NEURAL STEM CELLS - NEW PERSPECTIVES

Edited by **Luca Bonfanti**

Neural Stem Cells - New Perspectives
http://dx.doi.org/10.5772/56573
Edited by Luca Bonfanti

Contributors

Kaneyasu Nishimura, Luca Colucci-D\'Amato, MariaTeresa Gentile, Luca Bonfanti, Emilia Madarász, Caetana Carvalho, Bruno P Carreira, Ines Araujo, Ana Isabel Santos, Angelique Bordey, Manavendra Pathania, Shan Bian, Emmanuel Moyse, Young Gyu Chai, Nando Dulal Das, Verdon Taylor, Stefano Pluchino, Matteo Donega, Elena Giusto, Chiara Cossetti, Teri Belecky-Adams, Luciano Conti, Simona Casarosa, Jacopo Zasso, Joshua Goldberg

Published by InTech

Janeza Trdine 9, 51000 Rijeka, Croatia

Publishing Process Manager Iva Lipovic
Technical Editor InTech DTP team
Cover InTech Design team

Additional hard copies can be obtained from orders@intechopen.com

Neural Stem Cells - New Perspectives, Edited by Luca Bonfanti
p. cm.
ISBN 978-953-51-1069-9

For EU product safety concerns:
IN TECH d.o.o., Prolaz Marije Krucifikse Kozulić 3, 51000 Rijeka, Croatia,
info@intechopen.com or visit our website at intechopen.com.

Meet the editor

Luca Bonfanti, DVM, PhD, graduated (with honors) in 1987 at the School of Veterinary Medicine of the University of Turin, Italy. After a PhD in Functional Neuroanatomy and a postdoc at the Bordeaux University, he was appointed Assistant Professor in 1994 and Associate Professor in 2000 at the Turin University. He is now Professor of Veterinary Anatomy at the Department of Veterinary Sciences, and group leader at the Neuroscience Institute Cavalieri-Ottolenghi (NICO), a research center on brain repair. Since 1992 his research activity is devoted to study brain structural plasticity with particular reference to adult neurogenesis. In 2011 he was Visiting Professor at Northeastern University (Boston); he is receiving editor in the peer reviewed journal Frontiers in Neurogenesis.

Contents

Preface IX

Section 1 Neural Stem Cells as Progenitor Cells 1

Chapter 1 **Systems for ex-vivo Isolation and Culturing of Neural Stem Cells 3**
Simona Casarosa, Jacopo Zasso and Luciano Conti

Chapter 2 **Neural Stem Cell Heterogeneity 29**
Verdon Taylor

Chapter 3 **Diversity of Neural Stem/Progenitor Populations: Varieties by Age, Regional Origin and Environment 45**
Emília Madarász

Chapter 4 **Reactive Muller Glia as Potential Retinal Progenitors 73**
Teri L. Belecky-Adams, Ellen C. Chernoff, Jonathan M. Wilson and Subramanian Dharmarajan

Chapter 5 **Neural Stem Cell: Tools to Unravel Pathogenetic Mechanisms and to Test Novel Drugs for CNS Diseases 119**
Luca Colucci-D'Amato and MariaTeresa Gentile

Section 2 Neural Stem Cells and Neurogenesis 135

Chapter 6 **Postnatal Neurogenesis in the Subventricular Zone: A Manipulable Source for CNS Plasticity and Repair 137**
Manavendra Pathania and Angelique Bordey

Chapter 7 **Modulation of Adult Neurogenesis by the Nitric Oxide System 163**
Bruno P. Carreira, Ana I. Santos, Caetana M. Carvalho and Inês M. Araújo

Chapter 8 **A Vascular Perspective on Neurogenesis 199**
Joshua S. Goldberg and Karen K. Hirschi

Chapter 9 **Parenchymal Neuro-Glio-Genesis Versus Germinal Layer-Derived Neurogenesis: Two Faces of Central Nervous System Structural Plasticity 241**
Luca Bonfanti, Giovanna Ponti, Federico Luzzati, Paola Crociara, Roberta Parolisi and Maria Armentano

Section 3 Neural Stem Cells and Regenerative Medicine 269

Chapter 10 **A Survey of the Molecular Basis for the Generation of Functional Dopaminergic Neurons from Pluripotent Stem Cells: Insights from Regenerative Biology and Regenerative Medicine 271**
Kaneyasu Nishimura, Yoshihisa Kitamura, Kiyokazu Agata and Jun Takahashi

Chapter 11 **Systemic Neural Stem Cell-Based Therapeutic Interventions for Inflammatory CNS Disorders 287**
Matteo Donegà, Elena Giusto, Chiara Cossetti and Stefano Pluchino

Chapter 12 **Cell Adhesion Molecules in Neural Stem Cell and Stem Cell-Based Therapy for Neural Disorders 349**
Shan Bian

Chapter 13 **Neuroinflammation on the Epigenetics of Neural Stem Cells 381**
Nando Dulal Das and Young Gyu Chai

Chapter 14 **Primary Neural Stem Cell Cultures from Adult Pig Brain and Their Nerve-Regenerating Properties: Novel Strategies for Cell Therapy 397**
Olivier Liard and Emmanuel Moyse

Preface

During the last two decades stem cell biology has changed the field of basic research in life science as well as our perspective of its possible outcomes in medicine. At the beginning of the nineties, the discovery of neural stem cells in the mammalian central nervous system (CNS) made the generation of new neurons a real biological process occurring in the adult brain. Since then, a vast community of neuroscientists started to think in terms of regenerative medicine as a possible solution for incurable CNS diseases, such as traumatic injuries, stroke and neurodegenerative disorders. Nevertheless, in spite of the remarkable expansion of the field, the development of techniques to image neurogenesis in vivo, sophisticated in vitro stem cell cultures, and experimental transplantation techniques, no efficacious therapies capable of restoring CNS structure and functions through cell replacement have been convincingly developed so far. Deep anatomical, developmental, molecular and functional investigations have shown that new neurons can be generated only within restricted brain regions under the control of specific environmental signals. In the rest of the CNS, many problems arise when stem cells encounter the mature parenchyma, which still behaves as 'dogmatically' static tissue. More recent studies have added an additional level of complexity, specifically in the context of CNS structural plasticity, where stem cells lie within germinal layer-derived neurogenic sites whereas progenitor cells are widespread through the CNS.

Hence, two decades after the seminal discovery of neural stem cells, the real astonishing fact is the occurrence of such cells in a largely nonrenewable tissue. Still, the most intriguing question is which possible functional or evolutionary reasons might justify such oddity.

In other self-renewing tissues, such as skin, cornea, and blood, the role of stem cells in the tissue homeostasis is largely known and efficacious stem cell therapies are already available. The most urgent question is whether and how the potential of neural stem cells could be exploited within the harsh territory of the mammalian CNS. In this case, unlike other tissues, more intense and time-consuming basic research is required before achieving a regenerative outcome. The road of such research should travel through a better knowledge of several aspects which are still poorly understood, including the developmental programs leading to postnatal brain maturation, the heterogeneity of progenitor cells involved, the bystander effect that stem cell grafts exert even in the absence of cell replacement, and the cohort of stem cell-to-tissue interactions occurring both in homeostatic and pathological conditions.

In this book, the experience and expertise of many leaders in neural stem cell research are gathered with the aim of making the point on a number of extremely promising, yet unresolved, issues.

Luca Bonfanti DVM, PhD
Dept. of Veterinary Sciences, University of Turin
Neuroscience Institute Cavalieri-Ottolenghi (NICO)

Section 1

Neural Stem Cells as Progenitor Cells

Chapter 1

Systems for *ex-vivo* Isolation and Culturing of Neural Stem Cells

Simona Casarosa, Jacopo Zasso and Luciano Conti

Additional information is available at the end of the chapter

http://dx.doi.org/10.5772/55137

1. Introduction

During neural development, a relatively small and formerly considered homogeneous population of Neural Stem cells (NSCs) gives rise to the extraordinary complexity proper of the Central Nervous System (CNS). These represent populations of self-renewing multipotent cells able to differentiate into a variety of neuronal and glial cell types in a time- and region-specific manner throughout developmental stages and that account for a weak regenerative potential in the adult brain [1].

In the adult mammalian CNS, the presence of NSCs has been extensively investigated in two regions, the SVZ and the SGZ of the hippocampus, two specialized niches that control NSCs divisions in order to physiologically regulate their proliferative (symmetrical divisions) *vs* differentiative fate (asymmetrical divisions) [2].

In the early '90s it was shown that NSCs could be extracted from the developing and adult mammalian brain and expanded/manipulated/differentiated *in vitro* (Fig. 1).

This has represented a key step in the field, since the obtainment of *in vitro* NSC systems has been very useful in the last years in order to progress toward disclosure of the complex interplay of different extrinsic (signaling pathways) and intrinsic (transcription factors and epigenetics) signals that govern identity and functional properties of brain tissue-specific stem/progenitors [3]. Furthermore, it will also be a key step towards their exploitation for a better dissection of the molecular processes occurring in neurodegeneration [4]. Finally, NSC systems might represent major tools for the potential development of new cell-based and pharmacological treatments of neurodegenerative disorders and for assaying their toxicological effects [5].

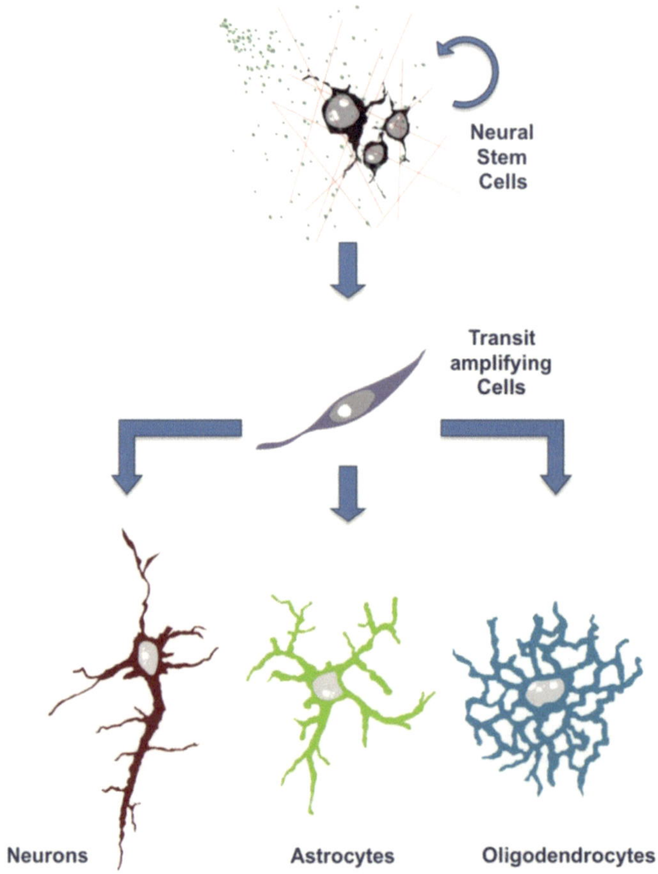

Figure 1. Process of NSC self-renewal and differentiation. NSCs are tri-potent cells. These cells during the differentiation process give rise to transiently dividing progenitors (transit amplifying progenitors) that subsequently undergo lineage restrictions toward neuronal, astrocytic and oligodendroglial mature cells.

Here we will review the functional properties of different *in vitro* NSC systems, providing also a direct comparison with NSCs present *in vivo*. Furthermore, we will discuss some of recent advancements in the development of *in vitro* systems that try to re-create *in vitro* some of the aspects of the physiological NSCs niches.

2. *In vivo* and *in vitro* developmental heterogeneity of NSCs populations

Vertebrate neural development starts with the process of neural induction, during and after gastrulation, which allows the formation of NEUROECTODERM from the dorsal-most part of the ectoderm. The molecular nature of the inductive signals that drive this process has been unveiled by studies in *Xenopus laevis*. These have shown that neural differentiation is promot-

http://dx.doi.org/10.5772/55137

ed by secretion of an array of BMP inhibitors, chordin, noggin and follistatin produced by an embryonic structure called "organizer" [6, 7]. The organizer also produces inhibitors of the Wnt signaling pathway, such as Dickkopf, frzb and cerberus [8]. Neural induction has shown a remarkable evolutionary conservation and a "default" model has been proposed, which states that ectodermal cells have an intrinsic predisposition to differentiate into neuroectoderm, unless inhibited by BMP signaling [9]. While in certain conditions this seems to be the case, in other assays positive inducers are needed, such as FGFs [10]. Finally, more recent studies show that inhibition of Activin/Nodal pathways also seems to be important for neural induction [8].

Progresses in cell culture technologies combined with a better understanding of these developmental progressions have allowed now to recapitulate these processes *in vitro* through neuralization of mouse and human pluripotent cells, i.e. Embryonic Stem cells derived from blastocyst stage (ESC; [11]) and reprogrammed cells (iPSC; [12, 13]), leading to the generation of populations of EARLY NEUROEPITHELIAL CELLS (Fig. 2). These cells give rise to all of the neural cells in the mature CNS thus denoting their extensive multipotential aptitude in terms of different cellular subtypes they can produce. Sox1 is the earliest identified marker of neural precursors in the mouse embryo and is present in dividing neural precursors from the NEURAL PLATE and NEURAL TUBE stages [3]. Studies on pluripotent cells support the "default" model for mammalian neural induction. *In vitro* studies have in fact shown that during neuronal differentiation, ESCs and iPSCs undertake gradual lineage restrictions analogous to those observed through *in vivo* fetal development, and a variety of distinctive progenitors can be generated. Accordingly, mouse and human pluripotent cells differentiate into sox 1 positive neuroepithelial cells (note that in human the earliest neuroepithelial marker is represented by pax6 that precedes sox1 expression) when grown in serum-free conditions in the absence of patterning signals [14-16]. ESCs and iPSCs neural induction can be enhanced by the addition of BMP-, Nodal- and Wnt-inhibitors, to minimize endogenous signals produced by ESCs/iPSCs themselves and recent studies have shown that paracrine signals (i.e. FGF4) are also needed for neurulation [17, 18].

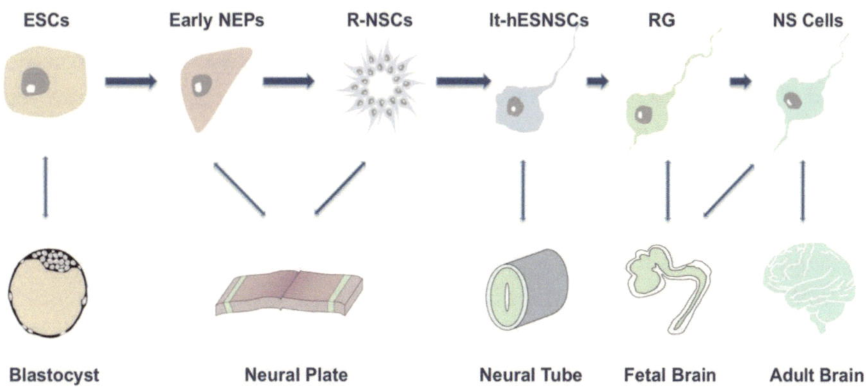

Figure 2. The different NSC populations that can be obtained in vitro correspond to stage-specific neural progenitors present at defined in vivo developmental stages.

Soon after neural induction process, pluripotent cell-derived neuroepithelial cells give rise to NEURAL ROSETTE structures, in which cells elongate and align radially, in a manner that mimics neural tube formation [19]. *In vivo*, the neural tube is formed after neurulation from the newly-induced neural plate and, as it closes, it is regionalized along the antero-posterior (A/P) axis (Fig. 3A), giving rise to four main areas: forebrain, midbrain, hindbrain and spinal cord. In amniotes, dorso-ventral (D/V) patterning takes place only after A/P patterning has occurred, after neural tube closure. The variety of neuronal cells that will be generated will have specific functions according to their position along these two axes.

Several evidences suggest that primary neural induction obtained by BMP inhibition generates anterior neural tissue, while to obtain tissue with posterior characteristics other molecules, known as "transformers", are needed. Three molecules with posteriorizing activities are known: retinoic acid (RA), Fgfs and Wnts [20, 21]. These signals are produced by the surrounding axial and paraxial mesoderm and endoderm, in addition two secondary signaling centers exist within the neural tube [22]. These are the Anterior Neural Ridge (ANR), located at the border between the forebrain and the non-neural ectoderm, and the isthmic organizer, located at the mid-hindbrain boundary. The ANR secretes the organizer molecules noggin and chordin, the resulting BMP signaling inhibition activates Fgf8, which in turn induces the expression of the transcription factor FoxG1 (Bf1), necessary for forebrain development [23]. The isthmic organizer is located at the boundary between the expression domains of the transcription factors Otx2 and Gbx2, and it is formed and maintained by an intricate regulatory network among these and other (En1/2, Pax2/5/8) transcription factors. The isthmic organizer secretes Fgf8, and the feedback loop that is set up assures the maintenance of the tissue identity [22]. RA and Wnts are produced by paraxial mesoderm with a high-posterior/low-anterior gradient and they are responsible for the patterning of midbrain, hindbrain and anterior spinal cord. Among the genes differentially regulated by varying concentrations of RA are the Hox genes, necessary for hindbrain and spinal cord A/P patterning [24, 25]. D/V patterning is mediated by signaling molecules secreted by the surrounding tissues (Fig. 3B). The overlying ectoderm produces TGFβ-family molecules that promote the formation of the roof plate in the dorsal neural tube, while the underlying notochord secretes SHH, that induces the ventral neural tube to become the floor plate. The roof plate and the floor plate in turn become a source of TGFβ and SHH, respectively. This creates two opposing gradients that provide positional information along the D/V axis, regulating the expression of key transcription factors. These will then act in a combinatorial manner to regulate the differentiation of specific neuronal and glial cell types in the correct position [26].

These *in vivo* studies have ultimately revealed that different neural progenitor populations can exist in a time and space-dependent manner and that their fate is greatly influenced by the interplay between specific extrinsic and intrinsic signaling molecules. ESCs- and iPSCs-derived neuroepithelial cells are able to perceive the positional information of patterning signals. These progenitors, when obtained in conditions that minimize endogenous signals, intrinsically acquire anterior identity, while they can be caudalized by the addition of FGFs, Wnts, RA [1, 19, 27, 28].

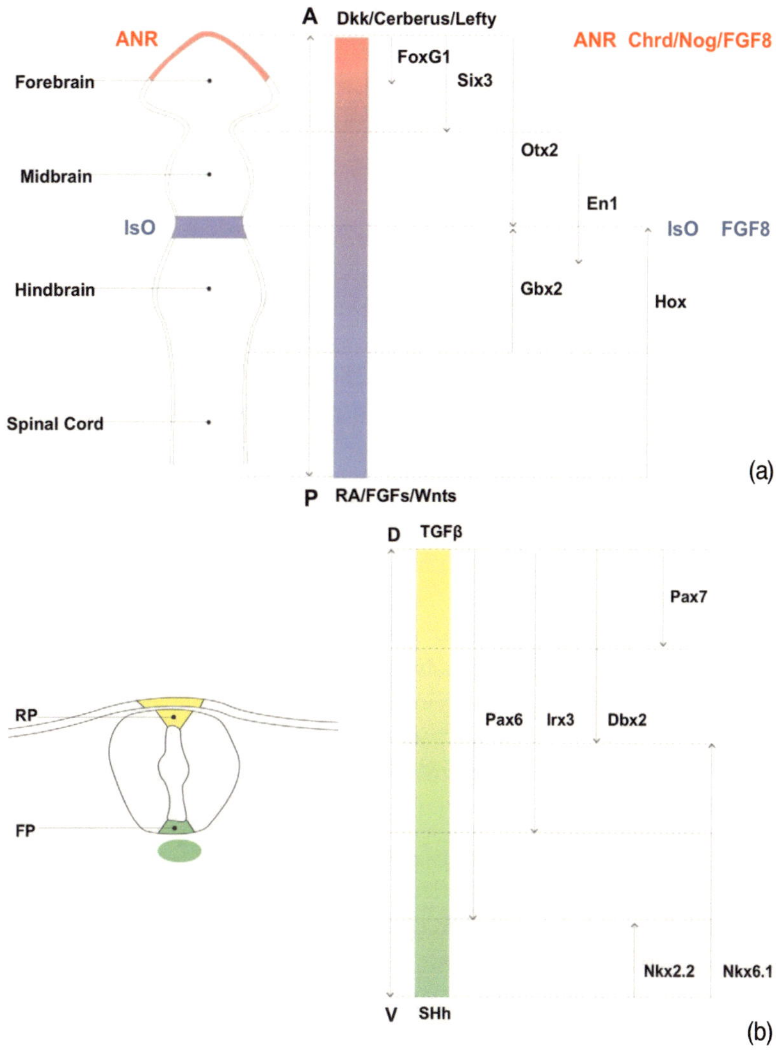

Figure 3. Regional patterning of the neural tube. Schematic diagrams showing antero-posterior (A) and dorso-ventral (B) patterning of the neural tube. The patterning process is driven by opposing gradients of signaling molecules that induce the expression of region-specific transcription factors in discrete areas. ANR: anterior neural ridge. IsO: Isthmic organizer. RP: roof plate. FP: floor plate.

Some studies have shown that NEUROEPITHELIAL CELLS cannot be maintained *in vitro* by the exposure to commonly used mitogens, i.e. basic fibroblast growth factor (FGF-2) and epidermal growth factor (EGF). These indeed convert these cells into radial glia populations characterized by a limited potentiality in neuronal sub-types they can give rise to. Nonetheless, it has been shown that a neuroepithelial population that grows in rosette-like structures (termed "R-NSCs") can be generated *in vitro* from human and mouse pluripotent cells when exposed to SHH/FGF8 signalling coupled to a N-Cadherin/Forse-1 cell sorting-based protocols [19]. These cells can be maintained *in vitro* for some passages by exposure to SHH and Notch agonists while showing a rostral BF1[+] neuroepithelial identity evocative of the signalling that *in vivo* are required for the induction of the anterior neuroepithelium. R-NSCs are characterized by a comprehensive differentiation potential toward CNS and PNS fates, supporting the idea that the R-NSCs represent neural precursors of the neural plate stage.

Another population of hESC-derived Sox1 positive self-renewing neuroepithelial cells named "lt-hESNSCs", has been described [29]. These cells can be grown as a nearly homogeneous population exhibiting clonogenicity and stable neurogenic potential. Remarkably, they can be maintained for many *in vitro* passages in the presence of FGF-2 and EGF and they preserve some properties of the R-NSCs, such as rosette-like growth, the expression of Bf1 and sensitivity to instructive signals that stimulate their conversion into distinct neuronal subpopulations. Molecular analyses have shown that lt-hESNSCs partly maintain rosette properties, possibly embodying an intermediate developmental stage between rosette-organized neuroepithelial cells and radial glia (see below).

As development proceeds, neuroepithelial cells lose sox1 expression and convert themselves into another transitory stem cell type, the so-called "RADIAL GLIA" (RG). This rapidly constitutes the main progenitor cell population in late development and early postnatal life while disappearing at later postnatal and adult stages [30, 31]. Large numbers of RG cells are found in primary cell cultures from dissociated E10.5-18.5 CNS tissue. Different populations of RG, characterized by lineage heterogeneity, with both regional and temporal varieties, give rise to sequential waves of neurogenesis, gliogenesis and oligodendrogenesis that build up the CNS. The *in vivo* developmental heterogeneity of RG has been also revealed by *in vitro* primary cultures studies that have shown a temporal constraint from neurogenesis to gliogenesis from RG isolated at initial or later developmental periods, respectively [32, 33].

The transition of neuroepithelial cells to RG cells is well recapitulated *in vitro* during neural differentiation of pluripotent cells. RG populations can be efficiently generated from ESCs/iPSCs using differentiation protocols that differ in major aspects between them. Bibel and collaborators generated transient (not expandable) populations of homogeneous RG cells that mature into glutamatergic neurons, as occurring during cortical development [34]. A different population of ESCs/iPSCs-derived RG cells can be obtained by exposing neuroepithelial cells to EGF and FGF-2. These rapidly lose Sox1 expression and acquire RG markers as BLBP and RC2 giving rise to RG-like cells which can be long term expanded in monolayer and at homogeneousness [35]. This conversion is dependent on Notch activity and on the exposure to EGF and FGF-2 [19, 35]. These self-renewing RG cells (called "NS cells") retain the marker signature of RG and the full capacity for tri-lineage neural differentiation, although their neuronal differentiation is limited to the GABAergic lineage [36-38]. These results indicate that pluripotent cells can be differentiated into distinct subtypes of RG – a non self-renewing type with aptitude to generate glutamatergic neurons, and a subtype that self-renews and exhibits a GABAergic differentiation. Such radial glial subtypes can also be found in the developing CNS *in vivo* although RG expansion *in vivo* is restricted to a defined time window.

Along with RG, a further immature population of cells with neuronal-restricted potential is represented by the BASAL PROGENITORs (BPs) that are located in the subventricular zone (SVZ) and can be generated both by neuroepithelial cells and RG [39, 40]. *In vitro* studies on

http://dx.doi.org/10.5772/55137

BPs are less comprehensive. Transitory induction of neurogenic Tbr2-positive BPs has been reported during the differentiation of ESCs to glutamatergic cortical neurons [27]. It has also been shown that BPs can be isolated from a subgroup of RG populations characterized by a high immunoreactivity for prominin that can make neurons only indirectly through the generation of BPs [41].

At the end of neurogenesis (in mice approximately at birth), neurogenic RG cells are exhausted and the remaining RG convert into astrocytes. The presence of stem cells has been reported in two regions of the adult mammalian brain, the SVZ and the SGZ of the hippocampus. Fate-mapping studies have shown that these adult NSC populations are represented by the type B astrocytes that directly derive from subpopulations of fetal RG cells. Therefore, RG and type B astrocytes appear to form a continuous lineage with stem cell potential [2]. These *in vivo* studies find a parallel indirect proof from the fact that *in vitro* adult-derived NSCs reacquire fetal characteristics, such as radial glia markers.

3. *In vitro* systems for NSCs isolation and expansion

The study of different types of stem cells has greatly beneficed from *in vitro* approaches that allow the reduction the intrinsic complexity of tissues. In order to allow stable maintenance *in vitro*, cells have to be immortalized, a procedure that blocks the progression of developmental programmes by pushing the cells to remain in enduring proliferation. Immortalization can be achieved by means of various methods, most usually by viral transduction of immortalizing oncogenes such as c-myc or SV40 Large T Antigen. Several immortalized murine and human NSC lines have been reported and, interestingly, it has been shown that they maintain many equivalences to non-immortalized lines, exhibiting neglectable signs of transformation both *in vivo* or *in vitro* [42-45]. Nevertheless, the physiological relevance of these lines might be weakened by the expression of potentially transforming oncogenes.

In the developing CNS, exponential cell division occurs only for brief developmental windows and NSCs represent transient populations. In the brain, NSC division is rigorously regulated by many factors of the "NICHE". The niche represents the particular cellular microenvironment that provides the appropriate milieu to support self-renewal and that controls the balance between symmetrical proliferative (producing two stem cells) and asymmetric cell divisions (generating one stem cell and one committed progenitor). Accordingly, for a stem cell to give rise to a clonal cell line, the physiological hindrances to continuous cell division have to be bypassed. However, until few years ago, it has been extremely difficult to stably propagate homogenous cultures of NSCs without oncogene-mediated immortalization procedures.

In the last two decades, oncogene-free procedures based on the use of soluble factors for selection and expansion of NSCs have been developed, permitting long-term maintenance of NSCs. The first report was from Reynolds and Weiss that in 1992 showed that the fetal and adult rodent brains contain cells competent for continuing *ex vivo* proliferation upon exposure to EGF and FGF-2 and that upon mitogen withdrawal exhibit tri-neural lineage differentiation [46, 47]. According to this procedure, freshly dissociated SVZ cells plated at low density (roughly 10^3-10^4 cells/cm^2) in the absence of cell adhesion substrates and in presence of EGF and/or FGF-2 have the tendency to loosely adhere to the plastic plate. Within few days, most of the cells die except a minor fraction of them that become smooth-edged and begin to proliferate while staying attached to the plate. Later, the progeny of these proliferating cells stick to each other forming sphere-shaped clones that detach from the plate thus floating in suspension giving rise to the so-called NEUROSPHERES. This assay, named "Neurosphere Assay" has thus been widely considered as a valuable method for isolating, enriching and maintaining embryonic and adult NSC populations *in vitro* [48]. Indeed, whereas NSCs in culture are characterized by the ability to considerably divide and self-renew thus giving rise to long-term expanding NSC lines, transit amplifying progenitors exhibit partial proliferative competence without self-renewal potential, and are eliminated during extensive sub-culturing. Notably, only a fraction of cells composing the neurosphere (commonly 1-10% for optimal cultures, although this value greatly differs depending on the age and on the brain area considered) are true stem cells, the remainder being differentiating progenitors at different stages, and even terminally differentiated neurons and glia [49]. Neurospheres can be sub-cultured by mechanical or enzymatic dissociation and by re-plating under the identical *in vitro* settings. As for the primary neurosphere culture, at every sub-culturing passage, differentiating/differentiated cells are supposed to die while the NSCs divide, generating secondary spheres that can then be further sub-cultured [50]. This procedure can be serially reiterated and, since each NSC gives rise to many NSCs by the time a neurosphere is generated, it ends in the expansion of the NSC population in culture.

Once established, neurosphere cultures can be expanded to obtain large amounts of cells that can then be cryopreserved. This permits the creation a pool of cells that can be later thawed and expanded for future experimentations. Nonetheless, several studies have shown that after few passages, the neurospheres greatly decrease their efficiency in neurogenic differentiation [51] and in the neuronal subtypes they can give rise to, mostly restricting their potential to the GABAergic lineage [52] (Fig. 4).

The accurate identification of the identity of the sphere-forming cell represents a key question. As committed progenitors are capable of only restricted proliferative capability and can generate only up to tertiary neurospheres, actually the designation of a cell as *bona fide* NSC should be retrospectively refereed only to a founder cell that self-renews extensively and can be propagated in long-term cultures. To this regard, it has been suggested that at least five sub-culturing passages are required to exclude the contribution of committed progenitors to the maintenance of the cell population. More rigorously, the assay should be performed with single dissociated cells (i.e. to plate a single cell per well) in order to avoid cell clustering and also fusion between neurospheres [53, 54].

Some researchers consider that three-dimensional organisation and the cellular milieu of the neurosphere as the *in vitro* equivalent of the *in vivo* neurogenic compartment [55, 56]. Although this view is a pure speculation, it is broadly accepted that the issue of the complexity of the

http://dx.doi.org/10.5772/55137

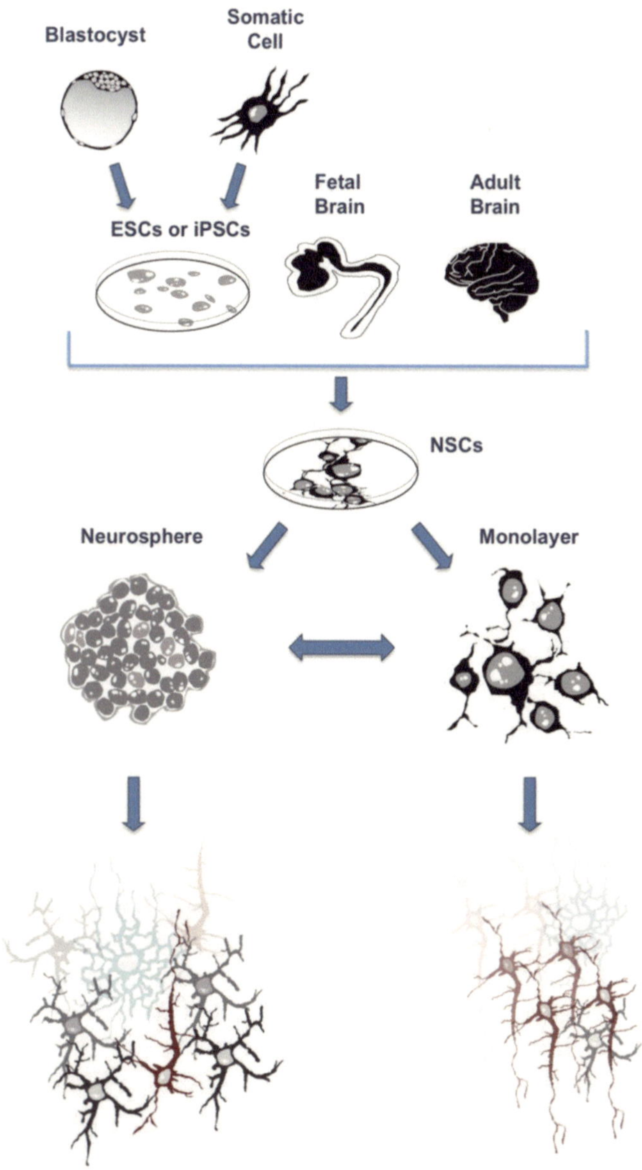

Figure 4. Neurospheres and monolayer NSCs can be obtained by different sources and have different neuronal differentiation efficiency. NSCs grown in monolayer and neurospheres can be derived from ESCs or iPS cells and from the germinative areas of the fetal and adult brain. The homogenous cellular composition of the NSCs grown in monolayer results in a higher neurogenic potential than neurospheres

neurosphere system represents a barrier for fine biochemical and molecular studies. The prospect of refining the neurosphere culture and of developing alternative *in vitro* systems, not only to enrich but also to select and clonally expand the *bona fide* stem cell population without losing the original prevalent neuronal fate, has been a recurrent issue in the stem cell field.

As an alternative to the neurosphere system, other researchers have developed monolayer-based methods [57]. In 1997, Gage and colleagues reported that progenitor cells with properties similar to NSCs from adult SVZ could be obtained from the adult hippocampus [58]. These hippocampal precursor cells propagate in monolayer and using *in vitro* procedures similar to the ones used for SVZ NSCs. Hippocampal precursors divide in response to FGF-2 and show tri-neural potential being able to differentiate into astroglia, oligodendroglia, and neurons *in vitro*. More recently, the optimization of novel and efficient strategies for the derivation and stable long-term propagation of NSCs from developing and adult neural tissue and from pluripotent cellular sources has been reported. It has been shown that transiently generated ESC-derived neural precursors, normally destined to differentiate to neuronal and glial cells, can be efficiently expanded as adherent clonal NSC lines in EGF and FGF-2 supplemented medium [19, 35]. In these growth conditions, cells undergo symmetrical division with neglectable accompanying differentiation, while shifting of the cultures to differentiative conditions prompts the cells to efficiently generate mature neurons, astrocytes and oligodendrocytes, thus indicating their NSC essence. The cells obtained by this procedure have been named Neural Stem (NS) cells. Notably, these results suggest that expansion of NS cells can occur in the absence of a complex cellular niche. Accordingly, NS cell expansion in monolayer conditions restrains spontaneous differentiation and permits proliferation of homogeneous *bona fide* NSCs.

Phenotypic characterization of NS cell cultures indicates a close similarity to forebrain RG [35]. Indeed, NS cells are homogenously immunopositive for nestin, SSEA1/Lex1, Pax6, prominin, RC2, vimentin, 3CB2, Glast, and BLBP, a set of markers diagnostic for neurogenic RG. NS cells keep their neurogenic potential after extensive expansion (over 100 passages), yet retaining the capability to produce a large proportion of mature neurons (Fig. 4). These results further indicate that the acquisition of RG properties endows the cells with a "niche" that traps them in a state of symmetric cell division. Significantly, NS cells do not represent a peculiarity of ESCs and iPSCs cell differentiation [35, 59]. In fact, similar lines can also be obtained from foetal or adult CNS and established from long-term expanded neurosphere cultures [35, 60, 61]. It is therefore possible that NS cells embody the resident NSC population within neurospheres. Further characterization of different mouse NS cell lines has demonstrated a close similarity in self-renewal, neuronal differentiation potential and molecular markers, independently from their origin. NS cells are not exclusive for mouse sources but it has indeed described the possibility to generate NS cells both from human fetal neural tissue and from human ESCs [62]. Interestingly, similar cells can be developed also from brain tumors and might serve as systems for find new targets in order to develop new therapeutic approaches [63, 64]. Similarly to NS cells, also lt-hESNSCs grow in monolayer and can be long-term expanded but differently from NS cells, they maintain sox 1 expression and a wide developmental competence [29, 65]. These aspects might be suggestive for some species-specific differences.

4. Influences of the *in vitro* systems on the molecular and biological properties of NSC lines

For brain tissue, founder NSCs existing during embryogenesis do not endure in adulthood but switch to a quiescent state following completion of development. Therefore, it might be expected that in order to achieve persistent propagation of NSCs *in vitro* it might not be merely sufficient to follow intrinsic programmed mechanisms but also modifications of the "Neural Stem Cells cellular "character" are required to adapt to the synthetic *in vitro* milieu might also be required. Indeed, the interaction of typical transient progenitor populations with the artificial *in vitro* environment (i.e. high levels of growth factor stimulation and/or different matrix or cell-cell interactions) may modify their transcriptional and epigenetic status, allowing them to be "turned" into NSC lines.

In this view, when coming to the nature of the NSCs, the crucial issue is if they do exactly represent a definite sub-population of NSC/progenitor existing *in vivo*. Currently, it is still not entirely understood if the accomplishment of the NSC status might be the effect of phenotypic alterations due to culture set and how physiologically relevant the consequent *in vitro* phenotype might be [3]. Thus, it is preferable to refer to *in vitro* expanded NSCs as NSC-like cells.

To this regard, the possibility that the mixture of mitogens may produce an artificial cell condition with a proper balance of key transcription factors able to suppress lineage commitment and allow self-maintaining divisions has to be considered. It has been shown that FGF-2 and EGF, two growth factors typically used for the *in vitro* maintenance of NSCs can alter the transcriptional and epigenetic phenotype. For example, expression of several genes can be directly stimulated *in vitro* in neural progenitors by exposure to FGF-2, suggesting that these genes might exert fundamental functions in the establishment of NSCs lines [66]. Similarly, foetal neural progenitors *in vitro* exposed to FGF-2, rapidly activate expression of Egfr (ErbB1) and Olig2, the latter being a bHLH transcription factor linked with the oligodendrocyte lineage and ventral CNS identity [66, 67]. Under expansion conditions with high levels of EGF and FGF-2, induction of Olig2 is required for the proliferation and self-renewal of neurosphere cells and NS cells, as demonstrated by analyses in which experimental interference with Olig2 expression severely decreases the amount and the quality of neurospheres [68]. Besides Olig2, it has been shown that acute exposure to FGF-2 induces neural progenitors to upregulate expression of a broad set of genes (for example CD44, GLAST, Olig1, Cdh20, Adam12 and Vav3) likely playing significant roles in the phenotype of the cells [69]. Likewise, EGF has been shown to deregulate expression of Dlx-2 in NS cells, NSC cultures and in transit-amplifying cells of the SVZ, inducing their switch into RG-like neurosphere-forming cells [51, 61, 69, 70]. Remarkably, stimulation of several of these genes (for instance Vav3 and CD44) occurs within few hours of FGF-2 exposure, possibly indicating that mitogen-mediated action is not suggestive for a physiological developmental progress but rather an acute transcriptional rearrangement [69].

NSCs *in vivo* have been shown to be tremendously heterogeneous in terms of transcriptional factors expression pattern, a feature predictable to confer a complex elaboration of positional signals [33]. To this regard, several reports have shown the occurrence *in vitro* of profound variations in the expression pattern of positional genes compared with primary precursors and progenitors *in vivo* thus leading to a mixed regional identity and limited neuronal differentiation. For example, neurospheres from the spinal cord have been shown to undergo upregulation of Olig2 and downregulation of the dorsal spinal cord transcription factors Pax3 and Pax7 [71]. Olig2 and Mash1 are also induced in E14 cortex or ganglionic eminence precursors, short- or long-term grown as neurospheres [72]. With some exceptions, a similar deregulation of the regional patterning is evident in the adherent NS cells and lt-hESNSCs cultures [29].

Importantly, this relaxation in the positional code might be related to a recurrent restriction in the competence to generate diverse neuronal subtypes. Indeed, NSCs have been reported to rapidly lose their original competence to generate site-specific neuronal subtypes when cultured *in vitro*, both in monolayer and in aggregation, in the presence of EGF and/or FGF-2, becoming mainly constrained to adopt a GABAergic fate [35, 52, 73, 74]. A notable exception is represented by the lt-hESNSCs [29], possibly indicating that for some reasons neuroepithelial cells derived from human pluripotent sources are more "predisposed" to long-term better preserve a broad neuronal sub-types developmental competence.

On the whole, these results might thus emphasize an artificial nature of cell culture, underlining the requirement for prudence in extrapolation of *in vitro* results to normal development or physiology without corresponding *in vivo* data [3]. Alternatively, this might be due to inadequate culture conditions that are not actually competent to preserve the molecular and biological properties of genuine NSCs.

5. Reconstruction of NSC niche *in vitro*

NSC niches present distinctive features leading to diverse ways to ensure neurogenesis. In the adult SVZ, three main immature neural populations lie adjacent to a layer of ependymal cells lining the lateral ventricle wall [2]. The Type B cells, representing the NSCs, reside interposed into the ependymal layer, displaying connections with both the ventricular wall and the blood vessels-network characterizing this niche. They are relatively quiescent but capable of giving rise to transit amplifying cells (Type C cells), a more rapidly dividing population that in turn generate the third population composed by neuroblasts (Type A cells) that migrate into glial tubes to reach the olfactory bulb. Besides these populations, a vital role for the maintenance of the niche is played by ependymal cells (Type E cells), astrocytes and endothelial cells. A comparable organization has been reported also for hippocampal SGZ niche although this exhibits a more planar structure [75, 76]. For a more detailed description of the neurogenic niches refer to of this book.

It emerges that both of these neurogenic niches are arranged to allow NSCs integration and to permit a strict responsiveness to signals from the *"external world"* (blood vessels and ventricles) and the *"neighboring world"* (newly generated neuroblasts, resident astrocytes and microglia, ECM components-forming scaffolds, etc.). All of these components harmoniously interact with each other providing both positive and negative signals and feedback that regulate NSCs activity.

Even though it is still a long way to fully understand the complex physiological context of a niche, researchers are now trying to reproduce *in vitro* at least some aspects of the dynamic *in vivo* environment. A better comprehension of the mechanisms underlying the NSC niche and the development of systems aimed at the reconstruction of this milieu will fill the gap between bi-dimensional (2D) simplified *in vitro* studies and the complex but physiological conditions of *in vivo* methods.

To this purpose, a synthetic NSC niche should recreate the complex interactions between NSCs and others cells, extracellular matrix, gradients of regulatory molecules and physical factors (Figure 5). In particular an ideal *in vitro* mimicked SVZ niche should contemplate the following minimal requirements:

1. presence of NSCs
2. production of the characteristic NSC niche-signaling molecules
3. presence of a basal lamina and extracellular matrix
4. autonomous production of cellular and molecular factors necessary for self-renewal and differentiation of resident stem cells
5. incorporation of extra-neural (i.e. endothelial cells) cells
6. spatial assembly reproducing the SVZ *in vivo* architecture.

In vitro generation of structures grossly simulating the SVZ NSC niche have been reported from mouse ESC-derived NSCs without the administration of mitogenic factors and complex physical scaffolds. In these studies, following a neuralization process with retinoic acid and plating the NSCs at high density on an entactin-collagen-laminin coated surface, heterogeneous multicellular aggregates appeared spontaneously, showing some of the characteristics postulated above, although a well-defined structural architecture was lacking [77]. In the last years, the development of new 3D culture systems that can allow to better reproduce *in vitro* structures in between standard monolayer culture and living organisms have been/are under investigation.

In this direction, standard culture methods involving petri dishes are being replaced with more accurate micro-scale devices, allowing procedures at the time and length scales of biological phenomena, enabling the control of multiple parameters, such as molecular and physical factors [78]. More attention is now focused on both the generation of morphogen-gradients, taking advantage of microfluidic systems, and three-dimensional extracellular matrix mimic-scaffolds in which multiple cells can be entangled allowing spatiotemporal control of the system and satisfying all of the features of a niche [79].

Microfluidic systems can reproduce a niche-like microenvironment permitting also the generation of concentration gradients of signaling molecules, often without the application of an external power source. Indeed, two different solutions can be introduced into the main channel of a microfluidic-chip by an osmotic pump. Since at this scale fluids mix only by diffusion, at the interface of the two solutions, diffusion generates a stable concentration

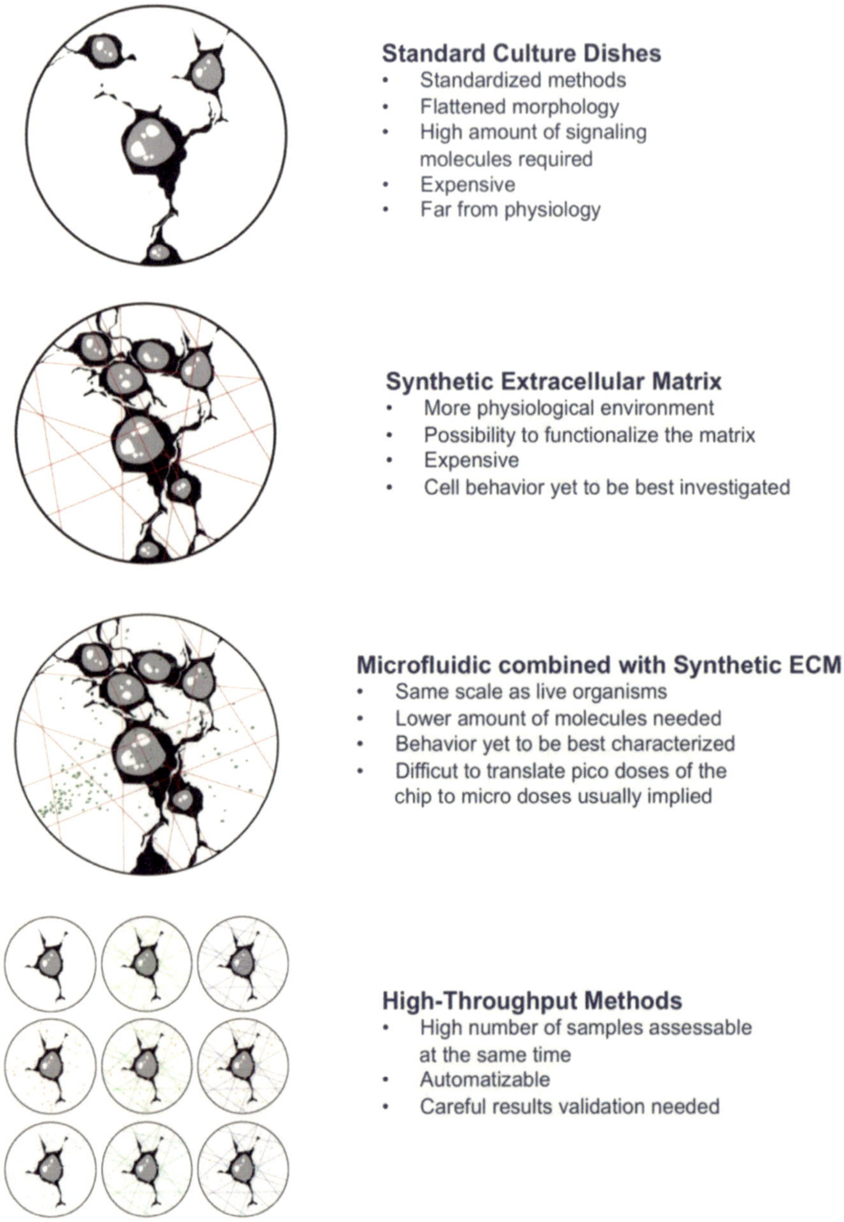

Figure 5. Schematic illustration of the different colture methods to reproduce *in vitro* the NSC niche.

gradient. To this regard, it has been shown that solutions of SHh, FGF8 or BMP4 are able to induce human ESC-derived NSCs neuronal differentiation, leading to the formation of a complex cellular network [80].

A fundamental impulse has come from the advance in the field of BIOMATERIALS. These have been greatly improved in the last few years, allowing now to finely control cell-matrix interactions, to direct cell migration and to permit the precise topographical administration of defined physical (both soluble or not) signals.

While it is quite difficult to modify only one variable with a naïve ECM component, the use of biomaterials has improved and simplified many experimental approaches. For example, when using natural matrices, decreasing the concentration of collagen leads to a decrease stiffness of the gel, nonetheless this also determines a decreasing in the concentration of adhesive ligands and an increase in diffusion, resulting in accumulation of variables to the system. This can be avoided with engineered biomaterials that enable isolation of individual variables, without varying others. Nowadays, synthetic biomaterials are greatly exploited to mimic the physical and mechanical features of the ECM. They allow to control a number of important parameters, including polymerization, degradation, and biocompatibility and to combine them with fully defined chemical components [81-87].

Another point of control allowed by new biomaterials is the possibility to incorporate cells releasing molecules or molecules *per se* as soluble factors, such as cytokines, NFs and GFs. Indeed, these molecules are constantly synthesized, secreted, transported, and depleted in NSC niches. To this regard, Zhang and colleagues have described a 16-residues peptide capable of self-assembly into membrane upon addition of a physiological concentration of salt [88]. Now commercially available as PuraMatrix™, it has been shown to support neurite outgrowth and synapse formation [89] and more recently to regulate murine and human NSCs growth and differentiation following adjunction of NSCs-active molecules [90-93].

Synthetic peptides can also be used in combination with a variety of polymers to provide materials with cell-adhesive, enzymatically degradable, and GFs-binding properties. Amino-acid sequences commonly include collagen-, laminin-, and fibronectin-cell-adhesive domains, these can be mixed together and with other bioactive motifs, such as proteolytically degradable sequences, to create a multifunctional peptide material with different physical properties. For instance, NSCs survival has been shown to be improved in a collagen hydrogel that incorporates laminin-derived adhesion motifs [94]. Peptides can also be used as structural components.

The reconstructions of a NSC niche can be translated to multiwell-based high-throughput methods for screening compounds that can positively regulate neurogenesis and thus be developed as potential therapeutic drugs. Protein-based microarrays have been developed and applied to diverse stem-cell populations [95-97]. These devices consist of robotically spotted GFs or ECM molecules in combinations, on cell repellent substrates in order to avoid cell migration, and cell fate changes are often analyzed via immunocytochemistry assays. Platforms like these have been used to analyze human NSCs differentiation and proliferation in response to combinations of ECM components, morphogens and other signaling proteins. A joint effect of Wnt and Notch pathways to maintain human NSCs in an undifferentiated state, a dose dependent activity of Notch ligands in shifting neuronal differentiation towards glial fate and a neurogenic effect of Wnt3A have thus been reported. Consequently, it is possible to highlight specific responses of single versus combination of stimuli in a high-throughput way [97].

These platforms are limited to adherent cells only and do not allow cell fates determination on single cells. The hydrogel microwell array, developed on micrometer-sized cavities, permits to analyze both adherent and nonadherent cells, trapped by gravitational sedimentation. The device has been used to analyze single cell-forming neurospheres, avoiding the usual merging events of neurosphere assay [98] and more recently it has been combined with robotic protein spotting to address the role of biochemical and biophysical factors on single nonadherent neural stem cell self-renewal [99].

6. Conclusions

Our knowledge of the neural progenitor identity and properties during development has been radically revolutionized by the possibility to isolate and expand NSCs *in vitro*. We have reviewed here the current and most commonly used *in vitro* methodologies to isolate, expand and functionally characterize NSC populations. The real identity and the potential lineage relationships between different types of stem/precursor cells isolated and cultured *in vitro* by these different methodologies represents a field of open and intense investigation.

In light of the complexity of the biological concerns governing stem cell maintenance and differentiation, significant progress will require a close coordination between *in vivo* and *in vitro* approaches. In this scenario, *in vitro* systems of NSCs shall allow a deep analysis at cellular level providing useful information to be further validate in the embryo and adult in order to identify relevance to normal physiology.

Establishment of *in vitro* settings necessarily results in disruption of the three-dimensional tissue structure, loss of specific cell-to cell contacts and modification of the extracellular environment and signaling. This might also lead to alteration of biological and molecular properties and acquisition of stem cell features by committed progenitors. Thus, although the versatility shown by NSC cultures *in vitro* can be envisaged as an advantage, extreme caution is necessary when considering the potential *in vivo* translation to developmental biology.

NSC biology holds tremendous potential for neurological therapy. It should be emphasized that the study of the intrinsic properties of NSCs and understanding the mechanisms of interaction between resident CNS cells and grafted NSCs will be mandatory for the development of new therapies able to slow the progression of neurodegenerative diseases.

Beside the therapeutical applications, NSCs systems present unique opportunities that are starting to be successfully explored for genetic or chemical screens in order to identify and optimize molecules/drugs that may allow a tight control on self-renewal and lineage specification of NSCs as well as their functional maturation, thus moving forward NSCs-based therapies.

We can anticipate that a rigorous characterization of the functional features of the NSC populations isolated and propagated by means of different cell culture systems shall allow us

http://dx.doi.org/10.5772/55137

to exploit the advantages offered by one method or the other, depending on the goal of our research.

Acknowledgements

Our apologies to all whose studies were not mentioned due to space limitations. We thank Riccardo Rossi for the creative illustrations used in the manuscript. L. Conti is supported by the Italian Ministry of Health; S. Casarosa is supported by the University of Trento and Cassa di Risparmio di Trento e Rovereto.

Author details

Simona Casarosa[1], Jacopo Zasso[2] and Luciano Conti[2]

*Address all correspondence to: luciano.conti@unimi.it

1 Centre for Integrative Biology, CIBIO, Via delle Regole, Mattarello (TN), Italy

2 Dipartimento di Scienze Farmacologiche & Biomolecolari, Università degli Studi di Milano, Via Balzaretti, Milan, Italy

References

[1] Grabel L. Developmental origin of neural stem cells: the glial cell that could. Stem Cell Rev. 2012;8(2):577-85. Epub 2012/02/22.

[2] Alvarez-Buylla A, Lim DA. For the long run: maintaining germinal niches in the adult brain. Neuron. 2004;41(5):683-6.

[3] Conti L, Cattaneo E. Neural stem cell systems: physiological players or in vitro entities? Nature reviews Neuroscience. 2010;11(3):176-87. Epub 2010/01/29.

[4] Pollard S, Conti L, Smith A. Exploitation of adherent neural stem cells in basic and applied neurobiology. Regenerative medicine. 2006;1(1):111-8. Epub 2007/05/01.

[5] Negri-Cesi P, Colciago A, Pravettoni A, Casati L, Conti L, Celotti F. Sexual differentiation of the rodent hypothalamus: hormonal and environmental influences. The Journal of steroid biochemistry and molecular biology. 2008;109(3-5):294-9. Epub 2008/04/12.

[6] Hemmati-Brivanlou A, Melton DA. Inhibition of activin receptor signaling promotes neuralization in Xenopus. Cell. 1994;77(2):273-81. Epub 1994/04/22.

[7] Sasai Y, Lu B, Steinbeisser H, De Robertis EM. Regulation of neural induction by the Chd and Bmp-4 antagonistic patterning signals in Xenopus. Nature. 1995;377(6551): 757. Epub 1995/10/26.

[8] Levine AJ, Brivanlou AH. Proposal of a model of mammalian neural induction. Dev Biol. 2007;308(2):247-56. Epub 2007/06/26.

[9] Hemmati-Brivanlou A, Melton D. Vertebrate neural induction. Annu Rev Neurosci. 1997;20:43-60. Epub 1997/01/01.

[10] Stern CD. Neural induction: 10 years on since the 'default model'. Curr Opin Cell Biol. 2006;18(6):692-7. Epub 2006/10/19.

[11] Smith AG. Embryo-derived stem cells: of mice and men. Annu Rev Cell Dev Biol. 2001;17:435-62.

[12] Takahashi K, Tanabe K, Ohnuki M, Narita M, Ichisaka T, Tomoda K, et al. Induction of pluripotent stem cells from adult human fibroblasts by defined factors. Cell. 2007;131(5):861-72.

[13] Takahashi K, Yamanaka S. Induction of pluripotent stem cells from mouse embryonic and adult fibroblast cultures by defined factors. Cell. 2006;126(4):663-76.

[14] Ying QL, Stavridis M, Griffiths D, Li M, Smith A. Conversion of embryonic stem cells into neuroectodermal precursors in adherent monoculture. Nat Biotechnol. 2003;21(2):183-6.

[15] Watanabe K, Kamiya D, Nishiyama A, Katayama T, Nozaki S, Kawasaki H, et al. Directed differentiation of telencephalic precursors from embryonic stem cells. Nature neuroscience. 2005;8(3):288-96. Epub 2005/02/08.

[16] Smukler SR, Runciman SB, Xu S, van der Kooy D. Embryonic stem cells assume a primitive neural stem cell fate in the absence of extrinsic influences. J Cell Biol. 2006;172(1):79-90. Epub 2006/01/05.

[17] Kunath T, Saba-El-Leil MK, Almousailleakh M, Wray J, Meloche S, Smith A. FGF stimulation of the Erk1/2 signalling cascade triggers transition of pluripotent embryonic stem cells from self-renewal to lineage commitment. Development. 2007;134(16): 2895-902. Epub 2007/07/31.

[18] Chambers SM, Fasano CA, Papapetrou EP, Tomishima M, Sadelain M, Studer L. Highly efficient neural conversion of human ES and iPS cells by dual inhibition of SMAD signaling. Nat Biotechnol. 2009;27(3):275-80. Epub 2009/03/03.

[19] Elkabetz Y, Panagiotakos G, Al Shamy G, Socci ND, Tabar V, Studer L. Human ES cell-derived neural rosettes reveal a functionally distinct early neural stem cell stage. Genes Dev. 2008;22(2):152-65.

[20] Kiecker C, Niehrs C. A morphogen gradient of Wnt/beta-catenin signalling regulates anteroposterior neural patterning in Xenopus. Development. 2001;128(21):4189-201. Epub 2001/10/31.

[21] Kudoh T, Wilson SW, Dawid IB. Distinct roles for Fgf, Wnt and retinoic acid in posteriorizing the neural ectoderm. Development. 2002;129(18):4335-46. Epub 2002/08/17.

[22] Wurst W, Bally-Cuif L. Neural plate patterning: upstream and downstream of the isthmic organizer. Nature reviews Neuroscience. 2001;2(2):99-108. Epub 2001/03/17.

[23] Anderson RM, Lawrence AR, Stottmann RW, Bachiller D, Klingensmith J. Chordin and noggin promote organizing centers of forebrain development in the mouse. Development. 2002;129(21):4975-87. Epub 2002/10/25.

[24] Maden M. Retinoic acid and limb regeneration--a personal view. Int J Dev Biol. 2002;46(7):883-6. Epub 2002/11/29.

[25] Nordstrom U, Jessell TM, Edlund T. Progressive induction of caudal neural character by graded Wnt signaling. Nature neuroscience. 2002;5(6):525-32. Epub 2002/05/15.

[26] Hoch RV, Rubenstein JL, Pleasure S. Genes and signaling events that establish regional patterning of the mammalian forebrain. Semin Cell Dev Biol. 2009;20(4): 378-86. Epub 2009/06/30.

[27] Bertacchi M, Pandolfini L, Murenu E, Viegi A, Capsoni S, Cellerino A, et al. The positional identity of mouse ES cell-generated neurons is affected by BMP signaling. Cell Mol Life Sci. 2012 Oct 16.

[28] Eiraku M, Watanabe K, Matsuo-Takasaki M, Kawada M, Yonemura S, Matsumura M, et al. Self-organized formation of polarized cortical tissues from ESCs and its active manipulation by extrinsic signals. Cell stem cell. 2008;3(5):519-32. Epub 2008/11/06.

[29] Koch P, Opitz T, Steinbeck JA, Ladewig J, Brustle O. A rosette-type, self-renewing human ES cell-derived neural stem cell with potential for in vitro instruction and synaptic integration. Proceedings of the National Academy of Sciences of the United States of America. 2009.

[30] Pollard SM, Conti L. Investigating radial glia in vitro. Progress in neurobiology. 2007;83(1):53-67.

[31] Malatesta P, Hartfuss E, Gotz M. Isolation of radial glial cells by fluorescent-activated cell sorting reveals a neuronal lineage. Development. 2000;127(24):5253-63.

[32] Malatesta P, Hack MA, Hartfuss E, Kettenmann H, Klinkert W, Kirchhoff F, et al. Neuronal or glial progeny: regional differences in radial glia fate. Neuron. 2003;37(5): 751-64.

[33] Pinto L, Gotz M. Radial glial cell heterogeneity--the source of diverse progeny in the CNS. Progress in neurobiology. 2007;83(1):2-23.

[34] Bibel M, Richter J, Schrenk K, Tucker KL, Staiger V, Korte M, et al. Differentiation of mouse embryonic stem cells into a defined neuronal lineage. Nature neuroscience. 2004;7(9):1003-9.

[35] Conti L, Pollard SM, Gorba T, Reitano E, Toselli M, Biella G, et al. Niche-independent symmetrical self-renewal of a mammalian tissue stem cell. PLoS biology. 2005;3(9):e283.

[36] Goffredo D, Conti L, Di Febo F, Biella G, Tosoni A, Vago G, et al. Setting the conditions for efficient, robust and reproducible generation of functionally active neurons from adult subventricular zone-derived neural stem cells. Cell death and differentiation. 2008;15(12):1847-56.

[37] Spiliotopoulos D, Goffredo D, Conti L, Di Febo F, Biella G, Toselli M, et al. An optimized experimental strategy for efficient conversion of embryonic stem (ES)-derived mouse neural stem (NS) cells into a nearly homogeneous mature neuronal population. Neurobiology of disease. 2009;34(2):320-31. Epub 2009/02/25.

[38] Biella G, Di Febo F, Goffredo D, Moiana A, Taglietti V, Conti L, et al. Differentiating embryonic stem-derived neural stem cells show a maturation-dependent pattern of voltage-gated sodium current expression and graded action potentials. Neuroscience. 2007;149(1):38-52.

[39] Haubensak W, Attardo A, Denk W, Huttner WB. Neurons arise in the basal neuroepithelium of the early mammalian telencephalon: a major site of neurogenesis. Proceedings of the National Academy of Sciences of the United States of America. 2004;101(9):3196-201.

[40] Miyata T, Kawaguchi A, Saito K, Kawano M, Muto T, Ogawa M. Asymmetric production of surface-dividing and non-surface-dividing cortical progenitor cells. Development. 2004;131(13):3133-45.

[41] Pinto L, Mader MT, Irmler M, Gentilini M, Santoni F, Drechsel D, et al. Prospective isolation of functionally distinct radial glial subtypes--lineage and transcriptome analysis. Molecular and cellular neurosciences. 2008;38(1):15-42.

[42] Martinez-Serrano A, Bjorklund A. Immortalized neural progenitor cells for CNS gene transfer and repair. Trends Neurosci. 1997;20(11):530-8.

[43] Cacci E, Villa A, Parmar M, Cavallaro M, Mandahl N, Lindvall O, et al. Generation of human cortical neurons from a new immortal fetal neural stem cell line. Exp Cell Res. 2007;313(3):588-601.

[44] Cattaneo E, Conti L. Generation and characterization of embryonic striatal conditionally immortalized ST14A cells. Journal of neuroscience research. 1998;53(2):223-34. Epub 1998/07/22.

[45] De Filippis L, Lamorte G, Snyder EY, Malgaroli A, Vescovi AL. A novel, immortal, and multipotent human neural stem cell line generating functional neurons and oligodendrocytes. Stem Cells. 2007;25(9):2312-21.

[46] Reynolds BA, Tetzlaff W, Weiss S. A multipotent EGF-responsive striatal embryonic progenitor cell produces neurons and astrocytes. The Journal of neuroscience : the official journal of the Society for Neuroscience. 1992;12(11):4565-74.

[47] Reynolds BA, Weiss S. Generation of neurons and astrocytes from isolated cells of the adult mammalian central nervous system. Science. 1992;255(5052):1707-10.

[48] Pastrana E, Silva-Vargas V, Doetsch F. Eyes wide open: a critical review of sphere-formation as an assay for stem cells. Cell stem cell. 2011;8(5):486-98. Epub 2011/05/10.

[49] Chojnacki AK, Mak GK, Weiss S. Identity crisis for adult periventricular neural stem cells: subventricular zone astrocytes, ependymal cells or both? Nature reviews Neuroscience. 2009;10(2):153-63.

[50] Chojnacki A, Weiss S. Production of neurons, astrocytes and oligodendrocytes from mammalian CNS stem cells. Nat Protoc. 2008;3(6):935-40.

[51] Ciccolini F. Identification of two distinct types of multipotent neural precursors that appear sequentially during CNS development. Molecular and cellular neurosciences. 2001;17(5):895-907.

[52] Machon O, Backman M, Krauss S, Kozmik Z. The cellular fate of cortical progenitors is not maintained in neurosphere cultures. Molecular and cellular neurosciences. 2005;30(3):388-97.

[53] Singec I, Knoth R, Meyer RP, Maciaczyk J, Volk B, Nikkhah G, et al. Defining the actual sensitivity and specificity of the neurosphere assay in stem cell biology. Nature methods. 2006;3(10):801-6.

[54] Jessberger S, Clemenson GD, Jr., Gage FH. Spontaneous fusion and nonclonal growth of adult neural stem cells. Stem Cells. 2007;25(4):871-4.

[55] Campos LS. Neurospheres: insights into neural stem cell biology. Journal of neuroscience research. 2004;78(6):761-9.

[56] Bez A, Corsini E, Curti D, Biggiogera M, Colombo A, Nicosia RF, et al. Neurosphere and neurosphere-forming cells: morphological and ultrastructural characterization. Brain research. 2003;993(1-2):18-29.

[57] Johe KK, Hazel TG, Muller T, Dugich-Djordjevic MM, McKay RD. Single factors direct the differentiation of stem cells from the fetal and adult central nervous system. Genes Dev. 1996;10(24):3129-40.

[58] Palmer TD, Takahashi J, Gage FH. The adult rat hippocampus contains primordial neural stem cells. Molecular and cellular neurosciences. 1997;8(6):389-404.

[59] Castiglioni V, Onorati M, Rochon C, Cattaneo E. Induced pluripotent stem cell lines from Huntington's disease mice undergo neuronal differentiation while showing alterations in the lysosomal pathway. Neurobiology of disease. 2012;46(1):30-40. Epub 2012/01/10.

[60] Onorati M, Binetti M, Conti L, Camnasio S, Calabrese G, Albieri I, et al. Preservation of positional identity in fetus-derived neural stem (NS) cells from different mouse central nervous system compartments. Cellular and molecular life sciences : CMLS. 2011;68(10):1769-83. Epub 2010/10/29.

[61] Pollard SM, Conti L, Sun Y, Goffredo D, Smith A. Adherent neural stem (NS) cells from fetal and adult forebrain. Cereb Cortex. 2006;16 Suppl 1:i112-20.

[62] Sun Y, Pollard S, Conti L, Toselli M, Biella G, Parkin G, et al. Long-term tripotent differentiation capacity of human neural stem (NS) cells in adherent culture. Molecular and cellular neurosciences. 2008;38(2):245-58.

[63] Pollard SM, Yoshikawa K, Clarke ID, Danovi D, Stricker S, Russell R, et al. Glioma stem cell lines expanded in adherent culture have tumor-specific phenotypes and are suitable for chemical and genetic screens. Cell stem cell. 2009;4(6):568-80. Epub 2009/06/06.

[64] Conti L, Crisafulli L, Caldera V, Tortoreto M, Brilli E, Conforti P, et al. REST controls self-renewal and tumorigenic competence of human glioblastoma cells. PloS one. 2012;7(6):e38486. Epub 2012/06/16.

[65] Falk A, Koch P, Kesavan J, Takashima Y, Ladewig J, Alexander M, et al. Capture of neuroepithelial-like stem cells from pluripotent stem cells provides a versatile system for in vitro production of human neurons. PloS one. 2012;7(1):e29597. Epub 2012/01/25.

[66] Dromard C, Bartolami S, Deleyrolle L, Takebayashi H, Ripoll C, Simonneau L, et al. NG2 and Olig2 expression provides evidence for phenotypic deregulation of cultured central nervous system and peripheral nervous system neural precursor cells. Stem Cells. 2007;25(2):340-53.

[67] Ciccolini F, Svendsen CN. Fibroblast growth factor 2 (FGF-2) promotes acquisition of epidermal growth factor (EGF) responsiveness in mouse striatal precursor cells: identification of neural precursors responding to both EGF and FGF-2. The Journal of neuroscience : the official journal of the Society for Neuroscience. 1998;18(19): 7869-80.

[68] Ligon KL, Huillard E, Mehta S, Kesari S, Liu H, Alberta JA, et al. Olig2-regulated lineage-restricted pathway controls replication competence in neural stem cells and malignant glioma. Neuron. 2007;53(4):503-17.

[69] Pollard SM, Wallbank R, Tomlinson S, Grotewold L, Smith A. Fibroblast growth factor induces a neural stem cell phenotype in foetal forebrain progenitors and during

embryonic stem cell differentiation. Molecular and cellular neurosciences. 2008;38(3): 393-403.

[70] Doetsch F, Petreanu L, Caille I, Garcia-Verdugo JM, Alvarez-Buylla A. EGF converts transit-amplifying neurogenic precursors in the adult brain into multipotent stem cells. Neuron. 2002;36(6):1021-34.

[71] Gabay L, Lowell S, Rubin LL, Anderson DJ. Deregulation of dorsoventral patterning by FGF confers trilineage differentiation capacity on CNS stem cells in vitro. Neuron. 2003;40(3):485-99.

[72] Hack MA, Sugimori M, Lundberg C, Nakafuku M, Gotz M. Regionalization and fate specification in neurospheres: the role of Olig2 and Pax6. Molecular and cellular neurosciences. 2004;25(4):664-78.

[73] Parish CL, Castelo-Branco G, Rawal N, Tonnesen J, Sorensen AT, Salto C, et al. Wnt5a-treated midbrain neural stem cells improve dopamine cell replacement therapy in parkinsonian mice. J Clin Invest. 2008;118(1):149-60.

[74] Bithell A, Finch SE, Hornby MF, Williams BP. Fibroblast growth factor 2 maintains the neurogenic capacity of embryonic neural progenitor cells in vitro but changes their neuronal subtype specification. Stem Cells. 2008;26(6):1565-74.

[75] Suh H, Consiglio A, Ray J, Sawai T, D'Amour KA, Gage FH. In vivo fate analysis reveals the multipotent and self-renewal capacities of Sox2+ neural stem cells in the adult hippocampus. Cell stem cell. 2007;1(5):515-28. Epub 2008/03/29.

[76] Zhao C, Deng W, Gage FH. Mechanisms and functional implications of adult neurogenesis. Cell. 2008;132(4):645-60. Epub 2008/02/26.

[77] Lutolf MP, Blau HM. Artificial stem cell niches. Adv Mater. 2009;21(32-33):3255-68. Epub 2010/10/01.

[78] Cimetta E, Figallo E, Cannizzaro C, Elvassore N, Vunjak-Novakovic G. Micro-bioreactor arrays for controlling cellular environments: design principles for human embryonic stem cell applications. Methods. 2009;47(2):81-9. Epub 2008/10/28.

[79] Burdick JA, Vunjak-Novakovic G. Engineered microenvironments for controlled stem cell differentiation. Tissue engineering Part A. 2009;15(2):205-19. Epub 2008/08/13.

[80] Park JY, Kim SK, Woo DH, Lee EJ, Kim JH, Lee SH. Differentiation of neural progenitor cells in a microfluidic chip-generated cytokine gradient. Stem Cells. 2009;27(11): 2646-54. Epub 2009/08/28.

[81] Drury JL, Mooney DJ. Hydrogels for tissue engineering: scaffold design variables and applications. Biomaterials. 2003;24(24):4337-51. Epub 2003/08/19.

[82] Hunt NC, Grover LM. Cell encapsulation using biopolymer gels for regenerative medicine. Biotechnology letters. 2010;32(6):733-42. Epub 2010/02/16.

[83] Lee KY, Mooney DJ. Hydrogels for tissue engineering. Chemical reviews. 2001;101(7):1869-79. Epub 2001/11/17.

[84] Mather ML, Tomlins PE. Hydrogels in regenerative medicine: towards understanding structure-function relationships. Regenerative medicine. 2010;5(5):809-21. Epub 2010/09/28.

[85] Nuttelman CR, Rice MA, Rydholm AE, Salinas CN, Shah DN, Anseth KS. Macromolecular Monomers for the Synthesis of Hydrogel Niches and Their Application in Cell Encapsulation and Tissue Engineering. Progress in polymer science. 2008;33(2): 167-79. Epub 2008/02/01.

[86] Peppas NA, Hilt JZ, Khademhosseini A, Langer R. Hydrogels in biology and medicine: From molecular principles to bionanotechnology. Adv Mater. 2006;18(11): 1345-60.

[87] Sanchis J, Canal F, Lucas R, Vicent MJ. Polymer-drug conjugates for novel molecular targets. Nanomedicine (Lond). 2010;5(6):915-35. Epub 2010/08/26.

[88] Zhang S, Holmes T, Lockshin C, Rich A. Spontaneous assembly of a self-complementary oligopeptide to form a stable macroscopic membrane. Proceedings of the National Academy of Sciences of the United States of America. 1993;90(8):3334-8. Epub 1993/04/15.

[89] Holmes TC, de Lacalle S, Su X, Liu G, Rich A, Zhang S. Extensive neurite outgrowth and active synapse formation on self-assembling peptide scaffolds. Proceedings of the National Academy of Sciences of the United States of America. 2000;97(12): 6728-33. Epub 2000/06/07.

[90] Cunha C, Panseri S, Villa O, Silva D, Gelain F. 3D culture of adult mouse neural stem cells within functionalized self-assembling peptide scaffolds. International journal of nanomedicine. 2011;6:943-55. Epub 2011/07/02.

[91] Liedmann A, Rolfs A, Frech MJ. Cultivation of human neural progenitor cells in a 3-dimensional self-assembling peptide hydrogel. Journal of visualized experiments : JoVE. 2012(59):e3830. Epub 2012/01/20.

[92] Ortinau S, Schmich J, Block S, Liedmann A, Jonas L, Weiss DG, et al. Effect of 3D-scaffold formation on differentiation and survival in human neural progenitor cells. Biomedical engineering online. 2010;9:70. Epub 2010/11/13.

[93] Yla-Outinen L, Joki T, Varjola M, Skottman H, Narkilahti S. Three-dimensional growth matrix for human embryonic stem cell-derived neuronal cells. Journal of tissue engineering and regenerative medicine. 2012. Epub 2012/05/23.

[94] Nakaji-Hirabayashi T, Kato K, Iwata H. Improvement of neural stem cell survival in collagen hydrogels by incorporating laminin-derived cell adhesive polypeptides. Bioconjugate chemistry. 2012;23(2):212-21. Epub 2012/01/11.

[95] Brafman DA, Shah KD, Fellner T, Chien S, Willert K. Defining long-term maintenance conditions of human embryonic stem cells with arrayed cellular microenvironment technology. Stem cells and development. 2009;18(8):1141-54. Epub 2009/03/31.

[96] LaBarge MA, Nelson CM, Villadsen R, Fridriksdottir A, Ruth JR, Stampfer MR, et al. Human mammary progenitor cell fate decisions are products of interactions with combinatorial microenvironments. Integrative biology : quantitative biosciences from nano to macro. 2009;1(1):70-9. Epub 2009/12/22.

[97] Soen Y, Mori A, Palmer TD, Brown PO. Exploring the regulation of human neural precursor cell differentiation using arrays of signaling microenvironments. Molecular systems biology. 2006;2:37. Epub 2006/07/06.

[98] Cordey M, Limacher M, Kobel S, Taylor V, Lutolf MP. Enhancing the reliability and throughput of neurosphere culture on hydrogel microwell arrays. Stem Cells. 2008;26(10):2586-94.

[99] Gobaa S, Hoehnel S, Roccio M, Negro A, Kobel S, Lutolf MP. Artificial niche microarrays for probing single stem cell fate in high throughput. Nature methods. 2011;8(11): 949-55. Epub 2011/10/11.

Chapter 2

Neural Stem Cell Heterogeneity

Verdon Taylor

Additional information is available at the end of the chapter

http://dx.doi.org/10.5772/55676

1. Introduction

The concept of neurogenic neural stem cells in the brains of adult mammals including humans is now widely accepted. In rodents these cells have been studied extensively both in vitro and in vivo. Of the two primary neurogenic regions in the rodent brain, the subventricular zone of the lateral ventricle wall generates the most neurons of multiple phenotypes. The newly generated neurons in the subventricular zone migrate to the olfactory bulb replenishing neurons and reconstituting the local circuitry responsible for olfaction. The dentate gyrus of the hippocampus generates a single neuron type, glutamatergic granule cells. These newborn granule cells contribute to specific forms of memory by integrating into existent circuits (Shors et al., 2001; Clelland et al., 2009; Garthe et al., 2009). Over the last few years, what was once considered to be a homogeneous population of astrocytic stem cells in both neurogenic brain regions is now turning out to be a more complex mixture of cells. Heterogeneous populations of cells with stem cell properties are being discovered in both the subventricular zone and dentate gyrus. This heterogeneity combined with potential diversity in signals forming the local niches could provide a situation where these multiple neural stem cell subpopulations contribute of tissue homeostasis and regeneration.

2. Neurogenesis in the subventricular zone

The lateral walls of the forebrain ventricles contain stem cells that generate neuronal subpopulations of the olfactory bulb throughout life (Reynolds and Weiss, 1992; Morshead et al., 1994; Doetsch et al., 1999b; Gage, 2000; Mirzadeh et al., 2008). Although much remains to be learnt about the neurogenic process and the fate determinants controlling maintenance, proliferation and differentiation of stem and progenitors cells in the subventricular zone, morphological, immunological and lineage tracing has recently uncovered a striking heterogeneity in the putative stem cell pool. In the first sections of this chapter I will look at some of the key findings and experiments identifying the stem cells and following their fate. I will also ask the question of whether single neural stem cells are multipotent in vivo and look at some for the experimental data addressing this and also cover emerging experimental data showing heterogeneity within the stem cell pool.

3. The subventricular zone and its progenitors

Continued neurogenesis from cells within the subependymal layer of the lateral ventricle wall implies stem cells as a driving force and a regulatory niche. Ultrastructural electronmicroscopic analysis has been instrumental in defining the morphological differences among cells within the subependymal layer of the ventricle wall (Doetsch et al., 1997; Doetsch et al., 1999a; Mirzadeh et al., 2008). Combining electromicroscopy with functional regeneration of the neurogenic niche, astrocytes have been shown to be primary progenitors of the subventricular zone (Doetsch et al., 1999b; Doetsch et al., 1999a; Doetsch et al., 2002). The subventricular zone astrocytes are defined as B-cells. B-cells have a polarized morphology extending an apical process and sensory cilium that projects between the ependymal call (E-cells) lining the lateral ventricle. These B-cell projections organize the E-cells into characteristic pinwheel structure (Mirzadeh et al., 2008). This is likely to be an important structural and signaling center in the stem cell niche. Based on their ultrastructural characteristics and location the B-cell population can be divided into two. B1-cells have their cell body between the chains of neuroblasts (A-cells) and the ependymal lining. B1-cells are quiescent and, based on thymidine incorporation assays and electronmicroscopic analysis, they rarely divide. B2-cells are more displaced towards the parenchyma of the underlying striatum and unsheathe the migrating chains of neuroblasts on route to the olfactory bulb (Doetsch et al., 1997). Unlike the structurally related B1-cells, B2-cells divide more prevalently. C-cells are the committed progeny of the B-cells, likely generated by asymmetric cell division, and they are mitotically highly active but undergo a limited number of divisions before differentiating. The progeny of the transient amplifying C-cells, the A-cells, migrate in chains through tubes formed by B-cells to the olfactory bulb. In adulthood, interneurons of the granule cell layer are the major newborn neuron type in the olfactory bulb, and together with periglomerular neurons, reform local circuits. In addition to neurons of the olfactory bulb, oligodendrocytes are also continuously generated in the subventricular zone and migrate to the corpus callosum. These oligodendrocytes are the product of Olig2-positive transient amplifying cells (a second type of C-cell). The relationship between the neurogenic C-cells and those that generate oligodendrocytes is hotly debated, as is whether they are the products of the same multipotent neural stem cells in the subventricular zone.

4. Heterogeneity within the subventricular zone neural stem cell pool

The mechanisms controlling the fate of progenitors in the subventricular zone remain unclear. The niche and its local interactions, morphogens and growth factors are one potential mode by

http://dx.doi.org/10.5772/55676

which the differentiation potential of the neural stem cells is controlled (Basak and Taylor, 2009). Assuming that all stem cells with the subventricular zone have the same potential, local differences within the niche or signals interpreted by committed progenitors *en route* to their final destination would be responsible for determining the multiple neuronal fates. The ectopic grafting of stem cells into the subventricular zone indicates some degree of plasticity within the neural stem cell population and suggest niche specific signals as fate determinants (Suhonen et al., 1996). However, even with the same niche, some neural stem cells seem to have autonomous fates and be heterogeneous in their potential (Kohwi et al., 2007; Merkle et al., 2007). By using homochronic/heterochronic transplantation experiments it has been shown that progenitor cells at different ontogenetic stages are intrinsically directed toward specific lineages (De Marchis et al., 2007). In addition, neuroblasts in the rostral migratory stream are also heterogeneous and may be committed to specific neuronal fates even before reaching the olfactory bulb (Hack et al., 2005; Kohwi et al., 2005). Thus, rather than being universally plastic, the neural stem cell pool may be made up of many stem cells with restricted potentials. This is also supported by region specific, viral-mediated genetic labeling of the subventricular neural stem cells in juvenile mice which show diversity in neuronal progeny generated rather than generating all neuron types (Merkle et al., 2007). Granule cells, the major neuron subtypes to be generated during adulthood, are produced from all anteroposterior and dorsoventral locations in the subventricular zone. However, most granule cells are generated from the dorsal and ventral most aspects of the subventricular zone (Merkle et al., 2007). Within this regionalization, the granule neurons generated from the dorsal subventricular zone migrate to a more superficial location in the granule cell layer of the olfactory bulb while those generated ventrally settle deeper in the granule cell layer (Merkle et al., 2007). This regional specification can also be mapped to the location of the stem cells during early postnatal development indicating not only a regional but also a developmentally-regulated fate specification (Merkle et al., 2007). Similarly, periglomerular neurons that migrate to the outer layer of the olfactory bulb also show a region-specific origin. Dorsal regions of the subventricular zone generate the majority of the thymidine hydroxylase-positive neurons whereas Calbindin-positive periglomerular neurons are generated preferentially from the ventral subventricular zone (Merkle et al., 2007). Calretinin-positive periglomerular and granule cells are generated from the medial wall of the lateral ventricle. As this region produces proportionally fewer granule cells in total this suggests that the niche of the medial wall directs the fate of neural stem cells towards Calretinin neuron generation. Although these findings do not rule out niche specific programming of multipotent cell fate, heterotopic transplantation strongly suggests that stem cells retain their differential potential when grafted into a different axial location (Merkle et al., 2007).

5. Mitotically active or quiescent neural stem cells

For many years mitotic inactivity or quiescence has been viewed as a primary stem cell trait. However, recent data in many systems including the intestine and blood suggest that stem cell may not need to be quiescent and some can divide frequently to drive the generation of new cells (Wilson et al., 2008; Essers et al., 2009; Fuchs, 2009; Li and Clevers, 2010). These active stem cells are the force behind tissue homeostasis and may reside side-by-side with quiescent stem cells that rarely if ever divide but that could be responsible for tissue regeneration. Ultrastructural cellular analysis of the subventricular zone implied that even within the B-cell compartment, B1 cells rarely if ever divide whereas B2 cells are detected in cell cycle (Doetsch et al., 1997). This raised the possibility that in the adult brain stem cells may also either be able to adopt different fates or, different neural stem cells exist which show strikingly different mitotic potential. More recently, mitotically active cells in the subventricular zone were show to be in close proximity to blood vessels suggesting a mitotic influence of the endothelium or blood-born factors (Shen et al., 2008; Tavazoie et al., 2008). This is particularly intriguing as endothelial cells express the Notch ligand Jagged1 and can active neural stem cells regulating maintenance and proliferation both in vitro and in vivo thus implying that activated neural stem cells my have a vascular contribution to their niche (Shen et al., 2004; Nyfeler et al., 2005).

In summary of current and past data, the heterogeneous mitotic activity among neural stem cells suggests at least two potential scenarios. Either individual cells are able to transit between a quiescent and an activated state, or, that there are different stem cells, some which are quiescent and rarely divide, and others that are more mitotically active, dividing frequently and driving the production of new neurons destined for the olfactory bulb. A similar situation of active and dormant stem cells is present in the crypts of the large intestine where previously identified slow or rarely dividing stem cells in the +4 position seem to be the cells responsible for regenerating the epithelial lining of the gut. Conversely, mitotically active cells that are interdigitated with paneth cells at the base of the crypt replenish the epithelial cells lining the villi (Li and Clevers, 2010).

6. Active and quiescent stem cells show differences in Notch signaling

Notch signaling regulates cell fate in many cell systems and across species (Artavanis-Tsakonas et al., 1999; Louvi and Artavanis-Tsakonas, 2006). Lateral signaling between neighboring cells presenting Notch ligands and expressing receptors classically results in binary fate decisions, often in cells undergoing cell division. Notch signaling is active in the subventricular zone and multiple ligands are present on B, C and E cells providing the potential for lateral signaling (Stump et al., 2002; Nyfeler et al., 2005; Imayoshi et al., 2010). Genetic ablation of Notch signaling in stem cells of the subventricular zone results in precocious differentiation and neurogenesis (Imayoshi et al., 2010; Basak et al., 2012). This in turn results in a loss of neural stem cells and a subsequent long-term suppression of neurogenesis. This is a "classical" role for Notch in the regulation of cell fate, whereby loss of Notch signaling during what should be an asymmetric neural stem cell division results in both daughter cells adopting a differentiated cell fate and a concomitant loss of stem cell self-renewal. However, the ablation of Notch from B-cells also results in quiescent B1-cells entering the cell cycle and the active neurogenic pool. This activation of cells that are normally in a mitotically inactive state contributes to a pulse of increased neuroblast production before extinction of the stem cells pool following inactivation of canonical Notch signaling (Imayoshi et al., 2010; Basak et al., 2012). Hence, Notch signaling through its canonical pathway not only regulates stem cell maintenance in the subventricular zone by repressing neuronal commitment of the stem cell but also suppresses mitotic activity of B1 cells. In addition, canonical Notch signaling is implicated in repressing the mitotic activity in ependymal cells lining the lateral ventricle during ischemic lesions (Carlen et al., 2009). Although the role of ependymal cells as stem cells is highly controversial, it remains possible that, under some degenerative/regenerative conditions, even these differentiated cells may be able to dedifferentiate or transdifferentiate to generate neuroblasts. How this regulation of proliferation function is controlled by Notch is unclear. However, analysis of Notch1 function in the subventricular zone suggests differential receptor usage by neural stem cell in different mitotic states. The Notch gene family contains four genes encoding highly related receptors. These receptors are able to bind all five canonical ligands. At least three Notchs, Notch1, Notch2 and Notch3 are expressed in the subventriuclar zone (Stump et al., 2002; Basak et al., 2012). Notch1, Notch2 and Notch3 are expressed by B-cells whereas Notch1 is also expressed by C-cells, A-cells and E-cells (Nyfeler et al., 2005; Carlen et al., 2009; Imayoshi et al., 2010; Basak et al., 2012). Genetic conditional inactivation of Notch1 from B-cells induces a loss of self-renewal during homeostatic neurogenesis. Notch1-deficient active stem cells fail to self-renew and spontaneously differentiate – similar to ablation of the canonical DNA-binding component of the pathway RBP-J in these cells. However, unlike when RBP-J is deleted, Notch1-deficiency in B1-cells does not result in spontaneous mitotic activity (Basak et al., 2012). The regulation of cell proliferation by Notch signaling has also been implicated in vitro where cultured neural stem cells lacking Notch1 fail to self-renew and differentiate and in the adult zebrafish quiescent progenitors proliferate when treated with the gamma-secretase inhibitor DAPT, which blocks Notch (Nyfeler et al., 2005; Chapouton et al., 2010). Conversely, B1-cells, although they express Notch1, do not seem to depend upon it for a quiescence signal. Thus, it is likely that molecular compensation or signal diversity between the Notch receptors is responsible for the quiescence of B1-cells. This remains to be examined in detail.

It has been difficult to identify and study active neural stem cell in the adult mouse subventricular zone due to an absence of selective markers. Most transgenes used to label neurogenic stem cells utilize the *Nestin*, *GLAST* or *Hes5* promoters (Mori et al., 2006; Balordi and Fishell, 2007; Lagace et al., 2007; Giachino and Taylor, 2009; Imayoshi et al., 2010; Bonaguidi et al., 2011; Basak et al., 2012). These promoters are all expressed by both quiescent and mitotic stem cells. However, combinations of transgenic reporter and surface expression of the Prominin1-associated glycoepitope CD133 and binding of the mitogen epidermal growth factor is able to select active from quiescent stem cells, C-cells and neuroblasts (Pastrana et al., 2009). Conversely, Inhibitor of DNA binding protein 1 (Id1) a target of transforming growth factor-β signaling, is expressed predominantly by quiescent B1-cells. Transgenic mice expressing green fluorescent protein under the control of the Id1 promoter label quiescent B1-cells in the subventricular zone (Nam and Benezra, 2009). These Id1-positive GFAP-positive B1 cells are relatively rare and divide infrequently to generate neuroblasts likely by asymmetric cell division. Interestingly, mitotic activity of subventricular zone neural stem cells requires Id proteins with loss of function resulting in a loss of self-renewal and neurogenesis (Nam and Benezra, 2009). It remains to be shown whether and how the quiescent and active stem cells in the subventricular zone are related to each other or whether they fulfill distinct functions for example homeostatic neurogenesis and regeneration. It is likely that the elucidation of the diverse neural stem cells in the subventricular zone is going to require the combination of different markers and genetic tools (Beckervordersandforth et al., 2010).

7. The hippocampus continually generates neurons which participate in memory formation

In contrast to the subventricular zone where proliferation and neurogenesis are eradicated soon after birth in humans, the dentate gyrus of the human hippocampus, like in rodents, continues to generate neurons from mitotically active progenitors cells all the way into adulthood. The cellular composition of the neurogenic niche in the dentate gyrus has been studied extensively (Seri et al., 2001; Kempermann et al., 2004; Steiner et al., 2006; Steiner et al., 2008). However, the identity and regulation of neural stem cells in the dentate gyrus remains unclear.

8. The progenitor pool in the dentate gyrus is morphologically and functionally heterogeneous

Self-renewing neural stem cells in the subgranular zone of the adult hippocampal dentate gyrus (also referred to as Type-1 cells) produce intermediate progenitor cells (IPs, Type-2a cells), NeuroD1 and Doublecortin-positive neuroblasts (Type-2b) and subsequently granule neurons (Seri et al., 2001; Kempermann et al., 2004; Steiner et al., 2006). Type-1 neural stem cells have their cell bodies in the subgranular zone and extend a long process through the granule cell layer to the overlaying molecular layer. Type-2 cells are transient intermediate progenitors. They also have their cell body in the subgranular zone but lack a long radial process and have a more rounded morphology with short stubby processes (Seri et al., 2001; Steiner et al., 2006). Neuroblasts by contrast extend a leading process and migrate into the granule cell layer. Whereas radial Type-1 cells are quiescent, Type-2 cells divide readily expanding the progenitor pool. Previous Bromodeoxyuridine labeling experiments suggested that Type-2a cells, which express proneural transcription factors, are the major proliferative progenitor in the adult dentate gyrus (Steiner et al., 2006). In addition, retroviral-labeling experiments showed that neuroblasts that have extended a radial process, exit cell cycle, and only go through one or two cell divisions (Seri et al., 2001). However, recent genetic labeling and lineage tracing of stem cells in the dentate gyrus revealed that Ascl1-positive Type-2a cells do not undergo symmetric cell divisions but generate an addition intermediate cell type, Tbr2-positive Type-2 cells (recently referred to as Type-2ab cells) (Bonaguidi et al., 2012; Lugert et al., 2012). The Tbr2-positive cells divide frequently to amplify the progenitor pool and increase the number of neurons generated from each stem cell division (Lugert et al., 2012).

9. Multiple stem cell populations in the dentate gyrus

The classical view of stem cells in the adult dentate gyrus implicates the quiescent radial glial like Type-1 cells as the primary progenitor. However, retroviral labeling is commonly used to examine neurogenesis in the dentate and to label cells that continue to generate multiple neurons over time (Seri et al., 2001; Suh et al., 2007). As retroviral integration and thus viral gene expression are dependent upon cells passing through the cell cycle, some long-term neurogenic stem cells in the dentate must be mitotically active. Radial Type-1 cells are rarely labeled in these retroviral experiments suggesting that other cells that lack a radial process must also display self-renewing and long-term neurogenic stem cell potential (Suh et al., 2007). This is also supported by lentiviral labeling experiments driving reporter expression from the *Sox2* promoter (Suh et al., 2007). Expression of the transcription factor Sox2 is associated with progenitor cells of the brain and required for their maintenance by regulating Notch, Sonic Hedgehog expression and Wnt activity (Steiner et al., 2008; Favaro et al., 2009; Kuwabara et al., 2009). A population of none radial stem cells with horizontally orientated processes has been identified by Cre-recombinase mediated lineage tracing (Suh et al., 2007). These horizontal cells display stem cell characteristics but are clearly distinct from the previously described Type-1 and Type-2 cells. Horizontal Type-1 neural stem cells like radial Type-1 stem cells in the dentate gyrus have active Notch signaling and are labeled with a Notch signal reporter allele *Hes5::GFP* (Ables et al., 2010; Ehm et al., 2010; Lugert et al., 2010). However, although they express Nestin they do not express the astrocytic protein GFAP. Hence, there remains some debate and despite their similarity in morphology to Type-2 cells, horizontal Type-1 stem cells do not express classic Type-2 cell markers including the proneural transcription factor Ascl1 – a Notch repressed target gene – Tbr2 or Doublecortin (Steiner et al., 2006; Lugert et al., 2010). In addition, horizontal Type-1 cells are more mitotically active than their radial counterparts (Lugert et al., 2010). Therefore, the horizontal *Hes5*-positive cells likely represent the Sox2-positive population of stem cells and those stem cells commonly traced and analyzed by retroviral labeling. Although the relationship between the radial and horizontal stem cells is not clear, horizontal cells rarely generate radial Type-1 cells in viral lineage tracing experiments. Interestingly, activated neurogenic stem cells in the dentate gyrus express Sox1, which, like Sox2 and Sox3, is a member of the SoxB1 family. *Sox1*, like *Hes5*, is expressed by radial and horizontal Type-1 cells (Venere et al., 2012). Lineage tracing shows that Sox1-positive Type-1 cells include the active neural stem cells and support that neurogenic stem cells in the dentate gyrus may switch between active and inactive states (Lugert et al., 2010; Venere et al., 2012).

10. Radial and horizontal hippocampal stem cells respond selectively to external cues

The classical view is that radial Type-1 stem cells divide infrequently to generate transient amplifying progenitors through asymmetric cell divisions. However, as described above there are additional progenitors in the hippocampal dentate gyrus that can function as stem cells. Hence, the question arises what are the functions of these multiple putative neural stem cells? Do they both contribute to neurogenesis in the adult hippocampus and are they in a lineage relationship with each other? Genetic labeling experiments suggest that both radial and horizontal stem cells may be functionally distinct or at least they respond differently to different pathophysiological cues (Lugert et al., 2010).

Analysis of hippocampal neurogenesis has shown it to be a dynamic process that diminishes with age but can be stimulated and modulated by physiology and pathology (Kuhn et al., 1996; Kempermann et al., 1998; Ben Abdallah et al., 2008; Fabel and Kempermann, 2008; Parent and Murphy, 2008; Steiner et al., 2008; Zhao et al., 2008). Voluntary physical exercise induces increased proliferation and generation of immature neurons. These neurons do not readily integrate into the dentate gyrus but the increased proliferation of the stem cells is significant (Fabel and Kempermann, 2008). Notch signaling also controls neural stem cell maintenance and differentiation within the dentate gyrus (Breunig et al., 2007; Ables et al., 2010; Ehm et al., 2010; Lugert et al., 2010; Lugert et al., 2012). Loss of Notch activity results in the loss of neural stem cells and their precocious differentiation culminating in a loss of neuron production (Ables et al., 2010; Ehm et al., 2010; Lugert et al., 2010). Genetic labeling of neural stem cells in the dentate that display Notch signaling has uncovered diversity in stem cell responses to pathophysiology (Lugert et al., 2010). Physical exercise stimulates proliferation of the radial type1 cells but not the horizontal stem cells (Lugert et al., 2010). Running induces the radial cells to enter the active stem cell pool without expanding the total stem cell population. This suggests that radial cells in physically active animals undergo asymmetric cell divisions to generate committed progenitors that increase the number of newborn neurons whilst maintaining the Type-1 stem cell pool through self-renewal. This also implies that radial stem cells respond to stimuli generated by increased physical activity that are not seen or are not interpreted in the same way by the horizontal stem cells. These findings seem, at first glance, to contradict previous experiments where Nestin expressing progenitors were labeled and suggested that radial Type-1 cells do not proliferate significantly in running mice (Steiner et al., 2008). It is likely that the differences in result reflect the different experimental paradigms used to identify the stem cells of the dentate gyrus. Where as *Hes5* expression identifies a smaller population of cells more restricted to the stem cell pools in the subgranular cell layer, the *Nestin* promoter is expressed by stem cells and more committed progenitors (Bonaguidi et al., 2012). Hence, it remains possible that the different labeling techniques and the extent of cell labeling could effect the quantification and interpretation.

11. Selective loss of active stem cells in the hippocampus of aged mice

Neurogenesis in the mammalian brain diminishes dramatically after birth, even in the dentate gyrus where neurons are continuously generated throughout life. This reduced neurogenesis is associated with a loss of mitotic cells (Kuhn et al., 1996; Kempermann et al., 1998; Ben Abdallah et al., 2008; Steiner et al., 2008). Whereas some reports have suggested an irreversible loss of neural stem cells in the dentate gyrus due to exit from the stem cell pool and differentiation into astrocytes (Encinas et al., 2011; Encinas and Sierra, 2012), others suggest that the stem cells are not lost but become dormant with age (Lugert et al., 2010; Bonaguidi et al., 2011; Venere et al., 2012). Hence, the reason for the substantial reduction in neuron production remains unclear but may be caused by a culmination of physiological changes.

Genetic lineage tracing of *Nestin* expressing cells revealed that, parallel to the reduced number of neurons generated from the labeled stem cells, radial Type-1 cells in the aged mouse brain enter cell cycle and, following a few cell divisions, differentiate into polymorphic astrocytes that lose radial morphology and presumably stem cell potential (Encinas et al., 2011). This "deforestation" or expenditure of the stem cells likely contributes to the reduction in mitotic progenitors and neurons (Encinas and Sierra, 2012). Surprisingly, in a parallel study using the same genetic tools, clonal analysis indicated that *Nestin* expressing stem cells within the subgranular layer can undergo prolonged neurogenesis. In addition, these clonal experiments revealed an additional degree of heterogeneity within the stem cell population of the dentate gyrus. Some labeled Type-1 cells remained quiescent over many months and failed to generate any viable offspring. Other Type-1 cells divided and generated clones of cells that included progenitors, neurons and astrocytes indicating multipotency (Bonaguidi et al., 2011). Partially supporting the proposal that some *Nestin*-expressing Type-1 cells may exit the stem cell pool, clones were found that contained only differentiated cells. Taken together these data indicate heterogeneity within the stem cells pools and it seems that a combination of entry of stem cells into a dormant state coupled with a partial loss of some progenitors may contribute to the age related decline in neurogenesis (Bonaguidi et al., 2011; Encinas et al., 2011).

In contrast, neural stem cells in the dentate gyrus labeled by Notch activity and Sox2 expression remain in the aged dentate gyrus (Lugert et al., 2010; Bonaguidi et al., 2011; Lugert et al., 2012). Interestingly however, the proportion of the cells that are mitotically active, which is predominantly the horizontal population, are lost. Hence, even in aged mice the number of stem cells remains relatively constant but their mitotic activity reduces and actively proliferating cells are lost, become quiescent, or dormant (Lugert et al., 2010; Bonaguidi et al., 2011; Lugert et al., 2012). This is similar to findings that *Sox1*–positive stem cells remain long-term neurogenic and can enter and exit the active stem cells pools (Venere et al., 2012).

A loss of stem cells in the dentate gyrus would suggest that the neurogenic process cannot be rescued or reversed in aged animals. However, physical exercise and pathological stimulation both stimulate proliferation, neural stem cell activation and under some conditions increased numbers of newly generated neurons (Rao et al., 2005; van Praag et al., 2005; Hattiangady et al., 2008; Jessberger and Gage, 2008; Rao et al., 2008; Zhao et al., 2008). Hence, although loss of stem cells could contribute to the age-related decline in neuron production, some cells with stem cell potential remain even in the dentate gyrus of old mice and these can be activated to proliferate and generate new cells (Lugert et al., 2010; Venere et al., 2012). It still remains unclear whether radial Type-1 cells in old mice enter the cell cycle during physical exercise or whether the few remaining horizontal cells could reactivate in the aged brain or whether a distinct cell population, previously not studied or labeled with the tools and techniques current available, replenishes the neural stem cell pools.

12. Seizures induce neural stem cell proliferation in the hippocampus

Chronic temporal lobe epilepsy is associated with an increase production of neurons in the dentate gyrus (Parent, 2007; Scharfman and Gray, 2007). Conversely, acute seizures dramatically induce abnormal production of neurons in the dentate gyrus, which may contribute to chronic epilepsy. Whether generation of new neurons in the hippocampus of patients with epilepsy is a result of the disease or contributes to the cause is not clear. In mice, experimentally induced seizures effect neuron production at multiple levels and not least by disproportionately increasing the number of neuroblasts (Type-3 cells) (Jessberger et al., 2005). Both the radial Type-1 and the horizontal stem cells are activated in response to experimentally induced seizures (Huttmann et al., 2003; Lugert et al., 2012; Venere et al., 2012). However, the proportion of radial cells that enter cell cycle is rather modest and the population is not expanded suggesting that their divisions generate more committed progenitors. The horizontal stem cells respond more homogeneously to seizures. The majority of them enter the cell cycle and the total number increases significantly (Lugert et al., 2010). The increase in horizontal cells could be the result of symmetric cell division but also generation of horizontal cells from the radial stem cell pool. Although current tools and techniques have not been able to address the mechanism, the increase in mitotically active stem cells following chronic seize and the differential response of the different stem cell pools has important implications for the cause and progression of temporal lobe epilepsy in humans.

13. Future perspectives

In the future it will be a major challenge to elucidate the heterogeneity within the stem cells pools and to address there cellular function. This will include understanding how these different populations and cell states are regulated and whether their functions are controlled by distinct niche signals or genetic and epigenetic mechanisms. Only the detailed analysis of neural stem cells in the adult brain could uncover their functions in homeostasis, aging and disease. This would raise the exciting possibility that specific neural stem cell subtypes could be directly targeted for therapy.

Author details

Verdon Taylor

Embryology and Stem Cell Biology, Department of Biomedicine, University of Basel, Basel, Switzerland

References

[1] Ables, J. L., Decarolis, N. A., Johnson, M. A., Rivera, P. D., Gao, Z., Cooper, D. C., Radtke, F., Hsieh, J. and Eisch, A. J. (2010). Notch1 is required for maintenance of the reservoir of adult hippocampal stem cells. J Neurosci. 30, 10484-10492.

[2] Artavanis-Tsakonas, S., Rand, M. D. and Lake, R. J. (1999). Notch signaling: cell fate control and signal integration in development. Science 284, 770-776.

[3] Balordi, F. and Fishell, G. (2007). Mosaic removal of hedgehog signaling in the adult SVZ reveals that the residual wild-type stem cells have a limited capacity for self-renewal. J Neurosci 27, 14248-14259.

[4] Basak, O. and Taylor, V. (2009). Stem cells of the adult mammalian brain and their niche. Cell Mol Life Sci. 66, 1057-1072.

[5] Basak, O, Giachino, C, & Fiorini, E. MacDonald, H. R. and Taylor, V. (2012). Neurogenic subventricular zone stem/progenitor cells are Notch1-dependent in their active but not quiescent state J Neurosci., 32, 5654-5666.

[6] Beckervordersandforth, R., Tripathi, P., Ninkovic, J., Bayam, E., Lepier, A., Stempfhuber, B., Kirchhoff, F., Hirrlinger, J., Haslinger, A., Lie, D. C., Beckers, J., Yoder, B., Irmler, M. and Gotz, M. (2010). In vivo fate mapping and expression analysis reveals molecular hallmarks of prospectively isolated adult neural stem cells. Cell Stem Cell 7, 744-758.

[7] Ben Abdallah N. M., Slomianka, L., Vyssotski, A. L. and Lipp, H. (2010). Early age-related changes in adult hippocampal neurogenesis in C57 mice. Neurobiol Aging. 31, 151-161.

[8] Bonaguidi, M. A., Song, J., Ming, G. L. and Song, H. (2012). A unifying hypothesis on mammalian neural stem cell properties in the adult hippocampus. Curr Opin Neurobiol. 22, 754-761. doi: 710.1016/j.conb.2012.1003.1013. Epub 2012 Apr 1013.

[9] Bonaguidi, M. A., Wheeler, M. A., Shapiro, J. S., Stadel, R. P., Sun, G. J., Ming, G. L. and Song, H. (2011). In vivo clonal analysis reveals self-renewing and multipotent adult neural stem cell characteristics. Cell. 145, 1142-1155.

[10] Breunig, J. J., Silbereis, J., Vaccarino, F. M., Sestan, N. and Rakic, P. (2007). Notch regulates cell fate and dendrite morphology of newborn neurons in the postnatal dentate gyrus. Proc Natl Acad Sci U S A 104, 20558-20563.

[11] Carlen, M., Meletis, K., Goritz, C., Darsalia, V., Evergren, E., Tanigaki, K., Amendola, M., Barnabe-Heider, F., Yeung, M. S., Naldini, L., Honjo, T., Kokaia, Z., Shupliakov, O., Cassidy, R. M., Lindvall, O. and Frisen, J. (2009). Forebrain ependymal cells are Notch-dependent and generate neuroblasts and astrocytes after stroke. Nat Neurosci. 12, 259-267.

[12] Chapouton, P., Skupien, P., Hesl, B., Coolen, M., Moore, J. C., Madelaine, R., Kremmer, E., Faus-Kessler, T., Blader, P., Lawson, N. D. and Bally-Cuif, L. (2010). Notch activity levels control the balance between quiescence and recruitment of adult neural stem cells. J Neurosci. 30, 7961-7974.

[13] Clelland, C. D., Choi, M., Romberg, C., Clemenson, G. D., Jr., Fragniere, A., Tyers, P., Jessberger, S., Saksida, L. M., Barker, R. A., Gage, F. H. and Bussey, T. J. (2009). A functional role for adult hippocampal neurogenesis in spatial pattern separation. Science. 325, 210-213.

[14] De Marchis, S., Bovetti, S., Carletti, B., Hsieh, Y. C., Garzotto, D., Peretto, P., Fasolo, A., Puche, A. C. and Rossi, F. (2007). Generation of distinct types of periglomerular olfactory bulb interneurons during development and in adult mice: implication for intrinsic properties of the subventricular zone progenitor population. J Neurosci. 27, 657-664.

[15] Doetsch, F., Garcia-Verdugo, J. M. and Alvarez-Buylla, A. (1997). Cellular composition and three-dimensional organization of the subventricular germinal zone in the adult mammalian brain. J Neurosci 17, 5046-5061.

[16] Doetsch, F., Garcia-Verdugo, J. M. and Alvarez-Buylla, A. (1999a). Regeneration of a germinal layer in the adult mammalian brain. Proc Natl Acad Sci U S A 96, 11619-11624.

[17] Doetsch, F., Caille, I., Lim, D. A., Garcia-Verdugo, J. M. and Alvarez-Buylla, A. (1999b). Subventricular zone astrocytes are neural stem cells in the adult mammalian brain. Cell 97, 703-716.

[18] Doetsch, F., Petreanu, L., Caille, I., Garcia-Verdugo, J. M. and Alvarez-Buylla, A. (2002). EGF converts transit-amplifying neurogenic precursors in the adult brain into multipotent stem cells. Neuron 36, 1021-1034.

[19] Ehm, O., Goritz, C., Covic, M., Schaffner, I., Schwarz, T. J., Karaca, E., Kempkes, B., Kremmer, E., Pfrieger, F. W., Espinosa, L., Bigas, A., Giachino, C., Taylor, V., Frisen, J. and Lie, D. C. (2010). RBPJkappa-dependent signaling is essential for long-term maintenance of neural stem cells in the adult hippocampus. J Neurosci. 30, 13794-13807.

[20] Encinas, J. M. and Sierra, A. (2012). Neural stem cell deforestation as the main force driving the age-related decline in adult hippocampal neurogenesis. Behav Brain Res. 227, 433-439.

[21] Encinas, J. M., Michurina, T. V., Peunova, N., Park, J. H., Tordo, J., Peterson, D. A., Fishell, G., Koulakov, A. and Enikolopov, G. (2011). Division-coupled astrocytic differentiation and age-related depletion of neural stem cells in the adult hippocampus. Cell Stem Cell. 8, 566-579.

[22] Essers, M. A., Offner, S., Blanco-Bose, W. E., Waibler, Z., Kalinke, U., Duchosal, M. A. and Trumpp, A. (2009). IFNalpha activates dormant haematopoietic stem cells in vivo. Nature. 458, 904-908.

[23] Fabel, K. and Kempermann, G. (2008). Physical activity and the regulation of neurogenesis in the adult and aging brain. Neuromolecular Med 10, 59-66.

[24] Favaro, R., Valotta, M., Ferri, A. L., Latorre, E., Mariani, J., Giachino, C., Lancini, C., Tosetti, V., Ottolenghi, S., Taylor, V. and Nicolis, S. K. (2009). Hippocampal development and neural stem cell maintenance require Sox2-dependent regulation of Shh. Nat Neurosci.

[25] Fuchs, E. (2009). The tortoise and the hair: slow-cycling cells in the stem cell race. Cell. 137, 811-819.

[26] Gage, F. H. (2000). Mammalian neural stem cells. Science 287, 1433-1438.

[27] Garthe, A., Behr, J. and Kempermann, G. (2009). Adult-generated hippocampal neurons allow the flexible use of spatially precise learning strategies. PLoS One. 4, e5464. Epub 2009 May 5467.

[28] Giachino, C. and Taylor, V. (2009). Lineage analysis of quiescent regenerative stem cells in the adult brain by genetic labelling reveals spatially restricted neurogenic niches in the olfactory bulb. Eur J Neurosci 30, 9-24.

[29] Hack, M. A., Saghatelyan, A., de Chevigny, A., Pfeifer, A., Ashery-Padan, R., Lledo, P. M. and Gotz, M. (2005). Neuronal fate determinants of adult olfactory bulb neurogenesis. Nat Neurosci. 8, 865-872.

[30] Hattiangady, B., Rao, M. S. and Shetty, A. K. (2008). Plasticity of hippocampal stem/progenitor cells to enhance neurogenesis in response to kainate-induced injury is lost by middle age. Aging Cell 7, 207-224.

[31] Huttmann, K., Sadgrove, M., Wallraff, A., Hinterkeuser, S., Kirchhoff, F., Steinhauser, C. and Gray, W. P. (2003). Seizures preferentially stimulate proliferation of radial glia-like astrocytes in the adult dentate gyrus: functional and immunocytochemical analysis. Eur J Neurosci 18, 2769-2778.

[32] Imayoshi, I., Sakamoto, M., Yamaguchi, M., Mori, K. and Kageyama, R. (2010). Essential roles of Notch signaling in maintenance of neural stem cells in developing and adult brains. J Neurosci 30, 3489-3498.

[33] Jessberger, S. and Gage, F. H. (2008). Stem-cell-associated structural and functional plasticity in the aging hippocampus. Psychol Aging 23, 684-691.

[34] Kempermann, G., Kuhn, H. G. and Gage, F. H. (1998). Experience-induced neurogenesis in the senescent dentate gyrus. J Neurosci. 18, 3206-3212.

[35] Kempermann, G., Jessberger, S., Steiner, B. and Kronenberg, G. (2004). Milestones of neuronal development in the adult hippocampus. Trends Neurosci 27, 447-452.

[36] Kohwi, M., Osumi, N., Rubenstein, J. L. and Alvarez-Buylla, A. (2005). Pax6 is required for making specific subpopulations of granule and periglomerular neurons in the olfactory bulb. J Neurosci. 25, 6997-7003.

[37] Kohwi, M., Petryniak, M. A., Long, J. E., Ekker, M., Obata, K., Yanagawa, Y., Rubenstein, J. L. and Alvarez-Buylla, A. (2007). A subpopulation of olfactory bulb GABAergic interneurons is derived from Emx1- and Dlx5/6-expressing progenitors. J Neurosci. 27, 6878-6891.

[38] Kuhn, H. G., Dickinson-Anson, H. and Gage, F. H. (1996). Neurogenesis in the dentate gyrus of the adult rat: age-related decrease of neuronal progenitor proliferation. J Neurosci. 16, 2027-2033.

[39] Kuwabara, T., Hsieh, J., Muotri, A., Yeo, G., Warashina, M., Lie, D. C., Moore, L., Nakashima, K., Asashima, M. and Gage, F. H. (2009). Wnt-mediated activation of NeuroD1 and retro-elements during adult neurogenesis. Nat Neurosci 12, 1097-1105.

[40] Lagace, D. C., Whitman, M. C., Noonan, M. A., Ables, J. L., DeCarolis, N. A., Arguello, A. A., Donovan, M. H., Fischer, S. J., Farnbauch, L. A., Beech, R. D., DiLeone, R. J., Greer, C. A., Mandyam, C. D. and Eisch, A. J. (2007). Dynamic contribution of nestin-expressing stem cells to adult neurogenesis. J Neurosci. 27, 12623-12629.

[41] Li, L. and Clevers, H. (2010). Coexistence of quiescent and active adult stem cells in mammals. Science 327, 542-545.

[42] Louvi, A. and Artavanis-Tsakonas, S. (2006). Notch signalling in vertebrate neural development. Nat Rev Neurosci 7, 93-102.

[43] Lugert, S., Vogt, M., Tchorz, J. S., Muller, M., Giachino, C. and Taylor, V. (2012). Homeostatic neurogenesis in the adult hippocampus does not involve amplification of Ascl1(high) intermediate progenitors. Nat Commun., 670.

[44] Lugert, S., Basak, O., Knuckles, P., Haussler, U., Fabel, K., Gotz, M., Haas, C. A., Kempermann, G., Taylor, V. and Giachino, C. (2010). Quiescent and active hippocampal neural stem cells with distinct morphologies respond selectively to physiological and pathological stimuli and aging. Cell Stem Cell 6, 445-456.

[45] Merkle, F. T., Mirzadeh, Z. and Alvarez-Buylla, A. (2007). Mosaic organization of neural stem cells in the adult brain. Science 317, 381-384.

[46] Mirzadeh, Z., Merkle, F. T., Soriano-Navarro, M., Garcia-Verdugo, J. M. and Alvarez-Buylla, A. (2008). Neural stem cells confer unique pinwheel architecture to the ventricular surface in neurogenic regions of the adult brain. Cell Stem Cell. 3, 265-278.

[47] Mori, T., Tanaka, K., Buffo, A., Wurst, W., Kuhn, R. and Gotz, M. (2006). Inducible gene deletion in astroglia and radial glia--a valuable tool for functional and lineage analysis. Glia 54, 21-34.

[48] Morshead, C. M., Reynolds, B. A., Craig, C. G., McBurney, M. W., Staines, W. A., Morassutti, D., Weiss, S. and van der Kooy, D. (1994). Neural stem cells in the adult

mammalian forebrain: a relatively quiescent subpopulation of subependymal cells. Neuron 13, 1071-12082.

[49] Nam, H. S. and Benezra, R. (2009). High levels of Id1 expression define B1 type adult neural stem cells. Cell Stem Cell. 5, 515-526.

[50] Nyfeler, Y., Kirch, R. D., Mantei, N., Leone, D. P., Radtke, F., Suter, U. and Taylor, V. (2005). Jagged1 signals in the postnatal subventricular zone are required for neural stem cell self-renewal. EMBO J 24, 3504-3515.

[51] Parent, J. M. and Murphy, G. G. (2008). Mechanisms and functional significance of aberrant seizure-induced hippocampal neurogenesis. Epilepsia 49 Suppl 5, 19-25.

[52] Pastrana, E., Cheng, L. C. and Doetsch, F. (2009). Simultaneous prospective purification of adult subventricular zone neural stem cells and their progeny. Proc Natl Acad Sci U S A. 106, 6387-6392.

[53] Rao, M. S., Hattiangady, B. and Shetty, A. K. (2008). Status epilepticus during old age is not associated with enhanced hippocampal neurogenesis. Hippocampus 18, 931-944.

[54] Rao, M. S., Hattiangady, B., Abdel-Rahman, A., Stanley, D. P. and Shetty, A. K. (2005). Newly born cells in the ageing dentate gyrus display normal migration, survival and neuronal fate choice but endure retarded early maturation. Eur J Neurosci 21, 464-476.

[55] Reynolds, B. A. and Weiss, S. (1992). Generation of neurons and astrocytes from isolated cells of the adult mammalian central nervous system. Science 255, 1707-1710.

[56] Seri, B., Garcia-Verdugo, J. M., McEwen, B. S. and Alvarez-Buylla, A. (2001). Astrocytes give rise to new neurons in the adult mammalian hippocampus. J Neurosci 21, 7153-7160.

[57] Shen, Q., Wang, Y., Kokovay, E., Lin, G., Chuang, S. M., Goderie, S. K., Roysam, B. and Temple, S. (2008). Adult SVZ stem cells lie in a vascular niche: a quantitative analysis of niche cell-cell interactions. Cell Stem Cell. 3, 289-300.

[58] Shen, Q., Goderie, S. K., Jin, L., Karanth, N., Sun, Y., Abramova, N., Vincent, P., Pumiglia, K. and Temple, S. (2004). Endothelial cells stimulate self-renewal and expand neurogenesis of neural stem cells. Science 304, 1338-1340.

[59] Shors, T. J., Miesegaes, G., Beylin, A., Zhao, M., Rydel, T. and Gould, E. (2001). Neurogenesis in the adult is involved in the formation of trace memories. Nature 410, 372-376.

[60] Steiner, B., Zurborg, S., Horster, H., Fabel, K. and Kempermann, G. (2008). Differential 24 h responsiveness of Prox1-expressing precursor cells in adult hippocampal neurogenesis to physical activity, environmental enrichment, and kainic acid-induced seizures. Neuroscience. 154, 521-529.

[61] Steiner, B., Klempin, F., Wang, L., Kott, M., Kettenmann, H. and Kempermann, G. (2006). Type-2 cells as link between glial and neuronal lineage in adult hippocampal neurogenesis. Glia. 54, 805-814.

[62] Stump, G., Durrer, A., Klein, A. L., Lutolf, S., Suter, U. and Taylor, V. (2002). Notch1 and its ligands Delta-like and Jagged are expressed and active in distinct cell populations in the postnatal mouse brain. Mech Dev 114, 153-159.

[63] Suh, H., Consiglio, A., Ray, J., Sawai, T., D'Amour, K. A. and Gage, F. H. (2007). In vivo fate analysis reveals the multipotent and self-renewal capacities of Sox2+ neural stem cells in the adult hippocampus. Cell Stem Cell 1, 515-528.

[64] Suhonen, J. O., Peterson, D. A., Ray, J. and Gage, F. H. (1996). Differentiation of adult hippocampus-derived progenitors into olfactory neurons in vivo. Nature. 383, 624-627.

[65] Tavazoie, M., Van der Veken, L., Silva-Vargas, V., Louissaint, M., Colonna, L., Zaidi, B., Garcia-Verdugo, J. M. and Doetsch, F. (2008). A specialized vascular niche for adult neural stem cells. Cell Stem Cell. 3, 279-288.

[66] van Praag, H., Shubert, T., Zhao, C. and Gage, F. H. (2005). Exercise enhances learning and hippocampal neurogenesis in aged mice. J Neurosci 25, 8680-8685.

[67] Venere, M., Han, Y. G., Bell, R., Song, J. S., Alvarez-Buylla, A. and Blelloch, R. (2012). Sox1 marks an activated neural stem/progenitor cell in the hippocampus. Development. 139, 3938-3949. doi: 3910.1242/dev.081133. Epub 082012 Sep 081119.

[68] Wilson, A., Laurenti, E., Oser, G., van der Wath, R. C., Blanco-Bose, W., Jaworski, M., Offner, S., Dunant, C. F., Eshkind, L., Bockamp, E., Lio, P., Macdonald, H. R. and Trumpp, A. (2008). Hematopoietic stem cells reversibly switch from dormancy to self-renewal during homeostasis and repair. Cell. 135, 1118-1129.

[69] Zhao, C., Deng, W. and Gage, F. H. (2008). Mechanisms and functional implications of adult neurogenesis. Cell 132, 645-660.

Chapter 3

Diversity of Neural Stem/Progenitor Populations: Varieties by Age, Regional Origin and Environment

Emília Madarász

Additional information is available at the end of the chapter

http://dx.doi.org/10.5772/55678

1. Introduction

Criteria of „neural stemness" characterize a large number of terminally non-differentiated neural tissue cells. Neural stem/progenitor cells capable for asymmetric mitoses (resulting in a similar and a differently committed daughter cell which may adopt neuronal or glial phenotypes in further development) are present during the entire life-span of vertebrates and have been found in almost all regions of the brain. With the advancement of neural tissue genesis and maturation, more and more stem/progenitor-like cells adopt "quiescent" states, but can be activated by appropriate (yet not properly understood) stimuli. Besides asymmetric (stem cell specific) division, these cells can multiply by symmetric mitoses resulting in identical progenies. Self-renewal and symmetric multiplication are responsible for maintaining or expanding stem/progenitor populations at the actual stage of neural commitment. Expanded pools of cells with similar, but yet flexible developmental potential can provide the desired number and type of cells for genesis, maintenance and repair of the nervous tissue. Except the ontogenetically and phylogenetically "oldest" pioneer and/or large projection-type neurons [1], the majority of neural tissue cells are produced through successive stem/progenitor stages [2]. The extreme cellular diversity of the mature CNS implies huge diversity in the precursor populations. Accordingly, a large number of neural stem/progenitor populations exist in different stages of neural cell fate commitment and display different cellular characteristics, developmental capability and flexibility. "Quiescent" and actively proliferating stem cells, transient amplifying progenitor populations and migrating or resident progenitor/precursor cells reside at various "niches" including the "professional" neurogenic zones, migratory routes and the neural parenchyma, as well. Drifts in cell biological features and differentiation potential of stem/progenitor/precursor cells are implemented by the advancement of development, by the position along the body axes and by the physiological or pathophysiological changes of the local environment including neighbouring cells, extracellular matrix, metabolite-, oxygen- and growth factor-supply and activity-driven ionic composition.

With improvement of methodologies, different populations of NS cells have been isolated, propagated and investigated *in vitro*. Despite of rapidly accumulating data, however, the similarities and differences among various NS populations are not properly understood either *in vivo* or *in vitro*. This chapter intends to give an insight in the intrinsic varieties of NSC populations. As examples, fairly distinct features of in vitro propagated early embryonic and adult-derived mouse neural stem/progenitor populations will be presented.

2. Diversity of neural stem cells from early embryonic tissue genesis to adult-hood cell production

2.1. Neural stem cells in the neural plate and in primary geminative zones

The earliest neural stem populations, the cylindrical neuroepithelial cells of the early embryonic neural plate (Fig.1) appear in the embryonic disc soon after embedding. As it was learned from studies on chick embryos, the anterior epiblast cells acquire „pre-neural" fate [3] as it is indicated by Sox2 expression [4] under the influence of morphogen gradients established by two early organizers, the anterior visceral endoderm (AVE) [5] and the primitive streak and later by the node. The different morphogen concentrations result in different activation of positional (region-specific) genes, predestinating regions for different future fate in both chick and mammalian embryos [6,7] (Fig.1). Accordingly, despite of the apparent morphological homogeneity [8], neuroepithelial cells are not identical from the very beginning of neural tissue genesis.

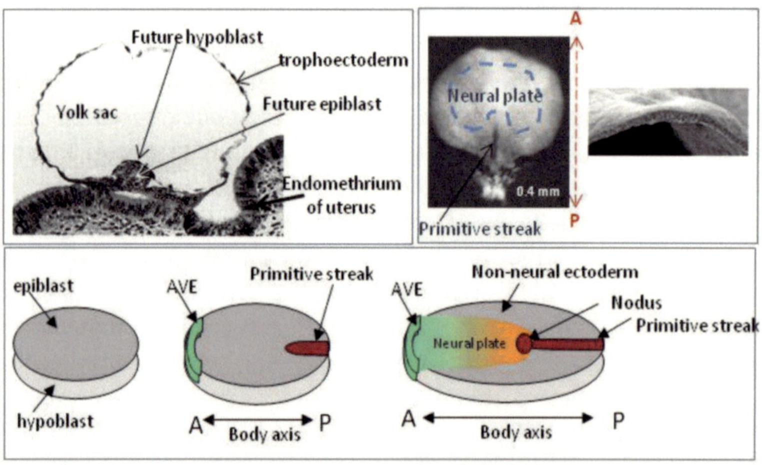

Figure 1. Scheme of formation and regionalization of the neural plate. The neural plate forms from the anterior epiblast in response to morphogens produced by the anterior visceral entoderm (AVE), the primitive streak and nodus

and the non-neural ectodermal ridge of the epiblast. The positions along the anterior (A) and posterior (P) body axes - imply regional differences among neuroectodermal cells.

During neural folding, a single layer of proliferating pre-neural stem cells called *radial neuroepithel cells* compose the neural plate and rapidly enlarge the neural primordium. These cells will give rise to primary neural stem cells, the *embryonic radial glia* (Fig.2), during the formation and closing of the neural tube (7-10 pcd in mouse). While the anterior cells of the epiblast of vertebrates were shown to generate neurons, if separated from their natural environment, ("default neural fate" [9]), or by blocking mesoderm-inducing morphogen actions, the mechanisms behind the transition of radial neuroepithel cells to radial glia cells are not properly understood.

Embryonic radial glial cells line the lumen of the neural tube in a single cell layer and compose the primary neural germinative zone (ventricular zone; VZ). The first neurons are generated in this layer by asymmetric mitoses of radial glial cells (Fig.2). Such division produces a daughter cell, which preserves radial glia characteristics and another one, a neuronal precursor, which looses proliferation capability and migrates outward from the lumen along a radial glia neighbour [10]. These early neuronal precursors will develop mainly to large, projecting type neurons [1]. Immediate cell-to cell interactions including Notch/Delta signalling [11] and growth factor – receptor signals (as neuregulin – ErbB; [12]) play inevitable roles in the determination and maintenance of asymmetric commitment of the two daughter cells.

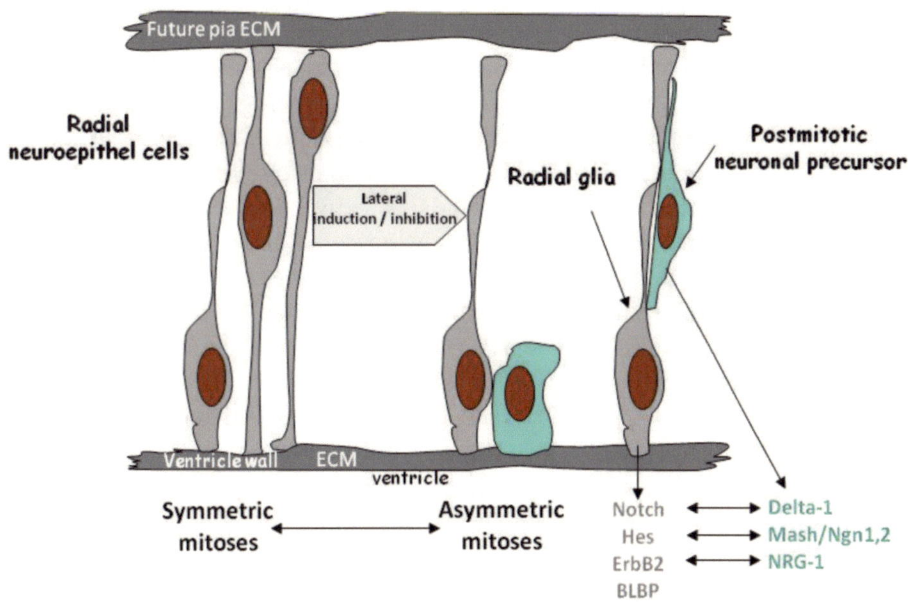

Figure 2. Formation of the primary neurogenic zone from radial neuroepithelial cells and generation of the first postmitotic neuronal precursors by asymmetric division of radial glial cells in the ventricular zone.

The early radial glial cells express several "marker" features (Table 1), which together with cell shape and localization may identify them. In contrast to later neural stem/progenitor populations, radial neuroepithelial and early radial glial cells express Oct4 and nanog embryonic stem cell genes and the anterior epiblast-characterizing Otx2 and En1,2 "positional" genes (see ref. in Table 1). These cells span the whole thickness of the early neural tube, and can divide without changing their spanned shape [8].

These early neural stem cells proliferate rapidly (with cycle time of about 14-16 hours), but not continuously, and with both symmetric and asymmetric divisions. The symmetric mitoses assure the expansion and maintenance of neural stem populations at the given stage of commitment. Intermittent non-mitotic periods help the attachment and out-migration of neuronal precursors. During the formation and closure of the tube, the neural primordium is composed by a single neurogenic zone, but with heterogeneous cellular constituents [13] comprising asymmetrically and symmetrically dividing stem/progenitor cells, and migrating, differentiating neuronal precursors.

The cellular diversity of the early neural tube is further enhanced by the time-delay in development along the anterior – posterior and dorso-ventral body axes. The tube formation and closure proceed with a delay from the zone of brachial arches (the region for future hind brain) to both, rostral and caudal directions, and the production and maturation of neural cells on the ventral face always precede those on the dorsal part in each neural domain [14]. It means that even small samples of the tissue will contain diverse populations of (stem/amplifying/differentiating) neural cells.

The cellular heterogeneity is further increased by the ongoing regional specification, which results in well-distinguishable domains with characteristic gene expression patterns along both the anterio-posterior and dorso-ventral axes of the growing CNS [28, 29]. Differential expression of positional master genes results in diverging expression patterns of "down-hill" genes including those coding for adhesion receptors and extracellular matrix proteins, and leads to the formation of morphological boundaries between the expression domains: the developing CNS is composed by segments called neuromeres [7, 30]. Along the A-P body axis, transversal segments will delineate primary brain vesicles (prosencephalon, mesencephalon, rhombencephalon) first, and smaller neuromeres later on. Inside the larger segment boundaries, smaller developmental / morphological entities will develop as the prosomeres in the telencephalon or rhombomeres in the hindbrain. In parallel with the anterior-posterior segmentation, dorso–ventral specification will identify longitudinal domains in the developing CNS. Dorso–ventral specification is initiated by the early lateral–medial determination of the neural plate, namely, the midline expression of sonic hedgehog and the lateral expression of bone morphogenic (BMP) proteins (Fig.3). With tube formation, progenies of midline plate cells become the ventral-most tube cells composing the notoplate. Derivatives of lateral-most neural plate cells will compose the neural crest and the closing dorsal lip of the tube, the later roof plate. Notoplate cells produce Shh, while roof plate cells produce BMPs and Wnt morphogenic factors, establishing dorso-ventral morphogen gradients throughout the developing CNS (for a recent review: [31]).

http://dx.doi.org/10.5772/55678

	„Marker" features	References
cell shape	Elongated, bipolar shape; Adherent junctions at the lumen-face; laminin receptors at the apical and basal ends;	[10, 8]
cellular motion	interkinetic nuclear migration; mitotic nuclei at the lumen face; cell division with spanned shape	[15, 8]
intermediate filament proteins	vimentin (characterizes many non-differentiated cells)	[refs in 1]
	GFAP (characterizes astrocytes; expression by radial glia cells is species-dependent)	
	nestin (expressed by many progenitor-type cells including muscle progenitors)	
	RC (1,2) nestin-associated protein-epitope; radial glia marker in the CNS	
master-gene expression	Oct3/4, nanog (embryonic stem cell/pluripotency genes; expressed in the epiblast and sporadically in the forming neural plate)	[16]
	Otx2 (anterior positional gene; expressed in the anterior epiblast and in the neural plate; later restricted to the forebrain regions)	[17, 18]
	En1,2 (engrailed positional gene; expressed in the anterior epiblast, in the neural plate; later restricted to isthmus and mesencephalic, cerebellar areas)	[19, 20]
	Sox2 (codes for a HMG box transcription factor; expressed generally by CNS neural stem cells)	[4,21,22]
	Pax6 (codes for a paired-box transcription factor; plays important roles in CNS patterning and eye development; expressed by multiple neural stem/ progenitor cells)	[23, 24]
	Hes1,3,5 (code for a basic helix-loop-helix transcription factor repressing tissue-specific cell differentiation)	[25]
Cell surface receptors and transporters	Notch1,3 (protein product: cell surface receptor with cleavable intracellular transcription fragment (NICD)	[11]
	ErbB (Neuregulin Trk receptor);	[12]
	GLAST (Glutamate-aspartate transporter; astrocytes also express)	[26]
Intracellular transporter	BLBP (brain lipid binding protein; Intracellar transporter of hydrophobic molecules including small nuclear ligand molecules)	[27]

Table 1. Radial glia "marker" features for characterization of radial glial cells

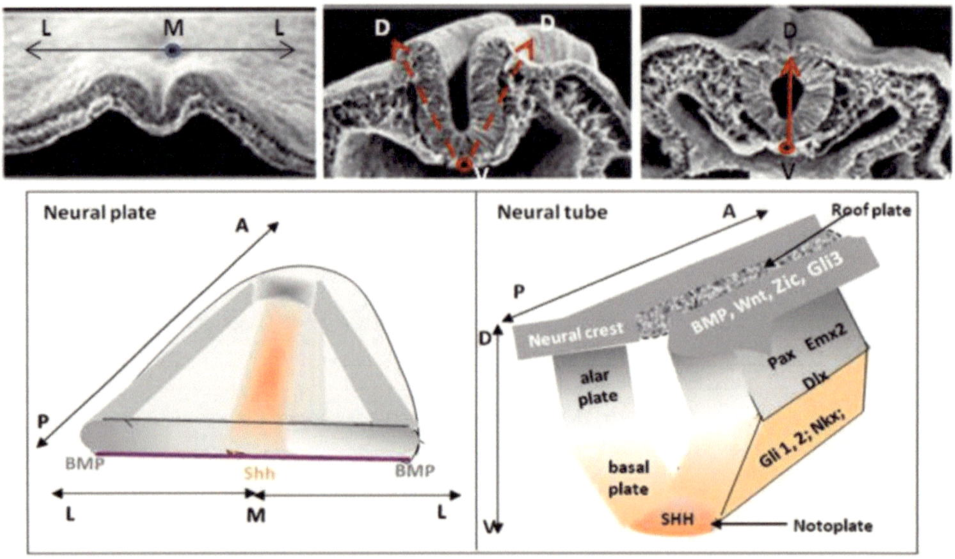

Figure 3. Schematic presentation of the establishment of dorso–ventral specification (longitudinal segmentation) in the neural tube. The lateral-medial (L-M) regionalization turns into dorso-ventral (D-V) regionalization when the neural plate folds and forms the neural tube. BMP: bone morphogenic proteins; SHH: sonic hedgehog protein; Zic, Gli-3, Pax6, Emx2, Dlx, Nkx are examples of genes expressed in different dorso-ventral domains.

The position along the antero-posterior and dorso–ventral axes determines the identity of future brain regions and seems to predict the phenotype of neurons through the orchestrated expression of defined region-specific, pro-neural and neuron-specific genes [28]. Boundaries between embryonic segments provide routes for the elongation of pioneer axons and delineate the paths for future fiber tracks.

For the time being, it is not clear whether the early embryonic position can (or how far can) determine the intrinsic developmental potential of individual cells. There are contradictory results on the "positional memory" of stem/progenitor cells if removed from their original position [32]. Our own data [33] indicate that neural stem-like cells isolated from early embryonic (E9) mouse forebrain do not display regional "memory" after in vitro propagation. In the course of in vitro neuron formation they express divergent region-specific genes, those not expressed in overlapping regions of the developing CNS. The data suggest that permanent presence of region-specifying factors is required for maintaining regional commitment, at least in case of early embryonic neural stem cells. This finding, however, does not compromise the fact that in vivo, neural stem/progenitor cells of the primary germinative (VZ) zone display important molecular and cell biological differences.

2.2. Diversity of stem/progenitor cells in the secondary germinative zones of the developing brain

With the thickening of the neural wall, increasing number of proliferation-capable cells loose contact either with the lumen-face (outer progenitors; OSVZ progenitors) or the pial surface (short progenitors; SNPs) or both (inner progenitors; ISVZ progenitors), and accordingly, change morphology and developmental characteristics [34, 8]. The number of primary radial glial cells decreases gradually, while their derivatives dividing at a distance from the luminar surface generate a novel germinative zone, the subventricular zone (SVZ). While a few SNP cells remain in contact with the luminar surface and thought to preserve "ancient" (primary) stem cell properties, the layer lining the ventricle wall transforms to the future ependyma and neural cell production is transposed to the secondary germinative layer, the SVZ. The SVZs along the entire neuraxis generate large number of progenitors and precursors including smaller projecting-type neurons, interneurons, astrocytes, oligodendrocytes and produces also enough stem cells to maintain an appropriate neurogenic capacity (for review see [1]). Novel precursors migrate from the SVZs to final destinations along defined routes in the developing neural parenchyma. While the lineage relations among different SVZ progenitors are not fully explored [8, 35], it is clear that evolutionary new SVZ progenitors provide the cell generating capacity for the enlarged interneuron and glia populations of the avian and mammalian CNS and provide cell-pools for the enormous expansion of the cerebellar cortex and the mammalian forebrain. With the appearance of SVZs, the developing CNS comprises a large variety of coexisting neural stem/progenitor cells both, in multiple neurogenic zones and inside / around the cell migratory routes.

While secondary (SVZ) germinative zones form all along the wall of brain ventricles and the spinal canal, the time of formation and the cell productivity of these zones are not uniform along the neuraxis. At most parts of the developing CNS, SVZs generate neuronal and glial precursors for local tissues, and the cell-generating activity decreases with the maturation of the local neural parenchyma. Such zones, still containing "resting" stem/progenitor cells, can restore cell-generating activity in response to tissue loss or specific (not yet properly understood) physiological/pathophysiological stimuli. The most productive secondary germinative zones, including the ventral forebrain SVZ [36], the external germinative layer (EGL) of the cerebellum [37] and the subgranular zone (SGZ) of the hippocampus [38], however, generate huge amounts of novel neural cells and function for a relatively long time, in case of the ventral forebrain SVZ and the SGZ, for the entire life-time.

The SVZ at the latero-ventral wall of the forebrain vesicles forms in a relatively early period (second embryonic week in mouse) of forebrain development [36]. It derives from the primary germinative (VZ) zone of the highly regionalized subpallial ganglionic eminences (LGE, MGE and CGE,) [39]. The enhanced cell production with preserved regional characteristics results in the formation of a variety of neuronal progenitors/precursors for developing subcortical and cortical tissue formations [39]. This zone generates neurons for the striatum and globus pallidus, provides the vast majority of interneurons of the cortex [40, 41], produces small connecting neurons for the ventral amygdala complex [42, 43] and, exclusively in humans, also for the dorsal thalamus [44]. Large number of forebrain oligodendrocytes also derives from the ventral SVZ complex [45, 46].

In contrast to the early appearance of the ventral forebrain SVZ, the external germinative layer (EGL) of the cerebellum appears around birth [47]. Secondary stem/progenitor cells from the dorsal lip of the IVth ventricle migrate on the top of developing Purkinje cells, and produce basket and stellar neurons and billiards of cerebellar granule cells [37] for the cerebellar cortex. The third "professional" secondary neurogenic zone of the mammalian brain is the subgranular zone (SGZ) of the hippocampus [38], which derives from the ventricular (primary germinative) zone of the most dorsal forebrain structure, the hem [48, 49]. In addition to astrocytes and oligodendrocytes, SGZ produces a single type of projecting neurons, the granule cells of the dentate gyrus.

2.3. Neural stem/progenitor cells in the adult mammalian brain

At most sites, the subventricular zones cease producing neurons and reduce their glia-production after a short postnatal period of normal development, except the external germinative layer (EGL) of the cerebellum, where cerebellar granule cells are generated for at least 10 days after birth in mouse and 1-2 years in human [47], the SVZ of the ventral forebrain [50] and SGZ of the hippocampus [51], where neuro- and gliogenic capacity is maintained for the entire life-time. The adult SVZ generates neuronal precursors mainly for regularly remodelling forebrain circuits, as the higher voice centre of singing birds [52] or the olfactory bulb in rodents [53]. Incorporation of adult SVZ-derived novel neurons, however, had been reported also in the neocortex, piriform cortex, olfactory tubercle and amygdala in rodents and in primates, as well [42, 43,54, 55]. In contrast to the SVZ, available data indicate that SGZ produces precursors solely to the dentate gyrus. Both, the SVZ [46] and the SGZ [51, 56], however, contain consecutive descendent populations of progenitor cells, which seem to mutually regulate each other's generation [57].

The routes of migration and integration of newly generated "adult" neuronal precursors into olfactory and hippocampal networks have been explored in many details [51, 53, 58, 59, 60], and the exploration has been highly facilitated by the use of multiple transgenic reporter mouse straits (reviewed in [35]). Novel neurons are integrated by functioning neuronal networks and tune the physiological parameters of functional circuits in both, the olfactory bulb [61] and the hippocampus [59, 60]. As it is widely accepted, continuous neurogenesis in the adult SVZ and SGZ plays organic roles in the physiological performance of the olfactory bulb and the hippocampus of mammals. In case of SVZ-derived precursors, however, further studies should explore the mechanisms behind the detachment of novel progenitors from the common migratory paths and the acquirement of different mature phenotypes.

Outside of adult neurogenic zones and migratory routes, developing novel neurons have been described in many regions of the adult brain, both in rodents (for ref. see [62]) and primates [54]. Local neuron formation was found at the striatum [63], hypothalamus [64], associate cortex [54, 65], ventral forebrain [43] and the substantia nigra of the midbrain (for review see [66]. Beside sporadic neuron formation, ongoing astroglia production and low-rate but permanent oligodendroglia replacement [67] also indicate that differentiation-capable neural stem/progenitor cells are scattered throughout the adult nervous tissue. Many of these progenitors might derive from continuously active ("professional") adult neurogenic zones [1, 45]. Quiescent cells with stem/progenitor capabilities however are present in the wall of brain ventricles and spinal canal [68] along the entire neuraxis, in large fibre bundles [69, 70] and in the functioning CNS parenchyma [71], as well. In response to stimuli, many quiescent tissue-resident neural cells can re-enter the cell cycle [72], can renew themselves and generate different – more than one - types of mature neural cells in vivo [69, 71, 73].

While such cells are regarded as neural stem cells according to „neural stemness" criteria, the time of birth, the route of migration, the conditions for settlement and quiescence are far from clear. The multiplicity of CNS derived stem/progenitor cells has been indicated also by the diversity of neural stem/progenitor cells investigated in vitro [74]; the data argue for a massive presence of tissue-resident neural cells with "stemness" properties throughout the CNS. To find single scattered quiescent cells in the healthy brain tissue or to determine their lineage relations are hard tasks, even with the use of transgenic reporter mice [35]. These neural stem / progenitor cells may have different origin, may represent different stages of neural cell differentiation and may adopt different cell physiological characteristics depending on the host environment. As they serve as resident progenitor pools for tissue repair and limited remodelling, it is of primary importance to explore their physiological characteristics and environmental requirements in order to understand local neural tissue reactions to physiological stimuli and various pathophysiological effects.

3. Neural stem cells in vitro: Experimental data on diverse clones

In vivo lineage analyses can hardly determine whether divergent fate decision could be made by a single cell or diversity was resulted by selective interactions inside a group of mixed cells [75]. Sophisticated cell isolation/propagation methods helped to prove the existence of multipotent neural stem cells. In vitro studies can describe neural stem cell features appearing under artificial conditions and can predict the *potential* phenotypes which can be acquired by the investigated cells. The in vivo fate and the phenotype manifested under in vivo conditions, however, can not be forecasted from in vitro data. For the time being, our knowledge on local microenvironmental factors governing the fate of differentiating neural cells or our understanding on characteristics and cell physiological requirements of differentiating cells do not allow extrapolating in vitro data to in vivo cell fates.

Neural stem/progenitor cells had been isolated and investigated in many laboratories using different methods and divergent source materials including developing or adult brain tissues and embryonic stem (ES) cells. Excellent recent reviews summarize the collected data including the expression of various gene clusters, the inducibility of different neuronal phenotypes and the growth factor requirements of various neural stem/progenitor preparations (for a recent review see [76]). Recent knowledge on differentiation-dependent changes in requirements for O_2 and adhesive environment has been also summarized [77]. The diverse experimental approaches, however, hinder comparative characterization of neural stem/progenitor populations derived from different CNS regions and developmental ages.

Based on more than 10 different neural stem cell clones isolated and characterized in our laboratory, this chapter presents a comparative summary on important differences displayed by one-cell derived neural stem cell populations isolated from different ages and/or from different regions of the mouse brain.

3.1. Lessons from primary brain cell cultures

From the middle of the last century, "primary" neural cell cultures gave a statistical insight into the cellular composition of many studied brain area and gained high importance in understanding the cellular characteristics of neural tissue constituents. It became clear that neuron-enriched cultures can be easily initiated from early stages of development (embryonic brain material), while tissue samples from later stages of brain maturation provide mainly glial cultures.

As terminally differentiated neurons and oligodendrocytes do not divide and display poor regeneration potential, primary neural cell cultures gave a hint on the presence and distribution of non-differentiated cells in the brain tissue, which served as precursors for the majority of cultured neurons and oligodendroglia cells. The in vitro preservation of many neuronal characteristics of the source region including the size, shape, neurotransmitter phenotypes and specific vulnerability of neurons [78] indicates either a relatively stable fate commitment of in vitro surviving progenitors/precursors or continued production of factors, which help to maintain some region-specific features.

In case of astrocytes, usually identified solely by GFAP expression, it was clear that in vitro propagation results in a "juvenile", "de-differentiated" phenotype, fairly distinct from in vivo astroglial cells [79]. In purified (90-95% GFAP-positive) mouse astroglial cultures many cells express nestin, and a few of them express also SSEA-1 stem cell antigen (Fig.4). After longer (2 -3 weeks) propagation, "epithel dome"-like structures of rapidly dividing cells and GFAP-negative process-bearing O2A bipotential glial progenitor cells [69] often appear in purified astroglial cultures (Fig.4) [80,81].

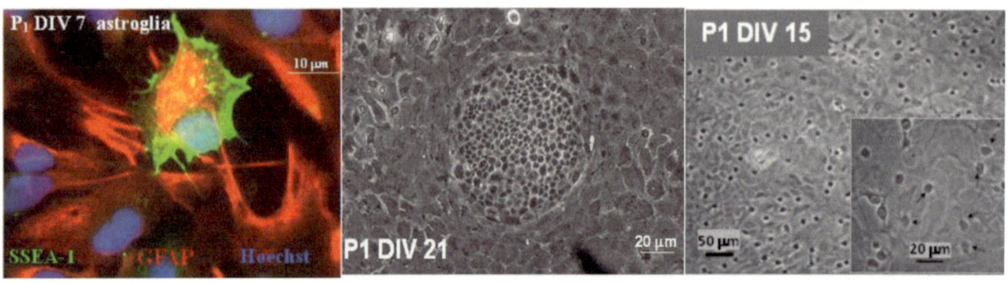

Figure 4. Non-differentiated cells in purified cultures of newborn (P1) mouse forebrain astroglial cells. Left: a SSEA-1 and GFAP double-immureactive cell on the 7th day in vitro. Middle: an "epithel dome"-like cellular expansion on the 21st day in vitro. Right: O2A-type progenitor cells in a 15-day old hypothalamic astrocyte culture.

As long-term propagation selects for rapidly proliferating cells, the size of such populations can be enlarged and one-cell derived clones may also emerge from such cells. Proliferation-based cloning from long-term propagated cultures, however, favours the selection for tumorigen, transformed cells and also, hinders the identification of in vivo origin of selected cells.

If primary neural cultures are prepared form early embryonic (E9-12 mouse) forebrain vesicles, clusters of rapidly proliferating non-differentiated cells occur with high frequency. From early embryonic source tissue, the larger frequency and rapid proliferation of non-differentiated cells allow isolating "almost primary" neural stem/progenitor cells after short-term in vitro propagation. Using the "serial splitting" method [82], we established and cloned various neural stem cell lines from E9 - E11.5 mouse forebrain vesicles from, both, p53$^{-/-}$ [83] and wild-type embryos.

While non-differentiated cells from the early embryonic (E9-12) brain attach readily to poly-L-lysine or collagen coated surfaces in the presence of serum, cells isolated from more developed (E14-15) mouse forebrain form aggregates under such conditions indicating that the surface of neighbouring cells provide better adherence than the provided artificial surface. The majority of cells exist inside the aggregates (e.g. in completely unknown microenvironments), for a 2-3 day period before large-scale migration starts from the aggregates.

Rapid attachment of freshly seeded cells from "older" forebrain suspensions was achieved, if pre-patterned integrin-ligands were provided as adherent surfaces. Brush-like peptide-conjugates built on a poly-L-lysine backbone and carrying rigid integrin-ligands (cyclic RGDfC pentapeptides) at the N-termini of regularly spaced poly-D/L lanine "spacer" side-chains (AK-c[RGDfC]) [84] proved to support the attachment and serum-free propagation of a number of non-differentiated cells including mouse and human ES cells (manuscript in preparation). As neurons do not attach to the peptide-coated surface and differentiated glial cells do not proliferate in serum–free conditions, non-differentiated cells of brain cell suspensions could be selectively propagated under serum-free conditions on the peptide-coated surface (Fig.5). From forebrain suspension of transgenic mouse embryos (E14.5) carrying the human GFAP-promoter-driven GFP construct [85], GFP-labelled but GFAP-negative cells grew rapidly on AK-c(RGDfC)-coated surfaces indicating that the surface-attached populations comprised neural progenitor-type cells (Fig.5) [74].

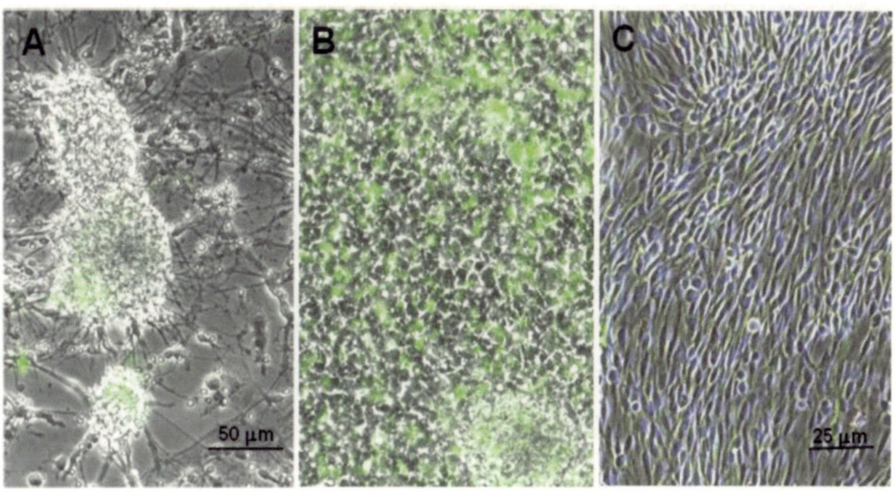

Figure 5. Embryonic (E 14.5) mouse$^{hGFAP\text{-}GFP}$ forebrain cells on the 3rd day on poly-L-lysine coated surface with serum (A) and on AK-c(RGDfC) coated surface without serum (B). After two passages on AK-c(RGDfC) sureum and in serum-free conditions, the cultures were composed by GFP-expressing morphologically homogeneous, surface-attached cells (C).

The "selective adhesion and serum free propagation" method allowed isolating and cloning cell lines from various parts of the embryonic (E 14-15) and adult (P50-75) mouse brains (Table 2).

3.2. Comparative characterization of neural stem/progenitor clones

Our first neural stem cell clones (NE-4C *[ATTC CRL 2925]* and NE-7C2) were established 15 years ago, from early embryonic (E9) anterior brain vesicles of p53-deficient mouse embryos [82,86]. The lack of p53 tumour suppressor protein did not prevent the in vitro formation of postmitotic neuronal precursors and later neurons [82], as did not hinder normal neural development of transgenic animals despite of increased tumour frequency in aging animals [83]. For control, however, similar cell lines had been established from wild-type mouse embryos (WNE cell lines) and were proved to display similar characteristics including morphology, chromosome-stability, cell cycle-time, regional gene expression and the schedule of in vitro neuron and glia formation. NE-4C, NE-7C2 and WNE cells showed similarities to P19 EC and embryoid body-derived ES cells in many aspects [33,87], and showed marked differences if compared to AK-c(RGDfC) adherent stem/progenitor cells isolated either from late embryonic or adult brain regions (Table 2). All clones shown on Table 2 can be propagated in vitro without differentiation or changing phenotype, and can generate neurons, astrocytes and oligodendrocytes in response to appropriate inducing stimuli.

<table>
<tr><th>Cell clone</th><th>tissue origin</th><th>age of source tissue</th><th>method of isolation</th></tr>
<tr><td>NE-4C
(ATTC;CRL-2925)</td><td rowspan="2">anterior brain vesicles;
$p53^{-/-}$ mouse embryo</td><td rowspan="2">E 9</td><td rowspan="3">Serial splitting;
serum-supported growth on
poly-L-lysine</td></tr>
<tr><td>NE-7C2</td></tr>
<tr><td>WNE</td><td>forebrain</td><td>E 11.5</td></tr>
<tr><td>RGl-1</td><td>whole forebrain</td><td rowspan="3">E 14.5</td><td rowspan="11">AK-c(RGDfC) adherence;
serum-free growth on AK-c(RGDfC) coated surfaces in the presence of EGF</td></tr>
<tr><td>A2</td><td>ventral forebrain</td></tr>
<tr><td>C4</td><td>dorsal forebrain</td></tr>
<tr><td>HC_A</td><td>hippocampus</td><td rowspan="8">P 50-75</td></tr>
<tr><td>SVZ_I</td><td rowspan="4">ventral forebrain SVZ</td></tr>
<tr><td>SVZ_K</td></tr>
<tr><td>SVZ_T</td></tr>
<tr><td>SVZ_M</td></tr>
<tr><td>CTX_H</td><td>parietal cortex</td></tr>
<tr><td>MES_D</td><td>dorsal mesencephalon (colliculus superior)</td></tr>
</table>

Table 2. Neural stem/progenitor clones investigated in the presented studies

Multiple differences were found among the characterized cell lines indicating important diversity of neural cells, those fulfilling the criteria of "stemness".

3.2.1. Morphological and cell biological characteristics of cloned stem/progenitor cell populations

Early embryonic neuroectoderm-derived (NE-4C, NE-7C2 and WNE) cells display epithel morphology (Fig.6), and produce uniform monolayers. These cells divide continuously (16-hour cycle-time in average) on poly-L-lysine in 10% FCS containing medium, but can not be propagated in serum-free conditions on AK-c(RGDfC) coated surfaces. NE cells express nestin intermedier filamentum protein [82,86] and two-third of the cells carries the mouse stem cell antigen, SSEA-1 [88]. In non-induced stage, the cells express Oct4, carry *blbp* and *glast* mRNAs (proteins could not be demonstrated), but do not express *pax6, mash1* or *neurogenin*s and do not produce GFAP and RC2 proteins [87].

In contrast to NE cells, embryo-derived AK-c(RGDfC)-adherent cells (RGl-1, A2, C4) proliferate on AK-c(RGDfC)-coated surfaces in serum-free, EGF (20 ng/ml) containing medium (cycle-time: 18-20 hrs) (Fig.6). The cells show elongated morphology, display nestin, RC2 and Sox2 immunreactivity, but are immune-negative for GFAP. They transcribe large amount of *blbp, glast, pax6, olig2* and also *gfap* mRNAs [74]. All clones express *mash1*, while neurogenin (*ngn2*) is not transcribed in the ventral forebrain-derived A2 clone. *Oct4* or *nanog* expression was never detected. According to the above characteristics, these cells were identified as embryonic radial glia-like (eRGl) cells.

Adult-derived AK-c(RGDfC) adherent cells were prepared from the ventral forebrain SVZ (clones: SVZ_I, SVZ_K, SVZ_M, SVZ_T) from the hippocampus (HC_A), from the parietal cortex far from corpus callosum (CTX_H) and also from the dorsal midbrain parenchyma (colliculus superior), far from the wall of the ventricle (MES_D). As it was expected, AK-c(RGDfC) adherent cells were much less frequent in the adult brain-derived cell suspensions, than in the embryonic preparations. The selective adhesivity and EGF-supported proliferation, however, resulted in several clones from the adult mouse brain. Isolation of CTX_H and MES_D clones indicates that cells with stem/progenitor features are present in the non-neurogenic adult brain parenchyma at a frequency high enough to sort them out easily.

Regardless of origin, adult-derived clones display GFAP-, nestin- and RC2 immunpositivity, and carry immunocytochemically detectable Sox2 and Pax6 proteins [74]. None of the cloned cells express *Oct4* or *nanog* pluripotency genes. According to these features, the cells are regarded as adult-derived RGl cells (aRGl). Adult RGl cells grow in two dimensional clusters, where elongated cells line up along each other (Fig.6 right panel) and divide with a cycle-time of 20-22 hours in average. By today, eRGl and aRGl clones are over 50-80 passages without pheno- or genotypic changes; they preserved 2n (40 chromosome) euploidity, morphological features and neural inducibility.

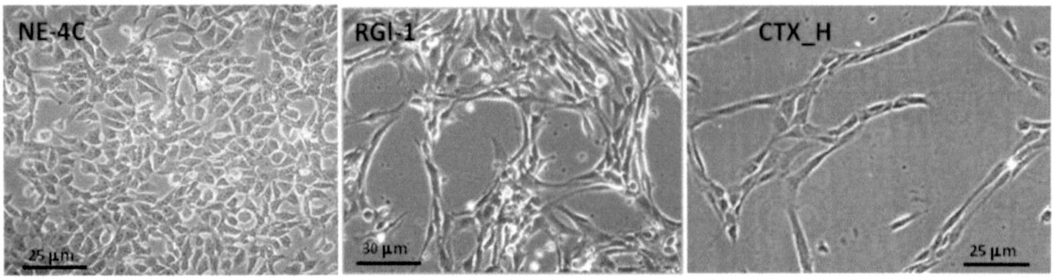

Figure 6. Phase-contrast view of NE-4C early embryonic neuroectodermal stem cells, the embryo-derived radial glia like RGl-1 cells and the adult cortex-derived CTX_H cells.

		NE-4C	RGl-1; A2	HC_A; SVZ_M
Cytochemical markers	nestin	+	+	+
	RC2	-	+	+
	GFAP	-	-	+
	βIII-tub	-	-	-
Gene expression	Oct 4	+	-	-
	Nanog	+	-	-
	Sox2	+	+	+
	Pax6	-	+	+
	Blbp	+	+	+
	GLAST	+	+	+
	Ngn2	-	+	+
	Mash1	-	+	+
	Math2	-	-	-
Induction of neural differentiation		retinoic acid	EGF withdrawal (neurons); serum (astrocytes); PDGF+FGF+forskolin/T3+AA (oligodendrocytes)	

Table 3. Similarities and differences between NE and RGl cells

Beside marked differences in shape, proliferation activity, pluripotency gene expression and GFAP and RC2 immunreactivity, RGl cells display different electrophysiological properties in comparison to early embryonic neuroectodermal (NE) cells. While both NE and RGl cells exist in multiple gap junction coupling, and consequently, display symmetric passive conductance in response to current injections [88], voltage-dependent K-currents (K_{DR}) can be registered in RGl cells [74], but not in NE cells.

3.2.2. Generation of neural tissue-type cells by NE and RGl type neural stem/progenitor cells

Studies on neuron and glia formation revealed further important differences between early embryonic neuroectoderm (NE) derived and embryonic or adult radial glia like stem/progenitor cells (Table 3).

Neural differentiation of NE cells is induced by retinoic acid (all-trans retinoic acid; RA; 10^{-8}-10^{-6} M) [82, 86, 89]. A short (6-hour) RA treatment initiate aggregation of cells and the rate of aggregation increases at higher RA-concentrations and/or longer treatment. Forced aggregation of cells without RA, however, does not induce neural development, and, if initial aggregation is prevented, RA-treatment alone results in severe cell decay without differentiation [89]. It seems, that RA primes the cells for intercellular inductive signalling, what takes places inside RA-primed aggregates. After RA-priming, neural differentiation proceeds along an apparently stable program (Fig.7), in the absence of RA, in both serum-containing and serum-free culture conditions. RA-priming is required also for the formation of astrocytes from NE cells, even if glia genesis starts only 7-10 days after RA-priming [87].

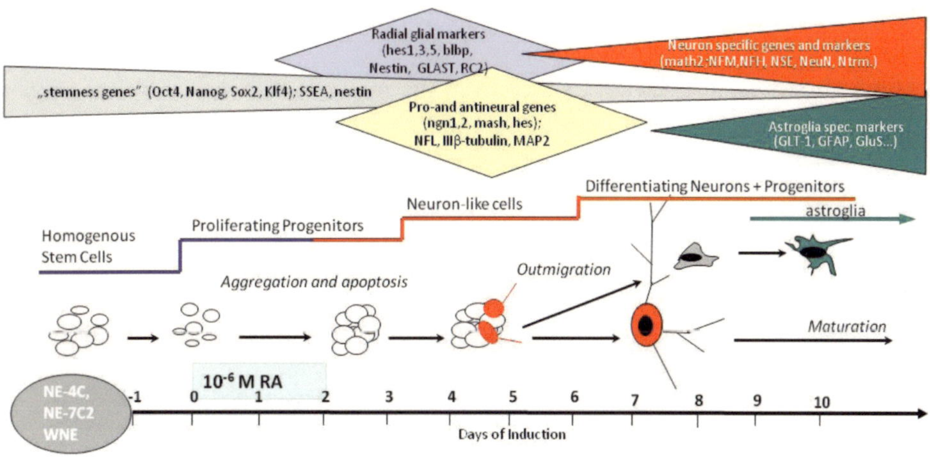

Figure 7. The scheme of RA-primed neural differentiation of NE cells.

20 hours after RA-priming, RC2 radial glia marker protein appears in the aggregated NE cells, and the first IIIβ-tubulin- and MAP2-positive neuronal precursors appear on the 3rd– 4rd day of induction, inside the aggregates. Expression of the proneural bHLH transcription factor *ngn2* is detected soon after induction (24-48 hours), increases during the first 5 days and decreases thereafter, when the "neuron-specific" *math2* transcription starts and increases gradually. The proportion of neurons reaches 50-55% of total cells during the second week of induction, and the first GFAP-positive cells appear around the 9th day of induction. Non-differentiated stem/progenitor cells persist during the entire period of neural differentiation as groups of SSEA-1 immunreactive epithelioid cells (Fig.8.) [33, 82, 87].

In contrast to NE cells, tissue-type differentiation of RGl cells is not induced by RA-treatment, regardless of the age and region of the source tissue [74, 90]. In RGl cells, neuronal differen-

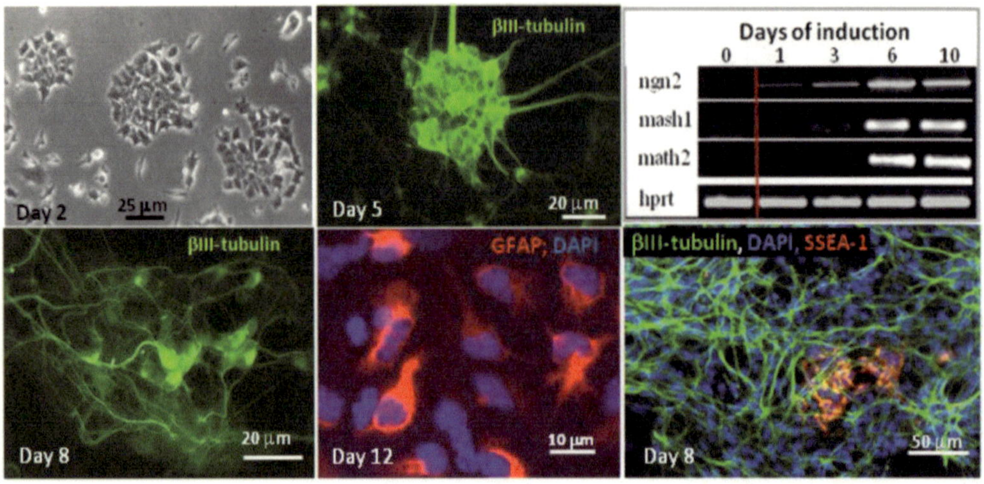

Figure 8. Immunocytochemical characterization of differentiating NE cells cells and the expression of some developmental neuronal "master" genes in the course of in vitro induced neural development. Developmental stages are indicated by days after RA-priming.

tiation is induced by withdrawal of EGF from the medium of dense cultures. Large-scale astroglia formation is initiated by adding serum (10% FCS), and oligodendroglia production is achieved by treating the cells with FGF, PDGF and forskolin followed by treatment with thyroid hormone (T3) and ascorbic acid, according to the protocol of Glaser [91] (Fig.9.).

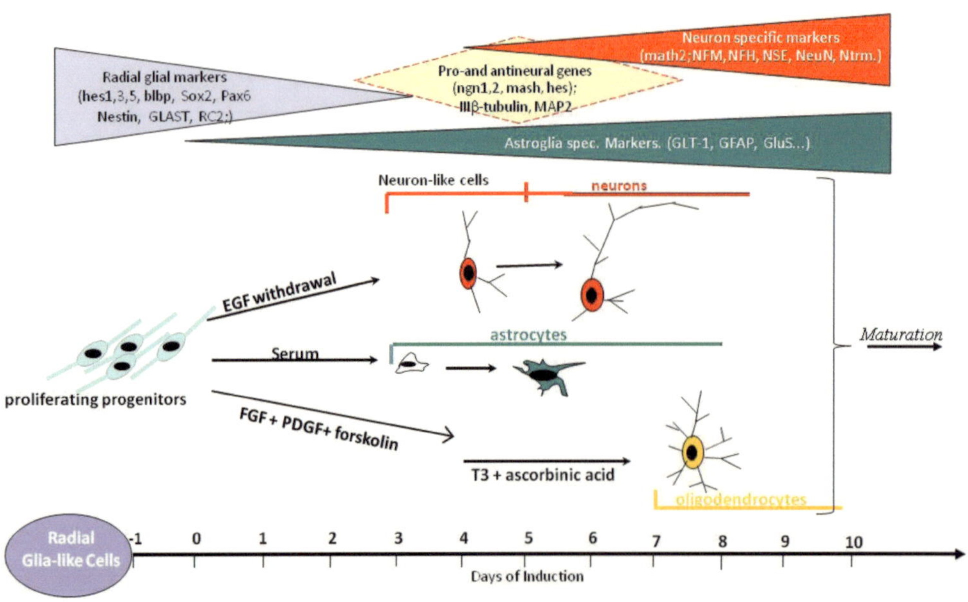

Figure 9. The scheme of in vitro neural tissue-type differentiation of RGl cells.

Regardless of origin, RGl cells give rise to neurons in a 5-day period after withdrawal of EGF, to astrocytes in 3 days after treatment with serum, and to oligodendrocytes 8 days after the onset of oligodendroglia induction (Fig.9 and 10).

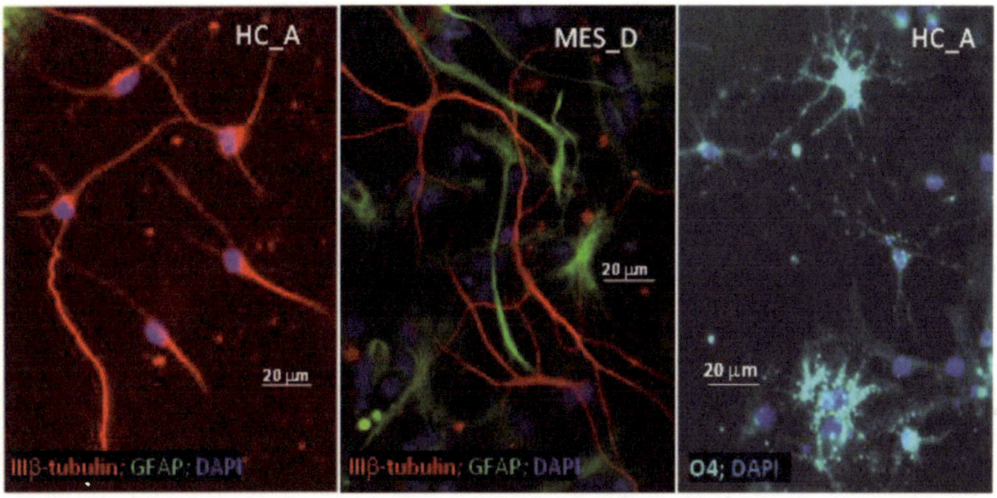

Figure 10. Neurons and oligodendrocytes generated by adult hippocampus- and mesencaphalon-derived (HC_A and MES_D) RGl cells are hown after neuron-specific IIIβ-tubulin and oligodendroglia-indicating O4 immunostaining, respectively.

RGl cells generate neurons in smaller proportion than NE cells; the ratio of neurons among total cells does not exceed 20-30%, and the yield varies among the different clones. Significant differences among the RGl clones were found also in the rate of oligodenroglia production [74]. The time-course of formation of "mature" phenotypes, however, is shorter in RGl cells than in NE cells.

3.2.3. RA-production and RA-responsiveness of NE and RGl cells

The retinoid-dependent and independent neuron formation is a striking difference between NE and RGl clones, respectively [90]. During development, RA regulates the formation of the neural tube and interferes with the patterning of the future hindbrain and spinal cord [92]. Important morphogenic roles of retinoids were demonstrated also in the developing forebrain [93], and the presence and production of RA were described in the developing lateral ganglionic eminence (LGE) [94,95]. RA-responsive cells persist in the main neural stem cell niches of the postnatal rodent brain including the SVZ and the dentate gyrus [96]. Depletion of RA in adult mice results in impaired neuronal differentiation in the dentate gyrus [97,98] and RA synthesis seems to regulate proliferation and gene transcription of at least a subset of neural stem cells in the SVZ [99,100]. As retinoid signalling seems to influence the fate of neural stem cells throughout life, the fundamental differences in RA-responsiveness between NE and RGl type neural stem/progenitor cells have been investigated in details.

Non-differentiated *NE* cells do not produce detectable retinoids [101], but differentiating daughter cells produce considerable amounts of RA [87, 90]. In contrast, RGl cells produce well-detectable amount of RA and retinoid production increases further with the advancement of neuronal differentiation. Endogenous RA production, on the other hand, does not influence the neuron formation by RGl cells: treatment with retinoic acid receptor (RAR) antagonist (AGN193109) does not result in significant changes in the number or morphology of RGl-derived neurons [90]. Expression of genes coding components of the retinoid metabolism and signalling revealed significant differences between NE and RGl cells. NE and RGl cells express non-identical sets of retinaldehyde-dehydrogenases (RALDHs) and nuclear retinoid receptor subunits. Moreover, some defined retinoid transporters (as STRA6), and catabolising enzymes (as CYP26s) are not expressed by RGl cells [90]. In accordance with Haskell and LaMantia [99], we concluded that retinoid metabolism and responsiveness are distinctive characteristics of defined subtypes of neural stem/progenitor populations.

3.2.4. *"Regional memory"*

NE-4C cells, in non-induced „stem cell stage", express only *otx2* and *en* from the investigated "region-specific", positional genes. In differentiated NE-4C cultures, however, many positional genes including *hoxb2* are actively transcribed (Table 4) [33] indicating that these early embryonic stem cells are "open" for a variety of "position-determined" development. Accordingly, glutamatergic, GABAergic and also serotonin producing neurons develop from NE-4C cells.

Non-induced RGl cells, on the other hand, express *gbx2*, *dlx2*, *emx2*, but not *hoxb2* or *nkx2.1* (Table 4), regardless of origin. The pattern of investigated genes indicates an "anterior to hindbrain" (lack of active *hoxb2*), and a "not caudo-ventral" (lack of *nkx2.1* expression) origin, but does not distinguish between forebrain regions and between forebrain and mesencephalon derivatives.

Clone gene	NE	RGl									
	NE-4C	Embryonic (E14.5)			Adult (P 50-75)						
		A2	C4	RGl-1	HC_A	CTX_H	MES_D	SVZ_I	SVZ_K	SVZ_T	SVZ_M
	Gene expression										
Emx2	-	+	+	+	+	+	+	+	+	+	+
Nkx2.1	-	-	-	-	-	-	-	-	-	-	-
Gbx2	-	ni	ni	+	+	+	+	+	+	+	+
Dlx2	-	+	+	+	+	+	+	+	+	ni	ni
Hoxb2	-	ni	ni	-	-	-	-	-	-	ni	ni
Ngn2	-	-	+	+	+	+	+	+	-	+	+
Mash1	-	+	+	+	+	+	+	+	+	+	+

(ni: non investigated)

Table 4. Positional gene expression by non-induced stem/progenitor cells.

The neurotransmitter phenotype of neuronal progenies, on the other hand, still indicates preservation of some fine region-restricted commitment. All RGl cells give rise to GABA-producing, VGAT-positive neurons. In adult-derived clones, vesicular glutamate transporter

http://dx.doi.org/10.5772/55678

(*vglut*) expression and VGLUT-immunpositivity [102] however was found only in the hippocampus-derived HC_A clone. As HC_A neurons express only *vglut1* [74], it seems that these cells preserve some regional, neurotransmitter-related identity, even after long-term in vitro propagation. Similarly, all adult SVZ-derived clones give rise to tyrosin-hydroxylase(TH)-expressing neurons [74]. This observation might be in accord with the known production of TH-positive olfactory neuron-precursors [103] in the adult SVZ.

The embryo-derived RGl-1 cells, on the other hand, express both, *vglut1*, *vglut2* and also TH. The finding may indicate a less advanced stage of commitment of these cells, but needs further studies.

Available sporadic results do not allow deciding on the preservation or loss of regional identity of in vitro propagated stem/progenitor cells. As clones comprise progenies of single cells, for far-range conclusions we need further studies on statistically sufficient number of clones from each brain region.

4. Distinct neural stem/progenitor stages require distinct environmental conditions

One cell derived, cloned populations are useful subjects to show the types of cells which can or can not derive from a given stem cell population, but neither of such studies can predict what sort of phenotypes can be manifested in vivo. In the course of in vivo tissue genesis, differentiating cells and their environment change in an orchestrated way: the conditions either help tissue integration or kill the differentiating cells assuring the survival of the right types and number of cells at the right places.

In our experience, NE-4C cells give rise to neurons with large frequency if implanted into early embryonic brain vesicles, but produce non-differentiating, tumour-like expansions in the subcortical regions of newborn mice. Moreover, the same cells diminish from the adult mouse forebrain, except the forebrain SVZ, where some cells reside for longer than 3 weeks [104]. The data demonstrate that the fate of NE stem/progenitor cells is strictly governed by environmental factors provided by the host tissue, and which are far from being explored.

In addition to initial requirements for stem/progenitor survival, the successive progenies need different environment and change basic physiological demands including O_2 -supply [105; 77]. Under hypoxic conditions, NE-4C stem cells survive and proliferate, but do not generate neurons. Soon after the onset of in vitro neuronal differentiation (48 hrs after RA-priming), however, hypoxia severely impairs cell survival [105] indicating that the basic metabolic characteristics of cells change soon after neural fate decision. The in vitro results were supported by data obtained on the sporadic neuron formation by NE-4C cells implanted into the adult mouse forebrain in response to hyperbaric oxygenation [105].

5. Conclusions

The presented results on some selected NS populations demonstrate that neural stem/progenitor cells derived from different brain regions and different ages may display significant diversity in many aspects including cell physiological and developmental features. The selected clones represent derivatives of a single stem cell and had been propagated under fairly artificial in vitro conditions; accordingly, none of them can represent *"the" stem cells in general* of the region of origin. From the data obtained on 3 different NE and 10 RGl cell clones, however, we could conclude that early embryonic neuroectoderm-derived stem cells display distinct *in vitro* features from those isolated by selective adhesion from later CNS tissues. In case of RGl cells, the adhesive preference-based selection and EGF-supported growth, imply a strong selection. Cells corresponding to such selection could be isolated from fairly distinct brain regions including areas not listed among "professional" neurogenic zones. *Correspondingly, RGl cells may not represent a defined lineage of stem/progenitor population, rather a common stage of neural progenitor-succession, which may appear in many lineages.* The integrin-based adhesion, EGF-supported proliferation, voltage-dependent K^+-flux, RA-production and RA-insensitivity may be characteristics of stem/progenitor cells in a defined stage of tissue integration.

While many NS populations have been characterized in vitro [76], were implanted into different brain regions of healthy and disease-model animals and have been considered also as potential tools for clinical cell therapy (for critical review: [106, 107]), the basic physiology of such cells, their needs for survival in successive developmental stages as well as the harmonization of such needs with conditions provided by the host tissue are rarely taken into account. As a further task, physiological demands of various NS populations and descending progenies should be explored and the roles of basic physiological factors in tuning differentiation should be understood.

Acknowledgements

I thank my colleagues, especially to Zsuzsanna Környei and Károly Markó at the Cellular and Developmental Neurobiology Unit of IEM-HAS, and Katalin Schlett at the Dept. of Physiology and Neurobiology, Eötvös L. Univ., for common work, thinking and discussions. I wish to thank the Nikon Microscopy Center at IEM, Nikon Austria GmbH and Auro-Science Consulting Ltd for kindly providing microscopy support. I apologise to Authors whose excellent publications were not referred here because of space limitations. The presented studies were supported by National Science Foundation (OTKA) (K 68939, K106191), National Innovation Office (Bio_Surf) and Eu Fp6 (BioDot) grants.

http://dx.doi.org/10.5772/55678

Author details

Emília Madarász

Institute of Experimental Medicin of Hungarian Academy of Sciences, Laboratory of Cellular and Developmental Neurobiology, Hungary

References

[1] Rakic P. Evolution of the neocortex: Perspective from developmental biology 2009. Nat Rev Neurosci., 10(10): 724–735

[2] Kriegstein A, Alvarez-Buylla A. 2009. The glial nature of embryonic and adult neural stem cells. Annu Rev. Neurosci 2009, 32:149-184.

[3] Albazerchi A., Stern C.D. 2007. A role for the hypoblast (AVE) in the initiation of neural induction, independent of its ability to position the primitive streak. Developmental Biology 301: 489–503

[4] Chapman SC, Schubert FR, Schoenwolf GC, Lumsden A. 2003. Anterior identity is established in chick epiblast by hypoblast and anterior definitive endoderm. Development;130:5091-5101

[5] Beddington, R. S. and Robertson, E. J. 1999. Axis development and early asymmetry in mammals. Cell 96,195 -209.

[6] Schoenwolf, G.C., 2001. Cutting, pasting and painting: experimental embryology and neural development. Nat. Rev. Neurosci. 2, 763–771.

[7] Shimamura K, Hartigan DJ, Martinez S, Puelles L, Rubenstein JL. 1995. Longitudinal organization of the anterior neural plate and neural tube. Development 121:3923–33

[8] Fietz S. A. and Huttner W.B. 2011. Cortical progenitor expansion, self-renewal and neurogenesis—a polarized perspective. Current Opinion in Neurobiology, 21:23–35

[9] Hemmati-Brivanlou A, Melton DA. 1994. Inhibition of activin receptor signaling promotes neuralization in Xenopus. Cell. 77: 273-81.

[10] Sidman RL, Rakic P. 1973. Neuronal migration with special reference to developing human brain: a review.. Brain Res., 62: 1-35.

[11] Artavanis-Tsakonas S, Rand MM, Lake RJ. 1999. Notch signaling: cell fate control and signal integration in development. Science 284: 770-776.

[12] Britsch S.The neuregulin-I/ErbB signaling system in development and disease. Adv Anat Embryol Cell Biol. 2007;190:1-65.

[13] Gal JS, Morozov YM, Ayoub AE, Chatterjee M, Rakic P, Haydar TF. 2006. Molecular and morphological heterogeneity of neural precursors in the mouse neocortical proliferative zones. J.Neurosci. 26: 1045-56

[14] Jacobson M., 1978. Developmental Neurobiology. Plenum Press

[15] Sauer FC. 1935. Mitosis in the neural tube.. J.Comp.Neurol. 62: 377-405

[16] Hart A.H., Hartley L., Ibrahim M, Robb L. 2004. Identification, Cloning and Expression Analysis of the Pluripotency Promoting Nanog Genes in Mouse and Human. Devl.Dynamics 230:187–198

[17] Kimura C, Shen MM, Takeda N, Aizawa S, Matsuo I. 2001. Complementary functions of Otx2 and Cripto in initial patterning of mouse epiblast. Dev Biol 235:12–32.

[18] Acampora D, Gulisano M, Broccoli V, Simeone A. 2001. Otx genes in brain morphogenesis. Prog Neurobiol. 64(1):69-95.

[19] Klingensmith J, Ang S-L, Bachiller D, Rossant J. 1999. Neural Induction and Patterning in the Mouse in the Absence of the Node and Its Derivatives. Developmental Biology 216, 535–549

[20] Nakamura H, Katahira T, Matsunaga E, Sato T. 2005Isthmus organizer for midbrain and hindbrain development. Brain Res Brain Res Rev.; 49(2):120-6.

[21] Sasai Y. 1998. Identifying the missing links: genes that connect neural induction and primary neurogenesis in vertebrate embryos. Neuron. 21: 455-458

[22] Ellis P, Fagan M B, Magness S T, Hutton S, Taranova O, Hayashi S, McMahon A, Rao M. 2004. SOX2, a Persistent Marker for Multipotential Neural Stem Cells Derived from Embryonic Stem Cells, the Embryo or the Adult. Dev Neurosci, 26:148–165

[23] Callaerts P, Halder G, Gehring WJ. 1997. PAX-6 in development and evolution. Annu Rev Neurosci.;20:483-532.

[24] Götz M, Stoykova A, Gruss P. Pax6 controls radial glia differentiation in the cerebral cortex. Neuron. 1998;21:1031–1044.

[25] Kageyama R, Nakanishi S; 1997. Current Opinion in Genetics & Development, 7:659-665

[26] Regan MR, Huang YH, Kim YS, Dykes-Hoberg MI, Jin L, Watkins AM, Bergles DE, Rothstein JD. Variations in promoter activity reveal a differential expression and physiology of glutamate transporters by glia in the developing and mature CNS. J Neurosci. 2007 Jun 20;27(25):6607-19.

[27] Liu R-Z, Mita R, Beaulieu M, Gao Z, Godbout R. Fatty acid binding proteins in brain development and disease. Int. J. Dev. Biol. 54: 1229-1239 (2010) doi: 10.1387/ijdb.092976rl

[28] Guillemot F., 2007. Spatial and temporal specification of neural fates by transcription factor codes. Development 134, 3771-3780

[29] Hébert JM, Fishell G. 2008. The genetics of early telencephalon patterning: some assembly required. Nat Rev Neurosci; 9(9):678-85.

[30] Puelles L, Rubenstein JL. 2003. Forebrain gene expression domains and the evolving prosomeric model. Trends Neurosci. 26(9):469-76.

[31] Kiecker and Lumsden 2012. The Role of Organizers in Patterning the Nervous System. Annu. Rev. Neurosci. 35:347–67.

[32] Onorati M, Binetti M, Conti L, Camnasio S, Calabrese G, Albieri I, Di Febo F, Toselli M, Biella G, Martynoga B, Guillemot F, Consalez GG, Cattaneo E. 2011. Preservation of positional identity in fetus-derived neural stem (NS) cells from different mouse central nervous system compartments. Cell Mol Life Sci. 68(10):1769-83.

[33] Varga B., Hadinger N, Gócza E., Dulberg V., Demeter K., Madarasz E.and Herberth B. 2008. Generation of diverse neuronal subtypes in cloned populations of stem-like cells. BMC Devl.Biol., 8: 89-107

[34] Hansen DV, Lui JH, Parker PR, Kriegstein AR, 2010. Neurogenic radial glia in the outer subventricular zone of human neocortex.Nature 2010, 464:554-561.

[35] Dhaliwal, J. and Lagace D. 2011. Visualization and genetic manipulation of adult neurogenesis using transgenic mice. Eur.J.Neurosci. 33: 1025-1036.

[36] García-Verdugo JM, Doetsch F, Wichterle H, Lim DA, Alvarez-Buylla A. 1998. Architecture and cell types of the adult subventricular zone: in search of the stem cells. J Neurobiol.; 36(2):234-48.

[37] Espinosa JS, Luo L. 2008. Timing neurogenesis and differentiation: insights from quantitative clonal analyses of cerebellar granule cells. J Neurosci. 28(10):2301-12.

[38] Kempermann G, Gage FH. 2000. Neurogenesis in the adult hippocampus.Novartis Found Symp.2000;231:220-235.

[39] Flames N, Pla R, Gelman D.M, Rubenstein J.L.R, Puelles L, Marın O. 2007. Delineation of Multiple Subpallial Progenitor Domains by the Combinatorial Expression of Transcriptional Codes. J. Neurosci., 27(36):9682–9695

[40] Anderson SA, Qiu M, Bulfone A, Eisenstat DD, Meneses J, Pedersen R, Rubenstein JLR. Mutations of the homeobox genes Dlx-1 and Dlx-2 disrupt the striatal subventricular zone and differentiation of late born striatal neuron. 1997. Neuron 19: 27-37.

[41] Anderson, SA., Mione, M., Yun, K., and Rubenstein, J. L. Differential origins of neocortical projection and local circuit neurons: role of Dlx genes in neocortical interneuronogenesis. 1999. Cereb Cortex 9: 646-54.

[42] Bernier PJ, Bedard A, Vinet J, Levesque M, Parent A. 2002. Newly generated neurons in the amygdala and adjoining cortex of adult primates. Proc Natl Acad Sci U S A 99:11464–11469.

[43] Marchis S, Fasolo A, Puche A.C. 2004. Subventricular Zone-Derived Neuronal Progenitors Migrate into the Subcortical Forebrain of Postnatal Mice. J. Comp. Neurol. 476:290–300

[44] Letinic K, Rakic P. 2001. Telencephalic origin of human thalamic GABAergic neurons. Nature Neurosci.;4:931–936.

[45] Richardson W.D, Kessaris N., Pringle N. 2006. Oligodendrocyte wars Nat. Rev. Neurosci. 7: 11–18

[46] Gonzalez-Perez O, Alvarez-Buylla A. 2011. Oligodendrogenesis in the subventricular zone and the role of epidermal growth factor. Brain Res Rev. 67(1-2):147-56.

[47] Donkelaar H. J. Lammens M., Wesseling P., Thijssen H. O. M., Renier W. O. 2003. Development and developmental disorders of the human cerebellum. J Neurol., 250 : 1025–1036

[48] Altman, J. and Bayer, S. A. 1990. Mosaic organization of the hippocampal neuroepithelium and the multiple germinal sources of dentate granule cells. J. Comp. Neurol. 301, 325-342.

[49] Galceran J, Miyashita-Lin E.M., Devaney E, John L. R. RubensteinJ.L.R, Grosschedl R. 2000. Hippocampus development and generation of dentate gyrus granule cells is regulated by LEF1. Development 127, 469-482

[50] Doetsch F, Garcia-Verdugo JM, Alvarez-Buylla A. Cellular composition and three-dimensional organization of the subventricular germinal zone in the adult mammalian brain. 1997. J.Neurosci., 17: 5046-5061.

[51] Zhao C, Teng EM, Summers RG Jr, Ming GL, Gage FH. 2006. Distinct Morphological Stages of Dentate Granule Neuron Maturation in the Adult Mouse HippocampusJ. Neurosci., 26(1):3–11

[52] Paton JA, Nottebohm FN. Neurons generated in the adult brain are recruited into functional circuits. Science. 1984;225(4666):1046-8.

[53] Lledo P.M., Alonso M, Grubb M.S, Adult neurogenesis and functional plasticityin neuronal circuits. Nat.Rev.Neurosci. 2006. 7: 179-193

[54] Gould E, Reeves A.J., Graziano M.S.A.,. Gross C. G. 1999. Neurogenesis in the Neocortex of Adult Primates. Science, 286: 548-552

[55] Bedard A, Levesque M, Bernier PJ, Parent A. 2002. The rostral migratory stream in adult squirrel monkeys: contribution of new neurons to the olfactory tubercle and involvement of the antiapoptotic protein Bcl-2. Eur J Neurosci 16:1917–1924.

[56] Hodge R.D., Thomas D. KowalczykT.D., Wolf S.A., Encinas J.M., Rippey C., Enikolopov G., Kempermann G., Hevner R.F.; 2008. Intermediate Progenitors in Adult Hippocampal Neurogenesis: Tbr2 Expression and Coordinate Regulation of Neuronal Output. J. Neurosci. 28(14): 3707-3717

[57] Li L, Clevers H. 2010. Coexistence of Quiescent and Active Adult Stem Cells in mammals. Science: 327 (5965): 542-545

[58] Lois C., Garcia-Verdugo J.M., Alvarez-Buylla, A.; 1996. Chain migration of neuronal precursors. Science 271: 978-981

[59] Zhao Ch, W. Deng, Gage F. (2008). Mechanisms and Functional Implications of Adult Neurogenesis. Cell 132: 645–660

[60] Marín-Burgin A, Mongiat L, Pardi M.B., Schinder A. F. (2012) Unique Processing During a Period of High Excitation/Inhibition Balance in Adult-Born Neurons. Science; 335(6073): 1238–1242

[61] Alonso M, Lepousez G, Sebastien W, Bardy C, Gabellec MM, Torquet N, Lledo PM. 2012. Activation of adult-born neurons facilitates learning and memory. Nat Neurosci. 2012 Jun;15(6):897-904.

[62] Fowler C.D., Liu Y., Wang Z. Estrogen and adult neurogenesis in the amygdala and hypothalamus. Brain Res Rev. 2008 March ; 57(2): 342–351.

[63] Collin T, Arvidsson A, Kokaia Z, Lindvall O. 2005. Quantitative analysis of the generation of different striatal neuronal subtypes in the adult brain following excitotoxic injury. Exp Neurol. 195(1):71-80.

[64] Migaud M, Batailler M, Segura S, Duittoz A, Franceschini I, Pillon D. 2010. Emerging new sites for adult neurogenesis in the mammalian brain: a comparative study between the hypothalamus and the classical neurogenic zones. Eur J Neurosci. 32(12): 2042-52.

[65] Bonfanti L, Nacher J., 2012. New scenarios for neuronal structural plasticity in non-neurogenic brain parenchyma: the case of cortical layer II immature neurons. Prog Neurobiol. 98(1):1-15.

[66] Geraerts M, Krylyshkina O, Debyser Z, Baekelandt V., 2007. Concise review: therapeutic strategies for Parkinson disease based on the modulation of adult neurogenesis. Stem Cells. 25(2):263-70.

[67] Baumann N., Pham-Dinh D. 2001. Biology of Oligodendrocyte and Myelin in the Mammalian Central Nervous System. Phys. Rev. 81: 871–927

[68] Meletis K, Barnabe-Heider F, Carlen M, Evergren E, Tomilin N, Shupliakov O, Frisen J. Spinal cord injury reveals multilineage differentiation of ependymal cells. PLoS Biol 6:e182, 2008.

[69] Raff MC, Miller RH, Noble M., 1983. A glial progenitor cell that develops in vitro into an astrocyte or an oligodendrocyte depending on culture medium. Nature. 303(5916): 390-396.

[70] Franklin R J. M. and Charles ffrench-Constant 2008. Remyelination in the CNS: from biology to therapy. Nat.Rev.Neurosci. 9(11):839-55

[71] Robel S, Berninger B, Gotz M. 2011. The stem cell potential of glia: lessons from reactive gliosis. Nat Rev Neurosci 12: 88–104,

[72] Ohira, K., 2011. Injury-induced neurogenesis in the mammalian forebrain. Cell Mol Life Sci. 68(10):1645-56.

[73] Decimo I., Bifari F., Krampera M., Fumagalli G. 2012. Neural stem cell niches in health and diseases. Curr. Pharm. Design 18: 1755-1783.

[74] Marko K, Kohidi T, Hadinger N, Jelitai M, Mezo G, Madarasz E. 2011. Isolation of radial glia-like neural stem cells from fetal and adult mouse forebrain via selective adhesion to a novel adhesive peptide-conjugate. PlosOne; I www.plosone.org 6 I 12 I e28538

[75] Ming,G. and Song H., 2011. Adult neurogenesis in the mammalian brain: Significant aswers and significant questions. Neuron 70: 687-702.

[76] Conti L., Cattaneo E.,2010. Neural stem cell systems: physiological players or in vitro entities? Nat.Rev.Neurosci. 11: 176-187

[77] Crossin, K., 2012. Oxygen levels and the regulation of cell adhesion in the nervous system: A control point for morphogenesis in development, disease and evolution? Cell Adh., Migr. 6 (1): 49–58.

[78] Chen J, Herrup K., 2008. Selective vulnerability of neurons in primary cultures and in neurodegenerative diseases. Rev Neurosci. 2008;19(4-5):317-26.

[79] Juurlink BH, Hertz L.1985. Plasticity of astrocytes in primary cultures: an experimental tool and a reason for methodological caution. Dev Neurosci.; 7(5-6):263-77.

[80] Madarász E, Kiss J, Bartók I. 1984. Cell production and morphological pattern formation in primary brain cell cultures. I. Pattern formation within the basal layer(s). Brain Res. 304: 339-349

[81] Madarász E, Theodosis DT, Poulain DA. 1991. In vitro formation of type 2 astrocytes derived from postnatal rat hypothalamus or cerebral cortex. Neuroscience. 43: 211-221.

[82] Schlett K, Madarasz E. 1997. Retinoic acid induced neural differentiation in a neuroectodermal cell line immortalized by p53 deficiency. J Neurosci Res. 47: 405-415.

[83] Livingstone L.R., White A., Sprouse J., Livanos E., Jacks T., Tisty T.D. Altered cell cycle arrest and gene amplification potential acoompany loss of wild-type p53. 1992. Cell, 70: 923-935.

[84] Marko K., Ligeti M., Mezo G., Mihala K., Kutnyinszky E., Kiss E., Hudecz F., Madarasz E. 2008. A novel synthetic peptide polymer with cyclic RGD motifs supports serum-free attachment of anchorage-dependent cells. Bioconjugate Chemistry 19: 1757-1766

[85] Nolte C, Matyash M, Pivneva T, Schipke CG, Ohlemeyer C, et al. (2001) GFAP promoter-controlled EGFP-expressing transgenic mice: a tool to visualize astrocytes and astrogliosis in living brain tissue. Glia 33: 72-86.

[86] Varju,P., Schlett,K. Eisel,U., Madarász,E. Schedule of NMDA receptor subunit expression and functional channel formation in the course of in vitro induced neurogenesis. J.Neurochem. 2001. 77, 1444-1456

[87] Hádinger N, Varga B.v, Berzsenyi S, Környei Zs, Madarasz E, Herberth B. 2009. Retinoic acid regulates astrogliogenesis. Int.J.Devl.Neurosci. 27: 365-375

[88] Jelitai M., Anderova M., Chvatal A., Madarasz E., 2007.Electrophysiological characterization of neural stem/progenitor cells during in vitro differentiation: a study on an immortalized neuroectodermal cell line. J.Neurosci.Res. 85(8):1606-1617

[89] Tarnok K., Pataki A., Kovacs J., Schlett K., Madarasz E., 2002. Stage -dependent effects of cell-to cell connections on in vitro induced neurogenesis. Eur.J.Cell Biol.,. 81 (7): 403-412

[90] Orsolits B, Borsy A, Madarász E, Mészáros Z, Kőhidi T, Markó K, Jelitai M, Welker E, Környei Z. Retinoid machinery in distinct neural stem cell populations with different retinoid responsiveness. Stem Cell Devl. 2013 in press

[91] Glaser T, Pollard SM, Smith A, Brustle O (2007) Tripotential differentiation of adherently expandable neural stem (NS) cells. PLoS One 2: e298.

[92] Maden M. 2006. Retinoids and spinal cord development. J Neurobiol 66:726-38.

[93] Halilagic A, V Ribes, NB Ghyselinck, MH Zile, P Dolle and M Studer. (2007). Retinoids control anterior and dorsal properties in the developing forebrain. Developmental biology 303:362-75.

[94] Toresson H, A Mata de Urquiza, C Fagerstrom, T Perlmann and K Campbell. 1999. Retinoids are produced by glia in the lateral ganglionic eminence and regulate striatal neuron differentiation. Development 126:1317-26

[95] Chatzi C, T Brade and G Duester. 2011. Retinoic acid functions as a key GABAergic differentiation signal in the basal ganglia. PLoS Biol 9:e1000609.

[96] Wagner E, T Luo, Y Sakai, LF Parada and UC Drager. 2006. Retinoic acid delineates the topography of neuronal plasticity in postnatal cerebral cortex. The European journal of neuroscience 24:329-40.

[97] Bonnet E, K Touyarot, S Alfos, V Pallet, P Higueret and DN Abrous. 2008. Retinoic acid restores adult hippocampal neurogenesis and reverses spatial memory deficit in vitamin A deprived rats. PLoS One 3:e3487.

[98] Jacobs S, DC Lie, KL DeCicco, Y Shi, LM DeLuca, FH Gage and RM Evans. 2006. Retinoic acid is required early during adult neurogenesis in the dentate gyrus. Proc Natl Acad Sci U S A 103:3902-7.

[99] Haskell GT and AS LaMantia. 2005. Retinoic acid signaling identifies a distinct precursor population in the developing and adult forebrain. J Neurosci 25:7636-47.

[100] Wang TW, H Zhang and JM Parent. 2005. Retinoic acid regulates postnatal neurogenesis in the murine subventricular zone-olfactory bulb pathway. Development 132:2721-32.

[101] Kornyei Z, E Gocza, R Ruhl, B Orsolits, E Voros, B Szabo, B Vagovits and E Madarasz. 2007. Astroglia-derived retinoic acid is a key factor in glia-induced neurogenesis. FASEB J 21:2496-509.

[102] Fremeau RT, Jr., Troyer MD, Pahner I, Nygaard GO, Tran CH, et al. 2001. The expression of vesicular glutamate transporters defines two classes of excitatory synapse. Neuron 31: 247-260.

[103] Pignatelli A, Ackman JB, Vigetti D, Beltrami AP, Zucchini S, Belluzzi O (2009) A potential reservoir of immature dopaminergic replacement neurons in the adult mammalian olfactory bulb. Pflugers Arch 457(4):899–915.

[104] Demeter K, Herberth B, Duda E, Kúsz E, Jaffredo T, Herman J.-P, Madarász E. 2004. Fate of neuroectodermal cells implanted into the adult, newborn or embryonic brain. Exp.Neurol. 188(2):254-267

[105] Zadori A., Agoston V.A., Demeter K., Hadinger N., Varady L., Kohídi T, Gobl A., Nagy Z, Madarász E. 2011. Survival and differentiation of neural stem cells at different oxygen levels. Exp.Neurology 227, (1): 136-148

[106] Chiu A Y and Rao M.S. 2011. Cell-Based Therapy for Neural Disorders — Anticipating Challenges. Neurotherapeutics 8:744–752

[107] Lindvall O and Björklund A. 2011. Cell Therapeutics in Parkinson's Disease. Neurotherapeutics 8:539–548

Chapter 4

Reactive Muller Glia as Potential Retinal Progenitors

Teri L. Belecky-Adams, Ellen C. Chernoff,
Jonathan M. Wilson and Subramanian Dharmarajan

Additional information is available at the end of the chapter

http://dx.doi.org/10.5772/55150

1. Introduction

Regenerative medicine has become a driving force in the treatment of disease and injury over the last decade [1]. This is due to the accumulation of knowledge in several key areas; 1) the mechanisms of disease processes, 2) creation of stem cells/induced pluripotent stem cells that might be used for therapeutic purposes, and 3) factors that are necessary for the proper differentiation of specific cell types. In any tissue, it might be possible to regenerate lost cells from exogenous stem cells, endogenous stem or progenitor cells, or endogenous cells that can dedifferentiate, proliferate and re-differentiate. Several endogenous populations of cells localized to the eye have been shown to be capable of replacing some or all retinal cell types in various species; 1) an endogenous population of progenitor cells in the periphery of the eye referred to as the ciliary marginal zone (CMZ), 2) the retinal pigmented epithelium, 3) non-pigmented cells adjacent to peripheral retina, 4) $NG2^+$ glial progenitors of the optic nerve, and finally 5) Müller glia of the retina [2]. This chapter will focus specifically on the responsiveness of Müller glia to disease or injury to the retina with a special emphasis on signals that have been shown to lead to the injury response and changes to the extracellular matrix that play a role in dedifferentiation and proliferation.

2. Müller Glial cell basics

Müller Glia, named after their discoverer Heinrich Müller, were first described in 1851 [3]. Müller Glia are a unique blend of radial glia, astrocytes, and oligodendrocytes that span the width of the mature retina from the outer limiting memberane in the outer nuclear layer to the inner limiting membrane at the edge of the retina and vitreous humor [4]. Müller cells are one of three possible macroglial cells that can be found in the retina. Astrocytes also migrate into the retina from the optic nerve and some species also contain oligodendrocytes in the nerve fiber layer [5]. However, Müller glia are the only glial cells that are derived from retinal progenitors. Müller cells play a wide variety of roles in both the developing and mature retina. In order to consider the full effect of gliosis in the diseased or injured retina, we must first understand their function in the normal retina.

2.1. Retinal histogenesis

Lineage analysis of retinal progenitors using various techniques have shown that many retinal progenitors have the capacity to produce all retinal cell types [6-10]. Retinal cells undergo a stereotypical pattern of differentiation in which some cells leave the cell cycle (are born) very early in retinal histogenesis, such as cone photoreceptors, ganglion cells, and horizontal cells, while other cells are generated at later timepoints [6, 7, 9-12]. Müller glia are born in the group of cells that are generated late in the ontogenic period.

Vertebrate retinal cells are arranged in a specific fashion in both layers and in columns [6-8, 13-17]. Figure 1 shows the arrangement of mature retinal cells in the outer, inner and ganglion cell layers. However, some of the cells are also arranged in a columnar fashion. The later-born cells, which include the rods, bipolar, and subpopulation of the amacrine cells, all migrate along the radially arranged Müller glial cells. These cells remain in close contact with the Müller glia even as differentiation continues and are thought to comprise a metabolic and/or processing circuit within the retina [17]. The early-born cells are not a part of this columnar unit. Rather than relying on the Müller glia to migrate to the correct layer of the retina, these cells undergo nuclear translocation in the relatively thinner early retina [18, 19].

Müller glia also share properties that allow them to organize the laminar structure of the retina. Cultured Müller glia or Müller glial conditioned-medium are capable of organizing the retinal neurospheres into a layered pattern which closely resembles that seen in the mature retina [20, 21]. While these experiments suggest that there may be a secreted factor which may mediate the organizational properties of Müller glia, recent experiments done in zebrafish suggest that the apico-basal polarity that is inherent in the development of Müller glia is also a critical part of its organizational capacity [22]. A disrupted apical Müller glial cell process in zebrafish mutated in the P50 subunit of dynactin leads to a disruption in the normal laminar development of the retina [22]. In mice, disruption of the outer limiting membrane that is comprised of the apical Müller glial endfeet disrupts the placement of photoreceptors such that misplaced photoreceptor nuclei are found adjacent to the retinal pigmented epithelium, in a region where photoreceptor outer segments would normally be located [23].

2.2. Synapse formation

The role of astrocytes in synaptogenesis in the CNS has been established by many investigators [24-26]. Müller glial cells have been considered by many to be astrocyte-related cells (See Table 1), hence Müller glia may play some role in synapse formation and/or maintenance in the retina. This idea has been tested in zebrafish retina with somewhat contradictory results [27, 28]. While it appears that the Müller glial cell processes do not invade the outer plexiform later

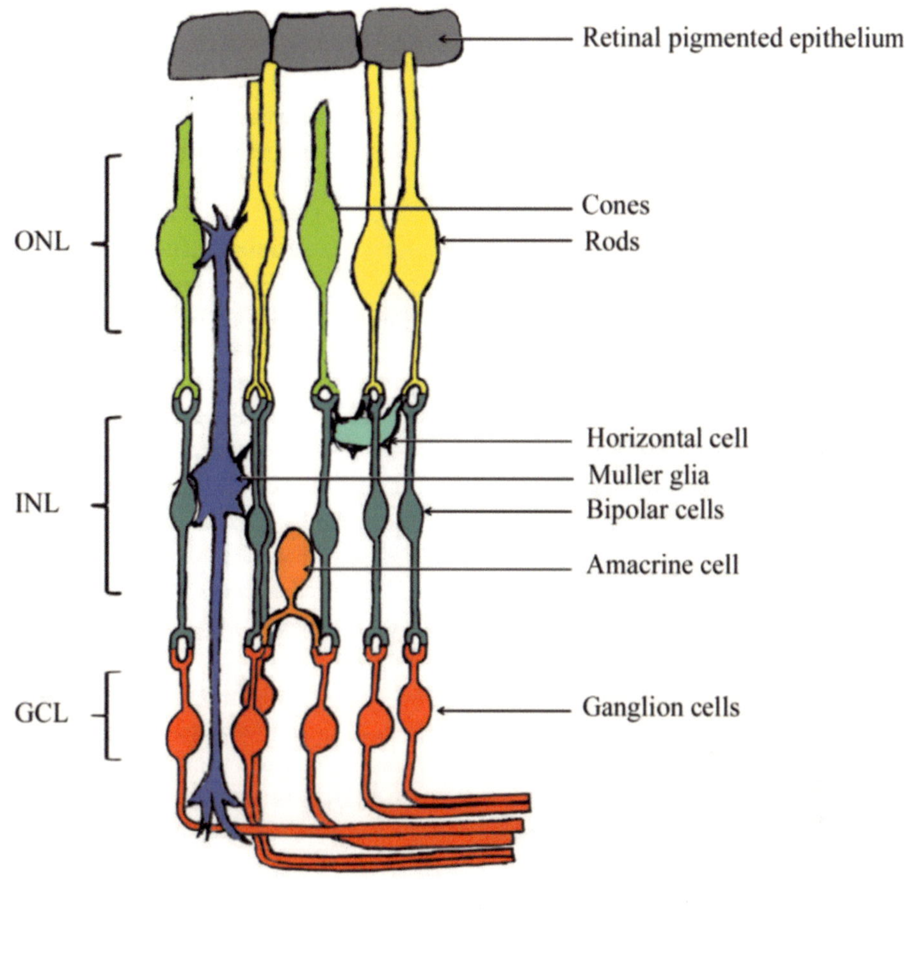

Figure 1. Organization of the mature retina: The retinal cells, which consist of neurons and glia, are organized into the outer nuclear layer (ONL), inner nuclear layer (INL) and the ganglion cell layer (GCL). The ONL consists of the rod and cone photoreceptor cells. The INL is made up of the horizontal cells, amacrine cells as well as the bipolar cells. The Müller glial cell bodies are also present in this layer. However, the processes of the Müller glial cells extend outward into the adjacent layers, extending throughout the thickness of the retina. The GCL is primarily consists of the ganglion cells which send out their axons out of the eye through the optic disc.

until after synapses have already formed and deletion of Müller glia during early retinal development does not affect cone synaptogenesis, a separate study examining the role of harmonin (USH1C) in zebrafish which is found in the retinal Müller glia, have disrupted ribbon synapses [27, 28]. Until this conflict can be resolved and the role of these cells have been investigated in other species, the role of Müller glia remains open.

	Astrocytes	Müller glia
Location	• Throughout the nervous system, including the retina and optic nerve [226]	• Found exclusively in the inner nuclear layer of the retina with the process spanning the entire width of the retina [227, 228]
Origin	• Originate from the glial restricted neural stem cells or the bipotent O2A progenitor cell type [229]	• Originate from the neural retinal progenitor cells [227, 228]
Morphology	• Have a stellate or star like morphology [226]	• Have a radial morphology [227]
Functions	• Scaffolding for migration of developing neurons [230] • Aid in the formation of synapses [231-233] • Aid in the formation of the blood brain barrier [234] • Serve as a source of nourishment and energy reserve for the neurons by providing glucose and storing excess glucose in the form of glycogen [235, 236] • Possess various channels and transporter (Na^+/K^+ channels, aquaporins etc.) which aid in the maintenance of homeostasis, pH levels and removal of toxic metabolites [237, 238] • Possess transporters for neurotransmitters (such as GABA, glycine, glutamate)which aid in clearance and release of these molecules into the synaptic space which can affect synaptic transmission[232, 239]	• Serve as a scaffolding for retinal organizations [227] • Help direct light through the retinal layers to the photoreceptor cells [240] • Help in recycling photopigments [241] • Aid in the formation of the blood retinal barrier [242] • Similar to astrocytes serve as a source of nourishment and energy reserve in the form of glucose and lactate respectively [227] • Help in maintenance of homeostasis and removal of toxic metabolites in a manner similar to astrocytes [115] • Neurotransmitter receptors (AMPA, GluR4, NMDA, GABA-A etc.), transporters and modulators (GLAST, GS, GAT etc.) help in neutransmitter recycling and also aid in glia-neuron communication [115]
Changes during reactive gliosis	• Changes in gliosis based on extent of injury which ranges from mild to moderate to severe [243] • Cells hypertrophy (particularly by increasing the expression of GFAP), change in morphology and upregulate various markers [244] • Increase proliferation and in severe cases form the "glial scar" [245]	• Similar to astrocytes following retinal damage, cells hypertrophy, change morphology and upregulate various markers [246] • Based on the ability or the lack of cells to proliferate, Müller cell gliosis is referred to as non conservative or conservative gliosis, respectively [115] • Glial scar is not a prominent feature of gliosis of the Müller glia [114, 115]

	Astrocytes	Müller glia
Stem cell potential	Following injury – • Cells dedifferentiate and have the potential to re-enter cell cycle [111] • Begin to express proteins associated to neural stem cells or radial glia (NG2, BLBP, nestin, DSD1, CD15) [99]	Following retinal injury – • Müller glial cells re-enter cell cycle and can proliferate [111, 114] • Following targeted ablation of photoreceptor and ganglion cells, regeneration of the respective cell types was observed from the Müller glia [111, 247]

Table 1. Comparison of Astrocyte and Müller glial Characteristics

2.3. Blood retinal barrier development and maintenance

The blood-brain barrier refers to the separation between the circulating blood and extracellular fluid found within the central nervous system. In the brain, this barrier is formed through the interactions between astrocytes and endothelial cells that form the vasculature [29]. In the eye, the blood-retinal barrier is maintained at two junctures; 1) an „outer barrier" in the form of the retinal pigmented epithelium (RPE), and 2) the „inner barrier" that is comprised of the endothelial cells of the retinal vasculature [30]. The endothelial cells of the retinal vasculature form tight junctions that are selectively permeable to hydrophobic molecules such as O_2, CO_2, and hormones, while restricting the entrance of bacteria and large or hydrophilic molecules (See Fig 2). Endothelial cells and pericytes that adhere to the outside of the endothelial cells are both encompassed by a basal lamina as well as the astrocytic endfeet. There is evidence that inner barrier is induced and maintained by both Müller glial and retinal astrocytic endfeet that ensheath retinal blood vessels [31]. The processes of retinal astrocytes, however, are limited to the never fiber and ganglion cell layer and can only interact with superficial vasculature near the inner surface of the retina [32]

Müller glia (as well as retinal astrocytes and retinal pigment epithelium) express factors that are critical to the formation of the deep plexus vasculature in the retina [33]. Angiogenesis is the result of a balance between the pro-angiogenic factor vascular endothelial growth factor (VEGF) and anti-angiogenic factor pigment-epithelium derived factor (PEDF) [33, 34]. The ratio of these factors carefully controls the growth of the deep plexus retinal vasculature. Not surprisingly, misregulation of these factors can lead to pathological neovascularization, a topic which will be covered later in this chaper. Many other interactions between Müller glia/ astrocytes and the vasculature have been proposed and/or documented. For instance, Paulson and Newman simulated a process whereby the activity of neurons indirectly regulated blood vessel dilation [35]. In a process referred to as siphoning, the Müller glia are proposed to take up K^+ released by active neurons and then release the K^+ at the endfeet that are in close proximity to the vasculature [35]. Thus the astrocyte can effectively redistribute the K^+ from the neuron, which may be some distance away from the nearest blood vessel, to a region immediately adjacent to the arteriole in a manner that is faster than would otherwise take place if the K^+ was undergoing simple diffusion. Further, this could also concentrate K^+ released over a wider area to the smaller area of the endfeet.

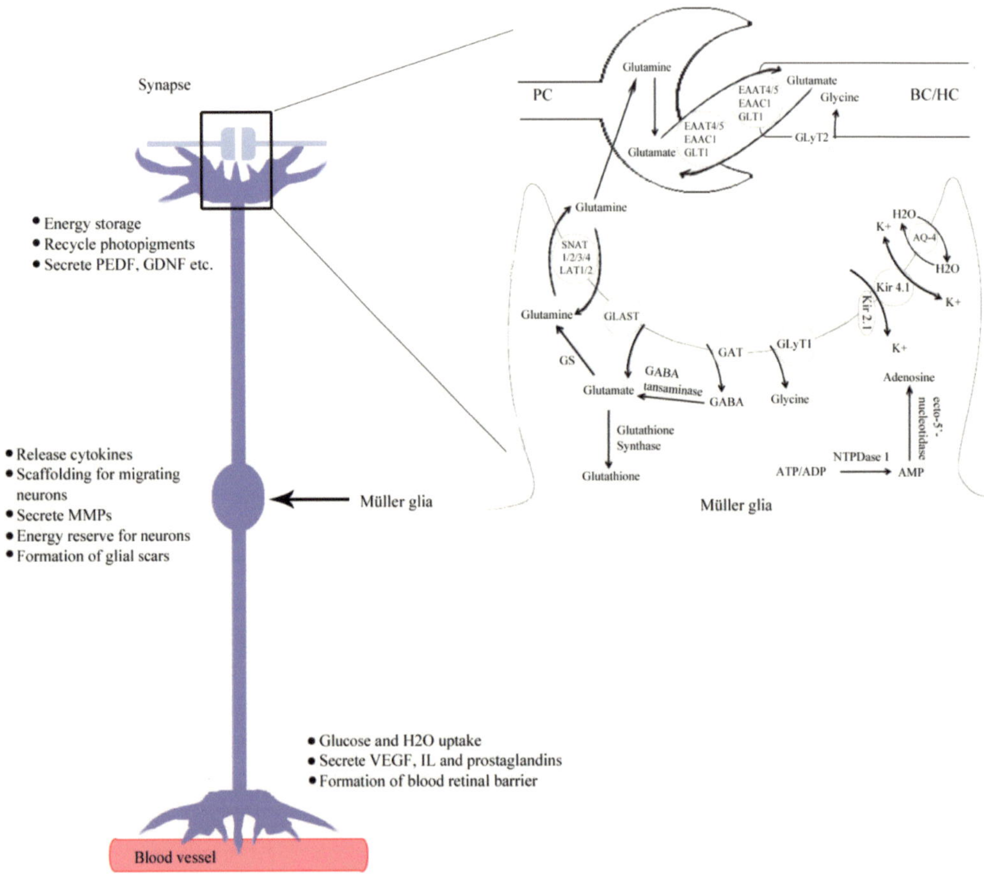

Figure 2. Aquaporin-4 (AQ-4), bipolar cell/ horizontal cell (BC/HC), excitatory amino acid transporter (EAAT), excitatory amino acid carrier (EAAC), glial cell derived neurotrophic factor (GDNF), glutamate transporter (GLT), glycine transporter(GlyT), glutamine synthetase (GS), interleukins (IL), L-type amino acid transporter (LAT), matrix metalloproteinases (MMP), photoreceptor cell (PC), pigment epithelium derived factor (PEDF), Na+ coupled neutral amino acid transporter (SNAT), glutamate aspartate transporter (GLAST), vascular endothelial growth factor (VEGF).

Müller glia are also known for releasing many growth factors, and many of these factors effect the endothelial cells. Transforming growth factor β1 (TGFβ1) is released by Müller glia and can increase the expression of tissue plasminogen activator inhibitor-1, which could potentially have the protective effect of reducing hemorhaging in the brain [36-38]. TGF-β1 also has been shown to have a morphological effect on cultured endothelial cells, inducing them to form capillary-like structures [39]. Mice with a loss of the integrin αVβ8 that is necesary for TGF-β activation within the retina also have abnormal superficial as well as deep plexus formation

[40]. Glial-derived neurotrophic factor (GDNF) and neurturin are also released by Müller glia and appear to enhance barrier function as measured by transendothelial resistance [41].

Communication between Müller glia and endothelial cells is not a one-way street. There also appear to be inductive signals released from the endothelial cells that effect Müller glial differentiation/function. It is well established that leukemia inhibitory factor (LIF) is secreted from endothelial cells and that it helps to induce astrocyte differentiation in optic nerve astrocytes [42, 43]. LIF and ciliary neurotrophic factor (CNTF) share a part of their receptor complex and intracellular signaling pathway; therefore it is not surprising to find that CNTF has also been shown to have effects on astrocyte development [44, 45]. Both CNTF and LIF are present in the developing retina and CNTF does increase the production of Müller glia [46]. However, an increase in the expression of LIF from the lens during retinogenesis inhibited the development of retinal vasculature and increased the expression of VEGF in retinal astrocytes and Müller glia [47]. Hence it is unclear whether LIF plays a role in Müller glial cell differentiation.

2.4. Metabolic coupling with neurons

The brain is a high energy consuming organ, using approximately 25% of the glucose present in the human body [48]. There is very tight coupling between the demand and supply in the central nervous system (CNS), and most of this expenditure is due to neuronal activity [48, 49]. However, neurons do not store much glycogen and therefore are reliant upon external sources to fuel their oxidative metabolism. In the retina, this need is met by both the Müller glia and retinal astrocytes. Glucose enters Müller glia via glucose transporter-1 (GLUT-1) and is phosporylated by hexokinase to produce glucose-6-phosphate (Fig 2). From here, part of the glucose-6-phosphate is stored with the Müller glial cell body as glycogen and the rest is metabolized to various carbohydrate intermediates [50-52]. Neurons can use a variety of substrates to fuel their oxidative metabolism, including lactate, pyruvate, alanine, glutamine, and glutamate [53, 54]. Müller glia metabolize glucose and glycogen deposits predominantly to pyruvate and lactate which is released to the extracellular milieu by the monocarboxylate transporter MCT2 [55, 56]. Neurons can then take up pyruvate and use it directly in the Krebs cycle to compensate during times of low glucose [50, 57]. Lactate generated by Müller glia is converted by lactate dehydorgenase and pyruvate kinase to pyruvate to power the Krebs cycle [55].

Active neurons, in turn, release glutamate, NH_4^+, K^+, and CO_2, all of which are taken up by the Müller glial cells and are either disposed of or recycled [4]. Glutamate is an excitatory neurotoxin, even at low extracellular concentrations, and is tightly regulated by Müller glia in the retina [58]. Müller cells take up glutamate via the glutamate/aspartate transporter, GLAST, and NH_4^+ via an ammonia transporter (AMT) [59, 60]. In addition to transporting glutamate into Müller glial cells, the GLAST protein co-transports $3Na^+$ ions and one H^+ and counter-transports one K^+ [61]. The influx of Na^+ into the Müller cell activates the Na^+/K^+ ATPase which further stimulates glycolysis [4, 62]. Both the NH4+ and glutamate are use to create L-glutamine by glutamine synthetase [60, 63-65]. The glutamine produced by Müller glia is then transported back to neuronal cells to aid in the synthesis of neurotransmitters glutamate and GABA [54]. The presence of glutamate and NH_4^+ have a combined action of increasing glycolysis by the Müller glia, in part by increasing the expression levels of glutamine synthetase [54, 66, 67].

Müller glia also act as a sink for excess extracellular K^+ in the retina, which is taken up by the inwardly rectifying K^+ (Kir) channels and the Na^+/K^+ ATPase of the Müller cells [62]. This elevation of K^+ concentration increases the glycogenolysis in cultured Müller glia, tightly coupling the breakdown of glycogen to neuronal activity [17]. The K^+ is then disposed of by passing K^+ into the subretinal space, the vitreous body, or the blood. [68, 69]. Finally, carbonic anhydrase converts CO_2 to bicarbonate which is then released by way of the H^+/HCO_3^- exchanger into the vitreous or blood (Fig 2) [70-73].

2.5. Regulation of neurotransmission

In the retina, glutamate is the primary excitatory neurotransmitter [74]. Müller glia have transporters for a wide variety of transmitters, including glutamate, GABA, Glycine, D-serine, dopamine, and ATP [75, 76]. The Müller glia take up neurotransmitters and other neuroactive substances and convert them to substances that can be supplied to retinal neurons as neurotransmitters or neurotransmitter precursors (Fig 2). The modulation of neuronal excitability through regulation of neurotransmitter availability is thought to serve three functions; 1) termination of neuronal signaling, 2) prevention of neurotransmitter spread to adjacent synapses, and 3) prevention of neurotoxicity resulting from prolonged presence of a transmitter at a synapse [4, 75]. In this section, we will briefly cover transport of the major retinal neurotransmitters into Müller glia, processing of the transmitter by the Müller glia and transport of products back to retinal neurons.

Müller glia express several glutamate transporters, depending upon the species, including the previously mentioned GLAST protein (also known as excitatory amino acid transporter 1 or EAAT1). In humans for instance, the dominant transporter is EAAT1, but EAAT2 and 3 can also be found [77]. Glutamate is the most widely used neurotransmitter used by retinal neurons, including photoreceptors, bipolar and ganglion cells. Both the photoreceptors and the bipolar cells have graded potentials, hence the amount of neurotransmitter released is directly correlated to the amount of stimulus. In addition, photoreceptors are wired a little differently than other neurons that transduce sensory information; they release glutamate in the dark and less glutamate upon transduction of light signals. Hence, removal of glutamate from the synaptic region is critical for normal tranmission of light signals to take place. Knockdown and knockout studies in the retina have indicated that a loss of GLAST leads to a loss of the electroretinogram b-wave, primarily because it aids in signal processing between photoreceptors and bipolar cells, rather than any neurotoxicity associated with high levels of glutamate [78, 79]. Consistent with the idea that Müller glia are critical for clearing away glutamate released at synapses are studies in which clearance of D-aspartate was tracked first to Müller glia followed by a redistribution into other neuronal cell types of the retina [80]. Glutamate can be converted to glutamine by glutamine synthetase, and is then transported back to neurons as a precursor to glutamate [63, 81]. Loss of glutamine synthetase activity leads to a loss of glutamate content in retinal neurons which leads to functional blindness within 2 minutes [82, 83].

There are several other neurotransmitters used in the retina, such as GABA, glycine, and dopamine. Since the interactions of these neurotransmitters are not as heavily studied as glutamate, only their uptake mechanism and potential processing within Müller glia will be discussed here. GABA is used by horizontal and amacrine cells within the retina and termination of GABA activity is brought about through the uptake of GABA by Na^+/Cl^--dependent GABA transporters (GATs) found in presynaptic neurons, Müller glia, and retinal astrocytes [76, 84, 85]. After uptake into Müller glia, GABA can be converted to glutamine via glutamine synthetase and, as specified above, is returned to neurons to act as substrates for neurotransmitters [86]. Müller glia also express glutamate decarboxylase which catalyzes the decarboxylation of glutamate to GABA. It is unclear, however, whether GABA can be released by Müller glia [76].

Dopamine performs a large number of functions in the developing and mature retina that are well out of the scope of this chapter. A full discussion of this topic can be found elsewhere [87]. Both the transporter and enzymes necessary for converting tyrosine to dopamine are expressed in Müller glia [88]. Likewise, ATP also performs a large number of functions in the developing and mature retina [89-91]. Müller glia express a subset of the P2X and P2Y ATP receptors and they also have the ability to convert ATP to adenosine and release both ATP and adenosine into the intracellular space [91, 92].

Müller glia also carry glutamate, GABA, purinergic, glycine, dopaminergic, noradrenergic and cholinergic receptors [76]. In some instances these receptors have been shown to coordinate release of neurotransmitters by neurons with enzymatic activity and or gene regulation in the Müller glial cells. An excellent example of this coordination is the regulation of glutamate receptors on GLAST activity and expression of GLAST. When glutamate receptors are activated on Müller glial membranes it leads to an increase in intracellular Ca^{2+} and protein kinase C (PKC). The activation of metabotropic glutamate receptors in Müller cells leads to an increase in Ca^{2+} and protein kinase C, and phosphorylation of GLAST by PKC leads to an increase in transport of glutamate [82, 93]. The increased transport of glutamate through GLAST appears to regulate activation of mammalian target of rapamycin (mTOR), which activates DNA binding of the transcription factor activator protein-1 (AP-1) and an increase in GLAST mRNA [94].

2.6. Other

Müller glia perform a variety of other functions beyond those already mentioned. For instance, in addition to siphoning K^+ released by retinal neurons, the Müller glia are also responsible for the transport of water that accumulates in the tissue as the end product of ATP synthesis [95]. The movement of water is specifically coupled to the movement of Na^+ and K^+ and, like K^+, is released into the bloodstream. Müller cells are also involved in phagocytosis of debris in the retina and in the release of antioxidant glutathione [96, 97]

3. Properties that are similar to stem cells/astrocytes

Studies using reactive astrocytes have shown the potential to dedifferentiate into cells having neural progenitor or stem cell like properties (Table 1) [98, 99]. Following stimulation, these cells show activation of signaling pathways such as EGF, FGF, SHH and Wnts, previously shown to be associated with the neural stem cells [98, 100-102]. Similarly, activated Müller glial cells following retinal injury have also shown the capacity to dedifferentiate into retinal progenitor cells [103]. Studies in lower vertebrates such as fish, amphibians and birds have shown the presence of a stem cell niche in the ciliary marginal zone (CMZ) of the retina [104-107]. Mammals, however, do not have a CMZ [108]. In mammals, a small group of cells in the non-pigmented portion adjacent to the retina can proliferate up to postnatal day 21, but these cells are low in abundance and are not thought to generate many cells [103, 109]. It may be more feasible to generate many retinal progenitor cells from activated Müller glia. Expression profiling of proliferating Müller cells suggests a stem cell like role for these cells [110, 111]. Culture of the Müller cells in an enriched medium generated "multipotent neurospheres", elucidating the stem cell role of Müller cells *in vitro*. Further transplantation of enriched Müller glial cells into injured retina generated cells with neuron like characteristics [112]. Müller cells have been shown to dedifferentiate, proliferate and give rise to amacrine cells, bipolar cells, retinal ganglion neurons as well as the photoreceptor cells. [110, 111, 113]. One important factor aiding the transformation of the Müller glial cells is the membrane depolarization due to a reduction of potassium ion conductance, primarily due to downregulation of the Kir channels in the Müller cell [114]. The dowregulation of the Kir channels leads to a decrease in the p27kip1 cyclin kinase inhibitor, which is then succeeded by re-entry into cell cycle. The downregulation of the Kir channels pushes these cells towards the proliferative stage [115].

4. Response of Müller Glia to injury or disease states

When there is injury or disease within the CNS, astrocytes respond by entering a state referred to as reactive gliosis. Reactive gliosis is an ill-defined set of molecular changes that alters the homeostatic role of the cells and their interactions with neurons, vasculature, and the immune system. Reactive gliosis is thought to be the result of signals received from the injured or diseased tissue that begins a molecular cascade within the glial cells resulting in a change of state [103]. There are a mulitude of questions that have arisen as a result of our limited understanding of gliois, and investigators are currently working to answer these questions. Among them:

- Is reactive gliosis one condition, or a host of related conditions?
- What are the molecular triggers of gliosis?
- Do all the triggers that appear to be involved in gliosis converge on one or two pathways that mediate the changes in Müller glial state, or, are their multiple pathways that can mediate multiple changes?
- Do different signals mitigate mild, moderate or severe reactive gliosis? How are these forms of gliosis related?
- Can severe reactive gliosis be attenuated, even when triggers are chronically present?

- Can the reactive gliosis be used to „supply“ multipotent stem cells to the retina to replace dead or dying neurons?
- Can the multipotent stem cells that arise from Müller glial cells be directed in their differentiation in vivo and can the number of progenitor cells differentiating into cell types other than Müller glia be increased substantially?

There appears to be a continuum in the states of reactive gliosis, from mild to severe. In the mild to moderate forms of gliosis, the cells may hypertrophy and show some changes to their functionality, but, if the trigger is removed, the cells may revert back to their former condition without altering the tissue [116]. In the more severe forms of reactive gliosis, cells hypertrophy, lose functionality, form glial scars that are inhibitory to axonal regeneration and neuronal surivival, and may also proliferate [116, 117]. The severe state is marked by the persistance of these characteristics. Within the mammalian retina, both the Müller glia and retinal astrocytes display reactivity to injury and disease. In this section we will talk about triggers of Müller glia, evidence that BMP7 may also be a trigger, and the changes in retinal homeostasis that result from reactive gliosis in the retina.

5. Triggers of reactive gliosis

5.1. Known triggers

Müller glial reactivity can be found in every identified disease and injury that plagues the eye, including diabetic retinopathy, glaucoma, age-related macular degeneration, retinitis pigmentosa, and many many others [118-122]. In considering reactive gliosis, there appears to multiple levels of complexity. For instance, there are a wide range of factors which have been shown to trigger reactive gliosis in Müller glia (Figure 3 and Table 2). Some of these triggers can have concentration-dependent effects upon astrocytes [116]. Further, different triggers can lead to specific molecular and functional changes in the Müller glia that may correspond to the various aspects of reactive gliosis [123]. Not only are there multiple triggers, but there is heterogeneity in the response of Müller glia to the same factor [118].

5.2. Bone morphogenetic proteins in Müller cell gliosis

Studies in the injured spinal cord have indicated a role for another family of growth factors; the bone morphogenetic proteins (BMPs). The BMPs are members of the TGF-β superfamily of growth factors. The receptors include two basic types, Type I and Type II, both of which are serine-threonine kinases. Receptors from each type must form heterodimers in order for signaling to occur, although the Type I receptors are downstream of the Type II. There are two non-canonical signaling pathways, BMP-MAPK and FRAP-STAT, that have more recently been identified in addition to the canonical SMAD-related pathway [45, 124-129]. Three type I receptors have been associated with the BMPs, activin-like kinase 2 (ALK2), ALK 3 and ALK6. Accumulated evidence has shown that in regards to the Type I receptor, BMP 6 and 7 activate the ALK2 receptor preferentially, whereas BMPs 2 and 4

Growth Factors and Cytokines	
	Ciliary Neurotrophic Factor/Leukemia Inhibitory Factor [86, 248-251]
	Epidermal growth factor/HB-EGF [84, 87, 180]
	Fibroblast growth factor 2 [250, 252]
	Brain-derived neurotrophic factor [250]
Transduction Pathways and Transcription Factors	
	STAT3 [248, 253, 254]
	NF-κB [255, 256]
	Toll-like receptor 2 [257]
	TRPV1 (Vanilloid Receptor) [85]
	Gp130 [249]
	Epidermal growth factor receptor [87]
	Fibroblast growth factor receptor [179]
	MEK [179, 258]
Other	
	Oxidative Stress/Ischemia [38, 254, 255]
	ATP
	Glucose [88, 259]
	Amyloid Beta [260]
	Endothelins [261]
	Nitric Oxide

Table 2. Triggers of Müller Glia Cell Activation

activate either ALK3 or ALK6 [130]. In addition to the canonical SMAD pathway, ALK3 and 6 also activate the BMP-MAPK and FRAP-STAT pathways [45, 124, 129]. The BMPs have been shown to act as a gliosis trigger in penetrating spinal cord injuries, and a differential role for ALK3 and 6 receptors has been ascribed to various aspects of gliosis, including hypertrophy, inflammation, and tissue loss [131, 132]. While BMPs have been studied in retinal injury, primarily as a survival factor for retinal neurons, it has not been studied as a potential trigger for reactive gliosis in Müller glia [133].

My lab has investigated the role of BMP7 as a potential trigger for reactive gliosis in Müller glia and retinal astrocytes. We and others have documented changes in BMP expression and signaling following injury or disease in the retina and optic nerve [134]. We have determined expression levels of BMPs and BMP intracellular signaling pathway members in a diabetic mouse model, the Akita mouse model (Ins^{AKITA}). These mice contain a naturally occurring missense mutation in

http://dx.doi.org/10.5772/55150

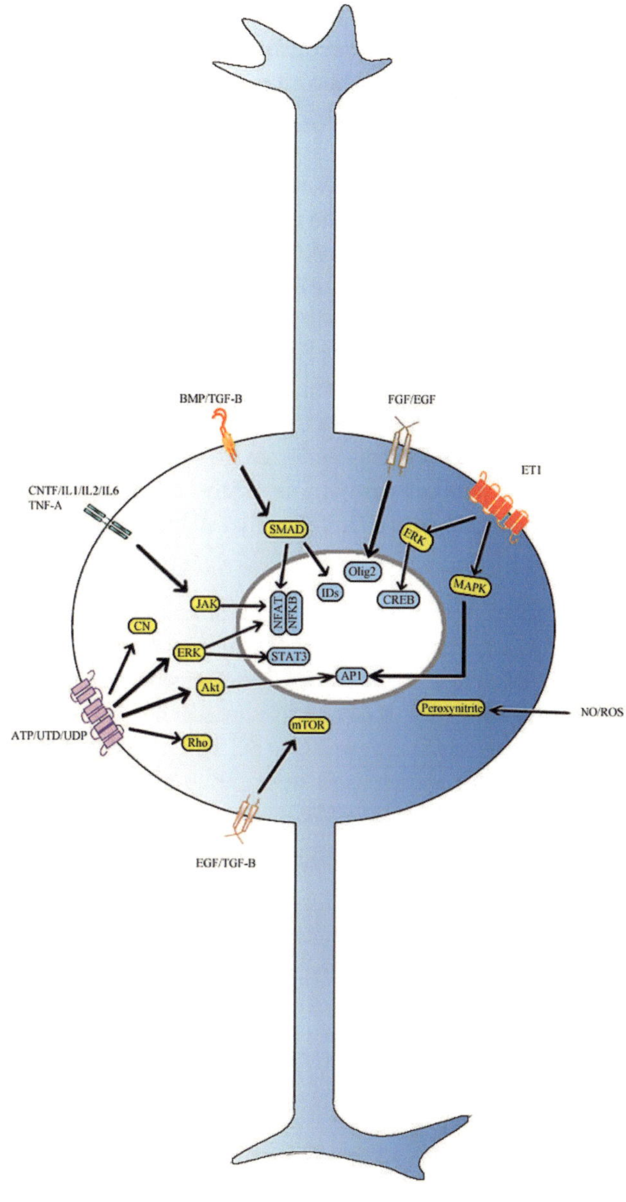

Figure 3. Schematic representation of various signaling mechanisms which trigger and regulate reactive gliosis in Müller glia. Growth factors such as TGF-β, BMP, EGF and CNTF; interleukins; as well as reactive oxygen species and free radicals are known factors to trigger gliosis in Müller glial cells. Activator protein-1 (AP1), adenosine triphosphate (ATP), bone morphogenetic protein (BMP), ciliary neurotrophic factor (CNTF), calcineurin (CN), cAMP response element binding protein (CREB), epidermal growth factor (EGF), endothelin 1 (ET1), extracellular-signal-regulated kinase (ERK), fibroblast growth factor (FGF), interleukin (IL), inhibitor of differentiation (ID), janus kinase (JAK), mitogen activated protein kinase (MAPK), mammalian target of rapamycin (mTOR), nuclear factor of activated T-cells (NFAT), nuclear factor kappa-light-chain-enhancer of activated B cells (NF-κB), nitric oxide (NO), reactive oxygen species (ROS), tumor necrosis factor-α (TNF-α), transforming growth factor-β (TGF-β), uridine triphosphate (UTP), uridine diphosphate (UDP).

the insulin 2 gene that causes a switch from a cysteine to a tyrosine residue at amino acid 96, removing one of the cysteines necessary for an intramolecular disulfide bond [135]. Heterozygous mice are severely insulin deficient and become diabetic at about 6 weeks of age [135]. For these studies we used two stages; mice that are 3 weeks of age have mild to no reactive gliosis, while 6 weeks of age has moderate gliosis. Levels of BMP expression were determined by reverse transcription – quantitative polymerase chain reaction (RT-qPCR) of RNA samples from 3 and 6 week old wild type and heterozygous mice. The graphs show changes in mRNA levels in the 3 and 6 week Ins^{AKITA} mice relative to levels of mRNA in wild type samples (Fig 4A, B). Further, genes that are known downstream targets of the BMP pathway, such as inhibitor of differentiation (ID) 1, 3, and MSX2 are also increased, consistent with an increase in BMP signaling (Fig 4A, B). To verify there was an increase in canonical BMP signaling, an increase in nuclear localization of phospho-SMAD1 (p-SMAD1,5,8) sections through wild-type and 6 week Ins^{AKITA} retina were immunolabeled for p-SMAD1,5,8 and glutamine synthetase (Fig 4C-N). The Ins^{AKITA} retina showed a clear increase in p-SMAD1,5,8 expression in the inner nuclear layer at 6 weeks of age, some of which was coincident with cells glutamine synthetase-expressing Müller glia (Fig 4L-N). There was also clear increase in p-SMAD1, 5, 8 in other cells of the inner nuclear layer and cells of the ganglion cell layer.

To test the role of BMPs in reactive gliosis in vivo, adult murine eyes were injected intravitreally with vehicle or BMP7 and analyzed 3 or 7 days post injection. At both 3 and 7 days post vehicle injection, there were the normal low levels of GFAP expression and moderate levels of glutamine synthetase in Müller glia (Fig 5A, B, G, H). A low level of the chondroitin sulfate proteoglycan, neurocan, is present throughout the retina (Fig 5C, I). Three days post BMP7 injection, no increase in GFAP was detected, but an increase in both glutamine synthetase and neurocan levels were detected (Fig 5D-F). Immunolabel of BMP7-injected eyes showed an increase of GFAP, glutamine synthetase and neurocan in comparson to vehicle-injected eyes (Fig 5J-L).

6. Characteristics of reactive gliosis in Müller Glial cells

Müller glia display many changes during reactive gliosis (Fig 6). We have grouped these changes into 6 broad categories; 1) hypertrophy, 2) loss of functionality, 3) neuroprotection, 4) inflammation, 5) proliferation, 6) remodeling.

6.1. Hypertrophy

Hypertrophy refers to the swelling of the Müller glial cell body and processes. The swelling is, in part, brought about by an increase in the expression of two type III intermediate filament genes, GFAP and vimentin. As with many changes that occur with reactive gliosis, upregulation of intermediate filaments and the ensuing hypertrophy has both good and bad characteristics associated with it. Hypertrophic glia help to form and maintain a barrier around injured tissue which helps to protect surrounding tissues from inflammatory signals [136, 137]. On one hand, there is evidence that the increased production of GFAP does not lead to diminished neuronal metabolism, eletrophysiology or visual function [138]. However, evidence from injured spinal cord indicates axonal regeneration and functional recovery was increased in GFAP/vimentin double-knockouts in comparison to wild type controls [139]. Further, the retinas of GFAP/vimentin double knockouts were also protected from retinal degeneration following retinal detachment, and integration and neurite extension from transplanted cells is also enhanced [140].

In addition to increased intermediate filament expression, hypertrophy is also the product of a loss of K^+ conductance into the blood stream as already covered in section 1.3, Müller glia take up K^+ released by retinal neurons and release it into the bloodstream. Water in the tissue, created through the process of oxidative synthesis of ATP, is removed through the pigmented epithelium and Müller glia. The movement of water is coupled to the movement of osmolytes, including Na^+ and K^+ ions, and are subsequently removed from the Müller cell bodies via release into the bloodstream [4]. Müller glia undergoing gliosis downregulate the K^+ channel, Kir4.1, that delivers K^+ to the vasculature, which uncouples the movement of K^+ and water into the blood. The end result is swelling of the Müller cell body.

6.2. Loss of functionality

Loss of functionality is a part of the general response of the cells to undergo dedifferentiation. However, the response of the Müller cells can vary depending upon the disease or injury present. A good example of this is the regulation of the glutamate transporter in disease and following mechanical injury. Downregulation of glutamate transporters is observed in glaucoma, ischemia and diabetic retinopathy, due to a decrease in the activity of the glutamate transporter GLAST. This in turn downregulates the activity of glutamine synthetase, an enzyme involved in glutamate recycling [141]. However, following mechanical nerve injury, as seen with the optic nerve crush model, glutamine synthetase was found to localize to the ganglion cell layer, aiding in the recycling of the excess glutamate released due to neuronal injury. [142].

The Kir channels (potassium channels) in the Müller glial cell membrane play an important role in the gliosis response as well. Decrease in conductance of the potassium ions due to down regulation of Kir 4.1 leads to an increase in potassium ions outside the membrane. This, in turn, decreases the transport of glutamate, glucose and water across the Müller glial cell surface. Consequently, an increase in the glutamate toxicity, decrease in glutathione synthase activity and osmotic swelling were observed in the retina, which contribute to the loss of glia/neuron interactions [97, 114, 120, 143-146].

There is also a reduction in the blood-retinal barrier function under hypoxic conditions. This appears to be driven by changes Müller cell expression of growth factors that regulate endothelial cell tight junctions. The balance between factors that increase endothelial cell tight junctions (PEDF, glial derived neurotrophic factor (GDNF), transforming growth factor Beta (TGFβ), thrombospondin, etc) and factors that decrease barrier function (VEGF, TNFα, FGF2, etc) is disrupted by reactive gliosis [34, 41, 147-153]. VEGF appears to be the dominant factor released from Müller glial cells in decrease of barrier function and angiogenesis that occurs in many forms of retinal injury and disease [153].

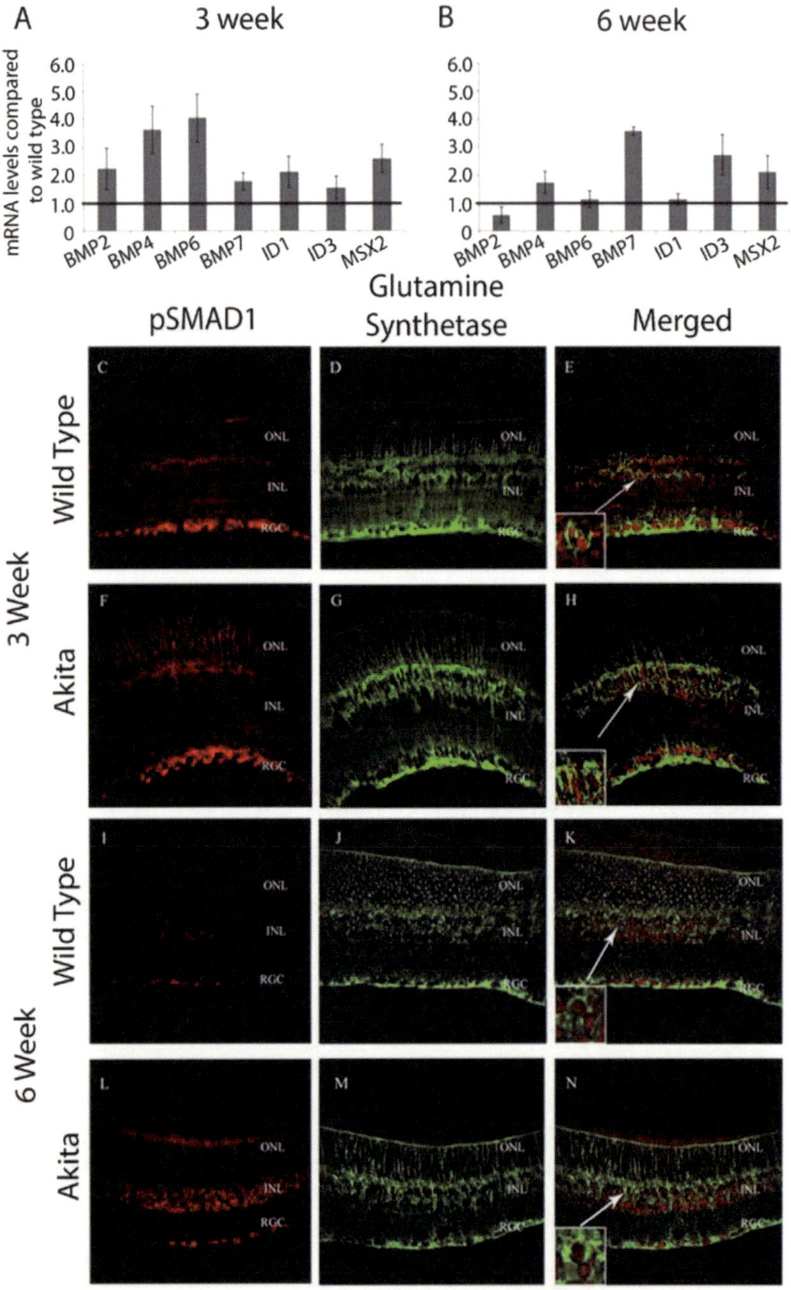

Figure 4. Analysis of retinas of the Ins2^{Akita} diabetic mouse shows increase in BMP signaling in the diseased eye when compared to the wild type eye. **A** and **B**: qPCR results analyzing the levels of various BMP molecules shown to be regulated during reactive gliosis and some of the targets of the canonical BMP signaling pathway, using RNA obtained from whole retinas in 3 week and 6 week diseased eye, respectively, normalized to their respective wild types. At the 3 week stage (**A**), when little or no gliosis is observed (data not shown) levels of BMP 2, 4 and 6 appear to be high. At the 6 week stage (**B**) when we do seen an increase in expression of GFAP, GS and neurocan (data not shown), there is

an increase in levels of BMP7 with a subsequent decrease in the levels of other BMP molecules, indicating a role for BMP7 in reactive gliosis in the diseased state. Immunohistochemistry was performed to determine the localization of phospho SMAD with glutamine synthetase in the retinas (**C – N**). The 3 week retinas show similar nuclear phospho SMAD levels in both the wild type and the Ins2^{Akita} (**C, E, F and H**). In the 6 week Ins2^{Akita}, there is a clear increase in the phospho SMAD levels in the inner nuclear layer nuclei (**L and N**) when compared to the wild type (**I and K**), possibly due to the increase in BMP7 shown previously (**B**).

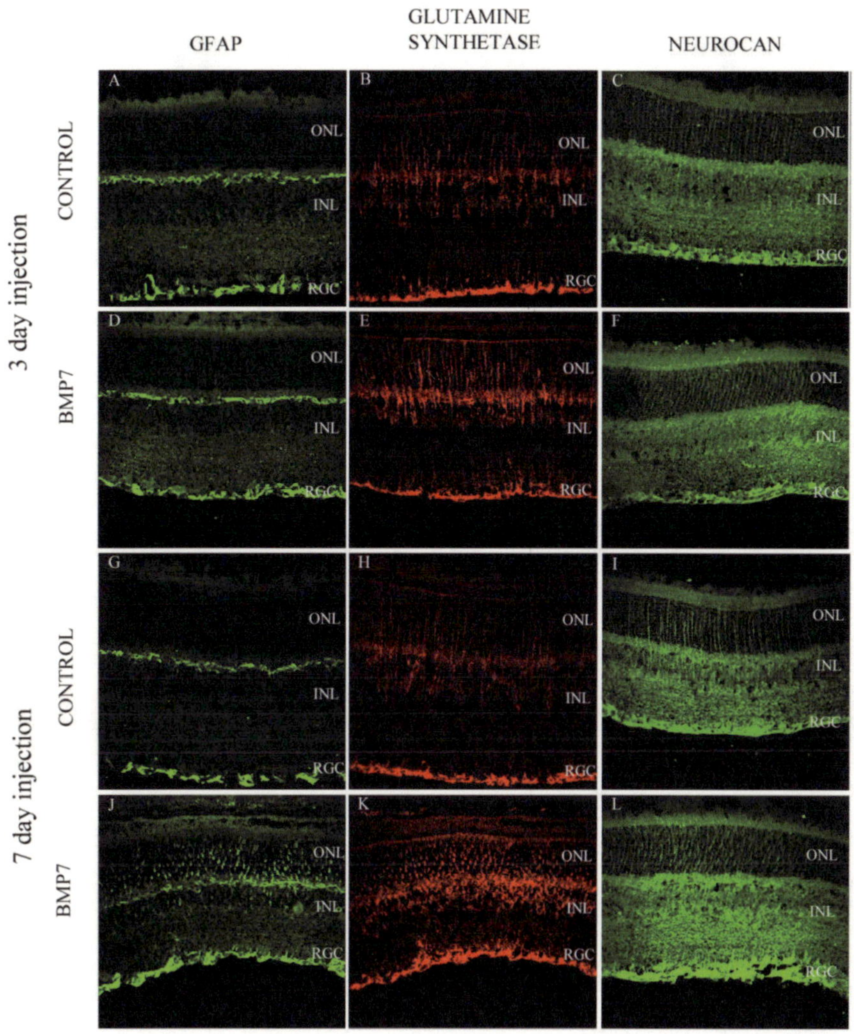

Figure 5. Effect of intra vitreal injections of BMP7 into normal mouse eyes – Retinal sections of eyes injected with either vehicle or BMP7 were analyzed 3 days (**A – F**) and 7 days (**G – L**) post injection via immunohistochemistry for the localization of glial fibrillary acidic protein (GFAP), glutamine synthetase (GS) and neurocan. Retinas isolated 3 days post injections do not show an increase in GFAP (**A and D**) or neurocan (**C and F**), although GS does seem to show an increase when compared to the vehicle injected eyes (**B and E**). Retinas isolated 7 days post injection did show a clear increase in GFAP (**G and J**), GS (**H and K**) and neurocan (**I and L**) in the BMP7 injected eyes compared to the control eyes, suggesting the BMP7 was able to trigger gliosis in these retinas.

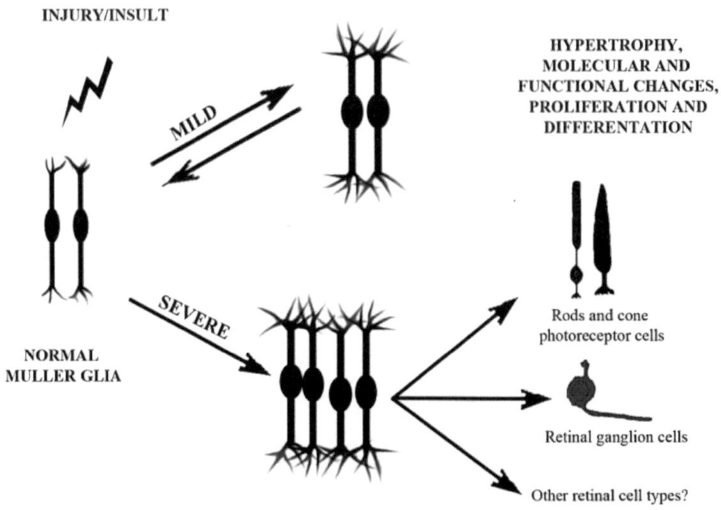

Figure 6. Schematic representation of reactive gliosis response in Müller glia depending on the extent of the injury. Mild changes in reactive gliosis comprises of hypertrophy of the cells due to an increase in glial fibrillary acidic protein and changes to the function and morphology of the cell, with little or no proliferation which has the potential to resolve once the stimulus subdues. Severe reactive gliosis occurs following tissue damage and induces Müller glial cell proliferation, overlapping of cell processes, hypertrophy, functional and morphological changes. Under severe gliosis conditions reactive Müller cells have shown the ability to dedifferentiate and give rise to some of the retinal cells types.

6.3. Neuroprotection

Reactive gliosis in Müller cells is a complex response dependent on the injury or disease. Diseases which lead to retinal degeneration such as retinal detachment, retinitis pigmentosa or physical damage to the retina elucidate such a response from the Müller cells to aid in neuoprotection and prevent apoptosis [114, 141]. A wide range of growth factors secreted by the reactive Müller cells, including bFGF, GDNF, CNTF, and VEGF [114, 141, 150, 154, 155]. Upregulation of CNTF and bFGF have been observed following mechanical injury, ischemia and NMDA mediated neuronal death [156-158]. These growth factors help to increase neuron survival and inhibit apoptosis, either directly as is the case for bFGF, or indirectly in the case of CNTF and GDNF [159, 160]. GDNF also upregulates GLAST, thereby, protecting neurons from excessive glutamate excitotoxicity [160]. VEGF is another factor which is upregulated following gliosis. Hypoxia as well as diabetes has shown to increase the VEGF secretion by Müller glial cells [161, 162]. VEGF may act directly by increasing the permeability of the endothelial cells [163]. VEGF may also be regulated by TGF-β released during hypoxia, which, along with other cytokines such as TNF-α, increase the expression of matrix metalloproteinases which can clear the basement membranes of these cells generating leaky vessels [38, 164].

Müller cells also protect retinal neurons from oxidative stress, excitotoxicity and from damaging reactive oxygen species via conversion of glutamate to glutamine as well as synthesis and release of antioxidants such as glutathione [165, 166]. However, concomitant with an increase in the antioxidant glutathione, during hypoxia, diabetic retinopathy, hyperglycemia and ischemia there is also an increase in the expression of inducible nitric oxide synthase and cyclooxygenase-2 [167, 168]. These enzymes can lead to production of nitric oxide, prostaglandins and superoxides which are detrimental to retinal neurons and may induce apoptosis in neural cells [169]. Nitric oxide also has a beneficial role as it increases blood flow by dilating blood vessels and prevents glutamate toxicity by closing *N*-methyl –D-aspartate (NMDA) receptors [170].

6.4. Inflammation

Müller cells also play a role in the inflammatory response observed in the retina, primarily seen in the diabetic retina. Under these conditions, the activated Müller cells begin expressing pro inflammatory cytokine interleukin-6 (IL-6) and IL-1B [171, 172]. They also increase expression of TNF-α which increases the expression of the chemokine IL-8 and MCP-8, and promotes infiltration of inflammatory cells [173]. The inflammatory response is further supported by the decrease in glutamate uptake in diabetic retinas. This increases the expression of glutaredoxin, which translocates NF-κB to the nucleus and increases the expression of pro-inflammatory proteins [141].

6.5. Proliferation

Dedifferentiation and proliferation of Müller glia is known to occur in many different species, including chick, fish, and even mammalians [108, 110, 174-177]. Several aspects of Müller cell proliferation are of interest here; 1) the molecular pathways that result in the release of the cells from their normally quiescent state, 2) extrinsic signals that are necessary for the proliferative reponse, and 3) directing progenitor cells to differentiation and integration into retinal tissue.

Several intracellular signaling pathways have been investigated to determine those that may be important for the proliferative response in dedifferentiating Müller glia. The FGF-MAPK pathway appears to be indespensible for the proliferative activity seen during reactive gliosis [178, 179]. The heparin binding epidermal growth factor (HBEGF)-MAPK pathway is be also induced in the Müller glia found in injured areas and appears to be associated with regeneration-associated genes [180]. Further, the HB-EGF pathway appears to be upstream of the WNT-β-catenin pathway, which has been very clearly associated wth re-entrance of Müller glia into the cell cycle [181]. More specifically, Müller glia that are poised to re-enter the cell cycle accumulate β-catenin in injured zebrafish retina, whereas those Müller cells that remain quiescent do not accumulate β-catenin [181]. Further, activation of the WNT/β-catenin pathway stimulates a loss of Müller glia and a concomitant increase in newly generated photoreceptors [181].

In order for Müller glia to re-enter and progress in the cell cycle, the cells would also have to suppress some of the cell cycle check-points that are responsible for the quiescent state of the cells. Inhibition of the cyclin kinase inhibitor p27 has been shown to play a pivitol role in the ability of Müller glia to re-enter the cell cycle. P27 regulates the cell cycle by blocking cell cycle progression into the S-phase, and hence is necessary for maintainance of the quiescent state [182]. Knock-out mice for p27 show many of the characteristics of reactive gliosis, including an increase in GFAP expression and proliferation and migration of cells into the subretinal space [138, 182-184].

6.6. Remodeling

There appears to be three elements of the retina which can undergo remodeling as a result of gliosis; 1) vasculature, 2) the Müller glia themselves, and 3) the extracellular matrix. The neovascularization is, for the most part, due to an imbalance between the antiangiogenic factor PEDF and and the angiogenic factor VEGF [162, 185-190]. Under hypoxic conditions, transcriptional activation of VEGF occurs by translocation of the newly stabilized hypoxia inducible factor-1α (HIF-1α) and it's partner HIF-1β to the nucleus where they bind to the hypoxia responsive element (HRE) in the 5′ flanking regions of the VEGF gene [191, 192]. VEGF is released into the extracellular mileu, where it penetrates the basal laminae and interacts with retinal endothelial cells. This interaction results in an increase in the release of a family of zinc-dependent endopeptidases called the matrix metalleoproteinases (MMPs), plasminogen activators, and other proteinases which degrade proteins, such as occludens, which necessary for the tight junction formation between endothelial cells [192-196]. The VEGF activates the MAPK pathway via phospholipase C-γ, which mediates proliferation of the endolthelial cells [197]. The MMPs also degrade the basal laminae, removing contact inhibition of the endothelial cells and permitting proliferation [38].

The Müller glia participate in remodeling themselves by extending hypertrophied processes into areas they are not typically found. For instance, processes can protrude into the subretinal space, plexiform layers, the vitreous, into occluded blood vessels, and even into the choroid [122, 198-203]. In some respects, the Müller glia are expanding into areas where degenerating neurons and/or axonal processes are found, such as the subretinal space or plexiform layers [204]. If these new processes persist, the end result is the formation of scar tissue, which can permanently block the reattachment of the retina, regeneration of outer segments or regeneration of synaptic contacts in the plexiform layers [118, 122, 205-209]. The extension of processes onto the vitreal surface of the retina results in the formation of periretinal membranes that may under epithelial to mesenchymal transformation into myofibrocytes that spread and become contractile [210]. The contractility leads to folds and/or deformations in the retina, causing visual distortions at the very least, and, at worst, can cause retinal detachments [211, 212]. Glial membranes/scars are a significant issue in the treatment of visual disorders in humans, occuring in appoximately 15% of retinal detachments [213].

The last element of the retina that undergoes remodeling during reactive gliosis is the extracellular matrix (ECM). During reactive gliosis, Müller glia upregulate expression of MMPs and the gene products are secreted and activated [196, 214-218]. Each MMP specifically targets and proteolytically cleaves one or more ECM molecules. The activity of MMPs is regulated by activators as well as inhibitors; the precursor molecules must be processed, either by already activated MMPs or by one of a variety of serine proteases and the MMPs can be inhibited by the tissue inhibitors of metalloproteinases (TIMPs) [219]. When activated, the MMPs degrade the existing ECM in preparation for replacement with an ECM that partially inhibits neurite outgrowth or supports abnormal neurite outgrowth [141]. In the normal adult retina heparin sulfate proeoglycans (HSPGs) are typically found on Müller glial endfeet and in retinal basal lamina, serving as a substrate for axonal outgrowth. The HSPG, via the HS chains, is also a ligand for the protein tyrosine phosphatase-sigma (PTP-σ), used in signaling in axons and growth cones in response to matrix cues. The HSPGs involved are agrin and collagen XVIII [220]. The HSPGs are lost in favor of the axonal outgrowth inhibitory molecules known as the chondroitan sulfate proteoglycans (CSPGs). The CSPGs include phosphacan, aggrecan, NG2, brevican, versican, and neurocan [221]. In addition to turning over the ECM, the degradation of the ECM also releases growth factors that are bound to the ECM, such as EGF, FGFs, BMPs, insulin, and VEGFs [219].

Müller glia can form new neurons in a process said to involve dedifferentiation of the Müller glia. Tenascin C (TNC), a matricellular protein, influences the dedifferentiation behavior of Müller glia in response to FGF2 in vitro, affecting the composition of the ECM. Sulfated chondroitin glycosaminoglycan chains in CSPG are the main target. Chondroitin sulfate increases in TNC-deficient mouse ECM [222]. The proteoglycan most affected by TNC is the CSPG Phosphacan/RPTPβ/ζ which bind to TNC [223]. TNC shows overlapping expression with phosphacan [224]. In a TNC knock out mouse TNC level rise. Studies using immunocytochemistry for phosphacan, Western Blots and PCR for mRNA levels show that it is the chondroitin sulfate chains that increase, not the amount of mRNA for CSPG core protein. Proliferation rates also increase in the TNC-deficient mice, but it is not clear if this affects exit from the cell cycle and differentiation [222].

SPARC(secreted protein, acidic and rich in cysteine)/osteonectin is also a matricellular. It interacts with growth factors and ECM forming a link that modulates the cell cycle and other cell behavior. SPARC remains expressed at significant levels in the adult CNS, moreso than in most normal adult tissues. SPARC is widely expressed in remodeling injured tissue and in morphogenesis in development [225]. In normal newborn and adult bovine retinas SPARC is found in ganglion cell soma and in ganglion cell axons, with higher expression in the adult tissue. SPARC is thought to have a function in maintaining healthy retinas and is localized to the ganglion cell layer (GCL), nerve fiber layer (NFL) and some retinal capillaries. Müller glia showed no immunoreactivity, but the GFAP-positive retinal astrocytes were SPARC-positive [225].

7. Conclusion

The evidence to date has shown that Müller glia undergo dedifferentiation and generate retinal progenitors that may be capable of differentiating into retinal neurons. Several potential problems have arisen that impact on the ability of those progentiors to effectively be used to regenerate large numbers of neurons following injury or during disease. Of the proliferating population that arise from dedifferentiated Müller glia, a very small percentage go on to become retinal neurons [4, 141]. The inability of the cells to differentiate into retinal neurons implies that either the signals and/or competence necessary for differentiation have been lost or there are signals present that direct progenitor cells away from differentiation into retinal neurons. Further, if the progenitor cells can be induced to differentiate, they will have to functionally integrate into the diseased or injured retina. This, in and of itself, will be a challenge if glial scars are present in the tissue as the glial scars will prevent ntegration by inhibiting migration, placement, and/or synapse formation. Clearly, investigators have been untangling which signaling pathways are critical for various aspects of reactive gliosis to occur. If signals that are necessary for proliferation can be separated from those necessary for glial scars to form, there is the possibility that therapeutic approaches could be engineered that will block scar formation while allowing proliferation to occur. There are many challenges ahead before the potential of Müller glia as a source for retinal regeneration can be realized.

Acknowledgements

This work was supported by grant 1R01EY019525 (TBA) and 1R15EY020816 (TBA) from the National Eye Institute and the American Health Assistance Foundation (TBA).

Author details

Teri L. Belecky-Adams[1,2], Ellen C. Chernoff[1,2], Jonathan M. Wilson[1,2] and Subramanian Dharmarajan[1,2]

*Address all correspondence to: tbadams@iupui.edu

1 Department of Biology, Indiana University-Purdue University Indianapolis, Indianapolis, IN, USA

2 Center for Regenerative Biology and Medicine, Indiana University- Purdue University Indianapolis, IN, USA

References

[1] Daadi MM. Novel paths towards neural cellular products for neurological disorders. Regenerative medicine. 2011 Nov;6(6 Suppl):25-30. PubMed PMID: 21999259. Epub 2011/10/26. eng.

[2] Wohl SG, Schmeer CW, Isenmann S. Neurogenic potential of stem/progenitor-like cells in the adult mammalian eye. Prog Retin Eye Res. 2012 May;31(3):213-42. PubMed PMID: 22353284. Epub 2012/02/23. eng.

[3] Sarthy VaR, Harris. The Retinal Muller Cell Structure and Function. Blakemore C, editor: Kluwer Academic Publishers; 2002.

[4] Bringmann A, Pannicke T, Grosche J, Francke M, Wiedemann P, Skatchkov SN, et al. Muller cells in the healthy and diseased retina. Prog Retin Eye Res. 2006 Jul;25(4): 397-424. PubMed PMID: 16839797. Epub 2006/07/15. eng.

[5] Miller RH. Regulation of oligodendrocyte development in the vertebrate CNS. Prog Neurobiol. 2002 Aug;67(6):451-67. PubMed PMID: 12385864.

[6] Wetts R, Fraser SE. Multipotent precursors can give rise to all major cell types of the frog retina. Science. 1988;239(4844):1142-5.

[7] Turner DL, Cepko CL. A common progenitor for neurons and glia persists in rat retina late in development. Nature. 1987 Jul 9-15;328(6126):131-6. PubMed PMID: 3600789. Epub 1987/07/09. eng.

[8] Holt CE, Bertsch TW, Ellis HM, Harris WA. Cellular determination in the Xenopus retina is independent of lineage and birth date. Neuron. 1988;1(1):15-26.

[9] Turner DL, Snyder EY, Cepko CL. Lineage-independent determination of cell type in the embryonic mouse retina. Neuron. 1990 Jun;4(6):833-45. PubMed PMID: 2163263. Epub 1990/06/01. eng.

[10] Young RW. Cell differentiation in the retina of the mouse. Anat Rec. 1985 Jun;212(2): 199-205. PubMed PMID: 3842042. Epub 1985/06/01. eng.

[11] Belecky-Adams T, Cook B, Adler R. Correlations between terminal mitosis and differentiated fate of retinal precursor cells in vivo and in vitro: analysis with the "window-labeling" technique. Dev Biol. 1996 Sep 15;178(2):304-15. PubMed PMID: 8812131. Epub 1996/09/15. eng.

[12] Zhang X, Serb JM, Greenlee MH. Mouse retinal development: a dark horse model for systems biology research. Bioinformatics and biology insights. 2011;5:99-113. PubMed PMID: 21698072. Pubmed Central PMCID: 3118678. Epub 2011/06/24. eng.

[13] Meller K, Tetzlaff W. Scanning electron microscopic studies on the development of the chick retina. Cell Tissue Res. 1976;170(2):145-59.

[14] Layer PG, Alber R, Mansky P, Vollmer G, Willbold E. Regeneration of a chimeric retina from single cells in vitro: cell-lineage-dependent formation of radial cell columns by segregated chick and quail cells. Cell Tissue Res. 1990 Feb;259(2):187-98. PubMed PMID: 2337919. Epub 1990/02/01. eng.

[15] Williams RW, Goldowitz D. Structure of clonal and polyclonal cell arrays in chimeric mouse retina. Proc Natl Acad Sci U S A. 1992 Feb 15;89(4):1184-8. PubMed PMID: 1741373. Pubmed Central PMCID: 48413. Epub 1992/02/15. eng.

[16] Fields-Berry SC, Halliday AL, Cepko CL. A recombinant retrovirus encoding alkaline phosphatase confirms clonal boundary assignment in lineage analysis of murine retina. Proc Natl Acad Sci U S A. 1992 Jan 15;89(2):693-7. PubMed PMID: 1731342. Pubmed Central PMCID: 48305. Epub 1992/01/15. eng.

[17] Reichenbach A, Stolzenburg JU, Eberhardt W, Chao TI, Dettmer D, Hertz L. What do retinal muller (glial) cells do for their neuronal 'small siblings'? Journal of chemical neuroanatomy. 1993 Jul-Aug;6(4):201-13. PubMed PMID: 8104418. Epub 1993/07/01. eng.

[18] Prada C, Puelles L, Genis-Galvez JM. A golgi study on the early sequence of differentiation of ganglion cells in the chick embryo retina. Anatomy and embryology. 1981;161(3):305-17. PubMed PMID: 7187824. Epub 1981/01/01. eng.

[19] Maslim J, Webster M, Stone J. Stages in the structural differentiation of retinal ganglion cells. J Comp Neurol. 1986 Dec 15;254(3):382-402. PubMed PMID: 3794013. Epub 1986/12/15. eng.

[20] Willbold E, Berger J, Reinicke M, Wolburg H. On the role of Muller glia cells in histogenesis: only retinal spheroids, but not tectal, telencephalic and cerebellar spheroids develop histotypical patterns. J Hirnforsch. 1997;38(3):383-96. PubMed PMID: 0009350510.

[21] Willbold E, Rothermel A, Tomlinson S, Layer PG. Muller glia cells reorganize reaggregating chicken retinal cells into correctly laminated in vitro retinae. Glia. 2000;29(1):45-57. PubMed PMID: 0010594922.

[22] Jing X, Malicki J. Zebrafish ale oko, an essential determinant of sensory neuron survival and the polarity of retinal radial glia, encodes the p50 subunit of dynactin. Development. 2009 Sep;136(17):2955-64. PubMed PMID: 19666822. Pubmed Central PMCID: 2723067. Epub 2009/08/12. eng.

[23] Rich KA, Figueroa SL, Zhan Y, Blanks JC. Effects of Muller cell disruption on mouse photoreceptor cell development. Exp Eye Res. 1995 Aug;61(2):235-48. PubMed PMID: 7556487. Epub 1995/08/01. eng.

[24] Ullian EM, Christopherson KS, Barres BA. Role for glia in synaptogenesis. Glia. 2004 Aug 15;47(3):209-16. PubMed PMID: 15252809. Epub 2004/07/15. eng.

[25] Ransom B, Behar T, Nedergaard M. New roles for astrocytes (stars at last). Trends Neurosci. 2003 Oct;26(10):520-2. PubMed PMID: 14522143. Epub 2003/10/03. eng.

[26] Risher WC, Eroglu C. Thrombospondins as key regulators of synaptogenesis in the central nervous system. Matrix Biol. 2012 Apr;31(3):170-7. PubMed PMID: 22285841. Epub 2012/01/31. eng.

[27] Phillips JB, Blanco-Sanchez B, Lentz JJ, Tallafuss A, Khanobdee K, Sampath S, et al. Harmonin (Ush1c) is required in zebrafish Muller glial cells for photoreceptor synaptic development and function. Disease models & mechanisms. 2011 Nov;4(6):786-800. PubMed PMID: 21757509. Pubmed Central PMCID: 3209648. Epub 2011/07/16. eng.

[28] Williams PR, Suzuki SC, Yoshimatsu T, Lawrence OT, Waldron SJ, Parsons MJ, et al. In vivo development of outer retinal synapses in the absence of glial contact. J Neurosci. 2010 Sep 8;30(36):11951-61. PubMed PMID: 20826659. Pubmed Central PMCID: 2946228. Epub 2010/09/10. eng.

[29] Abbott NJ. Astrocyte-endothelial interactions and blood-brain barrier permeability. J Anat. 2002 Jun;200(6):629-38. PubMed PMID: 12162730. Pubmed Central PMCID: 1570746.

[30] Cunha-Vaz J. The blood-ocular barriers. Surv Ophthalmol. 1979 Mar-Apr;23(5): 279-96. PubMed PMID: 380030. Epub 1979/03/01. eng.

[31] Tout S, Chan-Ling T, Hollander H, Stone J. The role of Muller cells in the formation of the blood-retinal barrier. Neuroscience. 1993 Jul;55(1):291-301. PubMed PMID: 8350991. Epub 1993/07/01. eng.

[32] Bussow H. The astrocytes in the retina and optic nerve head of mammals: a special glia for the ganglion cell axons. Cell Tissue Res. 1980;206(3):367-78. PubMed PMID: 6771013.

[33] Stone J, Itin A, Alon T, Pe'er J, Gnessin H, Chan-Ling T, et al. Development of retinal vasculature is mediated by hypoxia-induced vascular endothelial growth factor (VEGF) expression by neuroglia. J Neurosci. 1995 Jul;15(7 Pt 1):4738-47. PubMed PMID: 7623107. Epub 1995/07/01. eng.

[34] Eichler W, Yafai Y, Keller T, Wiedemann P, Reichenbach A. PEDF derived from glial Muller cells: a possible regulator of retinal angiogenesis. Exp Cell Res. 2004 Sep 10;299(1):68-78. PubMed PMID: 15302574. Epub 2004/08/11. eng.

[35] Paulson OB, Newman EA. Does the release of potassium from astrocyte endfeet regulate cerebral blood flow? Science. 1987 Aug 21;237(4817):896-8. PubMed PMID: 3616619. Pubmed Central PMCID: 2505270. Epub 1987/08/21. eng.

[36] Abukawa H, Tomi M, Kiyokawa J, Hori S, Kondo T, Terasaki T, et al. Modulation of retinal capillary endothelial cells by Muller glial cell-derived factors. Mol Vis. 2009;15:451-7. PubMed PMID: 19247458. Pubmed Central PMCID: 2647974. Epub 2009/02/28. eng.

[37] Ikeda T, Homma Y, Nisida K, Hirase K, Sotozono C, Kinoshita S, et al. Expression of transforming growth factor-beta s and their receptors by human retinal glial cells. Curr Eye Res. 1998 May;17(5):546-50. PubMed PMID: 9617551. Epub 1998/06/09. eng.

[38] Behzadian MA, Wang XL, Al-Shabrawey M, Caldwell RB. Effects of hypoxia on glial cell expression of angiogenesis-regulating factors VEGF and TGF-beta. Glia. 1998 Oct;24(2):216-25. PubMed PMID: 9728767. Epub 1998/09/05. eng.

[39] Ramsauer M, D'Amore PA. Contextual role for angiopoietins and TGFbeta1 in blood vessel stabilization. J Cell Sci. 2007 May 15;120(Pt 10):1810-7. PubMed PMID: 17502485. Epub 2007/05/16. eng.

[40] Arnold TD, Ferrero GM, Qiu H, Phan IT, Akhurst RJ, Huang EJ, et al. Defective retinal vascular endothelial cell development as a consequence of impaired integrin alphaVbeta8-mediated activation of transforming growth factor-beta. J Neurosci. 2012 Jan 25;32(4):1197-206. PubMed PMID: 22279205. Epub 2012/01/27. eng.

[41] Igarashi Y, Chiba H, Utsumi H, Miyajima H, Ishizaki T, Gotoh T, et al. Expression of receptors for glial cell line-derived neurotrophic factor (GDNF) and neurturin in the inner blood-retinal barrier of rats. Cell Struct Funct. 2000 Aug;25(4):237-41. PubMed PMID: 11129793. Epub 2000/12/29. eng.

[42] Mi H, Haeberle H, Barres BA. Induction of astrocyte differentiation by endothelial cells. J Neurosci. 2001 Mar 1;21(5):1538-47. PubMed PMID: 11222644. Epub 2001/02/27. eng.

[43] Sakimoto S, Kidoya H, Naito H, Kamei M, Sakaguchi H, Goda N, et al. A role for endothelial cells in promoting the maturation of astrocytes through the apelin/APJ system in mice. Development. 2012 Apr;139(7):1327-35. PubMed PMID: 22357924. Epub 2012/02/24. eng.

[44] Bonni A, Frank DA, Schindler C, Greenberg ME. Characterization of a pathway for ciliary neurotrophic factor signaling to the nucleus. Science. 1993 Dec 3;262(5139): 1575-9. PubMed PMID: 7504325. Epub 1993/12/03. eng.

[45] Rajan P, McKay RD. Multiple routes to astrocytic differentiation in the CNS. J Neurosci. 1998 May 15;18(10):3620-9. PubMed PMID: 9570793. Epub 1998/06/06. eng.

[46] Goureau O, Rhee KD, Yang XJ. Ciliary neurotrophic factor promotes muller glia differentiation from the postnatal retinal progenitor pool. Dev Neurosci. 2004;26(5-6): 359-70. PubMed PMID: 15855765. Epub 2005/04/28. eng.

[47] Ash J, McLeod DS, Lutty GA. Transgenic expression of leukemia inhibitory factor (LIF) blocks normal vascular development but not pathological neovascularization in the eye. Mol Vis. 2005;11:298-308. PubMed PMID: 15889014. Epub 2005/05/13. eng.

[48] Belanger M, Allaman I, Magistretti PJ. Brain energy metabolism: focus on astrocyte-neuron metabolic cooperation. Cell metabolism. 2011 Dec 7;14(6):724-38. PubMed PMID: 22152301. Epub 2011/12/14. eng.

[49] Sokoloff L. Relation between physiological function and energy metabolism in the central nervous system. J Neurochem. 1977 Jul;29(1):13-26. PubMed PMID: 407330. Epub 1977/07/01. eng.

[50] Kuwabara T, Cogan DG. Retinal glycogen. Arch Ophthalmol. 1961 Nov;66:680-8. PubMed PMID: 14460992. Epub 1961/11/01. eng.

[51] Poitry-Yamate C, Tsacopoulos M. Glial (Muller) cells take up and phosphorylate [3H]2-deoxy-D-glucose in mammalian retina. Neurosci Lett. 1991 Jan 28;122(2):241-4. PubMed PMID: 2027525. Epub 1991/01/28. eng.

[52] Poitry-Yamate CL, Poitry S, Tsacopoulos M. Lactate released by Muller glial cells is metabolized by photoreceptors from mammalian retina. J Neurosci. 1995 Jul;15(7 Pt 2):5179-91. PubMed PMID: 7623144. Epub 1995/07/01. eng.

[53] Zielke HR, Zielke CL, Baab PJ. Direct measurement of oxidative metabolism in the living brain by microdialysis: a review. J Neurochem. 2009 May;109 Suppl 1:24-9. PubMed PMID: 19393005. Pubmed Central PMCID: 2792919. Epub 2009/05/07. eng.

[54] Poitry S, Poitry-Yamate C, Ueberfeld J, MacLeish PR, Tsacopoulos M. Mechanisms of glutamate metabolic signaling in retinal glial (Muller) cells. J Neurosci. 2000 Mar 1;20(5):1809-21. PubMed PMID: 10684882. Epub 2000/02/24. eng.

[55] Tsacopoulos M, Poitry-Yamate CL, MacLeish PR, Poitry S. Trafficking of molecules and metabolic signals in the retina. Prog Retin Eye Res. 1998 Jul;17(3):429-42. PubMed PMID: 9695799. Epub 1998/08/08. eng.

[56] Lin RY, Vera JC, Chaganti RS, Golde DW. Human monocarboxylate transporter 2 (MCT2) is a high affinity pyruvate transporter. J Biol Chem. 1998 Oct 30;273(44): 28959-65. PubMed PMID: 9786900. Epub 1998/10/24. eng.

[57] Gohdo T, Ueda H, Ohno S, Iijima H, Tsukahara S. Heat shock protein 70 expression increased in rabbit muller cells in the ischemia-reperfusion model. Ophthalmic Res. 2001 Sep-Oct;33(5):298-302. PubMed PMID: 11586064. Epub 2001/10/05. eng.

[58] Izumi Y, Kirby CO, Benz AM, Olney JW, Zorumski CF. Muller cell swelling, glutamate uptake, and excitotoxic neurodegeneration in the isolated rat retina. Glia. 1999 Feb 15;25(4):379-89. PubMed PMID: 10028920. Epub 1999/02/24. eng.

[59] Otori Y, Shimada S, Tanaka K, Ishimoto I, Tano Y, Tohyama M. Marked increase in glutamate-aspartate transporter (GLAST/GluT-1) mRNA following transient retinal ischemia. Brain Res Mol Brain Res. 1994 Dec;27(2):310-4. PubMed PMID: 7898315. Epub 1994/12/01. eng.

[60] Derouiche A, Rauen T. Coincidence of L-glutamate/L-aspartate transporter (GLAST) and glutamine synthetase (GS) immunoreactions in retinal glia: evidence for coupling of GLAST and GS in transmitter clearance. J Neurosci Res. 1995 Sep 1;42(1): 131-43. PubMed PMID: 8531222. Epub 1995/09/01. eng.

[61] Kanai Y, Hediger MA. The glutamate/neutral amino acid transporter family SLC1: molecular, physiological and pharmacological aspects. Pflugers Arch. 2004 Feb; 447(5):469-79. PubMed PMID: 14530974. Epub 2003/10/08. eng.

[62] Tsacopoulos M, Magistretti PJ. Metabolic coupling between glia and neurons. J Neurosci. 1996 Feb 1;16(3):877-85. PubMed PMID: 8558256. Epub 1996/02/01. eng.

[63] Linser P, Moscona AA. Induction of glutamine synthetase in embryonic neural retina: localization in Muller fibers and dependence on cell interactions. Proc Natl Acad Sci U S A. 1979 Dec;76(12):6476-80. PubMed PMID: 42916. Pubmed Central PMCID: 411888. Epub 1979/12/01. eng.

[64] Tsacopoulos M, Poitry-Yamate CL, Poitry S. Ammonium and glutamate released by neurons are signals regulating the nutritive function of a glial cell. J Neurosci. 1997 Apr 1;17(7):2383-90. PubMed PMID: 9065499. Epub 1997/04/01. eng.

[65] Brew H, Attwell D. Electrogenic glutamate uptake is a major current carrier in the membrane of axolotl retinal glial cells. Nature. 1987 Jun 25-Jul 1;327(6124):707-9. PubMed PMID: 2885752. Epub 1987/06/01. eng.

[66] Bringmann A, Kuhrt H, Germer A, Biedermann B, Reichenbach A. Muller (glial) cell development in vivo and in retinal explant cultures: morphology and electrophysiology, and the effects of elevated ammonia. J Hirnforsch. 1998;39(2):193-206. PubMed PMID: 10022343. Epub 1999/02/18. eng.

[67] Germer A, Jahnke C, Mack A, Enzmann V, Reichenbach A. Modification of glutamine synthetase expression by mammalian Muller (glial) cells in retinal organ cultures. Neuroreport. 1997 Sep 29;8(14):3067-72. PubMed PMID: 9331915. Epub 1997/10/23. eng.

[68] Newman EA, Frambach DA, Odette LL. Control of extracellular potassium levels by retinal glial cell K+ siphoning. Science. 1984 Sep 14;225(4667):1174-5. PubMed PMID: 6474173. Pubmed Central PMCID: 2693189. Epub 1984/09/14. eng.

[69] Karwoski CJ, Lu HK, Newman EA. Spatial buffering of light-evoked potassium increases by retinal Muller (glial) cells. Science. 1989 May 5;244(4904):578-80. PubMed PMID: 2785716. Pubmed Central PMCID: 2562506. Epub 1989/05/05. eng.

[70] Newman EA. Acid efflux from retinal glial cells generated by sodium bicarbonate cotransport. J Neurosci. 1996 Jan;16(1):159-68. PubMed PMID: 8613782. Epub 1996/01/01. eng.

[71] Linser P, Moscona AA. Carbonic anhydrase C in the neural retina: transition from generalized to glia-specific cell localization during embryonic development. Proc Natl Acad Sci U S A. 1981 Nov;78(11):7190-4. PubMed PMID: 6118868. Pubmed Central PMCID: 349222. Epub 1981/11/01. eng.

[72] Sarthy PV, Bunt AH. The ultrastructure of isolated glial (Muller) cells from the turtle retina. Anat Rec. 1982 Feb;202(2):275-83. PubMed PMID: 7065426. Epub 1982/02/01. eng.

[73] Newman EA, Astion ML. Localization and stoichiometry of electrogenic sodium bicarbonate cotransport in retinal glial cells. Glia. 1991;4(4):424-8. PubMed PMID: 1657777. Epub 1991/01/01. eng.

[74] Wu SM, Maple BR. Amino acid neurotransmitters in the retina: a functional overview. Vision Res. 1998 May;38(10):1371-84. PubMed PMID: 9667005.

[75] Bringmann A, Pannicke T, Biedermann B, Francke M, Iandiev I, Grosche J, et al. Role of retinal glial cells in neurotransmitter uptake and metabolism. Neurochem Int. 2009 Mar-Apr;54(3-4):143-60. PubMed PMID: 19114072. Epub 2008/12/31. eng.

[76] de Melo Reis RA, Ventura AL, Schitine CS, de Mello MC, de Mello FG. Muller glia as an active compartment modulating nervous activity in the vertebrate retina: neurotransmitters and trophic factors. Neurochem Res. 2008 Aug;33(8):1466-74. PubMed PMID: 18273703. Epub 2008/02/15. eng.

[77] Kugler P, Beyer A. Expression of glutamate transporters in human and rat retina and rat optic nerve. Histochem Cell Biol. 2003 Sep;120(3):199-212. PubMed PMID: 12898275. Epub 2003/08/05. eng.

[78] Barnett NL, Pow DV. Antisense knockdown of GLAST, a glial glutamate transporter, compromises retinal function. Invest Ophthalmol Vis Sci. 2000 Feb;41(2):585-91. PubMed PMID: 10670492. Epub 2000/02/12. eng.

[79] Harada T, Harada C, Watanabe M, Inoue Y, Sakagawa T, Nakayama N, et al. Functions of the two glutamate transporters GLAST and GLT-1 in the retina. Proc Natl Acad Sci U S A. 1998 Apr 14;95(8):4663-6. PubMed PMID: 9539795. Pubmed Central PMCID: 22547. Epub 1998/05/16. eng.

[80] Pow DV, Barnett NL, Penfold P. Are neuronal transporters relevant in retinal glutamate homeostasis? Neurochem Int. 2000 Aug-Sep;37(2-3):191-8. PubMed PMID: 10812204. Epub 2000/05/17. eng.

[81] De Bock M, Culot M, Wang N, da Costa A, Decrock E, Bol M, et al. Low extracellular Ca(2+) conditions induce an increase in brain endothelial permeability that involves intercellular Ca(2+) waves. Brain Res. 2012 Jul 10. PubMed PMID: 22789903.

[82] Gonzalez MI, Lopez-Colom AM, Ortega A. Sodium-dependent glutamate transport in Muller glial cells: regulation by phorbol esters. Brain Res. 1999 Jun 12;831(1-2): 140-5. PubMed PMID: 10411993.

[83] Barnett NL, Pow DV, Robinson SR. Inhibition of Muller cell glutamine synthetase rapidly impairs the retinal response to light. Glia. 2000 Mar;30(1):64-73. PubMed PMID: 10696145. Epub 2000/03/01. eng.

[84] Nickerson PE, McLeod MC, Myers T, Clarke DB. Effects of epidermal growth factor and erythropoietin on Muller glial activation and phenotypic plasticity in the adult mammalian retina. J Neurosci Res. 2011 Jul;89(7):1018-30. PubMed PMID: 21484851.

[85] Leonelli M, Martins DO, Britto LR. TRPV1 receptors are involved in protein nitration and Muller cell reaction in the acutely axotomized rat retina. Exp Eye Res. 2010 Nov; 91(5):755-68. PubMed PMID: 20826152.

[86] Xue W, Cojocaru RI, Dudley VJ, Brooks M, Swaroop A, Sarthy VP. Ciliary neurotrophic factor induces genes associated with inflammation and gliosis in the retina: a gene profiling study of flow-sorted, Muller cells. PLoS One. 2011;6(5):e20326. PubMed PMID: 21637858. Pubmed Central PMCID: 3102695.

[87] Douglas MR, Morrison KC, Jacques SJ, Leadbeater WE, Gonzalez AM, Berry M, et al. Off-target effects of epidermal growth factor receptor antagonists mediate retinal ganglion cell disinhibited axon growth. Brain. 2009 Nov;132(Pt 11):3102-21. PubMed PMID: 19783665.

[88] Layton CJ, Becker S, Osborne NN. The effect of insulin and glucose levels on retinal glial cell activation and pigment epithelium-derived fibroblast growth factor-2. Mol Vis. 2006;12:43-54. PubMed PMID: 16446701.

[89] Dale N. Dynamic ATP signalling and neural development. J Physiol. 2008 May 15;586(10):2429-36. PubMed PMID: 18356200. Pubmed Central PMCID: 2464351.

[90] Leybaert L, Sanderson MJ. Intercellular Ca(2+) waves: mechanisms and function. Physiological reviews. 2012 Jul;92(3):1359-92. PubMed PMID: 22811430.

[91] Wurm A, Pannicke T, Iandiev I, Francke M, Hollborn M, Wiedemann P, et al. Purinergic signaling involved in Muller cell function in the mammalian retina. Prog Retin Eye Res. 2011 Sep;30(5):324-42. PubMed PMID: 21689780.

[92] Newman EA. Propagation of intercellular calcium waves in retinal astrocytes and Muller cells. J Neurosci. 2001 Apr 1;21(7):2215-23. PubMed PMID: 11264297. Pubmed Central PMCID: 2409971.

[93] Lopez-Colome AM, Salceda R, Fragoso G. Specific interaction of glutamate with membranes from cultured retinal pigment epithelium. J Neurosci Res. 1993 Mar 1;34(4):454-61. PubMed PMID: 8097266.

[94] Lopez-Colome AM, Martinez-Lozada Z, Guillem AM, Lopez E, Ortega A. Glutamate transporter-dependent mTOR phosphorylation in Muller glia cells. ASN neuro. 2012 Jul 23. PubMed PMID: 22817638. Pubmed Central PMCID: 3420017.

[95] Nagelhus EA, Horio Y, Inanobe A, Fujita A, Haug FM, Nielsen S, et al. Immunogold evidence suggests that coupling of K+ siphoning and water transport in rat retinal Muller cells is mediated by a coenrichment of Kir4.1 and AQP4 in specific membrane domains. Glia. 1999 Mar;26(1):47-54. PubMed PMID: 10088671.

[96] Francke M, Makarov F, Kacza J, Seeger J, Wendt S, Gartner U, et al. Retinal pigment epithelium melanin granules are phagocytozed by Muller glial cells in experimental retinal detachment. J Neurocytol. 2001 Feb;30(2):131-6. PubMed PMID: 11577251.

[97] Reichelt W, Stabel-Burow J, Pannicke T, Weichert H, Heinemann U. The glutathione level of retinal Muller glial cells is dependent on the high-affinity sodium-dependent uptake of glutamate. Neuroscience. 1997 Apr;77(4):1213-24. PubMed PMID: 9130799.

[98] Buffo A, Rite I, Tripathi P, Lepier A, Colak D, Horn AP, et al. Origin and progeny of reactive gliosis: A source of multipotent cells in the injured brain. P Natl Acad Sci USA. 2008 Mar 4;105(9):3581-6. PubMed PMID: ISI:000253846500072. English.

[99] Robel S, Berninger B, Gotz M. The stem cell potential of glia: lessons from reactive gliosis. Nat Rev Neurosci. 2011 Feb;12(2):88-104. PubMed PMID: ISI: 000286420100003. English.

[100] White BD, Nathe RJ, Maris DO, Nguyen NK, Goodson JM, Moon RT, et al. beta-Catenin Signaling Increases in Proliferating NG2+Progenitors and Astrocytes during Post-Traumatic Gliogenesis in the Adult Brain. Stem Cells. 2010 Feb;28(2):297-307. PubMed PMID: ISI:000275058400014. English.

[101] Liu B, Neufeld AH. Activation of epidermal growth factor receptors in astrocytes: From development to neural injury. J Neurosci Res. 2007 Dec;85(16):3523-9. PubMed PMID: ISI:000251778200003. English.

[102] Amankulor NM, Hambardzumyan D, Pyonteck SM, Becher OJ, Joyce JA, Holland EC. Sonic Hedgehog Pathway Activation Is Induced by Acute Brain Injury and Regulated by Injury-Related Inflammation. J Neurosci. 2009 Aug 19;29(33):10299-308. PubMed PMID: ISI:000269087300013. English.

[103] Fischer AJ, Bongini R. Turning Muller glia into neural progenitors in the retina. Mol Neurobiol. 2010 Dec;42(3):199-209. PubMed PMID: 21088932. Epub 2010/11/23. eng.

[104] Perron M, Kanekar S, Vetter ML, Harris WA. The genetic sequence of retinal development in the ciliary margin of the Xenopus eye. Dev Biol. 1998 Jul 15;199(2):185-200. PubMed PMID: ISI:000075365200002. English.

[105] Fischer AJ, Reh TA. Identification of a proliferating marginal zone of retinal progenitors in postnatal chickens. Developmental Biology. 2000;220(2):197-210.

[106] Fischer AJ, Bongini R. Turning Muller Glia into Neural Progenitors in the Retina. Mol Neurobiol. 2010 Dec;42(3):199-209. PubMed PMID: ISI:000284654000004. English.

[107] Otteson DC, Hitchcock PF. Stem cells in the teleost retina: persistent neurogenesis and injury-induced regeneration. Vision Res. 2003 Apr;43(8):927-36. PubMed PMID: ISI:000182254200009. English.

[108] Fischer AJ, Reh TA. Potential of Muller glia to become neurogenic retinal progenitor cells. Glia. 2003 Jul;43(1):70-6. PubMed PMID: 12761869. Epub 2003/05/23. eng.

[109] Kubota R, Hokoc JN, Moshiri A, McGuire C, Reh TA. A comparative study of neurogenesis in the retinal ciliary marginal zone of homeothermic vertebrates. Brain Res Dev Brain Res. 2002 Mar 31;134(1-2):31-41. PubMed PMID: 11947935.

[110] Fischer AJ, Reh TA. Muller glia are a potential source of neural regeneration in the postnatal chicken retina. Nature Neuroscience. 2001;4(3):247-52.

[111] Ooto S, Akagi T, Kageyama R, Akita J, Mandai M, Honda Y, et al. Potential for neural regeneration after neurotoxic injury in the adult mammalian retina. P Natl Acad Sci USA. 2004 Sep 14;101(37):13654-9. PubMed PMID: ISI:000223917900047. English.

[112] Das AV, Mallya KB, Zhao X, Ahmad F, Bhattacharya S, Thoreson WB, et al. Neural stem cell properties of Muller glia in the mammalian retina: regulation by Notch and Wnt signaling. Dev Biol. 2006 Nov 1;299(1):283-302. PubMed PMID: 16949068. Epub 2006/09/05. eng.

[113] Fischer AJ, Reh TA. Exogenous growth factors stimulate the regeneration of ganglion cells in the chicken retina. Dev Biol. 2002 Nov 15;251(2):367-79. PubMed PMID: 12435364. Epub 2002/11/19. eng.

[114] Bringmann A, Pannicke T, Grosche J, Francke M, Wiedemann P, Skatchkov SN, et al. Muller cells in the healthy and diseased retina. Prog Retin Eye Res. 2006 Jul;25(4): 397-424. PubMed PMID: ISI:000239884500003. English.

[115] Jadhav AP, Roesch K, Cepko CL. Development and neurogenic potential of Muller glial cells in the vertebrate retina. Prog Retin Eye Res. 2009 Jul;28(4):249-62. PubMed PMID: ISI:000269296600002. English.

[116] Sofroniew MV. Molecular dissection of reactive astrogliosis and glial scar formation. Trends Neurosci. 2009 Dec;32(12):638-47. PubMed PMID: 19782411. Pubmed Central PMCID: 2787735.

[117] Sofroniew MV, Vinters HV. Astrocytes: biology and pathology. Acta neuropathologica. 2010 Jan;119(1):7-35. PubMed PMID: 20012068. Pubmed Central PMCID: 2799634. Epub 2009/12/17. eng.

[118] Roesch K, Stadler MB, Cepko CL. Gene expression changes within Muller glial cells in retinitis pigmentosa. Mol Vis. 2012;18:1197-214. PubMed PMID: 22665967. Pubmed Central PMCID: 3365136.

[119] Robaszkiewicz J, Chmielewska K, Figurska M, Wierzbowska J, Stankiewicz A. Muller glial cells--the mediators of vascular disorders with vitreomacular interface pathology in diabetic maculopathy. Klinika oczna. 2010;112(10-12):328-32. PubMed PMID: 21469528.

[120] Lieth E, Barber AJ, Xu B, Dice C, Ratz MJ, Tanase D, et al. Glial reactivity and impaired glutamate metabolism in short-term experimental diabetic retinopathy. Penn State Retina Research Group. Diabetes. 1998 May;47(5):815-20. PubMed PMID: 9588455. Epub 1998/05/20. eng.

[121] Ganesh BS, Chintala SK. Inhibition of reactive gliosis attenuates excitotoxicity-mediated death of retinal ganglion cells. PLoS One. 2011;6(3):e18305. PubMed PMID: 21483783. Pubmed Central PMCID: 3069086.

[122] Sullivan R, Penfold P, Pow DV. Neuronal migration and glial remodeling in degenerating retinas of aged rats and in nonneovascular AMD. Invest Ophthalmol Vis Sci. 2003 Feb;44(2):856-65. PubMed PMID: 12556422.

[123] Fischer AJ, Omar G, Eubanks J, McGuire CR, Dierks BD, Reh TA. Different aspects of gliosis in retinal Muller glia can be induced by CNTF, insulin, and FGF2 in the absence of damage. Mol Vis. 2004 Dec 22;10:973-86. PubMed PMID: 15623987. Epub 2004/12/30. eng.

[124] Yamaguchi K, Nagai S, Ninomiya-Tsuji J, Nishita M, Tamai K, Irie K, et al. XIAP, a cellular member of the inhibitor of apoptosis protein family, links the receptors to TAB1-TAK1 in the BMP signaling pathway. EMBO J. 1999 Jan 4;18(1):179-87. PubMed PMID: 9878061. Pubmed Central PMCID: 1171113. Epub 1999/01/07. eng.

[125] Ishitani T, Ninomiya-Tsuji J, Nagai S, Nishita M, Meneghini M, Barker N, et al. The TAK1-NLK-MAPK-related pathway antagonizes signalling between beta-catenin and transcription factor TCF. Nature. 1999 Jun 24;399(6738):798-802. PubMed PMID: 10391247. Epub 1999/07/03. eng.

[126] Meneghini MD, Ishitani T, Carter JC, Hisamoto N, Ninomiya-Tsuji J, Thorpe CJ, et al. MAP kinase and Wnt pathways converge to downregulate an HMG-domain repressor in Caenorhabditis elegans. Nature. 1999 Jun 24;399(6738):793-7. PubMed PMID: 10391246. Epub 1999/07/03. eng.

[127] Kimura N, Matsuo R, Shibuya H, Nakashima K, Taga T. BMP2-induced apoptosis is mediated by activation of the TAK1-p38 kinase pathway that is negatively regulated by Smad6. J Biol Chem. 2000 Jun 9;275(23):17647-52. PubMed PMID: 10748100. Epub 2000/04/05. eng.

[128] Zhang D, Gaussin V, Taffet GE, Belaguli NS, Yamada M, Schwartz RJ, et al. TAK1 is activated in the myocardium after pressure overload and is sufficient to provoke heart failure in transgenic mice. Nat Med. 2000 May;6(5):556-63. PubMed PMID: 10802712. Epub 2000/05/10. eng.

[129] Rajan P, Panchision DM, Newell LF, McKay RD. BMPs signal alternately through a SMAD or FRAP-STAT pathway to regulate fate choice in CNS stem cells. J Cell Biol. 2003 Jun 9;161(5):911-21. PubMed PMID: 12796477. Pubmed Central PMCID: 2172962. Epub 2003/06/11. eng.

[130] Aoki H, Fujii M, Imamura T, Yagi K, Takehara K, Kato M, et al. Synergistic effects of different bone morphogenetic protein type I receptors on alkaline phosphatase induction. J Cell Sci. 2001 Apr;114(Pt 8):1483-9. PubMed PMID: 11282024. Epub 2001/04/03. eng.

[131] Sahni V, Mukhopadhyay A, Tysseling V, Hebert A, Birch D, McGuire TL, et al. BMPR1a and BMPR1b signaling exert opposing effects on gliosis after spinal cord injury. J Neurosci. 2010 Feb 3;30(5):1839-55. PubMed PMID: 20130193. Pubmed Central PMCID: 3093918. Epub 2010/02/05. eng.

[132] Fuller ML, DeChant AK, Rothstein B, Caprariello A, Wang R, Hall AK, et al. Bone morphogenetic proteins promote gliosis in demyelinating spinal cord lesions. Ann Neurol. 2007 Sep;62(3):288-300. PubMed PMID: 17696121. Epub 2007/08/19. eng.

[133] Ueki Y, Reh TA. Activation of BMP-Smad1/5/8 signaling promotes survival of retinal ganglion cells after damage in vivo. PLoS One. 2012;7(6):e38690. PubMed PMID: 22701694. Pubmed Central PMCID: 3368846.

[134] Zode GS, Clark AF, Wordinger RJ. Bone morphogenetic protein 4 inhibits TGF-beta2 stimulation of extracellular matrix proteins in optic nerve head cells: role of gremlin in ECM modulation. Glia. 2009 May;57(7):755-66. PubMed PMID: 19031438. Epub 2008/11/26. eng.

[135] Izumi T, Yokota-Hashimoto H, Zhao S, Wang J, Halban PA, Takeuchi T. Dominant negative pathogenesis by mutant proinsulin in the Akita diabetic mouse. Diabetes. 2003 Feb;52(2):409-16. PubMed PMID: 12540615. Epub 2003/01/24. eng.

[136] Lewis GP, Fisher SK. Up-regulation of glial fibrillary acidic protein in response to retinal injury: its potential role in glial remodeling and a comparison to vimentin expression. Int Rev Cytol. 2003;230:263-90. PubMed PMID: 14692684.

[137] Pekny M, Nilsson M. Astrocyte activation and reactive gliosis. Glia. 2005 Jun;50(4): 427-34. PubMed PMID: 15846805. Epub 2005/04/23. eng.

[138] Vazquez-Chona FR, Swan A, Ferrell WD, Jiang L, Baehr W, Chien WM, et al. Proliferative reactive gliosis is compatible with glial metabolic support and neuronal function. BMC Neurosci. 2011;12:98. PubMed PMID: 21985191. Pubmed Central PMCID: 3203081. Epub 2011/10/12. eng.

[139] Menet V, Prieto M, Privat A, Gimenez y Ribotta M. Axonal plasticity and functional recovery after spinal cord injury in mice deficient in both glial fibrillary acidic protein and vimentin genes. Proc Natl Acad Sci U S A. 2003 Jul 22;100(15):8999-9004. PubMed PMID: 12861073. Pubmed Central PMCID: 166427.

[140] Nakazawa T, Takeda M, Lewis GP, Cho KS, Jiao J, Wilhelmsson U, et al. Attenuated glial reactions and photoreceptor degeneration after retinal detachment in mice deficient in glial fibrillary acidic protein and vimentin. Invest Ophthalmol Vis Sci. 2007 Jun;48(6):2760-8. PubMed PMID: 17525210. Epub 2007/05/26. eng.

[141] Bringmann A, Wiedemann P. Muller glial cells in retinal disease. Ophthalmologica Journal international d'ophtalmologie International journal of ophthalmology Zeitschrift fur Augenheilkunde. 2012;227(1):1-19. PubMed PMID: 21921569.

[142] Chen H, Weber AJ. Expression of glial fibrillary acidic protein and glutamine synthetase by Muller cells after optic nerve damage and intravitreal application of brain-de-

rived neurotrophic factor. Glia. 2002 Apr 15;38(2):115-25. PubMed PMID: 11948805. Epub 2002/04/12. eng.

[143] Francke M, Faude F, Pannicke T, Bringmann A, Eckstein P, Reichelt W, et al. Electrophysiology of rabbit Muller (glial) cells in experimental retinal detachment and PVR. Invest Ophthalmol Vis Sci. 2001 Apr;42(5):1072-9. PubMed PMID: 11274088.

[144] Francke M, Faude F, Pannicke T, Uckermann O, Weick M, Wolburg H, et al. Glial cell-mediated spread of retinal degeneration during detachment: a hypothesis based upon studies in rabbits. Vision Res. 2005 Aug;45(17):2256-67. PubMed PMID: 15924940.

[145] Marc RE, Murry RF, Fisher SK, Linberg KA, Lewis GP. Amino acid signatures in the detached cat retina. Invest Ophthalmol Vis Sci. 1998 Aug;39(9):1694-702. PubMed PMID: 9699559.

[146] Sherry DM, Townes-Anderson E. Rapid glutamatergic alterations in the neural retina induced by retinal detachment. Invest Ophthalmol Vis Sci. 2000 Aug;41(9):2779-90. PubMed PMID: 10937598.

[147] Drescher KM, Whittum-Hudson JA. Herpes simplex virus type 1 alters transcript levels of tumor necrosis factor-alpha and interleukin-6 in retinal glial cells. Invest Ophthalmol Vis Sci. 1996 Oct;37(11):2302-12. PubMed PMID: 8843914.

[148] Aiello LP, Pierce EA, Foley ED, Takagi H, Chen H, Riddle L, et al. Suppression of retinal neovascularization in vivo by inhibition of vascular endothelial growth factor (VEGF) using soluble VEGF-receptor chimeric proteins. Proc Natl Acad Sci U S A. 1995 Nov 7;92(23):10457-61. PubMed PMID: 7479819. Epub 1995/11/07. eng.

[149] Roque RS, Rosales AA, Jingjing L, Agarwal N, Al-Ubaidi MR. Retina-derived microglial cells induce photoreceptor cell death in vitro. Brain Res. 1999;836(1-2):110-9. PubMed PMID: 0010415410.

[150] Yafai Y, Iandiev I, Wiedemann P, Reichenbach A, Eichler W. Retinal endothelial angiogenic activity: effects of hypoxia and glial (Muller) cells. Microcirculation. 2004 Oct-Nov;11(7):577-86. PubMed PMID: 15513867.

[151] Walshe TE, Saint-Geniez M, Maharaj AS, Sekiyama E, Maldonado AE, D'Amore PA. TGF-beta is required for vascular barrier function, endothelial survival and homeostasis of the adult microvasculature. PLoS One. 2009;4(4):e5149. PubMed PMID: 19340291. Pubmed Central PMCID: 2659748.

[152] Howe KL, Reardon C, Wang A, Nazli A, McKay DM. Transforming growth factor-beta regulation of epithelial tight junction proteins enhances barrier function and blocks enterohemorrhagic Escherichia coli O157:H7-induced increased permeability. Am J Pathol. 2005 Dec;167(6):1587-97. PubMed PMID: 16314472. Pubmed Central PMCID: 1613202.

[153] Simo R, Carrasco E, Garcia-Ramirez M, Hernandez C. Angiogenic and antiangiogenic factors in proliferative diabetic retinopathy. Current diabetes reviews. 2006 Feb; 2(1):71-98. PubMed PMID: 18220619.

[154] Bringmann A, Reichenbach A. Role of Muller cells in retinal degenerations. Frontiers in bioscience : a journal and virtual library. 2001 Oct 1;6:E72-92. PubMed PMID: 11578954. Epub 2001/10/02. eng.

[155] Eichler W, Yafai Y, Wiedemann P, Reichenbach A. Angiogenesis-related factors derived from retinal glial (Muller) cells in hypoxia. Neuroreport. 2004 Jul 19;15(10): 1633-7. PubMed PMID: 15232297.

[156] Honjo M, Tanihara H, Kido N, Inatani M, Okazaki K, Honda Y. Expression of ciliary neurotrophic factor activated by retinal Muller cells in eyes with NMDA- and kainic acid-induced neuronal death. Invest Ophthalmol Vis Sci. 2000 Feb;41(2):552-60. PubMed PMID: 10670488. Epub 2000/02/12. eng.

[157] Mervin K, Valter K, Maslim J, Lewis G, Fisher S, Stone J. Limiting photoreceptor death and deconstruction during experimental retinal detachment: the value of oxygen supplementation. American journal of ophthalmology. 1999 Aug;128(2):155-64. PubMed PMID: 10458170. Epub 1999/08/24. eng.

[158] Akiyama H, Nakazawa T, Shimura M, Tomita H, Tamai M. Presence of mitogen-activated protein kinase in retinal Muller cells and its neuroprotective effect ischemia-reperfusion injury. Neuroreport. 2002 Nov 15;13(16):2103-7. PubMed PMID: 12438934. Epub 2002/11/20. eng.

[159] Wahlin KJ, Adler R, Zack DJ, Campochiaro PA. Neurotrophic signaling in normal and degenerating rodent retinas. Exp Eye Res. 2001 Nov;73(5):693-701. PubMed PMID: 11747369. Epub 2001/12/19. eng.

[160] Koeberle PD, Baehr M. The upregulation of GLAST-1 is an indirect antiapoptotic mechanism of GDNF and neurturin in the adult CNS. Cell Death Differ. 2008 Mar; 15(3):471-83. PubMed PMID: ISI:000253239900006. English.

[161] Amin RH, Frank RN, Kennedy A, Eliott D, Puklin JE, Abrams GW. Vascular endothelial growth factor is present in glial cells of the retina and optic nerve of human subjects with nonproliferative diabetic retinopathy. Invest Ophthalmol Vis Sci. 1997 Jan;38(1):36-47. PubMed PMID: ISI:A1997WC89700005. English.

[162] Eichler W, Kuhrt H, Hoffmann S, Wiedemann P, Reichenbach A. VEGF release by retinal glia depends on both oxygen and glucose supply. Neuroreport. 2000 Nov 9;11(16):3533-7. PubMed PMID: 11095513.

[163] Murohara T, Horowitz JR, Silver M, Tsurumi Y, Chen D, Sullivan A, et al. Vascular endothelial growth factor/vascular permeability factor enhances vascular permeability via nitric oxide and prostacyclin. Circulation. 1998 Jan 6-13;97(1):99-107. PubMed PMID: 9443437.

[164] Majka S, McGuire PG, Das A. Regulation of matrix metalloproteinase expression by tumor necrosis factor in a murine model of retinal neovascularization. Invest Ophthalmol Vis Sci. 2002 Jan;43(1):260-6. PubMed PMID: 11773040. Epub 2002/01/05. eng.

[165] Kawasaki A, Otori Y, Barnstable CJ. Muller cell protection of rat retinal ganglion cells from glutamate and nitric oxide neurotoxicity. Invest Ophthalmol Vis Sci. 2000 Oct; 41(11):3444-50. PubMed PMID: 11006237. Epub 2000/09/28. eng.

[166] Huster D, Reichenbach A, Reichelt W. The glutathione content of retinal Muller (glial) cells: effect of pathological conditions. Neurochemistry international. 2000 Apr;36(4-5):461-9. PubMed PMID: 10733014. Epub 2000/03/25. eng.

[167] Du YP, Sarthy VP, Kern TS. Interaction between NO and COX pathways in retinal cells exposed to elevated glucose and retina of diabetic rats. Am J Physiol-Reg I. 2004 Oct;287(4):R735-R41. PubMed PMID: ISI:000223884500006. English.

[168] Goureau O, Regnier-Ricard F, Courtois Y. Requirement for nitric oxide in retinal neuronal cell death induced by activated Muller glial cells. J Neurochem. 1999 Jun;72(6): 2506-15. PubMed PMID: ISI:000080434800032. English.

[169] Brune B, von Knethen A. The role of nitric oxide and cyclooxygenase-2 in attenuating apoptosis. Journal of environmental pathology, toxicology and oncology : official organ of the International Society for Environmental Toxicology and Cancer. 2002;21(2):103-12. PubMed PMID: 12086396. Epub 2002/06/28. eng.

[170] Goldstein IM, Ostwald P, Roth S. Nitric oxide: a review of its role in retinal function and disease. Vision Res. 1996 Sep;36(18):2979-94. PubMed PMID: 8917798. Epub 1996/09/01. eng.

[171] Nakamura N, Hasegawa G, Obayashi H, Yamazaki M, Ogata M, Nakano K, et al. In creased concentration of pentosidine, an advanced glycation end product, and interleukin-6 in the vitreous of patients with proliferative diabetic retinopathy. Diabetes research and clinical practice. 2003 Aug;61(2):93-101. PubMed PMID: 12951277. Epub 2003/09/03. eng.

[172] Tezel G, Wax MB. Increased production of tumor necrosis factor-alpha by glial cells exposed to simulated ischemia or elevated hydrostatic pressure induces apoptosis in cocultured retinal ganglion cells. J Neurosci. 2000 Dec 1;20(23):8693-700. PubMed PMID: 11102475. Epub 2000/01/11. eng.

[173] Yoshida S, Yoshida A, Ishibashi T. Induction of IL-8, MCP-1, and bFGF by TNF-alpha in retinal glial cells: implications for retinal neovascularization during post-ischemic inflammation. Graefe's archive for clinical and experimental ophthalmology = Albrecht von Graefes Archiv fur klinische und experimentelle Ophthalmologie. 2004 May;242(5):409-13. PubMed PMID: 15029502. Epub 2004/03/19. eng.

[174] Bernardos RL, Barthel LK, Meyers JR, Raymond PA. Late-stage neuronal progenitors in the retina are radial Muller glia that function as retinal stem cells. J Neurosci. 2007 Jun 27;27(26):7028-40. PubMed PMID: 17596452.

[175] Fausett BV, Goldman D. A role for alpha1 tubulin-expressing Muller glia in regeneration of the injured zebrafish retina. J Neurosci. 2006 Jun 7;26(23):6303-13. PubMed PMID: 16763038.

[176] Karl MO, Hayes S, Nelson BR, Tan K, Buckingham B, Reh TA. Stimulation of neural regeneration in the mouse retina. Proc Natl Acad Sci U S A. 2008 Dec 9;105(49): 19508-13. PubMed PMID: 19033471. Pubmed Central PMCID: 2614791.

[177] Ooto S, Akagi T, Kageyama R, Akita J, Mandai M, Honda Y, et al. Potential for neural regeneration after neurotoxic injury in the adult mammalian retina. Proc Natl Acad Sci U S A. 2004 Sep 14;101(37):13654-9. PubMed PMID: 15353594. Pubmed Central PMCID: 518808.

[178] Fischer AJ, Scott MA, Ritchey ER, Sherwood P. Mitogen-activated protein kinase-signaling regulates the ability of Muller glia to proliferate and protect retinal neurons against excitotoxicity. Glia. 2009 Nov 1;57(14):1538-52. PubMed PMID: 19306360. Pubmed Central PMCID: 2775435. Epub 2009/03/24. eng.

[179] Fischer AJ, Scott MA, Tuten W. Mitogen-activated protein kinase-signaling stimulates Muller glia to proliferate in acutely damaged chicken retina. Glia. 2009 Jan 15;57(2):166-81. PubMed PMID: 18709648. Pubmed Central PMCID: 2774719. Epub 2008/08/19. eng.

[180] Wan J, Ramachandran R, Goldman D. HB-EGF is necessary and sufficient for Muller glia dedifferentiation and retina regeneration. Developmental cell. 2012 Feb 14;22(2): 334-47. PubMed PMID: 22340497. Pubmed Central PMCID: 3285435.

[181] Meyers JR, Hu L, Moses A, Kaboli K, Papandrea A, Raymond PA. ss-catenin/Wnt signaling controls progenitor fate in the developing and regenerating zebrafish retina. Neural development. 2012 Aug 24;7(1):30. PubMed PMID: 22920725.

[182] Dyer MA, Cepko CL. Control of Muller glial cell proliferation and activation following retinal injury. Nat Neurosci. 2000 Sep;3(9):873-80. PubMed PMID: 10966617.

[183] Koguchi K, Nakatsuji Y, Nakayama K, Sakoda S. Modulation of astrocyte proliferation by cyclin-dependent kinase inhibitor p27(Kip1). Glia. 2002 Feb;37(2):93-104. PubMed PMID: 11754208.

[184] Levine EM, Close J, Fero M, Ostrovsky A, Reh TA. p27(Kip1) regulates cell cycle withdrawal of late multipotent progenitor cells in the mammalian retina. Dev Biol. 2000;219(2):299-314. PubMed PMID: 0010694424.

[185] Ohno-Matsui K, Morita I, Tombran-Tink J, Mrazek D, Onodera M, Uetama T, et al. Novel mechanism for age-related macular degeneration: an equilibrium shift be-

tween the angiogenesis factors VEGF and PEDF. J Cell Physiol. 2001 Dec;189(3): 323-33. PubMed PMID: 11748590.

[186] Ogata N, Nishikawa M, Nishimura T, Mitsuma Y, Matsumura M. Unbalanced vitreous levels of pigment epithelium-derived factor and vascular endothelial growth factor in diabetic retinopathy. Am J Ophthalmol. 2002 Sep;134(3):348-53. PubMed PMID: 12208245.

[187] Duh EJ, Yang HS, Suzuma I, Miyagi M, Youngman E, Mori K, et al. Pigment epithelium-derived factor suppresses ischemia-induced retinal neovascularization and VEGF-induced migration and growth. Invest Ophthalmol Vis Sci. 2002 Mar;43(3): 821-9. PubMed PMID: 11867604.

[188] Miller JW, Adamis AP, Shima DT, D'Amore PA, Moulton RS, O'Reilly MS, et al. Vascular endothelial growth factor/vascular permeability factor is temporally and spatially correlated with ocular angiogenesis in a primate model. Am J Pathol. 1994 Sep; 145(3):574-84. PubMed PMID: 7521577. Epub 1994/09/01. eng.

[189] Boehm BO, Lang G, Feldmann B, Kurkhaus A, Rosinger S, Volpert O, et al. Proliferative diabetic retinopathy is associated with a low level of the natural ocular anti-angiogenic agent pigment epithelium-derived factor (PEDF) in aqueous humor. a pilot study. Hormone and metabolic research = Hormon- und Stoffwechselforschung = Hormones et metabolisme. 2003 Jun;35(6):382-6. PubMed PMID: 12920663.

[190] Boehm BO, Lang G, Volpert O, Jehle PM, Kurkhaus A, Rosinger S, et al. Low content of the natural ocular anti-angiogenic agent pigment epithelium-derived factor (PEDF) in aqueous humor predicts progression of diabetic retinopathy. Diabetologia. 2003 Mar;46(3):394-400. PubMed PMID: 12687338.

[191] Schofield CJ, Ratcliffe PJ. Oxygen sensing by HIF hydroxylases. Nat Rev Mol Cell Biol. 2004 May;5(5):343-54. PubMed PMID: 15122348.

[192] Penn JS, Madan A, Caldwell RB, Bartoli M, Caldwell RW, Hartnett ME. Vascular endothelial growth factor in eye disease. Prog Retin Eye Res. 2008 Jul;27(4):331-71. PubMed PMID: 18653375.

[193] Mignatti P, Tsuboi R, Robbins E, Rifkin DB. In vitro angiogenesis on the human amniotic membrane: requirement for basic fibroblast growth factor-induced proteinases. J Cell Biol. 1989 Feb;108(2):671-82. PubMed PMID: 2465298. Pubmed Central PMCID: 2115414.

[194] Unemori EN, Ferrara N, Bauer EA, Amento EP. Vascular endothelial growth factor induces interstitial collagenase expression in human endothelial cells. J Cell Physiol. 1992 Dec;153(3):557-62. PubMed PMID: 1447317.

[195] Lamoreaux WJ, Fitzgerald ME, Reiner A, Hasty KA, Charles ST. Vascular endothelial growth factor increases release of gelatinase A and decreases release of tissue inhibi-

tor of metalloproteinases by microvascular endothelial cells in vitro. Microvascular research. 1998 Jan;55(1):29-42. PubMed PMID: 9473407.

[196] Behzadian MA, Wang XL, Windsor LJ, Ghaly N, Caldwell RB. TGF-beta increases retinal endothelial cell permeability by increasing MMP-9: possible role of glial cells in endothelial barrier function. Invest Ophthalmol Vis Sci. 2001 Mar;42(3):853-9. PubMed PMID: 11222550.

[197] Takahashi T, Yamaguchi S, Chida K, Shibuya M. A single autophosphorylation site on KDR/Flk-1 is essential for VEGF-A-dependent activation of PLC-gamma and DNA synthesis in vascular endothelial cells. EMBO J. 2001 Jun 1;20(11):2768-78. PubMed PMID: 11387210. Pubmed Central PMCID: 125481.

[198] Fisher SK, Lewis GP. Muller cell and neuronal remodeling in retinal detachment and reattachment and their potential consequences for visual recovery: a review and reconsideration of recent data. Vision Res. 2003 Apr;43(8):887-97. PubMed PMID: 12668058.

[199] Wu KH, Madigan MC, Billson FA, Penfold PL. Differential expression of GFAP in early v late AMD: a quantitative analysis. Br J Ophthalmol. 2003 Sep;87(9):1159-66. PubMed PMID: 12928288. Pubmed Central PMCID: 1771844.

[200] Ramirez JM, Ramirez AI, Salazar JJ, de Hoz R, Trivino A. Changes of astrocytes in retinal ageing and age-related macular degeneration. Exp Eye Res. 2001 Nov;73(5): 601-15. PubMed PMID: 11747361. Epub 2001/12/19. eng.

[201] Bek T. Immunohistochemical characterization of retinal glial cell changes in areas of vascular occlusion secondary to diabetic retinopathy. Acta Ophthalmol Scand. 1997 Aug;75(4):388-92. PubMed PMID: 9374245.

[202] Bek T. Glial cell involvement in vascular occlusion of diabetic retinopathy. Acta Ophthalmol Scand. 1997 Jun;75(3):239-43. PubMed PMID: 9253965.

[203] Bek T. Capillary closure secondary to retinal vein occlusion. A morphological, histopathological, and immunohistochemical study. Acta Ophthalmol Scand. 1998 Dec; 76(6):643-8. PubMed PMID: 9881543.

[204] Bringmann A, Wiedemann P. Involvement of Muller glial cells in epiretinal membrane formation. Graefe's archive for clinical and experimental ophthalmology = Albrecht von Graefes Archiv fur klinische und experimentelle Ophthalmologie. 2009 Jul;247(7):865-83. PubMed PMID: 19415318.

[205] Anderson DH, Guerin CJ, Erickson PA, Stern WH, Fisher SK. Morphological recovery in the reattached retina. Invest Ophthalmol Vis Sci. 1986 Feb;27(2):168-83. PubMed PMID: 3943943.

[206] Fisher SK, Lewis GP, Linberg KA, Verardo MR. Cellular remodeling in mammalian retina: results from studies of experimental retinal detachment. Prog Retin Eye Res. 2005 May;24(3):395-431. PubMed PMID: 15708835.

[207] Zhu CL, Ji Y, Lee EJ, Grzywacz NM. Spatiotemporal pattern of rod degeneration in the S334ter-line-3 rat model of retinitis pigmentosa. Cell Tissue Res. 2012 Nov 10. PubMed PMID: 23143675.

[208] Marc RE, Jones BW. Retinal remodeling in inherited photoreceptor degenerations. Mol Neurobiol. 2003 Oct;28(2):139-47. PubMed PMID: 14576452.

[209] Sethi CS, Lewis GP, Fisher SK, Leitner WP, Mann DL, Luthert PJ, et al. Glial remodeling and neural plasticity in human retinal detachment with proliferative vitreoretinopathy. Invest Ophthalmol Vis Sci. 2005 Jan;46(1):329-42. PubMed PMID: 15623793.

[210] Guidry C. The role of Muller cells in fibrocontractive retinal disorders. Prog Retin Eye Res. 2005 Jan;24(1):75-86. PubMed PMID: 15555527.

[211] Girard P, Mimoun G, Karpouzas I, Montefiore G. Clinical risk factors for proliferative vitreoretinopathy after retinal detachment surgery. Retina. 1994;14(5):417-24. PubMed PMID: 7899716.

[212] Speicher MA, Fu AD, Martin JP, von Fricken MA. Primary vitrectomy alone for repair of retinal detachments following cataract surgery. Retina. 2000;20(5):459-64. PubMed PMID: 11039419.

[213] Miura M, Ideta H. Factors related to subretinal proliferation in patients with primary rhegmatogenous retinal detachment. Retina. 2000;20(5):465-8. PubMed PMID: 11039420.

[214] Hoffmann S, Friedrichs U, Eichler W, Rosenthal A, Wiedemann P. Advanced glycation end products induce choroidal endothelial cell proliferation, matrix metalloproteinase-2 and VEGF upregulation in vitro. Graefe's archive for clinical and experimental ophthalmology = Albrecht von Graefes Archiv fur klinische und experimentelle Ophthalmologie. 2002 Dec;240(12):996-1002. PubMed PMID: 12483322.

[215] Brown D, Hamdi H, Bahri S, Kenney MC. Characterization of an endogenous metalloproteinase in human vitreous. Curr Eye Res. 1994 Sep;13(9):639-47. PubMed PMID: 7805394.

[216] Grant MB, Caballero S, Tarnuzzer RW, Bass KE, Ljubimov AV, Spoerri PE, et al. Matrix metalloproteinase expression in human retinal microvascular cells. Diabetes. 1998 Aug;47(8):1311-7. PubMed PMID: 9703333.

[217] Steen B, Sejersen S, Berglin L, Seregard S, Kvanta A. Matrix metalloproteinases and metalloproteinase inhibitors in choroidal neovascular membranes. Invest Ophthalmol Vis Sci. 1998 Oct;39(11):2194-200. PubMed PMID: 9761302.

[218] Murphy AN, Unsworth EJ, Stetler-Stevenson WG. Tissue inhibitor of metalloproteinases-2 inhibits bFGF-induced human microvascular endothelial cell proliferation. J Cell Physiol. 1993 Nov;157(2):351-8. PubMed PMID: 7693724.

[219] Page-McCaw A, Ewald AJ, Werb Z. Matrix metalloproteinases and the regulation of tissue remodelling. Nat Rev Mol Cell Biol. 2007 Mar;8(3):221-33. PubMed PMID: 17318226. Pubmed Central PMCID: 2760082.

[220] Aricescu AR, McKinnell IW, Halfter W, Stoker AW. Heparan sulfate proteoglycans are ligands for receptor protein tyrosine phosphatase sigma. Mol Cell Biol. 2002 Mar; 22(6):1881-92. PubMed PMID: 11865065. Pubmed Central PMCID: 135600.

[221] Busch SA, Silver J. The role of extracellular matrix in CNS regeneration. Curr Opin Neurobiol. 2007 Feb;17(1):120-7. PubMed PMID: 17223033.

[222] Besser M, Jagatheaswaran M, Reinhard J, Schaffelke P, Faissner A. Tenascin C regulates proliferation and differentiation processes during embryonic retinogenesis and modulates the de-differentiation capacity of Muller glia by influencing growth factor responsiveness and the extracellular matrix compartment. Dev Biol. 2012 Sep 15;369(2):163-76. PubMed PMID: 22691363.

[223] Adamsky K, Schilling J, Garwood J, Faissner A, Peles E. Glial tumor cell adhesion is mediated by binding of the FNIII domain of receptor protein tyrosine phosphatase beta (RPTPbeta) to tenascin C. Oncogene. 2001 Feb 1;20(5):609-18. PubMed PMID: 11313993.

[224] Faissner A, Heck N, Dobbertin A, Garwood J. DSD-1-Proteoglycan/Phosphacan and receptor protein tyrosine phosphatase-beta isoforms during development and regeneration of neural tissues. Adv Exp Med Biol. 2006;557:25-53. PubMed PMID: 16955703.

[225] Yan Q, Sage EH, Hendrickson AE. SPARC is expressed by ganglion cells and astrocytes in bovine retina. J Histochem Cytochem. 1998 Jan;46(1):3-10. PubMed PMID: 10712100.

[226] Sofroniew MV, Vinters HV. Astrocytes: biology and pathology. Acta Neuropathol. 2010 Jan;119(1):7-35. PubMed PMID: ISI:000273174400003. English.

[227] Newman E, Reichenbach A. The Muller cell: A functional element of the retina. Trends Neurosci. 1996 Aug;19(8):307-12. PubMed PMID: ISI:A1996VA75000002. English.

[228] Reese BE. Development of the retina and optic pathway. Vision Res. 2011 Apr 13;51(7):613-32. PubMed PMID: ISI:000290061300003. English.

[229] Rompani SB, Cepko CL. A Common Progenitor for Retinal Astrocytes and Oligodendrocytes. J Neurosci. 2010 Apr 7;30(14):4970-80. PubMed PMID: ISI:000276414800017. English.

[230] Powell EM, Geller HM. Dissection of astrocyte-mediated cues in neuronal guidance and process extension. Glia. 1999 Mar;26(1):73-83. PubMed PMID: ISI: 000078911800008. English.

[231] Ullian EM, Sapperstein SK, Christopherson KS, Barres BA. Control of synapse number by glia. Science. 2001 Jan 26;291(5504):657-61. PubMed PMID: ISI: 000166616000045. English.

[232] Halassa MM, Fellin T, Haydon PG. The tripartite synapse: roles for gliotransmission in health and disease. Trends Mol Med. 2007 Feb;13(2):54-63. PubMed PMID: ISI: 000244641500002. English.

[233] Barres BA. Thrombospondins are astrocyte-secreted proteins that promote CNS synaptogenesis. Neuropsychopharmacol. 2004 Dec;29:S8-S. PubMed PMID: ISI: 000225588000026. English.

[234] Abbott NJ, Ronnback L, Hansson E. Astrocyte-endothelial interactions at the blood-brain barrier. Nat Rev Neurosci. 2006 Jan;7(1):41-53. PubMed PMID: ISI: 000234139600015. English.

[235] Pellerin L, Bouzier-Sore AK, Aubert A, Serres S, Merle M, Costalat R, et al. Activity-dependent regulation of energy metabolism by astrocytes: An update. Glia. 2007 Sep; 55(12):1251-62. PubMed PMID: ISI:000248678400006. English.

[236] Brown AM, Tekkok SB, Ransom BR. Energy transfer from astrocytes to axons: the role of CNS glycogen. Neurochem Int. 2004 Sep;45(4):529-36. PubMed PMID: ISI: 000222295500010. English.

[237] Obara M, Szeliga M, Albrecht J. Regulation of pH in the mammalian central nervous system under normal and pathological conditions: Facts and hypotheses. Neurochem Int. 2008 May;52(6):905-19. PubMed PMID: ISI:000255431200001. English.

[238] Simard M, Nedergaard M. The neurobiology of glia in the context of water and ion homeostasis. Neuroscience. 2004;129(4):877-96. PubMed PMID: ISI:000225726000004. English.

[239] Tsacopoulos M, Magistretti PJ. Metabolic coupling between glia and neurons. J Neurosci. 1996 Feb 1;16(3):877-85. PubMed PMID: ISI:A1996TQ57300002. English.

[240] Franze K, Grosche J, Skatchkov SN, Schinkinger S, Foja C, Schlid D, et al. Muller cells are living optical fibers in the vertebrate retina. P Natl Acad Sci USA. 2007 May 15;104(20):8287-92. PubMed PMID: ISI:000246599900019. English.

[241] Mata NL, Radu RA, Clemmons RS, Travis GH. Isomerization and oxidation of vitamin a in cone-dominant retinas: A novel pathway for visual-pigment regeneration in daylight. Neuron. 2002 Sep 26;36(1):69-80. PubMed PMID: ISI:000178295100008. English.

[242] Tout S, Chanling T, Hollander H, Stone J. The Role of Muller Cells in the Formation of the Blood-Retinal Barrier. Neuroscience. 1993 Jul;55(1):291-301. PubMed PMID: ISI:A1993LL16300024. English.

[243] Sofroniew MV. Molecular dissection of reactive astrogliosis and glial scar formation. Trends Neurosci. 2009 Dec;32(12):638-47. PubMed PMID: ISI:000272643900005. English.

[244] Ridet JL, Malhotra SK, Privat A, Gage FH. Reactive astrocytes: cellular and molecular cues to biological function. Trends Neurosci. 1997 Dec;20(12):570-7. PubMed PMID: ISI:A1997YJ58000010. English.

[245] Silver J, Miller JH. Regeneration beyond the glial scar. Nat Rev Neurosci. 2004 Feb; 5(2):146-56. PubMed PMID: ISI:000188601400017. English.

[246] Bringmann A, Reichenbach A. Role of Muller cells in retinal degenerations. Front Biosci. 2001 Oct;6:E77-E92. PubMed PMID: ISI:000171372600023. English.

[247] Dubois-Dauphin M, Poitry-Yamate C, de Bilbao F, Julliard AK, Jourdan F, Donati G. Early postnatal Muller cell death leads to retinal but not optic nerve degeneration in NSE-Hu-Bcl-2 transgenic mice. Neuroscience. 2000;95(1):9-21. PubMed PMID: 0010619458.

[248] Peterson WM, Wang Q, Tzekova R, Wiegand SJ. Ciliary neurotrophic factor and stress stimuli activate the jak-STAT pathway in retinal neurons and glia [In Process Citation]. J Neurosci. 2000;20(11):4081-90. PubMed PMID: 0010818143.

[249] Kirsch M, Trautmann N, Ernst M, Hofmann HD. Involvement of gp130-associated cytokine signaling in Muller cell activation following optic nerve lesion. Glia. 2010 May;58(7):768-79. PubMed PMID: 20091786.

[250] Wahlin KJ, Campochiaro PA, Zack DJ, Adler R. Neurotrophic factors cause activation of intracellular signaling pathways in Muller cells and other cells of the inner retina, but not photoreceptors. Invest Ophthalmol Vis Sci. 2000;41(3):927-36.

[251] Ueki Y, Wang J, Chollangi S, Ash JD. STAT3 activation in photoreceptors by leukemia inhibitory factor is associated with protection from light damage. J Neurochem. 2008 May;105(3):784-96. PubMed PMID: 18088375.

[252] Hollborn M, Jahn K, Limb GA, Kohen L, Wiedemann P, Bringmann A. Characterization of the basic fibroblast growth factor-evoked proliferation of the human Muller cell line, MIO-M1. Graefe's archive for clinical and experimental ophthalmology = Albrecht von Graefes Archiv fur klinische und experimentelle Ophthalmologie. 2004 May;242(5):414-22. PubMed PMID: 14963717.

[253] Nelson CM, Gorsuch RA, Bailey TJ, Ackerman KM, Kassen SC, Hyde DR. Stat3 defines three populations of Muller glia and is required for initiating maximal Muller glia proliferation in the regenerating zebrafish retina. J Comp Neurol. 2012 Aug 8. PubMed PMID: 22886421.

[254] Sasaki M, Ozawa Y, Kurihara T, Noda K, Imamura Y, Kobayashi S, et al. Neuroprotective effect of an antioxidant, lutein, during retinal inflammation. Invest Ophthalmol Vis Sci. 2009 Mar;50(3):1433-9. PubMed PMID: 18997089. Epub 2008/11/11. eng.

[255] Lin ST, Wang Y, Xue Y, Feng DC, Xu Y, Xu LY. Tetrandrine suppresses LPS-induced astrocyte activation via modulating IKKs-IkappaBalpha-NF-kappaB signaling pathway. Mol Cell Biochem. 2008 Aug;315(1-2):41-9. PubMed PMID: 18498042. Epub 2008/05/24. eng.

[256] Dvoriantchikova G, Barakat D, Brambilla R, Agudelo C, Hernandez E, Bethea JR, et al. Inactivation of astroglial NF-kappa B promotes survival of retinal neurons following ischemic injury. Eur J Neurosci. 2009 Jul;30(2):175-85. PubMed PMID: 19614983. Pubmed Central PMCID: 2778328. Epub 2009/07/21. eng.

[257] Shamsuddin N, Kumar A. TLR2 mediates the innate response of retinal Muller glia to Staphylococcus aureus. J Immunol. 2011 Jun 15;186(12):7089-97. PubMed PMID: 21602496. Pubmed Central PMCID: 3110513.

[258] Kinkl N, Sahel J, Hicks D. Alternate FGF2-ERK1/2 signaling pathways in retinal photoreceptor and glial cells in vitro. J Biol Chem. 2001 Nov 23;276(47):43871-8. PubMed PMID: 11571286.

[259] Zong H, Ward M, Madden A, Yong PH, Limb GA, Curtis TM, et al. Hyperglycaemia-induced pro-inflammatory responses by retinal Muller glia are regulated by the receptor for advanced glycation end-products (RAGE). Diabetologia. 2010 Dec;53(12): 2656-66. PubMed PMID: 20835858.

[260] Walsh DT, Bresciani L, Saunders D, Manca MF, Jen A, Gentleman SM, et al. Amyloid beta peptide causes chronic glial cell activation and neuro-degeneration after intravitreal injection. Neuropathol Appl Neurobiol. 2005 Oct;31(5):491-502. PubMed PMID: 16150120.

[261] Rattner A, Nathans J. The genomic response to retinal disease and injury: evidence for endothelin signaling from photoreceptors to glia. J Neurosci. 2005 May 4;25(18): 4540-9. PubMed PMID: 15872101.

Chapter 5

Neural Stem Cell: Tools to Unravel Pathogenetic Mechanisms and to Test Novel Drugs for CNS Diseases

Luca Colucci-D'Amato and MariaTeresa Gentile

Additional information is available at the end of the chapter

http://dx.doi.org/10.5772/55677

1. Introduction

1.1. Neurogenesis, stem cells and cellular models of diseases

In a study published in 1992, Weiss and Reynolds at University of Calgary, demonstrated for the first time that cells isolated from the brain of adult mice have the ability to proliferate in vitro and differentiate into neurons, astrocytes and oligodendrocytes using a specific cocktails of growth factors [1]. In 1999, similar stem/progenitor cells were isolated from the human adult periventricular subependymal zone and expanded in vitro [2].

Neural stem cells (NSC) are self-renewing, multipotent cells residing in the nervous system. NSC during development produce the enormous diversity of neurons, astrocytes and oligodendrocytes within the developing nervous system. However, accumulating evidence has clearly shown that a number of newborn neurons can be generated also from adult NSC, integrates into pre-existing neural circuits and is functional [3]. In the adult brain, neurogenesis is not a diffuse event and occurs in restricted regions, where classical developmental signals and morphogens such as, Bone Morphogenic Proteins (BMPs), ephrins, Noggin, Sonic hedgehog homolog (Shh), and Notch expression are maintained even after differentiation [4]. In particular, the Notch pathway is a highly conserved arbiter of cell fate decisions and is intimately involved in developmental processes [5]. Thus, besides its pivotal role in neural development, it is also involved in the control of neurogenesis, neuritic growth [6], neural stem cell maintenance [7], synaptic plasticity [8] and long term memory [9] both in the developing and adult brain.

In the adult brain the well-established restricted regions of neurogenesis, named niches, are the sub ventricular zone (SVZ) of the lateral ventricle wall and the dentate gyrus subgranular zone (SGZ) of the hippocampus [9]. Several reports describe that neurogenesis may also occur in other brain areas, including substantia nigra [10], and neocortex [11] even if these studies have not been confirmed by others.

In the currently prevalent view, primary adult SVZ NSCs *in vivo* are slowly dividing, long-term 5-bromo-2′-deoxyuridine (BrdU)-retaining progenitors that exhibit several common features of subventricular radial glia-like astrocytes and ventricular ependymal cells, including morphological characteristics and expression of the glial fibrillary acidic protein (GFAP) and the glycoprotein CD133, also known as Prominin-1.

Adult neurogenesis in either the SVZ or the SGZ is highly sensitive to environmental cues, physiological stimuli and neuronal activity, suggesting that the tailored addition of new neurons might serve specific neuronal functions. In fact, the complex dynamic equilibrium present in healthy adult CNS involves also the participation of functional NSC niches [12, 13].

The occurrence of major events breaking niches homeostasis may overcome their adaptive capacities and contribute or lead to disease. In CNS, various pathogenic events acting by different mechanisms may cause neural cell loss and chronic inflammation. Several agents and mediators sustaining these mechanisms also act on niche homeostasis and it is therefore expected that these two conditions may have a deep impact on NSC biology and niche properties. The neurogenesis within SVZ niche is differentially activated in various neurodegenerative pathologies. An increase in endogenous neurogenesis in the lateral ventricular walls occurs in several diseases including traumatic brain injury, vascular dementia, Huntington's disease, multiple sclerosis or epilepsy and physiological events such as ageing. [14-19]. In these pathophysiological conditions, a common feature leading to neurogenesis is represented by the upregulation of inflammatory cytokines production. In several cases it has been shown that the niche reaction to inflammation leads to generation of new neurons that integrate within the pre-existing circuitry [20].

In pathological conditions, such as stroke, NSC niches can appear also at unexpected ectopic sites as in brain parenchyma. In particular, part of the stroke-induced ectopic cells positive for doublecortin, a widely accepted marker of neurogenesis, originates from the SVZ resident NSCs that have migrated to the injured site where they differentiated into mature neurons [21] However, it is also possible that part of this ectopic pool may originate locally, leading to an injury-induced ectopic NSC niche formation [22]. Niche reaction to injury may thus proceed by amplification of pre-existing NSC niche and migration toward the damaged site or by activation of local and pre-existing quiescent NSC niches or by conversion of a normal tissue into a newly established NSC niche. The mechanisms and signals driving these reactions are many, some of them peculiar to the pathological condition and most of them endogenous to the NSC niche and already acting in physiological condition.

Altogether the findings described above lead to speculate that NSC may be therapeutically useful for CNS self-repair if properly mobilized and can supply cells for neuronal substitution in CNS injury or degeneration, providing potential therapeutic tools in CNS diseases. NSC, on the other hand, can be also useful for studying the mechanisms and molecules involved in neuronal differentiation, regulation of neurotransmitter biosynthesis and also physiopathology of brain diseases. Indeed, to date, deciphering the molecular and cellular basis of CNS disorders has been met with success in animal models for neurological disease. However, while the continued importance of animals in translational research is unquestionable, genetic and anatomical variation between rodents and humans have led to imperfect phenotypic correlations between genetic models and the human disease. Moreover, most neurodegenerative diseases are sporadic and depend on the complex interaction of genetic and environmental risk factors. Thus, it may be difficult to fully clarify these conditions in animals. For these reasons, it would be advisable if neurological disease cellular models should be developed and studied in concert with existing animal models. Recent advances in the areas of stem cell and reprogramming biology seem to provide a novel route to the generation of a wide variety of neural cell types for studying neurological diseases.

A critical point of neurological disease modelling using NSCs is the availability of reliable protocols that efficiently direct stem cells differentiation into the specific neural cell types affected in disorders of interest. Knowledge of the pathways that drives neural differentiation has led to rational approaches now routinely used to direct the differentiation of NSCs in vitro. It appears that a variety of neural phenotypes can be achieved, depending on the combination and timing of the inductive signals to which progenitors are exposed. Recently, several disease-related cell types have been generated in vitro by directed differentiation of NSCs such as spinal motor neurons, midbrain dopaminergic neurons, basal forebrain cholinergic neurons, cortical progenitors and oligodendrocytes [23-38]. Moreover, Lee and co-workers also generated sensory neurons and Schwann cells [39]. One of the first differentiation method has been developed for the production of spinal motor neurons. In particular, NSCs can be directed to differentiate into functional spinal motor neurons when challenged with retinoic acid and Sonic Hedgehog. Retinoic acid induces neuralization and caudalization of stem cells, while the induction of Sonic Hedgehog signalling converts spinal progenitors cells into motor neurons [40]. The NSCs derived spinal motor neurons, when transplanted in developing chick embryos, were able to form long axonal projections into skeletal muscle [27]. The ability to direct the differentiation of NSCs into specific motor neurons could have important implications for modelling of diseases such as amyotrophic lateral sclerosis (ASL) and understanding molecular mechanisms of selective vulnerability. Another clinically relevant neural subtype that has been generated in vitro from NSCs is dopaminergic neurons which could be used as cellular model of Parkinson Disease (PD). One of the strategies used to obtain dopaminergic neurons from NSCs rely on the combined activity of Sonic Hedgehog and FGF8, as first shown by Lee et al. [27]. Methodological improvements to enhance dopaminergic differentiation in vitro include co-culture with immortalized fetal astrocytes and dual inhibition of BMP/TGF beta signalling during neural induction. The in vivo functionality of dopaminergic neurons derived from NSCs has been demonstrated by transplantation assays into the striatum of 6-hydroxydopamine-lesioned parkinsonian rats showing partial recovery of motor function [41-46]. A recent study reported the controlled differentiation of NSCs to basal forebrain cholinergic neurons, the neuronal population affected in Azheimer's disease (AD) and associated with cognitive decline [33]. In particular, two methods have been applied to obtain basal forebrain cholinergic neurons from NSCs: the use of the diffusible ligand BMP9 and the transfection of two developmentally relevant transcription factors, Lhx8 and Gbx1, which were further shown to act downstream of BMP9 signalling. Basal forebrain cholinergic neurons obtained in these ways were also shown to stably engraft into murin hippocampal slice cultures and to generate electrophysiologically functional cholinergic synapses [33]. In addition, NSCs are also able to differentiate into glial cell types of the CNS. However, directed differentiation protocols and functional characterization of astrocytes derived from NSCs have yet to be reported. In contrast, there are already many approaches to obtain oligodendrocytes from NSCs [47-50]. The ability of these differentiated cells to myelinate axons has been demonstrated both in vitro by co-culture with rat hippoampal neurons [51] and in vivo by transplantation into the mouse model of dysmielination [52]. Moreover, studies evaluating NSCs-derived oligodendrocytes transplants into adult rats have reported improvement of locomotor recovery in both thoracic and spinal cord injury models [53].

Neural progenitors generated in vitro can also be directed towards cell types of the peripheral nervous system (PNS).

Disease modeling using pluripotent stem cells might greatly benefit if the genome of these cells could be readily modified. For instance, the generation of transgenic "reporter" cell lines using fluorescent reporter genes under the control of cell type-specific promoters could enable the purification, tracking, and functional characterization of disease cells models after directed differentiation. Indeed, most in vitro differentiation strategies result in a heterogenous population of differentiated cells, which can include progenitors and a variety of cellular intermediates. Thus, having the ability to prospectively identify, purify, and easily track the desired cell type by means of reporter-gene expression can facilitate downstream disease-specific assays, which could be hindered by the presence of other cell types. However, in spite of their self-renewal properties, NSCs cells can still be difficult to genetically manipulate. Various techniques for stem cell genetic modification have been reported, and these can result in random (i.e., transgenic) or targeted integration of the DNA construct [54]. Transgenic approaches include the use of plasmid transfection, lentiviral transduction, transposases, and bacterial artificial chromosomes (BAC) as DNA delivery systems [55-59]. In particular, Placantonakis and colleagues used BAC transgenesis of different reporter constructs to generate motor neuron lineages [59]. Some of the limitations associated with transgenic approaches include the possibility of insertional mutagenesis, transgene silencing or ectopic expression of the transgene due to position effects, and lack of faithful expression of a reporter transgene due to absence of regulatory elements in the promoter fragment driving its expression [56].

Development of methods to accurately correlate how a stem cell-derived neuron in culture correlates ontologically with the in vivo equivalent could be useful. In fact, neuronal dysfunction and degeneration as a result of the neurodegenerative process probably occurs much earlier than the initial neurological manifestations that characterize the disease. For example, prior to the onset of the motor component of the Parkinson disease (PD), a significant number of dopaminergic neurons have already been lost. Moreover, non-motor manifestations can predate motor manifestations by years. In PD, one of the earliest symptoms of disease may be olfactory and autonomic dysfunction and initial alpha-synuclein positive Lewy body pathology may occur in the dorsal motor nucleus of the glosso-pharyngeal and vagal nerves and anterior olfactory nucleus [60]. It would seem reasonable, in addition to studying midbrain dopaminergic neurons, to dissect the molecular causes that lead to degeneration of the motor component of PD, to obtain the cell types affected earliest in the disease. Methods to accelerate the time of the pathology in vitro will probably be important in adult-onset disease. Cellular stressors such as oxidative stress, growth factor withdrawal, starvation, selective neurotoxins, and heat shock may reveal differences in NSCs models. Dopaminergic neurons obtained from a neural stem cell line with a known mutation in LRRK2, typical of an early-onset PD, demonstrated increased proportions of caspase-3 activation suggesting a selective vulnerability when exposed to a variety of cellular stressors including hydrogen peroxide, proteosome inhibition, and 6-OHDA exposure [60].

In addition, with better methods to direct differentiation of pluripotent stem cells to non-neuronal cells, it may be possible to recapitulate the relevant microenvironments that are important in several neurodegenerative diseases [61]. In this regard, NSC cell-based co-culture models have provided informative models of the effects of glia in SOD-1-related Amyotrophic lateral sclerosis (ALS). For example, NSC-derived spinal motor neurons cultured in the presence of glial cells, either derived from SOD-1 transgenic mice or primary astrocytes genetically modified to express mutant SOD-1, are selectively vulnerable to the toxic effects [62, 63]. Furthermore, these culture models allowed subsequent searches for candidate mechanisms by which mutant glia exerted its effects and testing of drugs to rescue motor neuron death. Thus, future NSC-based models composed of titrated populations of neurons, glia, and skeletal muscle to create a functional motor unit in vitro would be informative for studies on several neurologic diseases.

2. Isolation and establishment of pluripotent neural cell line endowed with stemness properties

Here we describe briefly, the establishment of a novel neural cell line obtained from mesencephalic primary cultures generated from 11-day-old mouse embryos and showing staminal properties when appropriately cultured.

In particular, these cells were infected with a replication-defective retrovirus bearing c-*myc* and neomycin resistance genes. Neomycin-resistant clones were isolated and a stable cell line was further characterized and named mes-c-myc A1 (A1). The proto-oncogene c-*myc* was chosen because of its efficiency in immortalizing cells including neuroepithelial cells without neoplastic transformation. A1 cells deprived of serum and treated with cAMP show arrest of cell division, which is considered a prerequisite for neural cells to enter a differentiation program. Under both conditions, A1 cells express nestin and vimentin, markers of neural precursors, and the glial fibrillary acidic protein (GFAP), a marker of glial cells. A1 cells present numerous neuronal features, such as peripherin, neuron specific enolase (NSE), microtubule-associated protein 1 (MAP1), N-CAM. In addition, morphologically differentiated A1 cells show functional voltage-gated channels, a neuronal hallmark. Finally, these cells synthesize and accumulate GABA, the principal inhibitory neurotransmitter in CNS. A1 cells present features of neural progenitors and have a broad potential because they may represent a new tool in CNS developmental studies and could be useful in therapeutic applications.

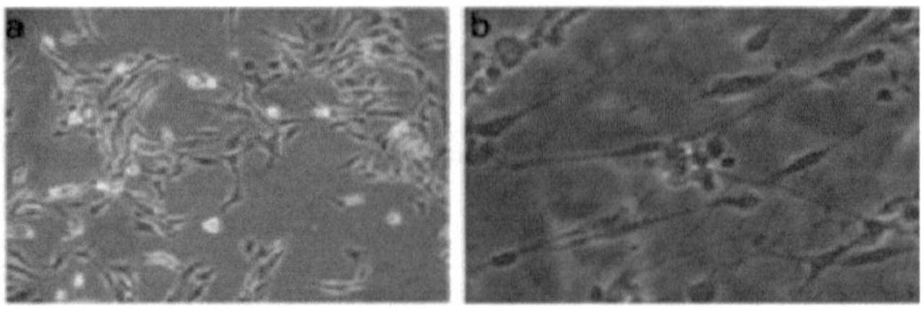

	a	b
Proliferation	+	-
neurite outgrowth	-	+
electric membrane activity	-	+
neuronal markers	+/-	++

Figure 1. a) Proliferating/undifferentiated A1-mes-c-myc cells and (b) unproliferating/differentiated A1 mes-cmyc cells with their phenotypic profiles. [64].

When cultivated in suspension in a serum free medium with the addition of epidermal growth factor (EGF) with or without basic fibroblast growth factor (bFGF), these cells are able to form neurosphere.

NSCs express EGF and bFGF receptors. The stimulation of these EGF and FGF cell-surface receptors promotes the activation of specific signal transduction pathways, such as the mitogen activated protein kinase (MAPK) and the phospholipase C (PLC) signal pathways [65]. Thus, A1 cells similarly to NSC derived from brain respond to such growth factors by generating neurospheres. Moreover, neurosphere are cellular clones derived from a single neural stem cell which divides to produce other NSC and/or other progenitors cells. Neurospheres can form not only from the clonal progeny of a single NSC, but also by the aggregation of cells into neurospheres [66]. For these reasons, the neurosphere assay (NSA) is one of the most frequently used methods to isolate, expand and also calculate the frequency of neural stem cells (NSCs). Furthermore, this serum-free culture system has also been employed to expand stem cells and determine their frequency from a variety of tumors and normal tissues. In our system, cells were counted and cultured in suspension ($2{,}5\times10^5$/ml) with neurosphere medium [MEM/F12, N_2, bFGF and EGF] in 25 cm^2 flasks with no substrate pre-treatment. Primary neurospheres formed after 4–5 days in vitro once every 7 days they were gently spun down [75 g), mechanically dissociated, and the medium was changed. To evaluate neural differentiation potential, neurospheres were dissociated and plated on poly-D-lysine precoated plates in neurosphere medium. After 6 days in culture, cells were kept for additional 3 days without b-FGF and EGF, and neuronal, astroglial and oligodendroglial differentiation was assessed by immunocytochemical analysis. Immunostaining for markers of neural (β-III-tubulin), astrocyte (GFAP) and oligodendrocyte (O4) cells clearly showed the presence of all three CNS cell types among the neurosphere derived cells. Further characterization of neuronal, astroglial, oligodendroglial and stem cells markers has been performed by real-time PCR. In particular, our analysis identified several stem cell genes and pluripotency-associated gene in neurosphere cells, including nestin, Nanog, and Sox2. By means of real-time PCR we also confirmed the presence of genes of neural cells such as β-III-tubulin and astrocyte such as GFAP. Finally, to definitely confirm the NSCs features of A1 mes-c-myc cells, we determined the self-renewal ability of secondary neurospheres. In particular, primary neurospheres were mechanically dissociated into single cells, counted and cultured in neurosphere medium in suspension in 25 cm^2 flasks again. As expected from NSCs, cells dissociated from primary neurosphere were able to generate secondary neurosphere.

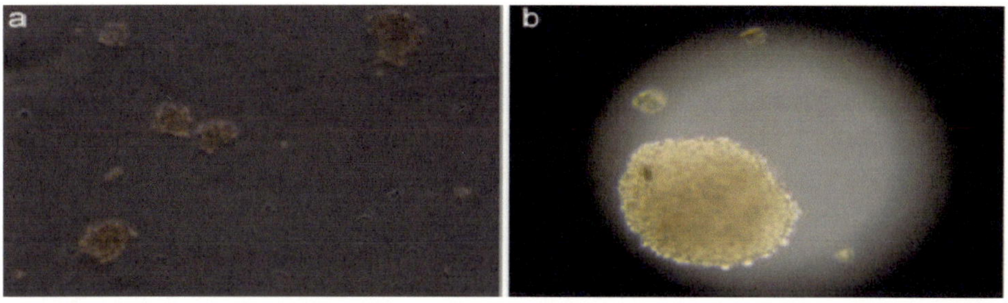

Figure 2. Neurospheres originated from A1mes-c-myc cells 2 days (a) and 10 days (b) after cultured in neurosphere medium.

It cannot be excluded that the contemporary maintainance of precursors and differentiated markers could be attributed to the persistent expression of the exogenous c-myc. However, in our model system the exogenous expression of c-myc does not impair block of proliferation elicited by serum deprivation and/or cAMP treatment. Moreover, it is generally accepted that c-myc is not oncogenic by itself likely because of its dual action as a promoter of cell division and as an apoptotic mediator in absence of growth factors. In addition, it has been reported that in genetically manipulated NSC clones there is a spontaneous and significant down-regulation of c-myc in concomitance with mitogen removal and the onset of differentiation [67,68]. Moreover, it is worth noting that c-myc is one of the genes used to generate iPS cells. Indeed, viral vectors have been used to transfer transcription factors, such as Oct4, Sox2, c-myc, Klf4, and nanog, to induce reprogramming of mouse fibroblasts, neural stem cells, neural progenitor cells, keratinocytes, B lymphocytes and meningeal membrane cells towards pluripotency [69-71]. Immortalization using the myc transcription factor has proven highly effective at extending the normal life span of human NSCs *in vitro* and maintaining a stable genotype and phenotype. Long term cell expansion with associated karyotype stability is a feature of myc immortalization. Traditionally thought of as a proto-oncogene, it has been recently reported that myc may be a 'stemness' gene driving rapid proliferation yet maintain multipotentiality in stem cells. Indeed, two NSC lines were recently immortalized from human progenitors by means of v-myc and c-myc [72], although useful to test drugs on molecular targets, the human origin of these cells may hamper the efficacy of transplants into mice due to species barrier.

3. NSC lines as model to test new therapeutic compounds

Given the potential of neural stem cells and their differentiated cell types to model keys aspects of neurological diseases, an obvious extension of this platform is represented by its use for drug discovery and predictive toxicology. Nearly 90% of new drugs tested in humans fail to ultimately come to clinical approval, with central nervous system disorders as a therapeutic area, among those with the highest rate of attrition. Arguably, these failures have resulted from a reliance on imperfect models used during preclinical development.

Neural stem cells (NSCs) are a good model to screen effective drugs that increase neurogenesis. Several classes of drugs have been reported to interfere with NSC homeostasis. Most of the interest has been focused on psychiatric drugs since the early discovery of links between neurogenesis and stress-related disorders. Several extensive reviews have already been published on this topic in the recent years. In particular, association between neurogenesis and depression has been the object of debate since the report of Malberg *et al.* showing, in 2000 by BrdU assay, that chronic antidepressant treatment significantly increased neurogenesis in the dentate gyrus of the hippocampus, and that the new cells became neurons [73-75]. These effects were in agreement with previous data showing that serotoninergic depletion decreased neurogenesis in the dentate gyrus and the SVZ [76]. The effects on neurogenesis were consistent with the time course for the therapeutic action of antidepressants. Chronic treatment with the classical antidepressant fluoxetine accelerated the maturation of immature neurons and enhanced a specific form of long term potentiation in the dentate gyrus; neurogenesis was also linked to the behavioral effect of fluoxetine measured by the novelty-suppressed feeding test [77, 78]. We have recently demonstrated, by means of western blotting and real time PCR, that A1 immortalized cell line is able to express serotonergic markers before and after differentiation. In particular, we demonstrated that A1 cells express the transcripts of the two rate-limiting enzymes necessary for the serotonin synthesis TPH1 and TPH2 and that the two enzymes were differently expressed in proliferating and differentiated cells. We also found that TPHs were modulated by fluoxetine and citalopram, two SSRI drugs widely used in therapy [79, 80].

In addition to synthetic chemicals, recent results show that some natural products also affect cell fate of NSCs [81]. Until recently, neuroprotective effects of natural products have been intensely studied. In particular, methanol extracts of Jeju native plants protected apoptosis induced by hydrogen peroxides. Visnagin, an active component extracted from the fruits of *Ammi visnaga*, which has been used as treatment for low blood-pressure, showed protective effects on kainic acid-induced mouse hippocampal cell death by reducing inflammation. BF-7 extracted from a sericultural product has significant protective effects on amyloid β peptide induced apoptosis through reduction of ROS generation and diminished caspase activity [82-85]. On this issue, also in our lab, we are in progress to test the biological effects of a natural compound on NSCs. In fact, in CNS, aberrant proliferation causes cancer whereas impaired survival of differentiated neurons induces neurodegenerative disorders. In order to find novel therapeutic targets able to inhibit aberrant cell proliferation and/or enhance differentiated cells survival, we analyzed properties of the aqueous extract of *Ruta graveolens* (*Ruta g.* a.e) on differentiated, non-proliferating and undifferentiated, proliferating neural cells. *Ruta g.* is currently used for its diuretic, sedative, and analgesic effects and recent studies described antiproliferative effects on different cancer cells.

In A1 cell system, *Ruta g.* a.e. induces increase of ERK 1/2 (ERKs) phosphorylation and death of A1 proliferating cells. In presence of the ERKs pathway inhibitor, *Ruta g.* a.e.-induced cell death decreases, indicating that ERKs is involved in the *Ruta g*. effect on A1 proliferating cells. Moreover, when *Ruta g*. a.e. is added, the number of differentiated A1 cells appears significantly higher as compared to control conditions and the analysis of the cell cycle showed an increased number of cells in G2/M phase in differentiated cells treated with *Ruta g*. a.e.

Thus, our data suggest that A1 cells could represent a model system of neural stem cell line able to allow a deep insight into the mechanisms of regulation neural gene expression and to identify novel therapeutic targets in the development of more useful drugs for the management of disorders of the CNS

4. Conclusion and future perspectives

Research in the area of stem cell biology and regenerative medicine, along with developmental and molecular neuroscience, will further our understanding of drug-induced effects (i.e. death, survival, neurotransmission) on neurons during their development. Moreover, in vitro models of stem cell-derived neural cell lines allow investigators, under control conditions and during intense neuronal growth, to delve deep into molecular mechanisms underlying the actions of various drugs and pathophysiological conditions at various developmental stages. In addition, since NSCs lines are capable to differentiate into non proliferating neuronal phenotypes, they represent a powerful tool to screen drugs exerting different effects according to the cell cycle.

In conclusion, the use of this models will likely lead to fewer pharmacological risks and/or identification of new compounds exerting biological effects on healthy and diseased neurons.

Author details

Luca Colucci-D'Amato and MariaTeresa Gentile

Department of Pharmaceutical, Biological and Enviromental Science and Technology, Second University of Naples, Caserta, Italy

References

[1] Reynolds, B. A, & Weiss, S. (1992). Generation of neurons and astrocytes from isolated cells of the adult mammalian central nervous system. Science , 255, 1707-1710.

[2] Kukekov, V. G, Laywell, E. D, Suslov, O, Davies, K, Scheffler, B, et al. (1999). Multipotent stem/progenitor cells with similar properties arise from two neurogenic regions of adult human brain. Exp Neurol , 156, 333-344.

[3] Brazel, C. Y, Romanko, M. J, Rothstein, R. P, & Levison, S. W. Roles of the mammalian subventricular zone in brain development. Progress in Neurobiology (2003). , 69, 49-69.

[4] Ehebauer, M, Hayward, P, & Arias, A. M. Notch, a universal arbiter of cell fate decisions. Science. (2006). Dec 1;, 314(5804), 1414-5.

[5] Sestan, N, Artavanis-tsakonas, S, & Rakic, P. Contact-dependent inhibition of cortical neurite growth mediated by notch signaling. Science. (1999). Oct 22;, 286(5440), 741-746.

[6] Hitoshi, S, Alexson, T, Tropepe, V, Donoviel, D, Elia, A. J, Nye, J. S, Conlon, R. A, Mak, T. W, Bernstein, A, & Van Der Kooy, D. Notch pathway molecules are essential for the maintenance, but not the generation, of mammalian neural stem cells. Genes Dev. (2002). Apr 1;, 16(7), 846-58.

[7] Poirazi, P, & Mel, B. W. Impact of active dendrites and structural plasticity on the memory capacity of neural tissue. Neuron. (2001). Mar;, 29(3), 779-96.

[8] Shors, T. J, Miesegaes, G, Beylin, A, Zhao, M, Rydel, T, & Gould, E. Neurogenesis in the adult is involved in the formation of trace memories. Nature. (2001). Mar 15;, 410(6826), 372-6.

[9] Soriano, E, Cobas, A, & Fairen, A. Neurogenesis of glutamic acid decarboxylase immunoreactive cells in the hippocampus of the mouse. I: Regio superior and regio inferior. J Comp Neurol (1989). , 281, 586-602.

[10] Zhao, M, Momma, S, Delfani, K, et al. Evidence for neurogenesis in the adult mammalian sub stantia nigra Proc Natl Acad Sci USA (2003). , 100, 7925-30.

[11] Gould, E, Reeves, A. J, Graziano, M. S, & Gross, C. G. Neurogenesis in the neocortex of adult primates. Science (1999). , 286, 548-52.

[12] Colucci-D'Amato LBonavita V, di Porzio U. The end of the central dogma of neurobiology: stem cells and neurogenesis in adult CNS. Neurol Sci. (2006). Sep;, 27(4), 266-70.

[13] Colucci-D'Amato Ldi Porzio U. Neurogenesis in adult CNS: from denial to opportunities and challenges for therapy. Bioessays. (2008). Feb;, 30(2), 135-45.

[14] Macas, J, Nern, C, Plate, K. H, & Momma, S. Increased Generation of Neuronal Progenitors after Ischemic Injury in the Aged Adult Human Forebrain. J Neurosci (2006). , 26, 13114-9.

[15] Yang, J, Jiang, Z, Fitzgerald, D. C, et al. Adult neural stem cells expressing IL-10 confer potent immunomodulation and remyelination in experimental autoimmune encephalitis. J Clin Invest (2009). , 119, 3678-91.

[16] Ekonomou, A, Ballard, C. G, Pathmanaban, O. N, et al. Increased neural progenitors in vascular dementia. Neurobiology of aging (2010). Feb 4

[17] Minger, S. L, Ekonomou, A, Carta, E. M, Chinoy, A, Perry, R. H, & Ballard, C. G. Endogenous neurogenesis in the human brain following cerebral infarction. Regen Med (2006). , 2, 69-74.

[18] Nait-oumesmar, B, Picard-riéra, N, & Kerninon, C. Baron-Van Evercooren A. The role of SVZ-derived neural precursors in demyelinating diseases: From animal models to multiple sclerosis. J Neurol Neurosci (2008). , 265, 26-31.

[19] Bonner, J. F, Connors, T. M, Silverman, W. F, Kowalski, D. P, Lemay, M. A, & Fischer, I. Grafted neural progenitors integrate and restore synaptic connectivity across the injured spinal cord. J Neurosci (2011). , 31, 4675-86.

[20] Kojima, T, Hirota, Y, Ema, M, et al. Subventricular Zone-Derived Neural Progenitor Cells Migrate Along a Blood Vessel Scaffold Toward The Post-stroke Striatum. STEM CELLS (2010). , 28, 545-54.

[21] Shimada, I. S, Peterson, B. M, & Spees, J. L. Isolation of Locally Derived Stem/Progenitor Cells From the Peri-Infarct Area That Do Not Migrate From the Lateral Ventricle After Cortical Stroke. Stroke (2010). e, 552-60.

[22] Nakagomi, T, Taguchi, A, Fujimori, Y, et al. Isolation and characterization of neural stem/progenitor cells from post-stroke cerebral cortex in mice. Eur J Neurosci (2009). , 29, 1842-52.

[23] Zuddas, A, Corsini, G. U, Barker, J. L, & Kopin, I. J. di Porzio U. Specific reinnervation of lesioned mouse striatum by grafted mesencephalic dopaminergic neurons European journal of Neuroscience 3; , 1, 72-85.

[24] Robert, J, Plunkett, M. D, Krysztof, S, Bankiewicz, M. D, & Alex, C. Cummins, Robert S. Miletich, M.D., Ph.D., Joan P. Schwartz, Ph.D., and Edward H. Oldfield, M.D.. ((1990). Long-term evaluation of hemiparkinsonian monkeys after adrenal autografting or cavitation alone. *Journal of Neurosurgery* 73:6, 918-926.

[25] Dimos, J. T, Rodolfa, K. T, Niakan, K. K, Weisenthal, L. M, Mitsumoto, H, Chung, W, Croft, G. F, Saphier, G, Leibel, R, Goland, R, et al. (2008). Induced pluripotent stem cells generated from patients with ALS can be differentiated into motor neurons. Science , 321, 1218-1221.

[26] Hu, B, & Zhang, Y. S.-C. ((2009). Differentiation of spinal motor neurons from pluripotent human stem cells. Nat. Protoc. , 4, 1295-1304.

[27] Lee, H, Shamy, G. A, Elkabetz, Y, Schofield, C. M, Harrsion, N. L, Panagiotakos, G, Socci, N. D, Tabar, V, & Studer, L. (2007b). Directed differentiation and transplantation of human embryonic stem cell-derived motoneurons. Stem Cells , 25, 1931-1939.

[28] Li, X, Du, J, Zarnowska, Z. -W, Pankratz, E. D, Hansen, M, Pearce, L. O, & Zhang, R. A. S.-C. ((2005). Specification of motoneurons from human embryonic stem cells. Nat. Biotechnol. , 23, 215-221.

[29] Cho, M. S, Lee, Y, Kim, E, Chung, J. Y, Cho, S, Kim, Y. H, Kang, D. -S, Lee, S. -M, Kim, H, Kim, M. -H, et al. (2008). Highly efficient and large-scale generation of functional dopamine neurons from human embryonic stem cells. Proc. Natl. Acad. Sci. USA , 105, 3392-3397.

[30] Hargus, G, Cooper, O, Deleidi, M, Levy, A, Lee, K, Marlow, E, Yow, A, Soldner, F, Hockemeyer, D, Hallett, P. J, et al. (2010). Differentiated Parkinson patient-derived induced pluripotent stem cells grow in the adult rodent brain and reduce motor asymmetry in Parkinsonian rats. Proc. Natl. Acad. Sci. USA , 107, 15921-15926.

[31] Nguyen, H. N, Byers, B, Cord, B, Shcheglovitov, A, Byrne, J, Gujar, P, & Kee, K. Schu ° le, B., Dolmetsch, R.E., Langston, W., et al. ((2011). LRRK2 mutant iPSC-derived DA neurons demonstrate increased susceptibility to oxidative stress. Cell Stem Cell , 8, 267-280.

[32] Roy, N. S, Cleren, C, Singh, S. K, Yang, L, Beal, M. F, & Goldman, S. A. (2006). Functional engraftment of human ES cell-derived dopaminergic neurons enriched by coculture with telomerase-immortalized midbrain astrocytes. Nat. Med. , 12, 1259-1268.

[33] Bissonnette, C. J, Lyass, L, Bhattacharyya, B. J, Belmadani, A, Miller, R. J, & Kessler, J. A. (2011). The controlled generation of functional Basal forebrain cholinergic neurons from human embryonic stem cells. Stem Cells , 29, 802-811.

[34] Eiraku, M, Watanabe, K, Matsuo-takasaki, M, Kawada, M, Yonemura, S, Matsumura, M, Wataya, T, Nishiyama, A, Muguruma, K, & Sasai, Y. (2008). Self-organized formation of polarized cortical tissues from ESCs and its active manipulation by extrinsic signals. Cell Stem Cell , 3, 519-532.

[35] Hu, B. Y, Du, Z. W, & Zhang, S. C. (2009). Differentiation of human oligodendrocytes from pluripotent stem cells. Nat. Protoc. , 4, 1614-1622.

[36] Kang, S. M, Cho, M. S, Seo, H, Yoon, C. J, Oh, S. K, Choi, Y. M, & Kim, D. W. (2007). Efficient induction of oligodendrocytes from human embryonic stem cells. Stem Cells , 25, 419-424.

[37] Keirstead, H. S, Nistor, G, Bernal, G, Totoiu, M, Cloutier, F, Sharp, K, & Steward, O. (2005). Human embryonic stem cell-derived oligodendrocyte progenitor cell transplants remyelinate and restore locomotion after spinal cord injury. J. Neurosci. , 25, 4694-4705.

[38] Nistor, G. I, Totoiu, M. O, Haque, N, Carpenter, M. K, & Keirstead, H. S. (2005). Human embryonic stem cells differentiate into oligodendrocytes in high purity and myelinate after spinal cord transplantation. Glia , 49, 385-396.

[39] Lee, G, Chambers, S. M, Tomishima, M. J, & Studer, L. (2010). Derivation of neural crest cells from human pluripotent stem cells. Nat. Protoc. , 5, 688-701.

[40] Wichterle, H, Lieberam, I, Porter, J. A, & Jessell, T. M. (2002). Directed differentiation of embryonic stem cells into motor neurons. Cell , 110, 385-397.

[41] Ben-hur, T, Idelson, M, Khaner, H, Pera, M, Reinhartz, E, Itzik, A, & Reubinoff, B. E. (2004). Transplantation of human embryonic stem cell-derived neural progenitors improves behavioral deficit in Parkinsonian rats. Stem Cells , 22, 1246-1255.

[42] Chambers, S. M, Fasano, C. A, Papapetrou, E. P, Tomishima, M, Sadelain, M, & Studer, L. (2009). Highly efficient neural conversion of human ES and iPS cells by dual inhibition of SMAD signaling. Nat. Biotechnol. , 27, 275-280.

[43] Cho, M. S, Lee, Y, Kim, E, Chung, J. Y, Cho, S, Kim, Y. H, Kang, D. -S, Lee, S. -M, Kim, H, Kim, M. -H, et al. (2008). Highly efficient and large-scale generation of functional dopamine neurons from human embryonic stem cells. Proc. Natl. Acad. Sci. USA , 105, 3392-3397.

[44] Cooper, O, Hargus, G, Deleidi, M, Blak, A, Osborn, T, Marlow, E, Lee, K, Levy, A, Perez-torres, E, Yow, A, & Isacson, O. (2010). Differentiation of human ES and Parkinson's disease iPS cells into ventral midbrain dopaminergic neurons requires a high activity form of SHH, FGF8a and specific regionalization by retinoic acid. Mol. Cell. Neurosci. , 45, 258-266.

[45] Lee, S. H, Lumelsky, N, Studer, L, Auerbach, J. M, & Mckay, R. D. (2000). Efficient generation of midbrain and hindbrain neurons from mouse embryonic stem cells. Nat. Biotechnol. , 18, 675-679.

[46] Chambers, S. M, Fasano, C. A, Papapetrou, E. P, Tomishima, M, Sadelain, M, & Studer, L. (2009). Highly efficient neural conversion of human ES and iPS cells by dual inhibition of SMAD signaling. Nat. Biotechnol. , 27, 275-280.

[47] Hatch, M. N, Nistor, G, & Keirstead, H. S. (2009). Derivation of high-purity oligodendroglial progenitors. Methods Mol. Biol. , 549, 59-75.

[48] Kang, S. M, Cho, M. S, Seo, H, Yoon, C. J, Oh, S. K, Choi, Y. M, & Kim, D. W. (2007). Efficient induction of oligodendrocytes from human embryonic stem cells. Stem Cells , 25, 419-424.

[49] Sharp, J, Frame, J, Siegenthaler, M, Nistor, G, & Keirstead, H. S. (2010). Human embryonic stem cell-derived oligodendrocyte progenitor cell transplants improve recovery after cervical spinal cord injury. Stem Cells , 28, 152-163.

[50] Lee, G, Kim, H, & Elkabetz, Y. Al Shamy, G., Panagiotakos, G., Barberi, T., Tabar, V., and Studer, L. ((2007a). Isolation and directed differentiation of neural crest stem cells derived from human embryonic stem cells. Nat. Biotechnol. , 25, 1468-1475.

[51] Giudice, A, & Trounson, A. (2008). Genetic modification of human embryonic stem cells for derivation of target cells. Cell Stem Cell , 2, 422-433.

[52] Di GiorgioF.P., Boulting, G.L., Bobrowicz, S., and Eggan, K.C. ((2008). Human embryonic stem cell-derived motor neurons are sensitive to the toxic effect of glial cells carrying an ALS-causing mutation. Cell Stem Cell , 3, 637-648.

[53] Lakshmipathy, U, Pelacho, B, Sudo, K, Linehan, J. L, Coucouvanis, E, Kaufman, D. S, & Verfaillie, C. M. (2004). Efficient transfection of embryonic and adult stem cells. Stem Cells , 22, 531-543.

[54] Siemen, H, Nix, M, Endl, E, Koch, P, Itskovitz-eldor, J, & Brustle, O. (2005). Nucleofection of human embryonic stem cells. Stem Cells Dev. , 14, 378-383.

[55] Ben-dor, I, Itsykson, P, Goldenberg, D, Galun, E, & Reubinoff, B. E. (2006). Lentiviral vectors harboring a dual-gene system allow high and homogeneous transgene expression in selected polyclonal human embryonic stem cells. Mol. Ther. , 14, 255-267.

[56] Xia, X, Zhang, Y, Zieth, C. R, & Zhang, S. C. (2007). Transgenes delivered by lentiviral vector are suppressed in human embryonic stem cells in a promoterdependent manner. Stem Cells Dev. , 16, 167-176.

[57] Lacoste, A, Berenshteyn, F, & Brivanlou, A. H. (2009). An efficient and reversible transposable system for gene delivery and lineage-specific differentiation in human embryonic stem cells. Cell Stem Cell , 5, 332-342.

[58] Wilber, A, Linehan, J. L, Tian, X, Woll, P. S, Morris, J. K, Belur, L. R, Mcivor, R. S, & Kaufman, D. S. (2007). Efficient and stable transgene expression in human embryonic stem cells using transposon-mediated gene transfer. Stem Cells , 25, 2919-2927.

[59] Placantonakis, D. G, Tomishima, M. J, Lafaille, F, Desbordes, S. C, Jia, F, Socci, N. D, Viale, A, Lee, H, Harrison, N, Tabar, V, & Studer, L. (2009). BAC transgenesis in human embryonic stem cells as a novel tool to define the human neural lineage. Stem Cells , 27, 521-532.

[60] Nguyen, H. N, Byers, B, Cord, B, Shcheglovitov, A, Byrne, J, Gujar, P, & Kee, K. Schu ° le, B., Dolmetsch, R.E., Langston, W., et al. ((2011). LRRK2 mutant iPSC-derived DA neurons demonstrate increased susceptibility to oxidative stress. Cell Stem Cell , 8, 267-280.

[61] Illeva, H, Polymenidou, M, & Cleveland, D. W. (2009). Non-cell autonomus toxicity in neurodegenerative disorders: ASL and beyond J Cell Biol , 187, 761-772.

[62] Di Giorgio FPBoulting GL., bobrowicz S., Eggan KC ((2008). Human embryonic stem cell derived moor neurons are sensitive to the toxic effect of glial cells carrying an ASL-causing mutation. Cell Stem Cell 3; , 637-648.

[63] Marchetto, M. C, Mutri, A. R, Mu, Y, Smith, A. M, Cesar, G. G, & Gage, F. H. (2008). Non-cell autonomus effect of human SOD-1 G37R astrocytes on motro neuronsderived from human embryonic stem cells Cell Stem Cell 3; , 649-657.

[64] Colucci-D'Amato GLTino A, Pernas-Alonso R, ffrench-Mullen JM., di Porzio U. Neuronal and glial properties coexist in a novel mouse CNS immortalized cell line. Exp Cell Res. (1999). Nov 1;, 252(2), 383-91.

[65] Vescovi, A. L, Reynolds, B. A, Fraser, D. D, & Weiss, S. bFGF regulates the proliferative fate of unipotent (neuronal) and bipotent (neuronal/astroglial) EGF-generated CNS progenitor cells. Neuron. (1993). Nov;, 11(5), 951-66.

[66] Lobo, M. V, Alonso, F. J, Redondo, C, López-toledano, M. A, Caso, E, Herranz, A. S, Paíno, C. L, Reimers, D, & Bazán, E. Cellular characterization of epidermal growth factor-expanded free-floating neurospheres. J Histochem Cytochem. (2003). Jan;, 51(1), 89-103.

[67] Learish, R. D, Bruss, M. D, & Haak-frendscho, M. Inhibition of mitogen-activated protein kinase kinase blocks proliferation of neural progenitor cells. Brain Res Dev Brain Res. (2000). Jul 30;, 122(1), 97-109.

[68] Chang, D. W, Claassen, G. F, & Hann, S. R. Cole MD The c-Myc transactivation domain is a direct modulator of apoptotic versus proliferative signals. Mol Cell Biol. (2000). Jun;, 20(12), 4309-19.

[69] De Filippis, L, Lamorte, G, Snyder, E. Y, Malgaroli, A, & Vescovi, A. L. A novel, immortal, and multipotent human neural stem cell line generating functional neurons and oligodendrocytes. Stem Cells. (2007). Sep;, 25(9), 2312-21.

[70] Yamanaka, S. Induced pluripotent stem cells: past, present, and future. Cell Stem Cell. (2012). Jun 14;, 10(6), 678-84.

[71] Patel, M, & Yang, S. Advances in reprogramming somatic cells to induced pluripotent stem cells. Stem Cell Rev. (2010). Sep;, 6(3), 367-80.

[72] De Filippis, L, & Ferrari, D. Rota Nodari L, Amati B, Snyder E, Vescovi AL. Immortalization of human neural stem cells with the c-myc mutant T58A. PLoS One. (2008). Oct 2;3(10):e3310.

[73] Kim, J. H, Auerbach, J. M, Rodriguez-gomez, J. A, Velasco, I, Gavin, D, Lumelsky, N, Lee, S. H, Nguyen, J, Sanchez-pernaute, R, Bankiewicz, K, & Mckay, R. Dopamine neurons derived from embryonic stem cells function in an animal model of Parkinson's disease. Nature (2002). , 418, 50-56.

[74] Wernig, M, Benninger, F, Schmandt, T, Rade, M, Tucker, K. L, Bussow, H, Beck, H, & Brustle, O. Functional integration of embryonic stem cell-derived neurons in vivo. J Neurosci (2004). , 24, 5258-5268.

[75] Malberg, J. E, Eisch, A. J, Nestler, E. J, & Duman, R. S. Chronic antidepressant treatment increases neurogenesis in adult rat hippocampus. J Neurosci. (2000). Dec 15;, 20(24), 9104-10.

[76] Benninghoff, J, Gritti, A, Rizzi, M, Lamorte, G, Schloesser, R. J, Schmitt, A, Robel, S, Genius, J, Moessner, R, Riederer, P, Manji, H. K, Grunze, H, Rujescu, D, Moeller, H. J, Lesch, K. P, & Vescovi, A. L. Serotonin depletion hampers survival and proliferation in neurospheres derived from adult neural stem cells. Neuropsychopharmacology. (2010). Mar;, 35(4), 893-903.

[77] Brezun, J. M, & Daszuta, A. Depletion in serotonin decreases neurogenesis in the dentate gyrus and the subventricular zone of adult rats. Neuroscience. (1999). , 89(4), 999-1002.

[78] Huang, W, Zhao, Y, Zhu, X, Cai, Z, Wang, S, Yao, S, Qi, Z, & Xie, P. Fluoxetine Upregulates Phosphorylated-AKT and Phosphorylated-ERK1/2 Proteins in Neural Stem Cells: Evidence for a Crosstalk between AKT and ERK1/2 Pathways. J Mol Neurosci. (2012). Jun 7.

[79] Di Lieto A, Leo D, Volpicelli F, di Porzio U, Colucci-D'Amato L. FLUOXETINE modifies the expression of serotonergic markers in a differentiation-dependent fashion in the mesencephalic neural cell line A1 mes c-myc. Brain Res. (2007). Apr 27;, 1143, 1-10.

[80] Gentile, M. T, Nawa, Y, Lunardi, G, Florio, T, & Matsui, H. Colucci-D'Amato L. Tryptophan hydroxylase 2 (TPH2) in a neuronal cell line: modulation by cell differentiation and NRSF/rest activity. J Neurochem. (2012). Sep 7.

[81] Zhuang, P, Zhang, Y, Cui, G, Bian, Y, Zhang, M, Zhang, J, Liu, Y, Yang, X, Isaiah, A. O, Lin, Y, & Jiang, Y. Direct stimulation of adult neural stem/progenitor cells in vitro and neurogenesis in vivo by salvianolic acid B. PLoS One. (2012). e35636

[82] Kwon, MS, & Lee, . . Neuroprotective effect of visnagin on kainic acid-induced neural cell death in the mice hippocampus. Korean J of Physiol Pharmacol. 2010; 14: 257-63

[83] Chae, H. S, Kang, Y. K, Shin, Y. K, Lee, H. J, Yu, J. L, Lee, K. G, Yeo, J. H, Kim, Y. S, Sohn, D. S, Kim, K. Y, Lee, W. B, Lee, S. H, & Kim, S. S. The role of BF-7 on neuroprotection and enhancement of cognitive function. Korean J of Physiol Pharmacol. (2004). , 8, 173-79.

[84] Kong, P. J, Kim, Y. M, Lee, H. J, Kim, S. S, Yoo, E. S, & Chun, W. neuroprotective effects of methanol extracts of Jeju native plants on hydrogen peroxide- induced cytotoxicity in SH-SY5Y human neuroblastoma cells. Korean J of Physiol Pharmacol (2007). , 11, 170-74.

[85] King, C. C. Culture and preparation of human embryonic stem cells for proteomics-based applications. Methods Mol Biol. (2010). , 584, 151-77.

Section 2

Neural Stem Cells and Neurogenesis

Chapter 6

Postnatal Neurogenesis in the Subventricular Zone: A Manipulable Source for CNS Plasticity and Repair

Manavendra Pathania and Angelique Bordey

Additional information is available at the end of the chapter

http://dx.doi.org/10.5772/55679

1. Introduction

Neurogenesis is the production of new nerve cells or neurons, a specialized class of cells that make up the functional components of the central nervous system (CNS). Throughout most of the CNS this process of neurogenesis is limited to the developmental period before birth, after which time no new cells are added to the pre-established circuitry. In mammals including humans however, neurogenesis persists into the early postnatal period in two discrete brain regions: the subgranular zone in the hippocampus and the subventricular zone (SVZ) lining the lateral ventricles [1-5]. It is unknown why neurogenesis continues in such discrete locations yet is excluded from most other brain regions. A finite number of neurons is thought to afford us a stable set of circuitry that is able to accumulate and assimilate experiential information throughout our lifetimes. However, this predetermined number of neurons is also our Achilles' heel as any accidental or pathological damage to the CNS often results in irreparable damage to neurons. Consequently, there is a great unmet need for endogenous sources of brain repair, for conditions such as neurodegenerative disorders, cognitive neurological impairments, epilepsy, and cancer. This reason is one of the primary driving forces behind the study of postnatal neurogenesis and it is hoped that once the mechanisms are understood, this process can be harnessed to provide therapeutic avenues for intractable neuropathologies. Considerable progress has been made in the last twenty years to unravel the mechanisms that both define and limit postnatal neurogenesis.

In this chapter we will limit our discussion to therapeutically relevant regulators of SVZ neurogenesis. We will first start with potentially manipulable intracellular and extracellular factors that have been found to control SVZ neurogenesis. Then, we will evaluate the signaling cascades downstream of neurotransmitter receptor activity that have also been shown to play regulatory roles within the SVZ. Finally, we will discuss the neuropathologies in which SVZ neurogenesis has been implicated.

In the two neurogenic regions, resident pools of slow-dividing, astrocyte-like stem cells generate highly proliferative transit amplifying progenitors that then give rise to fate-committed neuronal or glial precursors [1-4]. These precursors (called neuroblasts or glioblasts) then migrate to their final destinations and differentiate into postmitotic neurons, astrocytes and oligodendrocytes. In the case of neurogenesis, newborn neurons undergo additional maturation steps to develop an appropriate dendritic morphology, receive and form synapses, and survive an activity-dependent competitive process to integrate into the local circuitry [5]. Thus, neurogenesis may be broadly divided into two phases: an early phase that includes (a) stem cell proliferation, neuronal vs. glial fate commitment and migration; and a later phase that involves the (b) morphological and synaptic development and survival of newborn neurons. Throughout this chapter, astrocytic stem cells are defined as the self-renewing, multipotent cells present in the SVZ that express glial markers, including the glutamate aspartate transporter (GLAST), the intermediate filament proteins glial fibrillary acidic protein (GFAP) and nestin, and the carbohydrate Lewis X (Lex) [1;6]. Intermediate progenitors, also called transit amplifying progenitors, express epidermal growth factor receptor (EGFR) and mammalian achaete-scute homolog 1 (Ascl1/Mash1), while neuroblasts, or neuronal precursors, are defined by the presence of immature neuronal markers, including doublecortin (DCX) and βIII-tubulin (Tuj1) [7-10].

In the SVZ, astrocytic stem cells reside in-between the striatal parenchyma and the ependymal cell layer that lines the lateral ventricles [11]. These stem cells generate rapidly dividing transit amplifying progenitors, which give rise to neuroblasts that migrate by moving tangentially through the rostral migratory stream (RMS) and into the olfactory bulb (OB) [12-14]. During migration these proliferative cells travel in chains, and are ensheathed by specialized astrocytes [15]. In the OB the neuroblasts exit the RMS, change direction and migrate radially outward to differentiate into dopaminergic and GABAergic periglomerular and granule interneurons [16-20]. Granule cells form dendrodendritic reciprocal synapses with mitral and tufted neurons in the bulb and inhibit their activity to fine-tune their output, ultimately playing a role in olfactory discrimination and learning [21]. Periglomerular cells make one-way as well as reciprocal synapses with the apical dendrites of the mitral/tufted cells and the terminals of the olfactory nerves that converge into the glomeruli [20;22;23]. The progression from astrocytic stem cell to neuronal progenitor to synaptic integration requires tightly coordinated, complex regulation by a multitude of factors.

2. Progenitor proliferation, fate commitment and migration in the SVZ

2.1. Intracellular factors

SVZ neurogenesis is subject to tight regulation, confined to isolated microenvironments and sensitive to neuronal activity, stress, and aging. This control may be required to prevent network instability, maintain experience-driven memory and behavioral patterns, and prevent tumorigenesis. To this end, cell-intrinsic factors comprise a major component of this regulation and help coordinate neurogenesis by forming the bulwark of how cells make choices and adapt to changes in their environments. Importantly, these intracellular factors may represent clinically tractable opportunities for the treatment of neuropathologies.

In the postnatal SVZ, progenitor fate commitment is sequentially driven by a family of proneural proteins called the basic helix-loop-helix (bHLH) transcription factors, such as Mash1/Ascl1, neurogenin2 (Ngn2), Neuro-D1, Neuro-D2, Tbr1 and Tbr2, in a pattern similar to cortical and hippocampal neurogenesis [24;25]. This bHLH-driven fate commitment is thought to develop in a temporally successive fashion, from more broadly proneural proteins (Mash1/Ascl1) to more neuronal and subtype-specifying (Neuro-D1) [26]. Although more work needs to be done to fully flesh out the role of bHLH proteins in the postnatal SVZ, *in vitro* transient expression of bHLH proteins is sufficient to induce neuronal fate commitment.

Recently, a synthetic small molecule isoxazole 9 (Isx-9) has shown promise in enhancing hippocampal neurogenesis *in vivo*, by targeting a family of regulatory interactors of Mash1/Ascl1, the myocyte-enhancer family (Mef2) [27]. This study heralds future therapies that can target and harness specific intracellular pathways within the neurogenic niche to direct neurogenesis.

The bHLH proteins are thought to induce differentiation in part by activating cyclin-dependent kinase inhibitors that then induce cell cycle exit. This finding has been shown in culture and awaits confirmation *in vivo* [28]. Cdk inhibitors p27KIP1 and p19INK4d have also been shown to modulate proliferation in the SVZ. Mice lacking p27KIP1 show increased progenitor proliferation and a reduction in neuroblast number [29;30]. Furthermore, mice deficient for both p27KIP1 and p19INK4d display renewed proliferation of post-mitotic neurons and increased cell death- causing seizures, movement disorders, and death by postnatal day 18 [31]. This indicates that cdk inhibitors positively regulate cell cycle exit in SVZ progenitors and their absence prolongs or renews cell cycling. Cdk inhibitors mediate cell cycle exit by inhibiting phosphorylation of the retinoblastoma protein (Rb) [32]. Dephosphorylated Rb binds to and sequesters the E2F transcription factor that normally functions in the nucleus to positively regulate cell cycle progression, thereby promoting cell cycle exit. In the SVZ, E2F-deficient mice show reduced progenitor proliferation and neuroblast numbers, suggesting that E2F transcriptional activity is required for cell cycle progression and the maintenance of neurogenic ability [33]. Additionally, mice lacking the gene encoding the tumor suppressor p53, another cell cycle regulator, displayed enhanced proliferative capacity and increased differentiation into neurons and oligodendrocytes [34]. Taken together, these results outline a stereotyped program of cell-intrinsic mechanisms at work within the SVZ to regulate the proliferation and fate commitment of SVZ progenitors in the early steps of postnatal neurogenesis. The development of therapies that can target these pathways within neurogenic niches is the next step for endogenous sources of brain repair.

MicroRNAs (miRs) are short, non-coding, single-stranded RNA molecules approximately 19-23 nucleotides in length that regulate gene expression by binding to complementary elements in the untranslated regions of target mRNAs and inhibiting protein synthesis. They exert epigenetic control either to maintain the status quo in a cell, (*i.e.* to maintain tissue identity), or they act in dynamic processes occurring within cells to refine and sharpen transitional states, (*i.e.* facilitating the switch in expression profile occurring at active synapses that were previously silent) [35-37]. Their role in shaping the temporal dynamics and phenotypic outcome of gene regulatory networks means that the functions of particular miRs become especially resolvable in plastic processes like postnatal neurogenesis, and miRs have recently been implicated in regulating the fate commitment of SVZ progenitors [38]. Although as yet untested, one idea for how bHLH proteins are sequentially activated and repressed and the transitions from multipotent progenitor to post-mitotic neuron more sharply defined is the successive miR-mediated downregulation of targeted bHLH proteins. In the SVZ, microRNA-124 was shown to be upregulated upon neuronal differentiation in the SVZ, and overexpression increased the number of mature neurons at the cost of proliferative progenitors and neuroblasts, while knockdown decreased mature neuron production and increased the number of glia [38]. MiR-124 was shown to mediate these effects by antagonizing Sox9, a transcription factor that directs glial differentiation. MiR-124 has also been shown to target SCP1, a component of the REST/NRSF complex that represses neuronal genes in non-neuronal cells and PTBP1, a repressor of neuron-specific alternative splicing [39-42].

Together these results indicate that miR-124 is a critical regulator of neuronal development and tissue identity, acting early on in the shift from neuronal precursor to mature neuron. Synthetic miR anologues composed of locked nucleic acid technology (LNA) demonstrate a robust half-life and good tolerance in animal models and may prove a good strategy to induce neurogenesis in a therapeutic setting. Furthermore, transplantable cells stably expressing a complement of miRs may prove beneficial in cellular replacement strategies. Cell-intrinsic factors and their roles in the early phases of postnatal neurogenesis have only begun to become amenable to experimental dissection, and exciting developments are forthcoming.

2.2. Extracellular factors

Neurotrophic factors have long been implicated in the dynamic regulation of postnatal neurogenesis. In the SVZ, fibroblast growth factor 2 (FGF-2) has been shown to affect the early steps in neurogenesis, positively regulating progenitor proliferation and leading to an increase in the number of neurons migrating from the SVZ into the OB [43;44]. Furthermore, FGF-2 is known to interact with epidermal growth factor (EGF) receptor signaling in neuronal progenitors, wherein prior FGF exposure is necessary for progenitors to respond to EGF or transforming growth factor β (TGFβ), the endogenous ligand for EGF receptors [45-49]. EGF has been shown to also increase proliferation specifically in the SVZ transit amplifying progenitor population at the cost of neuronal differentiation. Exposure to EGF induces transit amplifying progenitors to increase their cycling and downregulate neurogenic markers [50]. Vascular endothelial growth factor (VEGF), an angiogenic protein that can produce neurotrophic effects, has been shown to stimulate proliferation and neuroblast production in the SVZ, while insulin-like growth factor-I (IGF-I), a growth factor implicated in mediating the positive effects of exercise on adult neurogenesis, has been also implicated in enhancing proliferation and migration in the postnatal SVZ [51;52].

Apart from modulating proliferation, diffusible factors have also been shown to affect the initial establishment of the neurogenic niche itself, by influencing stem cell self-renewal and cell fate decisions. The factors that regulate such homeostatic effects within the niche may comprise a separate class of growth factors and unlike FGF-2, VEGF and IGF-I, are not circulating systemically in the bloodstream at appreciable levels nor produced in an acute, dynamically regulated manner. Based on this hypothesis, sonic hedgehog (Shh), Wnts, bone morphogenic proteins (BMPs) and Notch/Delta are extracellular factors possibly acting in a more local fashion. Sonic hedgehog signaling is required for establishing and maintaining the quiescent pool of stem cells in the postnatal SVZ. Ablating Smoothened, the transmembrane protein required for hedgehog signaling, specifically in the SVZ results in the depletion of stem cells and proliferative transit amplifying progenitors by postnatal day 8 and the depletion of neuroblasts by postnatal day 30 [53]. The Notch-DSL (Delta/Serrate/LAG-2) pathway is a very highly conserved cell-cell signaling system that acts through single-pass transmembrane proteins. Binding of Notch to its ligand causes cleavage of an intracellular domain that translocates to the nucleus where it interacts with transcriptional regulators to initiate expression of target genes like Hes1 and Hes5. In mammals, Notch ligands like Delta-like and Jagged bind Notch on stem and proliferative cells to maintain self-renewal and prevent terminal differentiation. In the neonatal SVZ, retrovirally delivered activated Notch enhances the numbers of quiescent SVZ progenitors at the cost of migratory neuroblasts [54]. Another study shows that conditional ablation of Notch signaling in the ependymal cells reprograms these cells and enables them to leave their position in the epithelium, take on SVZ stem cell characteristics and differentiate into granule and periglomerular neurons in the OB [55]. BMPs, a family of growth factors within the transforming growth factor β superfamily, have been shown to instruct a glial lineage in SVZ stem cells, and noggin, a BMP antagonist secreted by the adjacent ependymal cells, blocks glial differentiation of stem cells in favor of neurogenesis [56]. However, deletion of Smad4, a downstream target of BMP signaling, instead of increasing neurogenesis has been shown to result in oligodendrocyte production and a neurogenic deficit [57]. Therefore, it seems that BMP signaling can have divergent effects in the SVZ.

Exciting justification for these disparate lines of research is now beginning to emerge. Small molecule inhibitors of glycogen-synthase kinase 3 (GSK-3) inhibitors have shown promise in enhancing neurogenesis in human neural progenitor cells [58]. GSK-3 is involved in the notch, Shh, Wnt/β-catenin and FGF signaling pathways and represents a movement of the field toward a more clinically oriented direction.

In the SVZ, neuroblasts have to migrate a greater distance than do their analogues in the dentate gyrus, to arrive at their ultimate destinations in the OB. Furthermore, the migration behavior exhibited by SVZ neuroblasts has two phases; it is tangential from the SVZ to the RMS-OB, and then becomes radial-like as the neuroblasts exit the RMS and begin synaptic integration into the granule cell layer. For our purposes in this review both phases of this migration are being considered within the "early" phase of neurogenesis, prior to dendritic arborization, reception of synaptic inputs, and survival. This feature of SVZ neurogenesis distinguishes it from hippocampal neurogenesis where the neuroblasts migrate very short distances, and has allowed for the isolation of a few important guidance molecules that enable this directed migration. Slit-1 and Slit-2 expression in the SVZ and septum is thought to repel neuroblasts away from the SVZ and towards the OB, while netrin expression in the mitral cells of the OB, and the coincident expression of netrin receptors neogenin and deleted in colorectal cancer (DCC) in neuroblasts may form a chemoattractive cue drawing neuroblasts towards the OB [59-61]. The secreted protein prokineticin2 (PK2) has also been shown to act as potent chemoattractant for SVZ neuronal progenitors. PK2 is expressed in the OB, and attracts SVZ cells to the OB through the two G-protein coupled prokineticin receptors (PRK1 and PRK2] [62]. Both ephrins and their Eph tyrosine kinase receptors are expressed in the SVZ and have been shown to play a role in both regulation of progenitor proliferation and neuroblast migration. EphA7, EphB2 and ephrin-B2 are associated with astrocytic progenitors in the SVZ, while ephrin-A2 is expressed in the neuroblasts. Ephrin-B2 and EphB2 signaling seem to positively regulate progenitor proliferation while also disrupting neuroblast migration in the postnatal SVZ, while ephrin-A2 seemed to positively regulate progenitor proliferation [63;64]. Tangential migration in the RMS has also been shown to involve the $\alpha6\beta1$ integrin [65]. Once in the OB, neuroblasts have to reorient from tangential to radial migration into the GCL. This process has been shown to involve the expression of tenascin-R, in both the GCL and internal plexiform layers of the OB, and reelin, expressed by the mitral cells [66-68].

There is now a large amount of information regarding the effects of neurotrophic factors on early stages of SVZ neurogenesis, all of which is not discussed here. Many exciting avenues are emerging for therapeutic intervention into neurodegenerative diseases and psychiatric illnesses and knowledge of how neurogenic niches are formed, maintained, and neuronal and glial programs directed is fundamental for devising a clinical paradigm of directed neurogenesis. However, these data on extracellular factor-based modulation of postnatal neurogenesis needs more critical validation within the context of *in vivo* experiments and behavioral analyses.

2.3. Dopamine

The SVZ is innervated by dopaminergic fibers originating in the substantia nigra, while the SGZ is innervated by dopaminergic fibers coming from the ventral tegmental area (VTA). Dopaminergic signaling has been shown to regulate progenitor proliferation through D2 receptors in both the SVZ and SGZ and D3 receptors in the SVZ [69-71]. In patients with Parkinson's Disease, SVZ proliferation is markedly reduced. This effect on proliferation has been shown to be mediated through the induction of EGF and CNTF secretion from SVZ stem cells in response to dopaminergic activity [72;73]. Dopaminergic deafferentiation reduces proliferation in the SVZ, and one study reports that this decrease in overall SVZ cell proliferation is nonetheless accompanied by an increase in numbers of cells expressing Pax6 in the dorsal SVZ. Pax6 is a transcription factor responsible for enabling a dopaminergic differentiation program in postnatally generated periglomerular neurons. Therefore, dopaminergic activity may not only affect proliferation but may also impact cell fate choice in the SVZ [74]. Drugs that potentiate dopamingergic signaling may represent one strategy to maintain neurongenesis postnatally. Dopaminergic regulation of postnatal neurogenesis is only beginning to be uncovered, and its role in neurogenesis has not been conclusively established as yet.

2.4. GABA

GABA, a major inhibitory neurotransmitter in the mature CNS, has a well-established role in the development of neuronal circuits in both the embryo and adult [8;75-77]. Its ambient release in the form of spillover from synaptic and extra-synaptic sources has led researchers to its role in regulating the functional integration of new neurons into both immature and mature networks. Owing to the initial abundance of the Cl^- importer NKCC1 and the low expression of the Cl^- exporter KCC2, internal Cl^- concentrations are higher in immature neurons than mature neurons [76]. The resultant high equilibrium potential for Cl^- in neuroblasts causes GABA, acting through $GABA_A$ receptors, to depolarize immature cells in the first few weeks after fate determination. The depolarizing effect of GABA on young neurons and progenitors has been shown to regulate key stages of neurogenesis such as proliferation, migration and morphogenesis, in both the embryonic and adult neurogenic zones [78-80].

Adult neurogenesis in the SVZ recapitulates the embryonic role for GABA [8;81;82]. In the SVZ, both astrocytic stem cells and their neuroblast progeny express $GABA_A$ receptors [5;83]. Electrophysiological evidence indicates that neuroblasts release GABA in a non-synaptic and non-vesicular fashion, and that this tonically activates $GABA_A$ receptors on SVZ astrocytic stem cells [84;85]. The SVZ astrocytes also express GABA transporters that may further regulate levels of ambient GABA within the niche. Pharmacological inhibition of $GABA_A$ receptors in SVZ slice-culture preparations increases mitotic activity within the SVZ [85;86]. Blocking GABA transporters or enhancing GABA release from neuroblasts on the other hand, slows the speed of their own migration, in a paracrine/autocrine fashion in the SVZ and RMS [87]. Furthermore, knocking down Na-K-2Cl cotransporter NKCC1 and thereby reducing GABA(A)-induced depolarization in the SVZ reduced proliferation, migration as well dendrite development [79;80]. These data together suggest that GABA has a role as a negative regulator of early stages of neurogenesis in the SVZ, where it reduces neuroblasts and SVZ astrocyte proliferation and decreases the speed of neuroblast migration. This is analogous to the role of $GABA_A$ activation in the developing cortex, where it also serves to limit proliferation of ventricular zone progenitors and migration of postmitotic neuroblasts [88-90].

GABA's role in regulating early phases of neurogenesis such as proliferation and migration has been examined more extensively in the SVZ where it has been shown to act as a negative regulator of early neurogenesis. Whether these effects are corroborated in the SGZ is as yet unknown. An attractive hypothesis explaining GABA's disparate roles in development, postnatal neurogenesis, and at the synapse is that neurotransmitter-based signaling may serve as a bridge that brings an activity-dependence to cell-autonomous and locally present instructive signals that drive neurogenesis and network plasticity. In this way, neuronal and metabolic activity may loop back onto the SVZ and SGZ.

2.5. Glutamate

During embryonic neurogenesis, glutamate signaling has been shown to influence proliferation, fate commitment, and migration of newborn neurons [88;89;91-93]. During postnatal neurogenesis, in the SVZ neuroblasts have been shown to express functional NMDA receptors as well as functional mGluR5 and GLU_{k5}-containing kainate receptors, using both electrophysiology and calcium imaging [94;95]. Evidence from our lab suggests glutamate released spontaneously from SVZ-RMS astrocytes generates phasic NMDA receptor activity in neuroblasts migrating towards the OB [96]. Both mGluR5 and GLU_{k5} activation have also been shown to mediate increases in intracellular Ca^{2+} transients in SVZ neuroblasts [95;97]. A mosaic of $GABA_A$, NMDA, mGluR5, and GLU_{k5} (now known as GluK2) receptor-expressing cells reside in the SVZ, where most cells express $GABA_A$ receptors in caudal SVZ and moving rostrally, a greater proportion of cells begin to express a combination of receptors. Ultimately, nearly half of all cells in the rostral RMS express all four types of receptors, indicating the continuing maturation of newborn cells along the SVZ-OB neurogenic axis. Mice lacking mGluR5, or in which mGluR5 was pharmacologically blocked, displayed a marked decrease in the number of proliferating cells in the SVZ [98]. This indicates a role for glutamate -acting tonically through metabotropic receptors- in positively regulating SVZ progenitor proliferation and antagonizing tonic $GABA_A$-ergic receptor-induced anti-mitotic activity; perhaps acting as a positive regulator of early neurogenic processes. Blocking GLU_{k5} in the RMS on the other hand, increased the speed of neuroblast migration, suggesting that tonic GLU_{k5}-mediated glutamatergic transmission decreases neuroblast clearance from the SVZ and acts in concert with GABA's effect on migration in the SVZ [97]. mGluR5 activity however, does not influence migration speed. It could be that GLU_{k5}-mediated signaling activated different Ca^{2+}-dependent intracellular cascades than mGluR5 signaling. It remains to be seen whether AMPA/kainate or NMDA receptor activity can have a positive effect on migration in the SVZ/RMS. These data together suggest that although glutamate receptor heterogeneity and the multiple intracellular pathways they may activate introduce ambiguity into what role glutamate may play in early neurogenesis in the SVZ, metabotropic glutamate receptor signaling enhances proliferation, while AMPA/kainate receptor signaling acts together with GABA to decrease migration of neuroblasts. More work is needed to fully flesh out the roles that the three different glutamate receptor families (NMDA, mGluR and AMPA/kainate) have in the SVZ. Work also is needed to elucidate how glutamate receptor heterogeneity parses among the different SVZ sublineages (Emx-1, Gsh2, Nkx2.1).

The diversity of glutamate receptors, the myriad intracellular pathways that they may activate and the many mechanisms by which levels of ambient glutamate are regulated suggests that glutamate, despite being nearly ubiquitously present, can have very specific and differential effects on SVZ cells. The data so far suggests glutamate may regulate the early phases of neurogenesis in manner that reflects this complexity. However, further work will involve clarifying some of the associated ambiguity surrounding glutamate availability, the receptor complement, the different intracellular pathways, and their effects on neurogenesis.

3. Morphogenesis, synaptogenesis and circuit integration in the OB

3.1. Intracellular factors

Later stages of neurogenesis include the survival, synaptic integration and dendritic elaboration of neuronal precursors within their target sites. CREB (cAMP response element binding) is a long-studied transcription factor known for underlying the later stages of synaptic plasticity and memory formation, as well as for linking neuronal activity to survival. In the postnatal SVZ-OB, CREB has been shown to be important in the survival and dendritic arborization of SVZ neuroblasts [99]. CREB phosphorylation is transient and parallels maturation, increasing during migration towards the OB and decreasing once radial migration and synaptic integration are completed. CREB-deficient mice show deficits in neuroblast survival in the OB, and CREB inhibition *in vitro* severely attenuates neurite outgrowth, suggesting that CREB positively modulates survival and dendritic elaboration in the OB and plays an important role in the later phases of SVZ neurogenesis. Data from our lab and others suggests that a CREB-regulated microRNA, miR-132, is involved in mediating some of the effects seen by impairing CREB activity in the SVZ. miR-132 expression is upregulated along the migratory route of the SVZ neuroblasts, peaking in the OB, and miR-132 overexpression enhances morphological complexity, spine density and survival of newborn neurons *in vivo* [100]. These data suggest that CREB and a CREB-regulated miRNA may form the basis of a structural plasticity program seen in SVZ postnatal neurogenesis. Intrinsic mechanisms regulating later stages of neurogenesis are some of the least elaborated aspects of postnatal neurogenesis. Additionally, with the emergence of inducible and conditional manipulation techniques, it has become possible to discretely assay the roles of many factors within the context of postnatal neurogenesis. Work is also emerging that utilizes the stop-flox-mediated overexpression of factors in a conditional and inducible manner. MiR-132 and other recently identified miRs that promote synapse maintenance because of their ease of delivery represent therapeutic strategies for the stable maintenance of newly generated neurons in disease states.

3.2. Extracellular factors

Later stages of postnatal neurogenesis have also been shown to be responsive to neurotrophic factor signaling. A single nucleotide polymorphism in the human brain derived neurotrophic factor (BDNF)-encoding gene (Val66Met) has been shown to correlate with mood disorders and memory deficits, and knock-in mice possessing the human SNP showed reduced activity-dependent BDNF secretion ultimately resulting in reduced survival of SVZ neuroblasts and impaired spontaneous olfactory discrimination [101]. In this study, activity-dependent BDNF signaling in the SVZ was shown to exert its effects on survival and olfactory function through TrkB receptors on neuroblasts. BDNF signaling and Trk receptor activity have been widely shown to have neurotrophic and synapse-potentiating effects in neurons and may represent a general strategy to promote the survival and maintenance of newly generated neurons.

Later stages of neurogenesis are poorly studied in the SVZ. Knowledge of molecules regulating the survival, synaptic integration and morphogenesis of newborn cells is more limited in comparison to the literature covering the DG. However, BDNF-signaling is an example of emerging data within the field that unites hypotheses between the two neurogenic niches. It is also interesting to note the sustained differences between the two niche microenvironments. NT-3-signaling is exclusive to the DG and may promote excitatory versus inhibitory neurogenic potential. It has also been suggested that the convergence of dopaminergic and serotonergic fibers defines the SVZ, while convergence of noradrenergic and serotonergic projections may define the SGZ [68].

3.3. GABA

GABA-mediated depolarization of immature neurons has been shown to be critical for synapse formation in the developing cortex [102]. Postnatally, following migration into the OB SVZ neuroblasts begin the process of integrating into the local circuitry by radially migrating out of the RMS core, elaborating a complex dendritic structure and establishing appropriate synapses [5;103]. A role for $GABA_A$ signaling in the initiation, elongation and stabilization of dendritic structures in immature neurons has been established in the OB. Specifically, it was discovered that ambient GABA-induced depolarization and Ca^{2+}-influx was necessary for the stabilization of emerging dendritic protrusions and enhanced the number and length of preexisting dendrites, in SVZ culture as well as OB slice preparations [104]. This effect was specific to the immature neuron population because six days following plating, KCC2 levels had increased sufficiently to block the depolarizing effects of GABA and the modulatory effects of GABA-depolarization on dendrite development. Furthermore, GABA activity promoted initiation and elongation of immature neuroblast dendrites in culture by stabilizing tubulin in its polymerized form. Knockdown of NKCC1 and prevention of GABA-mediated depolarization in immature neurons also resulted in dendritic morphological deficits. However, this effect was transient and dendritiric morphology recovered in adults [80].

This regulation by GABA in both the OB and DG helps shape neurogenesis as an activity-dependent process where GABA is involved in regulating later stages of postnatal neurogenesis orchestrating synapse formation and dendritic outgrowth. However, although GABA's role in the synaptic integration of postnatally generated neurons is becoming clearer, more work is needed to fully flesh out the internal mechanisms by which GABA activity leads to modulation of actions as disparate as proliferation, migration, synaptic integration, and dendritogenesis.

3.4. Glutamate

Glutamate has been shown to be important for neuroblast survival, dendritic development, and synaptogenesis in the developing CNS [105]. In the postnatal SVZ, spontaneous glutamate release from astrocytes onto neuroblasts results in phasic NMDAR activation that increases in frequency and amplitude upon migration towards the bulb. Genetic ablation of the NR1 subunit in migrating neuroblasts results in 60% of these NR1-deficient newborn neurons entering apoptosis, suggesting that NMDAR-dependent glutamatergic signaling is an important factor in regulating neuroblast survival and numbers of new neurons in the OB [106]. Once in the bulb, newly generated neurons begin to integrate synaptically into the local circuitry, generate action potentials and first establish GABAergic inputs followed by glutamatergic inputs ~4 weeks after birth [103]. Recently, newborn granule cells in the OB were shown to express a transient form of LTP in response to focal glutamatergic stimulation in the granule cell layer. This type of LTP was not present in mature granule cells and was observed in cells between 2 and 8 weeks old, implying that new neurons have a capacity for synaptic plasticity that is different from their mature counterparts [107]. Perhaps this sort of synaptic enhancement can help explain the positive effects olfactory learning has on SVZ neuroblast survival, as well as the negative effects anti-mitotic activity in the SVZ has on olfactory discrimination.

4. Responsiveness and involvement of postnatal neurogenesis in distinct neuropathologies

4.1. Alzheimer's disease

Alzheimer's disease (AD) is a late-onset neurological disease with a heritable component characterized by deposition of β-amyloid peptides (Aβ), formation of neurofibrillary tangles, reactive astrocytosis, activation of microglial cells and cholinergic deficits [108]. The effect of AD is a progressive neurodegeneration throughout the neocortex and hippocampus, and severe dementia [109]. SVZ neurogenesis is reduced in mouse models of AD and has also been shown to be diminished in postmortem tissue from human AD patients. Mice harboring familial mutations in amyloid precursor protein (APP) and presinilin 1 (PS1) show decreased proliferation in the SVZ [110;111]. Infusion of Aβ peptide into the lateral ventricles also decreases proliferation in the postnatal SVZ [112]. SVZ-derived neural progenitor cells from PS1 mutants and the APPSwe/PS1ΔE9 double-mutants showed decreased cycling *in vitro* [112; 113]. In postmortem tissue, AD patients showed decreased numbers of proliferative ($Ki67^+$) cells in the SVZ [114]. Anosmia or hyposmia, the inability or reduced ability to perceive smell, are predictive indicators of Alzheimer's progression in the clinic. However, it remains to be conclusively established whether altered SVZ neurogenesis is the cause of this disrupted olfaction in AD patients.

4.2. Parkinson's disease

Parkinson's disease (PD) develops due to the specific loss of dopaminergic neurons in the substantia nigra (SN) and results in impaired regulation of movement, mood, and motivation [69]. In mouse models of PD SVZ proliferation is reduced. This is thought to be due to the loss of dopaminergic inputs to the SVZ from the SN via the nigrostriatal pathway, as chemical ablation of these fibers results in decreased proliferation in the SVZ and decreased numbers of mature granule neurons in the OB [69;74]. This effect on proliferation was partially rescued with the application of the dopamine precursor levodopa. Furthermore, in postmortem tissue from human AD patients SVZ proliferation was reduced, as were numbers of immature neurons in the granule cell layer of the OB [69]. However, increases in numbers of periglomerular dopaminergic neurons have also been reported in mouse models of PD using chemical ablation, and in PD postmortem human tissue [74;115]. Because dopaminergic activity of periglomerular cells generally inhibits the transmission of olfactory information, it is thought that the decreased numbers of granule cells and the increased numbers of periglomerular cells together contribute to the hyposmia and disturbed olfaction seen in PD patients.

4.3. Huntington's disease

Huntington's disease (HD) is caused by expansions in CAG repeat elements in the gene encoding huntingin. This leads to aggregation of mutant huntingtin and neurodegeneration in the striatum [116]. In mouse models of HD there is little striatal neurodegeneration and consequently SVZ neurogenesis remains unchanged. However, in rat models of striatal degeneration SVZ proliferation is increased [117;118]. Some cells from the SVZ are seen to ectopically migrate into the damaged striatum and begin expressing markers of newborn neurons, although any functional recovery was not reported [118]. In HD patients an increase in proliferation in the SVZ is observed that corresponds with the number of CAG repeats, and SVZ cells are seen to migrate into the damaged striatum where they express both proliferative as well as mature neuronal markers. It remains to be seen whether the expression of both proliferative and mature markers in SVZ-derived cells within the HD-damaged striatum is symptomatic of HD or in fact, can contribute to functional recovery [119-121]. However, the potential for endogenous repair for HD can still be seen as promising, as newborn neurons would take a long time to develop huntingtin inclusions and in the meantime participate in the maintenance of striatal circuitry.

4.4. Ischemic stroke

A stroke results from either a hemorrhage or blocked cerebral arteries, leading to diminished local blood flow (ischemia) in a brain region and loss of neurons. In stroked tissue, the core infarcted area is distinguishable from the surrounding penumbral area by the exaggerated necrosis and little potential for regeneration. In the penumbral region on the other hand, neuronal regeneration has been demonstrated as it is perfused by collateral arteries and not wholly dependent on the occluded artery for oxygen. As ischemic stroke is one of the most frequent causes of mortality in industrialized countries, a lot of research has been undertaken to probe the capacity for regeneration in this condition. In rodent and primate models of stroke where the medial cerebral artery is occluded (MCAO), SVZ proliferation and the numbers of neurons in the OB are increased [122-124]. In addition, ectopic neurogenesis is also observed in the penumbral areas, such as the striatum [125]. Some groups have also reported ectopic neurogenesis in cortical regions following stroke (Gu 2000, sun 2003) but this has been denied by others [125]. In the stroked striatum, SVZ-derived cells differentiate into medium spiny GABAergic neurons which represent 90% of striatal neurons and are lost there, although once in the striatum many of the newborn neurons undergo cell death [125]. However, this finding is greatly encouraging for the continued study of an effective neuronal replacement strategy as a means to treat stroke damage in the CNS. As proof of this idea, ablating neurogenesis in mouse models of stroke greatly exacerbated cell death and postischemic sensorimotor deficits, suggesting that neurogenesis can account for some amelioration of stroke-induced damage [126]. In human stroke patients, increased proliferation has been observed in the ipsilateral SVZ and traces of ectopic neurogenesis were seen in the cortex [127-129]. The functional recovery that this observed increase in neurogenesis following stroke is able to accomplish remains to be validated, but it does suggest that some measure of recovery is endogenously possible and may be drawn out with continued research and more-tailored therapeutic intervention.

4.5. Epilepsy

Epilepsy has also been shown to alter SVZ neurogenesis. In rats, pilocarpine-induced seizures increased SVZ proliferation as well as expanded the extent of the RMS. Ectopic migration and

http://dx.doi.org/10.5772/55679

increased immature neurons were also observed [130]. In humans, increases proliferation and ectopic migration have also been observed in organotypic slice preparations [131]. These effects on neurogenesis seem to be symptomatic of epilepsy, whether they can be harnessed as a way to treat the damage caused by repeated seizure activity remains to be seen.

4.6. Precancerous lesions and cortical heterotopias

It has long been suggested that the SVZ is the source of origin for malignant gliomas. The prognosis for these cancers is very poor and for glioblastoma, the most common variant in adults, the median survival rate is only 9-12 months [132-134]. A few years ago, in a mouse model of malignant astrocytoma that included a p53 deletion and a conditional disruption of the neurofibromatosis type 1 (NF1) gene, researchers conclusively established that the originating tumorigenic mutation arises within the SVZ atrocyte-like stem cell [135]. Recently it was further shown that although the mutation arises in the neural stem cell, the cancer begins at a subsequent stage, when these stem cells have committed to the oligodendrocytic lineage [136]. Studies of the molecular characteristics of low-grade human astrocytomas suggested that most often in these conditions p53 is deleted and Ras signaling is elevated. Since NF1 is a negative regulator of Ras, its deletion would result in increased Ras activity and in conjunction with the p53 deletion more accurately model human astrocytomas. In mice, this genetic strategy produced astrocytomas with complete penetrance, suggesting that NF1 and p53 deletion are sufficient to induce cancer. When tumor development was closely followed in these mice, it was discovered to arise from the SVZ in nearly every instance before dispersing to other brain regions. These results demonstrate that therapeutic intervention utilizing SVZ neurogenesis must guarantee against the elevated risk of tumorigenesis, and that continued research is necessary to manipulate the proliferation, fate choice, migration and differentiation of SVZ progenitors. Perhaps one day a therapy can be conceived to induce cell death in progenitors that have become transformed into precancerous cell types.

It has also been recently demonstrated that SVZ dysfunction can contribute to the pathophysiology of neuropsychiatric conditions like tuberous sclerosis complex (TSC). TSC is caused by loss of either one of two tumor suppressor genes, *TSC1* and *TSC2*, which encode hamartin and tuberin, respectively. Mutations in these genes lead to hyperactivity of the mammalian target of rapamycin (mTOR) signaling pathway. Neurological symptoms of TSC include seizures, autism, psychiatric problems and the presence or subendymal nodules, heterotopias and giant, ectopically localized cells with both neuronal and glial characteristics. A recent study modeled TSC in mice by conditionally ablating *Tsc1* specifically in the postnatal SVZ. This produced ectopic migration and differentiation of neuronal precursors, resulting in hetertopias and micronodules containing neurons with a hypertrophic dendritic tree in aberrant locations. Furthermore, *Tsc1*-mutant cells were shown to be rerouted to forebrain structures where they differentiated into neurons and glia. This remarkable rerouting of SVZ cells to the cortex is thought to be occurring at a very low rate under normal circumstances, but becomes elevated when mTOR activity is pathogenically increased [137]. It is hypothesized that these ectopic cells in the cortex contribute to network malfunction in higher-order cognitive function. This research also opens up the exciting idea of actively rerouting cells to the cortex, or other desired brain regions, for directed endogenous circuit repair.

5. Conclusion

New neurons continue to be produced throughout life in two regions of the mammalian CNS and a plethora of research has accumulated demonstrating how this amazing propensity for plasticity is orchestrated and regulated. Postnatal SVZ neurogenesis has been shown to make important contributions to coordinated network activity in the OB as well as serving as a sensor for different neurological disease states. But most importantly, it continues to provide tantalizing potentials for a source of endogenous repair within the CNS.

Author details

Manavendra Pathania and Angelique Bordey*

*Address all correspondence to: angelique.bordey@yale.edu

Departments of Neurosurgery, and Cellular and Molecular Physiology, Yale University School of Medicine, New Haven, Connecticut, USA

References

[1] Doetsch F, Caille I, Lim DA, Garcia-Verdugo JM, Alvarez-Buylla A. Subventricular zone astrocytes are neural stem cells in the adult mammalian brain. Cell 1999 Jun 11;97(6):703-16.

[2] Imura T, Kornblum HI, Sofroniew MV. The predominant neural stem cell isolated from postnatal and adult forebrain but not early embryonic forebrain expresses GFAP. J Neurosci 2003 Apr 1;23(7):2824-32.

[3] Seri B, Garcia-Verdugo JM, McEwen BS, Alvarez-Buylla A. Astrocytes give rise to new neurons in the adult mammalian hippocampus. J Neurosci 2001 Sep 15;21(18): 7153-60.

[4] Alvarez-Buylla A, Lim DA. For the long run: maintaining germinal niches in the adult brain. Neuron 2004 Mar 4;41(5):683-6.

[5] Carleton A, Petreanu LT, Lansford R, Alvarez-Buylla A, Lledo PM. Becoming a new neuron in the adult olfactory bulb. Nat Neurosci 2003 May;6(5):507-18.

[6] Capela A, Temple S. LeX/ssea-1 is expressed by adult mouse CNS stem cells, identifying them as nonependymal. Neuron 2002 Aug 29;35(5):865-75.

[7] Platel JC, Gordon V, Heintz T, Bordey A. GFAP-GFP neural progenitors are antigenically homogeneous and anchored in their enclosed mosaic niche. Glia 2009 Jan 1;57(1):66-78.

[8] Bordey A. Adult neurogenesis: basic concepts of signaling. Cell Cycle 2006 Apr;5(7): 722-8.

[9] Parras CM, Galli R, Britz O, Soares S, Galichet C, Battiste J, et al. Mash1 specifies neurons and oligodendrocytes in the postnatal brain. EMBO J 2004 Nov 10;23(22): 4495-505.

[10] Tomita K, Moriyoshi K, Nakanishi S, Guillemot F, Kageyama R. Mammalian achaete-scute and atonal homologs regulate neuronal versus glial fate determination in the central nervous system. EMBO J 2000 Oct 16;19(20):5460-72.

[11] Doetsch F, Garcia-Verdugo JM, Alvarez-Buylla A. Cellular composition and three-dimensional organization of the subventricular germinal zone in the adult mammalian brain. J Neurosci 1997 Jul 1;17(13):5046-61.

[12] Altman J. Autoradiographic and histological studies of postnatal neurogenesis. IV. Cell proliferation and migration in the anterior forebrain, with special reference to persisting neurogenesis in the olfactory bulb. J Comp Neurol 1969 Dec;137(4):433-57.

[13] Lois C, Alvarez-Buylla A. Long-distance neuronal migration in the adult mammalian brain. Science 1994 May 20;264(5162):1145-8.

[14] De MS, Fasolo A, Shipley M, Puche A. Unique neuronal tracers show migration and differentiation of SVZ progenitors in organotypic slices. J Neurobiol 2001 Dec;49(4): 326-38.

[15] Lois C, Garcia-Verdugo JM, Alvarez-Buylla A. Chain migration of neuronal precursors. Science 1996 Feb 16;271(5251):978-81.

[16] Altman J, Das GD. Autoradiographic and histological studies of postnatal neurogenesis. I. A longitudinal investigation of the kinetics, migration and transformation of cells incorporating tritiated thymidine in neonate rats, with special reference to postnatal neurogenesis in some brain regions. J Comp Neurol 1966 Mar;126(3):337-89.

[17] Hinds JW. Autoradiographic study of histogenesis in the mouse olfactory bulb. I. Time of origin of neurons and neuroglia. J Comp Neurol 1968 Nov;134(3):287-304.

[18] Kishi K. Golgi studies on the development of granule cells of the rat olfactory bulb with reference to migration in the subependymal layer. J Comp Neurol 1987 Apr 1;258(1):112-24.

[19] Brill MS, Ninkovic J, Winpenny E, Hodge RD, Ozen I, Yang R, et al. Adult generation of glutamatergic olfactory bulb interneurons. Nat Neurosci 2009 Nov 1.

[20] Kosaka K, Toida K, Aika Y, Kosaka T. How simple is the organization of the olfactory glomerulus?: the heterogeneity of so-called periglomerular cells. Neurosci Res 1998 Feb;30(2):101-10.

[21] Lledo PM, Alonso M, Grubb MS. Adult neurogenesis and functional plasticity in neuronal circuits. Nat Rev Neurosci 2006 Mar;7(3):179-93.

[22] Kosaka T, Hataguchi Y, Hama K, Nagatsu I, Wu JY. Coexistence of immunoreactivities for glutamate decarboxylase and tyrosine hydroxylase in some neurons in the periglomerular region of the rat main olfactory bulb: possible coexistence of gamma-aminobutyric acid (GABA) and dopamine. Brain Res 1985 Sep 16;343(1):166-71.

[23] Kosaka K, Kosaka T. synaptic organization of the glomerulus in the main olfactory bulb: compartments of the glomerulus and heterogeneity of the periglomerular cells. Anat Sci Int 2005 Jun;80(2):80-90.

[24] Roybon L, Deierborg T, Brundin P, Li JY. Involvement of Ngn2, Tbr and NeuroD proteins during postnatal olfactory bulb neurogenesis. Eur J Neurosci 2009 Jan;29(2): 232-43.

[25] Farah MH, Olson JM, Sucic HB, Hume RI, Tapscott SJ, Turner DL. Generation of neurons by transient expression of neural bHLH proteins in mammalian cells. Development 2000 Feb;127(4):693-702.

[26] Roybon L, Mastracci TL, Ribeiro D, Sussel L, Brundin P, Li JY. GABAergic Differentiation Induced by Mash1 Is Compromised by the bHLH Proteins Neurogenin2, NeuroD1, and NeuroD2. Cereb Cortex 2009 Sep 18.

[27] Petrik D, Jiang Y, Birnbaum SG, Powell CM, Kim MS, Hsieh J, et al. Functional and mechanistic exploration of an adult neurogenesis-promoting small molecule. FASEB J 2012 Aug;26(8):3148-62.

[28] Bally-Cuif L, Hammerschmidt M. Induction and patterning of neuronal development, and its connection to cell cycle control. Curr Opin Neurobiol 2003 Feb;13(1): 16-25.

[29] Doetsch F, Verdugo JM, Caille I, Alvarez-Buylla A, Chao MV, Casaccia-Bonnefil P. Lack of the cell-cycle inhibitor p27Kip1 results in selective increase of transit-amplifying cells for adult neurogenesis. J Neurosci 2002 Mar 15;22(6):2255-64.

[30] Coskun V, Luskin MB. The expression pattern of the cell cycle inhibitor p19(INK4d) by progenitor cells of the rat embryonic telencephalon and neonatal anterior subventricular zone. J Neurosci 2001 May 1;21(9):3092-103.

[31] Zindy F, Cunningham JJ, Sherr CJ, Jogal S, Smeyne RJ, Roussel MF. Postnatal neuronal proliferation in mice lacking Ink4d and Kip1 inhibitors of cyclin-dependent kinases. Proc Natl Acad Sci U S A 1999 Nov 9;96(23):13462-7.

[32] Weinberg RA. The retinoblastoma protein and cell cycle control. Cell 1995 May 5;81(3):323-30.

[33] Yoshikawa K. Cell cycle regulators in neural stem cells and postmitotic neurons. Neurosci Res 2000 May;37(1):1-14.

[34] Gil-Perotin S, Marin-Husstege M, Li J, Soriano-Navarro M, Zindy F, Roussel MF, et al. Loss of p53 induces changes in the behavior of subventricular zone cells: implication for the genesis of glial tumors. J Neurosci 2006 Jan 25;26(4):1107-16.

[35] Hornstein E, Shomron N. Canalization of development by microRNAs. Nat Genet 2006 Jun;38 Suppl:S20-S24.

[36] Giraldez AJ, Cinalli RM, Glasner ME, Enright AJ, Thomson JM, Baskerville S, et al. MicroRNAs regulate brain morphogenesis in zebrafish. Science 2005 May 6;308(5723):833-8.

[37] Schratt GM, Tuebing F, Nigh EA, Kane CG, Sabatini ME, Kiebler M, et al. A brain-specific microRNA regulates dendritic spine development. Nature 2006 Jan 19;439(7074):283-9.

[38] Cheng LC, Pastrana E, Tavazoie M, Doetsch F. miR-124 regulates adult neurogenesis in the subventricular zone stem cell niche. Nat Neurosci 2009 Apr;12(4):399-408.

[39] Visvanathan J, Lee S, Lee B, Lee JW, Lee SK. The microRNA miR-124 antagonizes the anti-neural REST/SCP1 pathway during embryonic CNS development. Genes Dev 2007 Apr 1;21(7):744-9.

[40] Conaco C, Otto S, Han JJ, Mandel G. Reciprocal actions of REST and a microRNA promote neuronal identity. Proc Natl Acad Sci U S A 2006 Feb 14;103(7):2422-7.

[41] Makeyev EV, Zhang J, Carrasco MA, Maniatis T. The MicroRNA miR-124 promotes neuronal differentiation by triggering brain-specific alternative pre-mRNA splicing. Mol Cell 2007 Aug 3;27(3):435-48.

[42] Boutz PL, Stoilov P, Li Q, Lin CH, Chawla G, Ostrow K, et al. A post-transcriptional regulatory switch in polypyrimidine tract-binding proteins reprograms alternative splicing in developing neurons. Genes Dev 2007 Jul 1;21(13):1636-52.

[43] Mudo G, Bonomo A, Di L, V, Frinchi M, Fuxe K, Belluardo N. The FGF-2/FGFRs neurotrophic system promotes neurogenesis in the adult brain. J Neural Transm 2009 Aug;116(8):995-1005.

[44] Wagner JP, Black IB, DiCicco-Bloom E. Stimulation of neonatal and adult brain neurogenesis by subcutaneous injection of basic fibroblast growth factor. J Neurosci 1999 Jul 15;19(14):6006-16.

[45] Kuhn HG, Winkler J, Kempermann G, Thal LJ, Gage FH. Epidermal growth factor and fibroblast growth factor-2 have different effects on neural progenitors in the adult rat brain. J Neurosci 1997 Aug 1;17(15):5820-9.

[46] Jin K, Sun Y, Xie L, Batteur S, Mao XO, Smelick C, et al. Neurogenesis and aging: FGF-2 and HB-EGF restore neurogenesis in hippocampus and subventricular zone of aged mice. Aging Cell 2003 Jun;2(3):175-83.

[47] Jin K, Xie L, Childs J, Sun Y, Mao XO, Logvinova A, et al. Cerebral neurogenesis is induced by intranasal administration of growth factors. Ann Neurol 2003 Mar;53(3): 405-9.

[48] Lillien L, Raphael H. BMP and FGF regulate the development of EGF-responsive neural progenitor cells. Development 2000 Nov;127(22):4993-5005.

[49] Burrows RC, Wancio D, Levitt P, Lillien L. Response diversity and the timing of progenitor cell maturation are regulated by developmental changes in EGFR expression in the cortex. Neuron 1997 Aug;19(2):251-67.

[50] Doetsch F, Petreanu L, Caille I, Garcia-Verdugo JM, Alvarez-Buylla A. EGF converts transit-amplifying neurogenic precursors in the adult brain into multipotent stem cells. Neuron 2002 Dec 19;36(6):1021-34.

[51] Hurtado-Chong A, Yusta-Boyo MJ, Vergano-Vera E, Bulfone A, de PF, Vicario-Abejon C. IGF-I promotes neuronal migration and positioning in the olfactory bulb and the exit of neuroblasts from the subventricular zone. Eur J Neurosci 2009 Sep;30(5): 742-55.

[52] Jin K, Zhu Y, Sun Y, Mao XO, Xie L, Greenberg DA. Vascular endothelial growth factor (VEGF) stimulates neurogenesis in vitro and in vivo. Proc Natl Acad Sci U S A 2002 Sep 3;99(18):11946-50.

[53] Balordi F, Fishell G. Mosaic removal of hedgehog signaling in the adult SVZ reveals that the residual wild-type stem cells have a limited capacity for self-renewal. J Neurosci 2007 Dec 26;27(52):14248-59.

[54] Chambers CB, Peng Y, Nguyen H, Gaiano N, Fishell G, Nye JS. Spatiotemporal selectivity of response to Notch1 signals in mammalian forebrain precursors. Development 2001 Mar;128(5):689-702.

[55] Carlen M, Meletis K, Goritz C, Darsalia V, Evergren E, Tanigaki K, et al. Forebrain ependymal cells are Notch-dependent and generate neuroblasts and astrocytes after stroke. Nat Neurosci 2009 Mar;12(3):259-67.

[56] Lim DA, Tramontin AD, Trevejo JM, Herrera DG, Garcia-Verdugo JM, Alvarez-Buylla A. Noggin antagonizes BMP signaling to create a niche for adult neurogenesis. Neuron 2000 Dec;28(3):713-26.

[57] Colak D, Mori T, Brill MS, Pfeifer A, Falk S, Deng C, et al. Adult neurogenesis requires Smad4-mediated bone morphogenic protein signaling in stem cells. J Neurosci 2008 Jan 9;28(2):434-46.

[58] Lange C, Mix E, Frahm J, Glass A, Muller J, Schmitt O, et al. Small molecule GSK-3 inhibitors increase neurogenesis of human neural progenitor cells. Neurosci Lett 2011 Jan 13;488(1):36-40.

[59] Wu W, Wong K, Chen J, Jiang Z, Dupuis S, Wu JY, et al. Directional guidance of neuronal migration in the olfactory system by the protein Slit. Nature 1999 Jul 22;400(6742):331-6.

[60] Murase S, Horwitz AF. Deleted in colorectal carcinoma and differentially expressed integrins mediate the directional migration of neural precursors in the rostral migratory stream. J Neurosci 2002 May 1;22(9):3568-79.

[61] Nguyen-Ba-Charvet KT, Picard-Riera N, Tessier-Lavigne M, Baron-Van EA, Sotelo C, Chedotal A. Multiple roles for slits in the control of cell migration in the rostral migratory stream. J Neurosci 2004 Feb 11;24(6):1497-506.

[62] Ng KL, Li JD, Cheng MY, Leslie FM, Lee AG, Zhou QY. Dependence of olfactory bulb neurogenesis on prokineticin 2 signaling. Science 2005 Jun 24;308(5730):1923-7.

[63] Conover JC, Doetsch F, Garcia-Verdugo JM, Gale NW, Yancopoulos GD, Alvarez-Buylla A. Disruption of Eph/ephrin signaling affects migration and proliferation in the adult subventricular zone. Nat Neurosci 2000 Nov;3(11):1091-7.

[64] Holmberg J, Armulik A, Senti KA, Edoff K, Spalding K, Momma S, et al. Ephrin-A2 reverse signaling negatively regulates neural progenitor proliferation and neurogenesis. Genes Dev 2005 Feb 15;19(4):462-71.

[65] Emsley JG, Hagg T. alpha6beta1 integrin directs migration of neuronal precursors in adult mouse forebrain. Exp Neurol 2003 Oct;183(2):273-85.

[66] Saghatelyan A, de CA, Schachner M, Lledo PM. Tenascin-R mediates activity-dependent recruitment of neuroblasts in the adult mouse forebrain. Nat Neurosci 2004 Apr;7(4):347-56.

[67] Hack I, Bancila M, Loulier K, Carroll P, Cremer H. Reelin is a detachment signal in tangential chain-migration during postnatal neurogenesis. Nat Neurosci 2002 Oct; 5(10):939-45.

[68] Hagg T. Molecular regulation of adult CNS neurogenesis: an integrated view. Trends Neurosci 2005 Nov;28(11):589-95.

[69] Hoglinger GU, Rizk P, Muriel MP, Duyckaerts C, Oertel WH, Caille I, et al. Dopamine depletion impairs precursor cell proliferation in Parkinson disease. Nat Neurosci 2004 Jul;7(7):726-35.

[70] Miller FD, Gauthier-Fisher A. Home at last: neural stem cell niches defined. Cell Stem Cell 2009 Jun 5;4(6):507-10.

[71] Kim Y, Wang WZ, Comte I, Pastrana E, Tran PB, Brown J, et al. Dopamine stimulation of postnatal murine subventricular zone neurogenesis via the D3 receptor. J Neurochem 2010 Aug;114(3):750-60.

[72] O'Keeffe GC, Tyers P, Aarsland D, Dalley JW, Barker RA, Caldwell MA. Dopamine-induced proliferation of adult neural precursor cells in the mammalian subventricular zone is mediated through EGF. Proc Natl Acad Sci U S A 2009 May 26;106(21): 8754-9.

[73] Yang P, Arnold SA, Habas A, Hetman M, Hagg T. Ciliary neurotrophic factor mediates dopamine D2 receptor-induced CNS neurogenesis in adult mice. J Neurosci 2008 Feb 27;28(9):2231-41.

[74] Winner B, Geyer M, Couillard-Despres S, Aigner R, Bogdahn U, Aigner L, et al. Striatal deafferentation increases dopaminergic neurogenesis in the adult olfactory bulb. Exp Neurol 2006 Jan;197(1):113-21.

[75] Barker JL, Behar T, Li YX, Liu QY, Ma W, Maric D, et al. GABAergic cells and signals in CNS development. Perspect Dev Neurobiol 1998;5(2-3):305-22.

[76] Owens DF, Kriegstein AR. Is there more to GABA than synaptic inhibition? Nat Rev Neurosci 2002 Sep;3(9):715-27.

[77] Ben Ari Y. Excitatory actions of gaba during development: the nature of the nurture. Nat Rev Neurosci 2002 Sep;3(9):728-39.

[78] Marty A, Llano I. Excitatory effects of GABA in established brain networks. Trends Neurosci 2005 Jun;28(6):284-9.

[79] Mejia-Gervacio S, Murray K, Lledo PM. NKCC1 controls GABAergic signaling and neuroblast migration in the postnatal forebrain. Neural Dev 2011;6:4.

[80] Young SZ, Taylor MM, Wu S, Ikeda-Matsuo Y, Kubera C, Bordey A. NKCC1 knockdown decreases neuron production through GABA(A)-regulated neural progenitor proliferation and delays dendrite development. J Neurosci 2012 Sep 26;32(39): 13630-8.

[81] Ge S, Pradhan DA, Ming GL, Song H. GABA sets the tempo for activity-dependent adult neurogenesis. Trends Neurosci 2007 Jan;30(1):1-8.

[82] Esposito MS, Piatti VC, Laplagne DA, Morgenstern NA, Ferrari CC, Pitossi FJ, et al. Neuronal differentiation in the adult hippocampus recapitulates embryonic development. J Neurosci 2005 Nov 2;25(44):10074-86.

[83] Wang DD, Krueger DD, Bordey A. Biophysical properties and ionic signature of neuronal progenitors of the postnatal subventricular zone in situ. J Neurophysiol 2003 Oct;90(4):2291-302.

[84] Wang DD, Krueger DD, Bordey A. GABA depolarizes neuronal progenitors of the postnatal subventricular zone via GABAA receptor activation. J Physiol 2003 Aug 1;550(Pt 3):785-800.

[85] Nguyen L, Malgrange B, Breuskin I, Bettendorff L, Moonen G, Belachew S, et al. Autocrine/paracrine activation of the GABA(A) receptor inhibits the proliferation of neurogenic polysialylated neural cell adhesion molecule-positive (PSA-NCAM+) precursor cells from postnatal striatum. J Neurosci 2003 Apr 15;23(8):3278-94.

[86] Liu X, Wang Q, Haydar TF, Bordey A. Nonsynaptic GABA signaling in postnatal subventricular zone controls proliferation of GFAP-expressing progenitors. Nat Neurosci 2005 Sep;8(9):1179-87.

[87] Bolteus AJ, Bordey A. GABA release and uptake regulate neuronal precursor migration in the postnatal subventricular zone. J Neurosci 2004 Sep 1;24(35):7623-31.

[88] LoTurco JJ, Owens DF, Heath MJ, Davis MB, Kriegstein AR. GABA and glutamate depolarize cortical progenitor cells and inhibit DNA synthesis. Neuron 1995 Dec; 15(6):1287-98.

[89] Haydar TF, Wang F, Schwartz ML, Rakic P. Differential modulation of proliferation in the neocortical ventricular and subventricular zones. J Neurosci 2000 Aug 1;20(15): 5764-74.

[90] Behar TN, Li YX, Tran HT, Ma W, Dunlap V, Scott C, et al. GABA stimulates chemotaxis and chemokinesis of embryonic cortical neurons via calcium-dependent mechanisms. J Neurosci 1996 Mar 1;16(5):1808-18.

[91] Cameron HA, Hazel TG, McKay RD. Regulation of neurogenesis by growth factors and neurotransmitters. J Neurobiol 1998 Aug;36(2):287-306.

[92] Schlett K. Glutamate as a modulator of embryonic and adult neurogenesis. Curr Top Med Chem 2006;6(10):949-60.

[93] Komuro H, Rakic P. Modulation of neuronal migration by NMDA receptors. Science 1993 Apr 2;260(5104):95-7.

[94] Di GG, V, Caruso A, Cappuccio I, Ricci VL, Romeo S, Della RC, et al. The mGlu5 metabotropic glutamate receptor is expressed in zones of active neurogenesis of the embryonic and postnatal brain. Brain Res Dev Brain Res 2004 May 19;150(1):17-22.

[95] Platel JC, Dave KA, Bordey A. Control of neuroblast production and migration by converging GABA and glutamate signals in the postnatal forebrain. J Physiol 2008 Aug 15;586(16):3739-43.

[96] Platel JC, Dave KA, Gordon V, Lacar B, Rubio ME, Bordey A. NMDA receptors activated by subventricular zone astrocytic glutamate are critical for neuroblast survival prior to entering a synaptic network. Neuron 2010 Mar 25;65(6):859-72.

[97] Platel JC, Heintz T, Young S, Gordon V, Bordey A. Tonic activation of GLUK5 kainate receptors decreases neuroblast migration in whole-mounts of the subventricular zone. J Physiol 2008 Aug 15;586(16):3783-93.

[98] Giorgi-Gerevini V, Melchiorri D, Battaglia G, Ricci-Vitiani L, Ciceroni C, Busceti CL, et al. Endogenous activation of metabotropic glutamate receptors supports the proliferation and survival of neural progenitor cells. Cell Death Differ 2005 Aug;12(8): 1124-33.

[99] Giachino C, De Marchis S, Giampietro C, Parlato R, Perroteau I, Schutz G, et al. cAMP response element-binding protein regulates differentiation and survival of newborn neurons in the olfactory bulb. J Neurosci 2005 Nov 2;25(44):10105-18.

[100] Pathania M, Platel J-CJ, Teng Z-Q, Zhao X, Bordey A. A role for microRNA-132 in the postnatal subventricular zone and neurogenesis. Program No.531.2009 International Society for Stem Cell Research 7th Annual Meeting Poster Abstracts.Barcelona, Spain: International Society for Stem Cell Research, 2009.Online. 2009.

[101] Bath KG, Mandairon N, Jing D, Rajagopal R, Kapoor R, Chen ZY, et al. Variant brain-derived neurotrophic factor (Val66Met) alters adult olfactory bulb neurogenesis and spontaneous olfactory discrimination. J Neurosci 2008 Mar 5;28(10):2383-93.

[102] Wang DD, Kriegstein AR. GABA regulates excitatory synapse formation in the neocortex via NMDA receptor activation. J Neurosci 2008 May 21;28(21):5547-58.

[103] Belluzzi O, Benedusi M, Ackman J, LoTurco JJ. Electrophysiological differentiation of new neurons in the olfactory bulb. J Neurosci 2003 Nov 12;23(32):10411-8.

[104] Gascon E, Dayer AG, Sauvain MO, Potter G, Jenny B, De Roo M, et al. GABA regulates dendritic growth by stabilizing lamellipodia in newly generated interneurons of the olfactory bulb. J Neurosci 2006 Dec 13;26(50):12956-66.

[105] Nguyen L, Rigo JM, Rocher V, Belachew S, Malgrange B, Rogister B, et al. Neurotransmitters as early signals for central nervous system development. Cell Tissue Res 2001 Aug;305(2):187-202.

[106] Platel J-CJ, Dave K, Gordon V, Lacar B, Bordey A. NMDA-receptors are critical for SVZ postnatal neurogenesis. Program No.30.16/B21.2009 Neuroscience Meeting Planner.Chicago, IL: Society for Neuroscience, 2009.Online. 2009.

[107] Nissant A, Bardy C, Katagiri H, Murray K, Lledo PM. Adult neurogenesis promotes synaptic plasticity in the olfactory bulb. Nat Neurosci 2009 Jun;12(6):728-30.

[108] Savonenko A, Xu GM, Melnikova T, Morton JL, Gonzales V, Wong MP, et al. Episodic-like memory deficits in the APPswe/PS1dE9 mouse model of Alzheimer's disease: relationships to beta-amyloid deposition and neurotransmitter abnormalities. Neurobiol Dis 2005 Apr;18(3):602-17.

[109] Wenk GL. Neuropathologic changes in Alzheimer's disease. J Clin Psychiatry 2003;64 Suppl 9:7-10.

[110] Haughey NJ, Nath A, Chan SL, Borchard AC, Rao MS, Mattson MP. Disruption of neurogenesis by amyloid beta-peptide, and perturbed neural progenitor cell homeostasis, in models of Alzheimer's disease. J Neurochem 2002 Dec;83(6):1509-24.

[111] Rodriguez JJ, Jones VC, Verkhratsky A. Impaired cell proliferation in the subventricular zone in an Alzheimer's disease model. Neuroreport 2009 Jul 1;20(10):907-12.

[112] Veeraraghavalu K, Choi SH, Sisodia SS. Expression of familial Alzheimer's disease-linked human presenilin 1 variants impair enrichment-induced adult hippocampal neurogenesis. Neurodegener Dis 2010;7(1-3):46-9.

[113] Demars M, Hu YS, Gadadhar A, Lazarov O. Impaired neurogenesis is an early event in the etiology of familial Alzheimer's disease in transgenic mice. J Neurosci Res 2010 Aug 1;88(10):2103-17.

[114] Ziabreva I, Perry E, Perry R, Minger SL, Ekonomou A, Przyborski S, et al. Altered neurogenesis in Alzheimer's disease. J Psychosom Res 2006 Sep;61(3):311-6.

[115] Huisman E, Uylings HB, Hoogland PV. A 100% increase of dopaminergic cells in the olfactory bulb may explain hyposmia in Parkinson's disease. Mov Disord 2004 Jun; 19(6):687-92.

[116] Curtis MA, Eriksson PS, Faull RL. Progenitor cells and adult neurogenesis in neurodegenerative diseases and injuries of the basal ganglia. Clin Exp Pharmacol Physiol 2007 May;34(5-6):528-32.

[117] Curtis MA, Eriksson PS, Faull RL. Progenitor cells and adult neurogenesis in neurodegenerative diseases and injuries of the basal ganglia. Clin Exp Pharmacol Physiol 2007 May;34(5-6):528-32.

[118] Tattersfield AS, Croon RJ, Liu YW, Kells AP, Faull RL, Connor B. Neurogenesis in the striatum of the quinolinic acid lesion model of Huntington's disease. Neuroscience 2004;127(2):319-32.

[119] Curtis MA, Penney EB, Pearson AG, van Roon-Mom WM, Butterworth NJ, Dragunow M, et al. Increased cell proliferation and neurogenesis in the adult human Huntington's disease brain. Proc Natl Acad Sci U S A 2003 Jul 22;100(15):9023-7.

[120] Curtis MA, Penney EB, Pearson J, Dragunow M, Connor B, Faull RL. The distribution of progenitor cells in the subependymal layer of the lateral ventricle in the normal and Huntington's disease human brain. Neuroscience 2005;132(3):777-88.

[121] Curtis MA, Waldvogel HJ, Synek B, Faull RL. A histochemical and immunohistochemical analysis of the subependymal layer in the normal and Huntington's disease brain. J Chem Neuroanat 2005 Jul;30(1):55-66.

[122] Jin K, Minami M, Lan JQ, Mao XO, Batteur S, Simon RP, et al. Neurogenesis in dentate subgranular zone and rostral subventricular zone after focal cerebral ischemia in the rat. Proc Natl Acad Sci U S A 2001 Apr 10;98(8):4710-5.

[123] Zhang R, Zhang L, Zhang Z, Wang Y, Lu M, Lapointe M, et al. A nitric oxide donor induces neurogenesis and reduces functional deficits after stroke in rats. Ann Neurol 2001 Nov;50(5):602-11.

[124] Koketsu D, Furuichi Y, Maeda M, Matsuoka N, Miyamoto Y, Hisatsune T. Increased number of new neurons in the olfactory bulb and hippocampus of adult non-human primates after focal ischemia. Exp Neurol 2006 May;199(1):92-102.

[125] Arvidsson A, Collin T, Kirik D, Kokaia Z, Lindvall O. Neuronal replacement from endogenous precursors in the adult brain after stroke. Nat Med 2002 Sep;8(9):963-70.

[126] Jin K, Wang X, Xie L, Mao XO, Greenberg DA. Transgenic ablation of doublecortin-expressing cells suppresses adult neurogenesis and worsens stroke outcome in mice. Proc Natl Acad Sci U S A 2010 Apr 27;107(17):7993-8.

[127] Marti-Fabregas J, Romaguera-Ros M, Gomez-Pinedo U, Martinez-Ramirez S, Jimenez-Xarrie E, Marin R, et al. Proliferation in the human ipsilateral subventricular zone after ischemic stroke. Neurology 2010 Feb 2;74(5):357-65.

[128] Jin K, Wang X, Xie L, Mao XO, Zhu W, Wang Y, et al. Evidence for stroke-induced neurogenesis in the human brain. Proc Natl Acad Sci U S A 2006 Aug 29;103(35):13198-202.

[129] Macas J, Nern C, Plate KH, Momma S. Increased generation of neuronal progenitors after ischemic injury in the aged adult human forebrain. J Neurosci 2006 Dec 13;26(50):13114-9.

[130] Parent JM, Valentin VV, Lowenstein DH. Prolonged seizures increase proliferating neuroblasts in the adult rat subventricular zone-olfactory bulb pathway. J Neurosci 2002 Apr 15;22(8):3174-88.

[131] Gonzalez-Martinez JA, Bingaman WE, Toms SA, Najm IM. Neurogenesis in the postnatal human epileptic brain. J Neurosurg 2007 Sep;107(3):628-35.

[132] Lantos PL, Cox DJ. The origin of experimental brain tumours: a sequential study. Experientia 1976 Nov 15;32(11):1467-8.

[133] Vick NA, Lin MJ, Bigner DD. The role of the subependymal plate in glial tumorigenesis. Acta Neuropathol 1977 Sep 26;40(1):63-71.

[134] Sanai N, Alvarez-Buylla A, Berger MS. Neural stem cells and the origin of gliomas. N Engl J Med 2005 Aug 25;353(8):811-22.

[135] Zhu Y, Guignard F, Zhao D, Liu L, Burns DK, Mason RP, et al. Early inactivation of p53 tumor suppressor gene cooperating with NF1 loss induces malignant astrocytoma. Cancer Cell 2005 Aug;8(2):119-30.

[136] Liu C, Sage JC, Miller MR, Verhaak RG, Hippenmeyer S, Vogel H, et al. Mosaic analysis with double markers reveals tumor cell of origin in glioma. Cell 2011 Jul 22;146(2):209-21.

[137] Feliciano DM, Bordey A. Newborn cortical neurons: only for neonates? Trends Neurosci 2012 Oct 10.

Chapter 7

Modulation of Adult Neurogenesis by the Nitric Oxide System

Bruno P. Carreira, Ana I. Santos,
Caetana M. Carvalho and Inês M. Araújo

Additional information is available at the end of the chapter

http://dx.doi.org/10.5772/54982

1. Introduction

Nitric oxide (NO) is a gaseous free radical that acts as a second messenger having an important biological role in intercellular communication and in intracellular signaling in many tissues, including the brain (reviewed by [1]). NO is synthesized by the nitric oxide synthase (NOS) family of enzymes, which convert L-arginine to L-citrulline and NO. There are three different isoforms of NOS: a) neuronal NOS (nNOS or NOS I), b) endothelial NOS (eNOS or NOS III), and c) inducible NOS (iNOS or NOS II) (reviewed by [2]). Different members of the NOS family control different functions of NO (reviewed by [1]).

In the central nervous system (CNS), NO has been linked to the regulation of synaptic plasticity and cognitive functions, and it is also associated with the control of biological functions including sleep-wake cycle, appetite, body temperature, and modulation of hormone release as reviewed by [3]. In the last decade, there has been a growing interest in the study of the role of NO in neurogenesis, the process by which new neurons are formed in the brain. NO regulates neurogenesis in diverse ways, and the different NO synthases are important players in the different effects on neurogenesis. Under physiological conditions NO synthesized from nNOS acts as a negative regulator of neurogenesis [4-9], while in inflammatory conditions, such as neurodegenerative disorders or acute brain insults, a decrease in nNOS and an increase in iNOS expression may act as a mechanism to enhance neurogenesis [8,10-13]. In fact, depending on the source, NO has a pro-neurogenic effect either by promoting neural stem cell (NSC) proliferation, as recently described by our group [13,14], but also by favoring other steps of neurogenesis such as migration [15], differentiation and survival [10,16]. Although the exact molecular mechanisms underlying this dual effect of NO on neurogenesis are not fully clarified, the modulation of the NO system seems be a good target for the development of strategies to improve endogenous neurogenesis following brain damage.

In this chapter, we describe the use of two different strategies for the enhancement of endogenous neurogenesis using drugs that are linked to the nitrergic system: 1) Nitric oxide-releasing non-steroidal anti-inflammatory drugs (NO-NSAID); 2) Phosphodiesterase type 5 (PDE5) inhibitors. PDE5 inhibitors are suitable to be used in the clinic for the treatment of several pathologies, such as erectile dysfunction [17] and pulmonary arterial hypertension [18], while NO-NSAID are being studied as an alternative to NSAID in the treatment of systemic inflammatory conditions [19]. Although little is known about the use of these drugs for the treatment of CNS disorders, the evidence so far is encouraging. Several reports describe these drugs as a good strategy to promote regeneration of lesioned areas or to be used as an adjuvant approach in cell replacement therapies since they favor neurogenesis [20-22]. Thus, the design of therapeutic strategies using these drugs to efficiently enhance the different steps of neurogenesis, such as a) proliferation, b) migration, c) differentiation, d) integration and, e) survival of NSC in the injured CNS, seems to be a valuable therapeutic approach to improve brain repair.

2. Neurogenesis in the adult mammalian brain

The discovery of NSC in the adult mammalian brain had a strong contribution for the understanding of adult CNS plasticity. Two regions have been classically described as having the characteristics necessary for the maintenance of NSC: a) the subgranular zone (SGZ) of the dentate gyrus (DG) of the hippocampus [23,24], and b) the subventricular zone (SVZ) of the lateral ventricles [25]. NSC can be isolated from the SGZ or SVZ and cultured *in vitro*, since some of the characteristics of these regions can be kept in culture in the presence of growth factors such as the epidermal growth factor (EGF) [26] and/or basic fibroblast growth factor (bFGF) [27].

In vivo, from the SVZ and SGZ, NSC undergo a complex process leading ultimately to the formation of new neurons, a phenomenon referred to as neurogenesis, which enables the continuous production of neuronal cells throughout the adult life of mammals, including humans. Neurogenesis can be summarized into six main stages: 1) proliferation of precursor cells; 2) fate determination; 3) migration; 4) differentiation; 5) integration in the neuronal circuitry, and 6) long-term survival of functional newborn neurons. Each of these stages is tightly regulated locally, and numerous agents have been described to be responsible for the physiological regulation of neurogenesis, such as EGF, bFGF, Numb, Notch, Sox, Sonic hedgehog, Noggin, among others (for review see [28,29]).

When trauma occurs in the CNS, new needs arise for the brain, mainly for repair, and various signals are released from injured areas influencing neurogenic niches and the behavior of NSC, which can migrate to the affected sites. Brain damage may be a) acute, such as traumatic brain injury, ischemic stroke or prolonged brain seizures, or b) chronic, such as slow-progressing neurodegenerative diseases. All these conditions are fol-

http://dx.doi.org/10.5772/54982

lowed by an inflammatory response [30]. Indeed, several studies have shown that adult neurogenesis is influenced by various pathological conditions, as discussed previously [31]. Models of brain damage were used to demonstrate that neurogenesis may be favored following injury, particularly acute injury, which is generally accepted as an attempt of the brain to repair [31]. On the other hand, the neurogenic capacity is decreased in neurodegenerative diseases, such as Alzheimer's disease, Huntington's disease or Parkinson's disease [32-34].

However, several questions remain unclear about this issue, in particular: a) which factors regulate neurogenesis during inflammation; b) which signaling pathways are involved in the recruitment of NSC for the injured sites; c) how new neurons are integrated and are able to survive long term; d) how can neurogenesis be modulated to improve its efficiency in an inflammatory context. The search for the answers to some of these questions is the challenge of regenerative medicine and a major target by the scientific community nowadays.

3. Neuroinflammation

Neuroinflammation is a biological response to noxious stimuli affecting the CNS, such as stress, injury or infection by external pathogens [35,36]. The main role of the inflammatory response is that of providing an harmful environment for external agents that cause injury and to regain homeostasis, being mediated by the activation of two major groups of cells from the immune system: a) CNS resident cells - microglia and astrocytes, and b) hematopoietic system migrating cells - lymphocytes, monocytes and macrophages [37,38]. The activation of these cells is characterized by the release of different regulatory substances, including chemokines such as stromal derived factor (SDF)-1alpha, complement molecules, monocytes chemoattractant protein-1 (MCP-1), cytokines such as interferon (IFN)-gamma, tumor necrosis factor (TNF)-alpha, interleukine (IL)-1beta, IL-18 and IL-6, glutamate, reactive oxygen species (ROS) and reactive nitrogen species (RNS) like NO (for review see [31]). Although the main function of neuroinflammation is to protect the brain by promoting the removal of noxious stimuli and committed/dead cells, and thus reestablishing brain tissue homeostasis, neuroinflammation may also become deregulated and contribute to perpetuate secondary tissue damage, as reported previously [39]. In fact, the creation of a positive feedback loop through inflammation itself may result in neuronal loss and/or neuronal damage.

In short, neuroinflammation may have a dual effect on the cellular environment, being beneficial or detrimental, depending on the time and state of activation of inflammatory cells [40]. Accordingly, the inflammatory response has been linked to the mechanisms that lead to various CNS diseases, also affecting SVZ and SGZ niches, therefore compromising neurogenesis [41]. Whether this means that inflammation is always detrimental to neurogenesis, or whether it is harmful only when the homeostasis of SVZ and/or SGZ is compromised, will be discussed in the next section.

4. Neuroinflammation and neurogenesis

As mentioned in the previous section, it is now widely accepted that neuroinflammation modulates neurogenesis in different ways, either by increasing or, alternatively, decreasing it [42]. Depending on the severity and complexity of the inflammatory response, which can range from a mild acute to a chronic uncontrolled process, neurogenesis may be dually regulated. Factors such as a) the type of inflammatory stimuli, b) the type of inflammatory cells, c) the type of inflammatory mediator, d) the area of injured tissue and e) for how long the inflammatory cells, particularly microglia, remain activated, are decisive for the shift from a pro-neurogenic to an anti-neurogenic inflammatory status [43].

In this context, the involvement of a particular type of inflammatory cells such as microglia, considered by most authors as the "hallmark of neuroinflammation", seems to be of major importance in the modulation of neurogenesis. The main features of microglial cells are a) the expression of scavenger receptors, b) antigen presentation molecules (Major Histocompatibility Complex (MHC) class II), c) pattern-recognition receptors, and d) production of various cytokines and other inflammatory factors (ROS and RNS) [44]. For a long time, microglial cells were considered as the damaging agents of the inflammatory response, with a default response always leading to detrimental effects on neuronal surrounding environment. However, recent studies describe microglial activity to be plastic (for review see [42]). In fact, the plasticity of microglia seems to be a determining factor in this dual regulation of neurogenesis, since it can assume different morphologies and different phenotypes and subsequently release mediators along an inflammatory response that may influence the physiology of the NSC [45]. Apparently, microglial cells and factors released during inflammatory responses appear to have a dual role in neurogenesis [13,42].

Therefore, numerous studies have reported the involvement of different microglial-derived inflammatory mediators in the regulation of neurogenesis and/or neuroprotection [31,46,47]. Moreover, it has been reported that chronic microglial activation can stimulate one or more stages of neurogenesis, such as NSC proliferation, migration and differentiation, while the long-term survival of newborn neurons seems to be reduced in this context [31].

4.1. Anti-neurogenic role of inflammation

Neuroinflammation, in particular microglial activation, was initially described to be detrimental to neurogenesis [48,49]. Several studies have demonstrated microglia activation by lipopolysaccharide (LPS) to hinder neurogenesis in adult rats [48], by a mechanism mediated through TNF-alpha increased production [50,51]. Other studies have linked this anti-neurogenic effect of inflammation to the increased production of other proinflammatory mediators such as interleukins IL-1 beta and IL-6, or cytokines IFN-gamma and TNF-alpha [52-55]. In addition, ROS and RNS, in particular NO, have also been described as being involved in the detrimental effect of neuroinflammation in the

http://dx.doi.org/10.5772/54982

formation of new neurons in the adult brain of rodents [49,56,57]. In Table 1 we summarize the main findings concerning the effect of the most important proinflammatory mediators in neurogenesis.

In addition, the deleterious role of inflammation in neurogenesis was corroborated by numerous studies which demonstrated that neurogenesis can be restored when the inflammatory response is controlled by the administration of: a) antibiotics, such as minocycline [48,58,59], or b) non-steroidal anti-inflammatory drugs, such as indomethacin [48,49,60,61].

4.2. Pro-neurogenic role of inflammation

Contrary to initial observations, recent studies indicate that neuroinflammation may also support different stages of neurogenesis, thus favoring the formation of new neurons following injury to the CNS [44]. Thus, the inflammatory microenvironment is responsible for sending "activating signals" to NSC resident in neurogenic niches, such as SVZ or SGZ, that thereafter migrate to the injured areas where they differentiate and integrate the neuronal network [62,63]. In this context, microglial cells are described as central in the regulation of this process, suggesting an ambiguous role of microglia in the regulation of neurogenesis in inflammatory conditions [64].

Apparently, although microglia may be detrimental to neurogenesis in early stages of the inflammatory response after acute insults, prolonged inflammatory response, also referred as chronic inflammation, appears to have a protective effect by directing the replacement of damaged or lost cells [45,65-68]. Thus, it was shown in several studies that inhibition of microglial activation results in continuous production of new neurons from adult NSC [69,70]. Moreover, chronic activation of microglia is concomitant with long-term survival of newly formed neurons [71,72].

Several proinflammatory mediators have been related to the pro-neurogenic effect of inflammation, including: a) cytokines such as IFN-gamma or TNF-alpha, b) chemokines such as SDF-1alpha and its receptor CXCR4 [69,73], or c) trophic factors such as brain-derived neurotrophic factor (BDNF) and glial-derived neurotrophic factor (GDNF) involved in the removal of damaged synapses [72] (see Table 1). In addition to the pro-neurogenic effect, these studies also suggest a neuroprotective role of microglial cells for newborns cells.

4.3. Future studies

Overall, it seems clear that more knowledge about the crosstalk between inflammation and neurogenesis is lacking. For instance, it is necessary to better characterize the genetic and proteomics of the microglial response, as well as more targeted studies are needed to clarify how neuroinflammation modulates each of the neurogenic stages. Identifying which genes are expressed, and subsequently, what kind of proteins are present during an inflammatory response will allow the development of different strategies to control or mitigate the deleterious effects of neuroinflammation on neurogenesis.

Inflammatory mediator	Proliferation of NSC		Differentiation of NCS		Survival of NSC		Signaling pathway	References
	SVZ	DG	SVZ	DG	SVZ	DG		
IL-1 beta	-	-	0	0	+	+	SAPK/JNK	[74,75]
IL-6	-	-	-	-	-	-	JAK/STAT and MAPK	[48,56,67]
IFN-gamma	-	0	+	+	-	0	ERK 1/2 pathway	[52,64,76-78]
	+	0	0	0	=	0	NF-kappaB	[79,80]
TNF-alpha	+	0	0	-	0	-	0	[50]
	0	0	0	0	-	-	TNF-R1 and TNF-R2	[53,76]
	+	0	-	0	=	0	TNF-R1	[81]
	+	+	+	+	+	+	TNF-R2	[82]
	+	0	+	0	+	0	TNF-R1	[83]
	0	-	0	-	+	0	TNR-R1	[54]
	+	+	+	+	+	+	TNF-R2	[43,54]
SDF-1alpha	+	+	+	0	+	0	CXC-R4	[69,73,84]

The effects listed in Table 1 may not be direct. +, Increase; -, decrease; =, not changed; 0, not reported.

Table 1. Modulation of adult neurogenesis by inflammatory mediators.

5. Nitric oxide

NO is a short-lived gaseous free radical synthesized by different members of the nitric oxide synthase family of enzymes. NOS are present in most tissues of the body and convert L-arginine to L-citrulline and NO [2,85]. The NOS family of enzymes is characterized by the existence of three different isoforms in mammalian cells: a) neuronal NOS (nNOS, type I), is constitutively expressed in neurons, where it localizes to synaptic spines, and is activated by calcium/calmodulin following the activation of glutamate receptors; b) endothelial NOS (eNOS, type III), is constitutively expressed in endothelial cells and astrocytes, is regulated by phosphorylation/dephosphorylation and/or by calcium/calmodulin; and c) inducible NOS (iNOS, type II), which regulation is dependent on de novo synthesis of the enzyme, particularly in inflammatory conditions [2,86,87].

Involved in a variety of physiological processes, NO has been described as an important regulator of the activity of systems such as the cardiovascular, immune and nervous systems [88]. There are several biological functions that depend on NO formation, including the regulation of body temperature, appetite and sleep-wake cycle (for review see [3]). The main mechanism regulating NO activity is at the level of its synthesis. In the CNS, NO has a distinct action when compared to classical neurotransmitters, as it is synthesized on demand, diffusing from synaptic terminals, acting not only in NO-releasing cells, but also in neighboring cells [89]. Initially described as an intracellular messenger, NO has also been associated with synaptic plasticity, which is linked to cognitive function, neuronal development and modulation of hormone release [90]. In this context, the role of NO as an intracellular messenger is mediated by increasing cyclic guanosine 3', 5'-monophosphate (cGMP) levels, following the activation of N-Methyl-D-aspartate (NMDA)-type glutamate receptors [91]. Unlike classical neurotransmitters, which are stored in vesicles or released by exocytosis and further inactivated by re-uptake or enzymatic degradation, NO ends its action after reacting with intracellular substrates [1]. According to the literature, the action of NO in the brain has been associated with two different outcomes: a) regulation of physiological events, by its action as an intracellular messenger [90], or b) regulation of cell death mechanisms, due to its action as a cytotoxic agent [92,93]. This will be explored next in section 5.1.

5.1. Nitric oxide and neuroinflammation

As mentioned in the previous section, the action of NO in the CNS is characterized by the interaction with multiple intracellular targets, activating or inhibiting various signaling pathways. Thus, NO has been described to be involved in the regulation of several physiological functions, but also of various pathophysiological processes [2,85]. This dual action depends on the NOS isoform that catalyzes the formation of NO: a) nNOS and eNOS-derived NO is more involved in the regulation of physiological functions, and b) iNOS-derived NO is more involved in pathophysiological processes. In fact, iNOS is not normally expressed in the healthy brain, but in the presence of pro-inflammatory stimuli such as cytokines, external pathogens, such as bacteria or virus, or stress, such as hypoxia, iNOS may be expressed primarily on macrophages, astrocytes, microglia and endothelial cells [3,86,94,95], but also in neurons [96,97]. However, it should be mentioned that NO overproduction has also been linked to nNOS activation following persistent glutamate excitatory input during an inflammatory response, which has also been linked to iNOS expression [3]. Once expressed, iNOS continuously produces NO, and high levels are reached, in a process that can last for several days, having a cytotoxic effect by inhibiting mitochondrial respiratory chain enzymes, ultimately inducing apoptosis in target cells [95,98-101]. A key factor for the local effect of NO is the concentration achieved. Thus, in physiological conditions, NO concentrations could range from 0.1 to 100 nM, which is lower than those observed in inflammatory conditions, being less reactive. Accordingly, the action of NO is accomplished primarily by binding to the heme group of soluble guanylate cyclase (sGC), whose activation leads to the subsequent production of cGMP [102].

Increased levels of NO have been linked to oxidative and nitrosative stress phenomena, which have been described as involved in the development of several neurodegenerative disorders [2,85]. Thus, a massive release of NO can lead to the production of nitrogen dioxide (NO_2), after the direct reaction between NO and oxygen. NO_2 is a highly reactive nitrosative specie that can react with NO, producing dinitrogen trioxide (N_2O_3). Moreover, NO_2 can also oxidize or nitrate a wide variety of molecules, being the nitration of tyrosine to 3-nitrotyrosine a classical example [103]. N_2O_3, in turn, is involved in other phenomena such as nitrosation/nitrosylation, by reacting with amine or thiol groups, being a good example cysteine, which may be nitrosated to S-nitrosocysteine [103]. Furthermore, NO can also react with superoxide to produce peroxynitrite ($ONOO^-$), another extremely reactive molecule which can oxidize or nitrate other molecules, which has been described to be involved in the pathogenesis of several neurodegenerative diseases, such as Parkinson's disease, Alzheimer's disease, Huntington's disease, multiple sclerosis and amyotrophic lateral sclerosis [2,104-106]. Likewise, both S-nitrosylation and nitration lead to alterations in the function of proteins, which may be regarded as regulatory phenomena of its activity [103].Thus, understanding the involvement of these phenomena in the pathoetiology of disorders affecting the brain may highlight a potential therapeutic role in modulating these events.

5.2. Nitric oxide and neurogenesis

The involvement of NO in the regulation of neurogenesis is a matter of debate given the range of different observations reported in the literature. In fact, the role of NO in neurogenesis only recently has been identified [4,5,107]. Depending on the source and concentration attained locally in the brain, NO has a dual influence in the neurogenic process both by inhibiting or stimulating neurogenesis.

Based on the distribution of NO-producing cells in stem cell niches, several works have proposed to study the involvement of NO produced at the perivascular niche - which includes pericytes and smooth muscle fibroblasts, endothelial cells, microglia, glial progenitors and astrocytic endfeet – in the regulation of neurogenesis, thus reinforcing the involvement of NO signaling in angiogenesis and neurogenesis [4,108,109]. Indeed, the discovery of blood vessels expressing eNOS and neurons expressing nNOS, close to SVZ and SGZ neurogenic niches, was essential for the establishment of a causal relationship between NO and the formation of new neurons. Moreover, it was also shown that nNOS-derived NO is involved in the regulation of neurogenesis, particularly by regulating NSC function, so that a cytostatic function can be assigned to NO in the CNS [4,5,107]. Thus, NO production occurs in close proximity to the NSC. Other authors have shown that nitrergic neurons expressing nNOS are arranged in close relationship throughout the rostral migratory stream (RMS), also describing a regulatory action of NO in the migration of SVZ-derived progenitor cells along the RMS [110].

Although most studies initially performed reported NO as an anti-neurogenic agent in the normal adult brain, in hypoxic ischemic stress conditions its effect can be radically different favoring stem cell proliferation, as demonstrated in recent studies [4,12,13,108,111,112]. In fact, the oxygen tension environment appears to modulate the effect of perivascular NO in neurogenesis [112,113], which may vary from: a) pro-neurogenic action, dependent on the expression of eNOS and nNOS, in physiological condition [108]; b) to anti-neurogenic action, dependent on the expression of iNOS, in extreme environments such as hypoxic and ischemic tissue and/or tumors [12,13,114,115]. However, more studies should be conducted to clearly understand how NO produced by different cell types from the perivascular niche regulate neurogenesis. In fact, it still remains to clarify the limit of oxidative stress and other redox states that lead to the differential production of NO by each NOS isoforms - nNOS, eNOS and iNOS - so that one can describe its perivascular action for neurogenesis.

During development, NO is differentially and transiently produced in the brain [116-118]. Moreover, the differential cellular and subcellular localization of nNOS in the CNS may explain different functions of NO produced by nNOS [119]. In fact, in the cortex, there are two types of NOS neurons whose distribution is of particular interest due to the relationship between the sites of NO production and the sites of development of particular pathologies [119]. Furthermore, its pre- or postsynaptical expression influences nNOS functions [120,121]. In the adult olfactory bulb (OB), nNOS is highly expressed in developing neurons of the olfactory epithelium during embryogenesis [120,122] and in the periglomerular cells and granule cells in the OB in the adult [120,123], being necessary for the early postnatal development and for the glomerular OB organization, respectively [123,124]. Furthermore, following a lesion in the OB, nNOS expression is upregulated causing repopulation of this region [117,122,123]. Moreover, developing ependymal cells, which are in close association with SVZ-derived progenitor cells, also transiently express nNOS after birth, but its activity decreases with the maturation of the central canal [125], thus suggesting a role of NO synthesized by nNOS in the development of ependymal cells [125]. Ependymal cells, together with astrocytes, create an appropriate environment for neurogenesis [126].

NO production may be induced by neurotrophic factors, which results in an antiproliferative effect on target cells by inducing cell cycle arrest/exit favoring cell differentiation [127-129]. Most of the studies on the involvement of NO in adult neurogenesis characterized its effect on cell proliferation. However, the evaluation of survival and integration of newly-generated neurons in the neuronal circuitry is also important, since NO is known to be an important regulator of apoptosis [130]. In this context, different studies have shown that NO increases short-term survival of progenitor cell progeny in the DG of adult rats by inhibiting apoptosis after SE [131], and further preventing increases in the activity of caspase-3 [132].

5.2.1. Anti-neurogenic role of nitric oxide

The anti-neurogenic effect of NO has been attributed to its production via nNOS, as demonstrated in several studies using *in vitro* and *in vivo* experimental models. Thus, it was reported that nNOS-derived NO has an antiproliferative effect, and may be also involved in neuronal differentiation, survival and synaptic plasticity [4-6,133,134]. The anti-proliferative effect of NO was confirmed by several authors, which showed that the inhibition of NO production by intra-ventricular infusion of a NOS inhibitor or by the knockout of nNOS increase cell proliferation in the DG or in the olfactory subependymal zone of rodents [4,6,7,108]. Indeed, other studies were performed where the selective inhibition of nNOS with 7-nitroindazole (7-NI) was shown to greatly increase cell proliferation in the SVZ, RMS and OB of adult rats, but not in the DG [5]. Moreover, the inhibition of nNOS was also shown to increase neurogenesis and to reduce infarct size, following a stroke [135]. The presence of differentiated nitrergic neurons in the periphery of the neurogenic areas, mainly surrounding the SVZ, and its anatomical organization, contributes to this physiological downregulation of neurogenesis [5,110]. However, the inhibitory role of nNOS-derived NO in neurogenesis was also demonstrated in the DG, after cerebral ischemia [135]. In the DG, the neural precursors of the SGZ are in close proximity with the nitrergic neurons of the hilus, also suggesting a role for NO in the control of adult neurogenesis in this region [136]. These studies showed that chronic inhibition of nNOS increases neurogenesis, supporting the idea that, physiologically, NO produced by nNOS has an anti-neurogenic effect. Recently, several studies have suggested a mechanism for the negative effect of NO on neurogenesis in the SVZ. These authors suggested the inhibition of the EGF receptor [134] by a mechanism dependent on the nitrosylation of specific cysteine residues and the activation of the phosphoinositide 3-kinase (PI3-K)/Akt signaling pathway [9] as the main mechanisms by which NO negatively regulates neurogenesis in the SVZ (Table II). Furthermore, these authors described the antimitotic effect of NO as being related to the nuclear presence of the cyclin-dependent kinase inhibitor $p27^{Kip1}$ [9].

5.2.2. Neurogenic role of nitric oxide

The pro-neurogenic effect of NO has been reported in several studies using genetic or pharmacological approaches, showing that increased levels of iNOS after an insult to the brain are related to increased neurogenesis in the hippocampus, an event correlated with concurrent decreases in nNOS levels [8,137-139] (Table 2). However, in a study regarding the effect of NO on cell proliferation it was described the involvement of NO derived from both iNOS and nNOS in the enhancement of neurogenesis in the DG of adult rats, following seizures [140]. Other studies also showed that NO synthesized by iNOS following ischemia or by eNOS stimulates neurogenesis in the SVZ or DG, respectively [12,111]. Furthermore, we recently showed that the iNOS-derived NO promotes the proliferation of NSC in the hippocampus of adult rats following SE [13]. Following an injury, the concomitant neuroinflammation results in the activation of microglial cells, which continuously express iNOS [141]. This event leads to the production of large amounts of NO that was shown to be favorable to increasing neurogenesis following acute brain injuries.

Although some questions remain to be assessed, several studies have sought to explore the signaling pathways by which NO from inflammatory origin exerts its pro-neurogenic effect, namely in the regulation of proliferation. Recently, we have shown that supraphysiological concentrations of NO induce the proliferation of SVZ-derived NSC through the activation of at least two signaling pathways, in a biphasic manner: a) the mitogen-activated protein (MAP) kinase ERK 1/2 pathway, and/or b) the cGMP/cGMP-dependent kinase (protein kinase G; PKG) pathway. Thus, the proliferative effect of NO seems to be initially mediated by the direct activation of ERK1/2 signaling pathway [13]. The increased activation of the ERK 1/2 signaling pathway after exposure to NO, leads to the activation of several downstream targets, namely the kinase p90RSK, subsequently leading to decreased nuclear levels of its target $p27^{Kip1}$, allowing cell cycle progression and cell division [13]. Moreover, the activation of cGMP/cGMP-dependent kinase (PKG) pathway appears to be involved following longer periods of exposure to supraphysiological levels of NO [14]. NO involvement in the regulation of other stages of neurogenesis has also been investigated. NO released in inflammatory conditions is also involved in NSC differentiation into astrocytes, a process also referred to as astrogliogenesis, by a mechanism dependent on the activation of JAK/STAT-1 signal transduction pathway [142].

Taken together, these findings show that NO is an important regulator of neurogenesis. The effect of NO on neurogenesis seems to be dependent on the developmental period and of the source of NO. Furthermore, depending on the local concentration and surrounding molecular environment NO may regulate neurogenesis in various ways, either favoring it, or impairing it [136,143,144]. As discussed above, NO has concentration-dependent effects. Thus, under physiological conditions NO acts as a negative regulator of neurogenesis [4,5,107], whereas in inflammatory conditions a decrease in nNOS and an increase in iNOS can act as a mechanism to enhance neurogenesis [12,13,134,145,146]. However, the exact molecular mechanisms underlying this dual effect are not fully understood and more studies are needed to determine the downstream targets of NO, particularly to identify potential therapeutic targets and to assess whether modulation of these players is possible to improve the outcome of neurogenesis. Most of the drugs used in studies for the characterization of NO involvement in neurogenesis are therapeutically used with other purposes unrelated to brain injury recovery. So, its implementation as a therapeutic strategy to modulate neurogenesis should be explored. Next, some of the most promising pharmacological approaches intended to modulate signaling pathways dependent on NO will be discussed.

NO source	Proliferation of NSC		Differentiation of NCS		Survival of NSC		Signaling pathway	References
	SVZ	DG	SVZ	DG	SVZ	DG		
nNOS	-	0	=	0	=	0	Nitrosylation of EGF receptor	[9]
	-	0	=	0	0	0	(PI3-K)/Akt pathway	[9,134]
	0	-	0	-	0	0	PSA-NCAM and CREB	[147]
	0	-	0	-	0	-	cAMP phosphorylation	[6]
eNOS	+	0	+	0	0	0	BDNF and VEGF	[148]
	0	+	0	+	0	=	VEGF	[111]
iNOS	+	0	0	0	=	0	ERK 1/2 pathway	[13]
							cGMP/PKG pathway	[14]
	+	+	+	+	0	=	NMDA receptor	[62,149,150]
	+	+	+	+	0	0	L-VGCC	[151]
	+	0	+*	0	=	0	JAK/STAT-1 pathway	[142]

The effects listed in Table II may not be direct. +, increase; -, decrease; =, not changed; 0, not reported; Polysialylated-neuronal cell adhesion molecule, PSA-NCAM; cAMP response element-binding, CREB; Brain-derived neurotrophic factor, BDNF; Vascular endothelial growth factor, VEGF; L-type voltage-gated Ca^{2+} channel, L-VGCC. * - astrogliogenesis.

Table 2. NO-dependent modulation of adult neurogenesis.

6. The nitric oxide system as a target to enhance endogenous neurogenesis

In physiological conditions, damaged cells and tissues are continuously being repaired in order to maintain homeostasis and normal function of the organism. A deregulation or malfunction of self-repair mechanisms could lead to the emergence of several pathologies as referred in sections 3 and 4. In the adult CNS, the major limitation that researchers face is the restricted ability for regeneration. Moreover, this process is even more limited during an inflammatory process as the surrounding environment is detrimental to the survival of newborn cells [43]. Acute brain lesions, such as stroke, spinal cord injury, trauma and seizures, which are accompanied by an inflammatory response, have a strong participation on neuronal loss [43]. Neuroinflammation is also a hallmark of chronic pathologies, such as Alzheimer's disease, Huntington's disease and Parkinson's disease [31]. Thus, to overcome the limited ability for brain repair and as an attempt to revert the loss of neurons following inflammation, some strategies have been studied. The most promising strategies include a) stimulation of endogenous neurogenesis, or b) transplantation of exogenous neural precursors/stem cells. Transplantation of exogenous stem cells is a complex approach with several disadvantages including ethical concerns. Furthermore, the risk of rejection and uncontrolled proliferation of grafted cells, which may lead to tumor formation, raises some concerns about its therapeutic applicability. However, although the potentiation of endogenous neurogenesis appears to be a better approach, with higher possibility for therapeutic application, some disadvantages/limitations should be taken into account, such as: a) low yield in the formation of new neurons, b) low rate of long-term survival of new neurons, and c) poor specificity for increasing neurogenesis in the target/lesioned tissue. Here we focus on the stimulation of endogenous neurogenesis by targeting the pre-existing pools of NSC, particularly in SVZ and SGZ niches, mainly by modulating the nitrergic pathways.

As discussed in section 5.2, NO has been widely described as a dual regulator of adult neurogenesis, being involved in the regulation of proliferation, migration, neuronal differentiation and survival of NSC (see Table II). The great majority of studies in the literature characterized the involvement of NO in the regulation of NSC proliferation. In fact, as reported by our group, NO from inflammatory origin has a proliferative effect in the SVZ and SGZ [13]. However, more studies about the involvement of NO in the regulation of migration, differentiation in functional neurons that must correctly integrate neuronal circuits and survival of the newly formed neurons must be performed in order to understand how these neurogenic steps are regulated in an inflammatory context. Although little is known about the *in vivo* applicability of this strategy, recent encouraging evidences are already in the literature where the pharmacological modulation of different players in the nitrergic system has been proved to promote neurogenesis. However, we believe that this approach in a regenerative context should not be considered as an isolated approach, but instead, it could be adjuvant to other strategies in order to ensure an efficacious therapy. Therefore, two different strategies should be considered to enhance neurogenesis: a) controlled increase in NO levels by using NO donors, particularly NO-NSAID and, b) prevention of cGMP degradation by the use of PDE5 inhibitors. Next, these therapeutic approaches for brain repair will be discussed.

6.1. Nitric oxide-releasing non-steroidal anti-inflammatory drugs

NO-releasing drugs have been widely used in several studies for the characterization of the involvement of NO in the regulation of different steps of endogenous neurogenesis. These pharmacological tools were essential to understand that NO-mediated effects on neurogenesis are time and concentration-dependent [14,142]. A wide variety of NO-releasing compounds are available, being the most common the diazeniumdiolates, also referred as NONOates (such as DEA/NO, SPER/NO or DETA/NO), that spontaneously release NO under physiological conditions [152]. NONOates were also used in numerous studies to investigate the effect of supraphysiological concentrations of NO on neurogenesis, thus mimicking high NO concentration achieved in the brain in inflammatory conditions [13,142]. However, NO-releasing drugs are chemically distinct, having different half-life times, releasing different amounts of NO *in vitro*. The major disadvantages of the use of these drugs lie in the inability to control the amount of NO released *in vivo*, and the incapacity to specifically release NO in the target tissue/ cells. Moreover, factors such as pH, temperature, some co-factors and light, are able to alter the release of NO by these compounds [152,153]. As described above, given that different amounts of NO have different effects on neurogenesis, it is essential to control the release of NO in order to keep it in levels that are beneficial to neurogenesis. Thus, it arises the need to develop new NO-releasing drugs in order to overcome these disadvantages.

More recently, a new class of NO-releasing compounds has been developed, NO-NSAID. These drugs are synthesized by adding a nitric oxide donating group to classical NSAID. Conventional NSAID are broad-spectrum compounds used worldwide due to their properties as analgesics, antipyretics and, at higher doses, anti-inflammatory. However, chronic use of NSAID is limited, mainly due to increased side effects in the gastrointestinal (GI) tract, cardiovascular system and kidneys (extensively reviewed by [154-156]). Traditional NSAID exert their effect by inhibiting both isoforms of cyclooxygenase enzyme (COX-1 and COX-2), thus blocking the synthesis of prostaglandins. The great majority of side effects associated to the use of these drugs are associated with the inhibition of COX-1 pathway, and subsequent decrease in gastroprotective prostaglandins.

To overcome these side effects and improve safety of NSAID, new drugs were designed a) coxibs, selective COX-2 inhibitors, and b) hybrid prodrugs, which include NO-NSAID. The latter drugs take advantage of some characteristics of NO such as its potent vasodilator effect, inhibition of leukocyte adherence to the gastric vascular endothelium and inhibition of caspase activity, thus mimicking the biological effects of prostaglandins in the GI tract [19,157,158]. Several *in vivo* studies have shown that NO released by NO-NSAID has a reduced GI toxicity profile compared to NSAID alone, without affecting the anti-inflammatory effectiveness [159-161]. In fact, low levels of conventional NO donors were shown to inhibit cell apoptosis *in vivo* by inhibiting caspase activity and, thus, sparing the gastric mucosa from the pro-apoptotic effect induced by TNF-alpha, an effect that seems to be dependent on cGMP formation [161-164]. In addition, NO released by NO-NSAID inactivates caspases, contributing to the gastric-sparing effect of these drugs. Moreover, these NO-donating drugs release NO in amounts that mimics *in vivo* NO production by constitutive NOS, which seems to be linked to a reduced toxicity when compared to the parent NSAID [19,165]. In addition, the relatively slow rate of NO release by NO-NSAID when compared to classic NO donors, such as sodium nitroprusside (SNP) [166], allows a more controlled release of NO and a long-lasting protective effect, which should be considered a major advantage in the use of these drugs. Since there is no massive burst in NO levels, excitotoxic events are prevented when compared to classical drugs.

Chronic inflammatory events were linked to Alzheimer's disease, where several pro-inflammatory mediators are released, such as the cytokines IL-1beta and TNF-alpha [167], and caspase enzymes are activated [168]. Chronic administration of NSAID appears to reduce the risk for developing Alzheimer's disease [169-172], also ameliorating impairment of cognitive functions in patients (reviewed in [173]). Other studies have been performed to study the effect of anti-inflammatory drugs in the treatment of acute brain lesions, such as *status epilepticus* and ischemia. In this context, anti-inflammatory drugs, such as indomethacin, have been described to reduce microglial activation and to promote NSC proliferation and improve migration and survival of newborn cells, thus restoring neurogenesis following cranial irradiation or focal ischemia [49,174]. Therefore, although the neuroprotective effects of NSAID in models of chronic brain inflammation have been recently described in the literature, the side effects of NSAID in other biological systems should not be ignored. Given the advantages of NO-NSAID, and given their ability to rapidly cross the blood-brain barrier (BBB) [175], NO-NSAID have been considered for the treatment of CNS disorders, particularly for the control of neuroinflammation that, as already discussed, may affect neurogenesis [165]. However, to date, little is known about the effect of NO-NSAID on neurogenesis following acute or chronic brain injury. Nevertheless, studies in models of chronic brain inflammation showed that chronic administration of NO-flurbiprofen significantly attenuated brain inflammation by decreasing the density and reactive state of microglial cells [176,177]. In this study, treatment with NO-flurbiprofen reduced brain inflammation and attenuated the effects of LPS-activated microglia in young and adult rats, but not in aged rats, which suggested this drug to be a possible therapeutic tool to be used in the onset of Alzheimer's disease, before the development of chronic inflammatory events associated with age [178]. Besides the reports that NSAID decrease the expression of iNOS in inflammatory cells, NO-flurbiprofen appears to upregulate the expression of this enzyme in LPS-activated microglial cells [179]. This effect leads to an even higher increase in NO production, which has been attributed to NO released from NO-flurbiprofen, since traditional NO donors lead to similar results. Interestingly, the activation of microglial iNOS following a brain insult enhances NSC proliferation in the SGZ following epileptic seizures, thus promoting neurogenesis [13].

Overall, the beneficial effects of NO-NSAID observed in experimental models of neurodegenerative diseases are encouraging for the development of strategies to control neuroinflammation and target endogenous neurogenesis by using these drugs [165]. However, further studies need to be conducted in order to understand the mechanisms and within which concentrations NO-derived from NO-NSAID may promote neurogenesis.

6.2. Phosphodiesterase 5 inhibitors

The main intracellular target of NO is the heme-containing enzyme sGC. Activation of sGC leads to an increased production of cGMP [102,180], which subsequently activates cGMP-dependent PKG [181,182]. PKG regulates various physiological events, such as synaptic plasticity or synthesis and release of neurotransmitters (reviewed by [183]). In physiological conditions, intracellular cGMP levels are controlled through cyclic nucleotide phosphodiesterases (PDE), enzymes that hydrolyze the 3'-phosphodiester bound of cyclic AMP (cAMP) or cGMP, originating their respective inactive monophosphates, 5′-AMP or 5′-GMP. PDE are ubiquitous enzymes classified in 11 families by their different substrate specificity, kinetic properties and cellular and subcellular distribution (extensively reviewed in [184]). As different PDE families present such a wide distribution among the tissues, including the brain, inhibition of one or more PDE has been studied as an approach for the treatment of several diseases, mainly by controlling the levels of the respective second messengers cAMP and/or cGMP.

cGMP-dependent physiological functions, may be regulated by controlling PDE type 5 isoenzyme activity, which specifically hydrolyzes cGMP. Thus, a good strategy to increase intracellular levels of cGMP may be through inhibition of this enzyme [185]. PDE5 is a widely expressed cytosolic enzyme, whose protein activity was found in the lung, vascular and tracheal smooth muscle, spleen, platelets, corpus cavernosum [186-188], being also highly present in several brain regions, including Purkinje cells and SVZ [189-191]. PDE5 and PDE5 inhibition have been extensively studied in the last decades and several PDE5 inhibitors have been developed. The most characterized PDE5 inhibitor is sildenafil, commercially available as Viagra, a drug used for the treatment of erectile dysfunction and pulmonary arterial hypertension. However, besides PDE5, sildenafil also inhibits PDE 1 and 6 with lower potency [192]. In order to overcome this issue, more selective PDE5 inhibitors were developed for the treatment of erectile dysfunction: vardenafil (Levitra), tadalafil (Cialis) and, more recently, avanafil (Stendra). In addition, a new compound with even higher selectivity for PDE5 was also developed, T0156 [193].

The decrease in cGMP levels appears to be one of the causes for the decreased neurogenesis in aging, which normally correlates with the development of neurodegenerative diseases [194]. Although neurogenesis is increased in early stages of neurodegenerative diseases, as a compensatory mechanism, the more advanced or severe stages are characterized by impairment of neurogenesis [195]. In the aged brain, there is a decrease in NO levels with a concomitant decline in cGMP levels, ultimately resulting in the abolishment of cell proliferation and impairments in learning and memory [194]. Targeting an enzyme specific for the hydrolysis of cGMP, such as PDE5, has been proven to be a good strategy to reverse this process and, thus, enhance neurogenesis following acute or chronic brain insults. In fact, PDE5 inhibitors are known to modulate several functions in the adult brain. Several reports showed that PDE5 inhibitors, such as sildenafil, have a neuroprotective role, by improving memory and learning [20,196-201]. Beyond the important role in memory and cognition, PDE5 inhibitors could also be used to target endogenous neurogenesis in the adult brain. In neurodegenerative diseases such as Alzheimer's disease, the progressive neurodegeneration results in cognitive dysfunction, with memory loss and motoneural impairment. The administration of PDE5 inhibitors has been studied as a possible therapy for this disease, due to their ability to reverse long-term memory deficits [202,203]. Sildenafil has also been described to improve symptoms of multiple sclerosis [22], while chronic administration of sildenafil or tadalafil appears to have an anxiolytic effect [204]. Moreover, following an acute injury, PDE5 inhibitors are described to enhance endogenous neurogenesis and neuronal function recovery in models of ischemic injury or stroke [205-208]. In addition, sildenafil was shown to stimulate SVZ-derived NSC proliferation, an effect that appears to be dependent on the activation of the PI3-K/Akt pathway [191].

Overall, apart from small differences in the selectivity for PDE5, the majority of PDE5 inhibitors present similar effects in increasing cGMP levels and subsequent activation of nitrergic pathways. In spite of the fact that inhibition of PDE5 does not have an anti-inflammatory effect as NO-NSAID, the neuroprotective effect of PDE5 inhibitors appears to be consensual. However, in the CNS, the effect of PDE5 inhibitors is highly dependent on their permeability to the BBB, and more studies need to be conducted in order to correctly characterize the kinetics on PDE5 inhibitors permeabilization into the CNS. Within this background, the modulation of PDE5 activity could be a good approach to control the levels of cGMP, which could be used in the treatment of several pathologies in which the levels of cGMP are altered. Although there are some studies focused on the stimulation of neurogenesis, the use of inhibitors for PDE5 deserves further investigation in order to clarify their role in controlling different stages of neurogenesis, including migration, differentiation, functionality and survival of newborn neurons, and further understand the mechanisms underlying these effects.

6.3. Other strategies

The involvement of NO in a wide-range of physiological processes and cell function makes it a desirable molecule to use in the clinics, being a major target of pharmaceutical industry. Besides the strategies mentioned above, many synthetic compounds with various chemical and biological modifications have been developed in order to overcome some limiting factors of NO such as its short half-life, the instability during storage and its potential toxicity. Thus, recent innovations in the field of nanotechnology of the profile of NO-donating drugs are being tested to increase the utility and the safety of these compounds in order to be used in biomedical applications, as described below.

There is a wide variety of NO donors that are capable of releasing NO spontaneously or in a controlled way to certain target tissues. The great challenge is how to release NO and to achieve an optimal concentration locally in the brain, thus promoting a therapeutic effect with minimum toxicity [209]. Recent investigations aim at incorporating NO donors into biopolymers mimicking endogenous production of NO at target sites [202]. Nanomaterials are delivery systems with many advantages and a promising therapeutic applicability. These new systems are advantageous due to their: a) small size; b) ability to target specific tissues or cells, having the capacity to cross several biological barriers, such as BBB, reaching tissues that are inaccessible to classic drugs; c) ability to accumulate high drug concentrations; d) enhancement of bioavailability and drug solubility; e) facilitation of drug administration; f) increase of drug circulation in the blood; g) reduction of the dose required to exert an efficient therapeutic effect; and e) decreased local toxicity and reduction of side effects (reviewed by [203]).

The application of nanomaterials to classic NO donors may be an alternative to improve their stability and to therapeutically deliver NO. Among the most studied nanosystems are liposomes and polymeric nanocarriers, such as micelles and hydrogels. Overall, this emergent field of study is of great interest since it allows the development of compounds that release NO in a controlled and sustained way. However, there is a lack of studies concerning the application of these strategies to the CNS. To date, none of these nanosystems is commercially available to target/improve endogenous neurogenesis and further studies are needed in order to develop effective NO-releasing drugs. By these strategies, NO levels in certain targets can be regulated overcoming the traditional limitations of classical NO donors, thus allowing the control of NO levels in specific regions of adult brain in an attempt to repair the lesioned brain.

7. Future directions

Most brain disorders have common features such as neurodegeneration and neuroinflammation. Understanding the mechanisms underlying the evolution of these pathologies, the factors that lead to their onset and the biology of neuronal injury is of extreme importance for the development of efficient therapies, thus allowing to act on risk groups in order to prevent their occurrence. Neurogenesis is an important mechanism of repair in the adult brain, being considered as a critical target to counteract the loss of neurons. As discussed above, two promising strategies could be considered to improve neurogenesis, which include a) transplantation of exogenous neural precursors/stem cells, or b) stimulation of endogenous neurogenesis. However, both strategies for increasing neurogenesis have been linked to an inflammatory response.

Transplantation of exogenous stem cells is a complex and invasive approach with several disadvantages, raising questions about its therapeutic applicability, such as: a) uncontrolled proliferation of grafted cells that may lead to tumor formation, b) the risk of rejection, and c) ethical concerns. However, potentiation of endogenous neurogenesis appears to be a better approach, although with some disadvantages/limitations, such as: a) low yield in the formation of new brain cells, b) low rate of long-term survival of newly generated neurons, and c) poor specificity for increasing local neurogenesis in the target/lesioned tissue. Overall, stimulation of endogenous neurogenesis appears to have higher possibilities for a therapeutic application although it is a less efficient strategy, it has been considered a safer approach when compared to the invasive transplantation of exogenous precursor/stem cells.

Knowing how the inflammatory response affects neurogenesis and the factors that are altered following brain lesion will allow the modulation of certain signaling pathways involved in the regulation of neurogenesis. In fact, the modulation of the nitrergic system could be beneficial for controlling neurogenesis following brain inflammation.

Nitric oxide, by its importance as a regulator of neurogenesis, appears as potential target for the enhancement of endogenous neurogenesis, thus, the development of selective drugs for modulation of the nitrergic signaling pathways is an increasing challenge to pharmaceutical companies. Currently, many strategies are under study for the treatment of CNS disorders, some of them targeting the nitrergic system. The development of NO-NSAID is of great interest as it combines the anti-inflammatory effect to the release of NO, thus reducing the deleterious effects of neuroinflammation and, simultaneously, taking advantage of the pro-neurogenic effect of NO [165]. Moreover, PDE5 inhibitors also seem to be a good strategy to improve neurogenesis, although they lack an anti-inflammatory effect when compared to NO-NSAID [210]. Although little is known about the applicability of this strategy in a regenerative context, recent encouraging evidences support that NO-NSAID and PDE5 inhibitors should be considered as therapeutic strategies to enhance neurogenesis as discussed in this chapter.

In spite of all the evidences showing the important role of the nitrergic system in the modulation of neurogenesis, further studies are needed. In fact, more studies regarding the regulation of migration, differentiation in functional neurons and survival of the newly generated cells must be performed in order to fully understand how these neurogenic events are regulated in an inflammatory context, given the large number of molecular players involved besides NO. Modulation of the nitrergic pathways in a regenerative context should be considered, not as an isolated approach, but instead, as an adjuvant strategy in order to ensure an efficacious therapy.

Acknowledgements

This work was supported by the Foundation for Science and Technology, (FCT, Portugal), project PEst-OE/EQB/LA0023/2011, COMPETE and FEDER (project PTDC/SAU-NEU/102612/2008). Bruno P. Carreira and Ana I. Santos are supported by FCT (fellowships SFRH/BPD/78901/2011 and SFRH/BD/77903/2011).

Author details

Bruno P. Carreira[1], Ana I. Santos[1,2,3], Caetana M. Carvalho[1] and Inês M. Araújo[1,2,3]

*Address all correspondence to: imaraujo@ualg.pt

1 Centre for Neuroscience and Cell Biology, Neuroendocrinology and Neurogenesis Group, University of Coimbra, Coimbra, Portugal

2 Regenerative Medicine Program, Department of Biomedical Sciences and Medicine, University of Algarve, Faro, Portugal

3 IBB-Institute for Biotechnology and Bioengineering, Centre of Molecular and Structural Biomedicine, University of Algarve, Faro, Portugal

References

[1] Moncada S, Bolanos JP: Nitric oxide, cell bioenergetics and neurodegeneration. J Neurochem 2006;97:1676-1689.

[2] Guix FX, Uribesalgo I, Coma M, Munoz FJ: The physiology and pathophysiology of nitric oxide in the brain. Prog Neurobiol 2005;76:126-152.

[3] Calabrese V, Mancuso C, Calvani M, Rizzarelli E, Butterfield DA, Stella AM: Nitric oxide in the central nervous system: Neuroprotection versus neurotoxicity. Nat Rev Neurosci 2007;8:766-775.

[4] Packer MA, Stasiv Y, Benraiss A, Chmielnicki E, Grinberg A, Westphal H, Goldman SA, Enikolopov G: Nitric oxide negatively regulates mammalian adult neurogenesis. Proc Natl Acad Sci U S A 2003;100:9566-9571.

[5] Moreno-Lopez B, Romero-Grimaldi C, Noval JA, Murillo-Carretero M, Matarredona ER, Estrada C: Nitric oxide is a physiological inhibitor of neurogenesis in the adult mouse subventricular zone and olfactory bulb. J Neurosci 2004;24:85-95.

[6] Zhu XJ, Hua Y, Jiang J, Zhou QG, Luo CX, Han X, Lu YM, Zhu DY: Neuronal nitric oxide synthase-derived nitric oxide inhibits neurogenesis in the adult dentate gyrus by down-regulating cyclic amp response element binding protein phosphorylation. Neuroscience 2006;141:827-836.

[7] Fritzen S, Schmitt A, Koth K, Sommer C, Lesch KP, Reif A: Neuronal nitric oxide synthase (nos-i) knockout increases the survival rate of neural cells in the hippocampus independently of bdnf. Mol Cell Neurosci 2007;35:261-271.

[8] Luo CX, Zhu XJ, Zhou QG, Wang B, Wang W, Cai HH, Sun YJ, Hu M, Jiang J, Hua Y, Han X, Zhu DY: Reduced neuronal nitric oxide synthase is involved in ischemia-induced hippocampal neurogenesis by up-regulating inducible nitric oxide synthase expression. J Neurochem 2007;103:1872-1882.

[9] Murillo-Carretero M, Torroglosa A, Castro C, Villalobo A, Estrada C: S-nitrosylation of the epidermal growth factor receptor: A regulatory mechanism of receptor tyrosine kinase activity. Free Radic Biol Med 2009;46:471-479.

[10] Zhang R, Zhang L, Zhang Z, Wang Y, Lu M, Lapointe M, Chopp M: A nitric oxide donor induces neurogenesis and reduces functional deficits after stroke in rats. Ann Neurol 2001;50:602-611.

[11] Lu D, Mahmood A, Zhang R, Copp M: Upregulation of neurogenesis and reduction in functional deficits following administration of deta/nonoate, a nitric oxide donor, after traumatic brain injury in rats. J Neurosurg 2003;99:351-361.

[12] Zhu DY, Liu SH, Sun HS, Lu YM: Expression of inducible nitric oxide synthase after focal cerebral ischemia stimulates neurogenesis in the adult rodent dentate gyrus. J Neurosci 2003;23:223-229.

[13] Carreira BP, Morte MI, Inacio A, Costa G, Rosmaninho-Salgado J, Agasse F, Carmo A, Couceiro P, Brundin P, Ambrosio AF, Carvalho CM, Araujo IM: Nitric oxide stimulates the proliferation of neural stem cells bypassing the epidermal growth factor receptor. Stem Cells 2010;28:1219-1230.

[14] Carreira BP, Morte MI, Lourenco AS, Santos AI, Inacio A, Ambrosio AF, Carvalho CM, Araujo IM: Differential contribution of the guanylyl cyclase-cyclic gmp-protein

kinase g pathway to the proliferation of neural stem cells stimulated by nitric oxide. Neurosignals 2012; epub Feb 23, 2012.

[15] Tegenge MA, Rockel TD, Fritsche E, Bicker G: Nitric oxide stimulates human neural progenitor cell migration via cgmp-mediated signal transduction. Cell Mol Life Sci 2010;68:2089-2099.

[16] Contestabile A: Regulation of transcription factors by nitric oxide in neurons and in neural-derived tumor cells. Prog Neurobiol 2008;84:317-328.

[17] Corbin JD: Mechanisms of action of pde5 inhibition in erectile dysfunction. Int J Impot Res 2004;16 Suppl 1:S4-7.

[18] Hemnes AR, Champion HC: Sildenafil, a pde5 inhibitor, in the treatment of pulmonary hypertension. Expert Rev Cardiovasc Ther 2006;4:293-300.

[19] Keeble JE, Moore PK: Pharmacology and potential therapeutic applications of nitric oxide-releasing non-steroidal anti-inflammatory and related nitric oxide-donating drugs. Br J Pharmacol 2002;137:295-310.

[20] Rutten K, Vente JD, Sik A, Ittersum MM, Prickaerts J, Blokland A: The selective pde5 inhibitor, sildenafil, improves object memory in swiss mice and increases cgmp levels in hippocampal slices. Behav Brain Res 2005;164:11-16.

[21] Koc, Kucukguzel SG: Medicinal chemistry and anti-inflammatory activity of nitric oxide-releasing nsai drugs. Mini Rev Med Chem 2009;9:611-619.

[22] Pifarre P, Prado J, Baltrons MA, Giralt M, Gabarro P, Feinstein DL, Hidalgo J, Garcia A: Sildenafil (viagra) ameliorates clinical symptoms and neuropathology in a mouse model of multiple sclerosis. Acta Neuropathol 2011;121:499-508.

[23] Seri B, Garcia-Verdugo JM, McEwen BS, Alvarez-Buylla A: Astrocytes give rise to new neurons in the adult mammalian hippocampus. J Neurosci 2001;21:7153-7160.

[24] Limke TL, Rao MS: Neural stem cells in aging and disease. J Cell Mol Med 2002;6:475-496.

[25] Doetsch F, Scharff C: Challenges for brain repair: Insights from adult neurogenesis in birds and mammals. Brain Behav Evol 2001;58:306-322.

[26] Doetsch F, Petreanu L, Caille I, Garcia-Verdugo JM, Alvarez-Buylla A: Egf converts transit-amplifying neurogenic precursors in the adult brain into multipotent stem cells. Neuron 2002;36:1021-1034.

[27] Gritti A, Parati EA, Cova L, Frolichsthal P, Galli R, Wanke E, Faravelli L, Morassutti DJ, Roisen F, Nickel DD, Vescovi AL: Multipotential stem cells from the adult mouse brain proliferate and self-renew in response to basic fibroblast growth factor. J Neurosci 1996;16:1091-1100.

[28] Ming GL, Song H: Adult neurogenesis in the mammalian central nervous system. Annu Rev Neurosci 2005;28:223-250.

[29] Ming GL, Song H: Adult neurogenesis in the mammalian brain: Significant answers and significant questions. Neuron 2011;70:687-702.

[30] Lowenstein DH, Parent JM: Brain, heal thyself. Science 1999;283:1126-1127.

[31] Whitney NP, Eidem TM, Peng H, Huang Y, Zheng JC: Inflammation mediates varying effects in neurogenesis: Relevance to the pathogenesis of brain injury and neurodegenerative disorders. J Neurochem 2009;108:1343-1359.

[32] Tatebayashi Y, Lee MH, Li L, Iqbal K, Grundke-Iqbal I: The dentate gyrus neurogenesis: A therapeutic target for alzheimer's disease. Acta Neuropathol 2003;105:225-232.

[33] Gil JM, Mohapel P, Araujo IM, Popovic N, Li JY, Brundin P, Petersen A: Reduced hippocampal neurogenesis in r6/2 transgenic huntington's disease mice. Neurobiol Dis 2005;20:744-751.

[34] Lazic SE, Grote H, Armstrong RJ, Blakemore C, Hannan AJ, van Dellen A, Barker RA: Decreased hippocampal cell proliferation in r6/1 huntington's mice. Neuroreport 2004;15:811-813.

[35] Nencini P, Sarti C, Innocenti R, Pracucci G, Inzitari D: Acute inflammatory events and ischemic stroke subtypes. Cerebrovasc Dis 2003;15:215-221.

[36] Schmidt OI, Heyde CE, Ertel W, Stahel PF: Closed head injury--an inflammatory disease? Brain Res Brain Res Rev 2005;48:388-399.

[37] Stoll G, Jander S: The role of microglia and macrophages in the pathophysiology of the cns. Prog Neurobiol 1999;58:233-247.

[38] Streit WJ, Walter SA, Pennell NA: Reactive microgliosis. Prog Neurobiol 1999;57:563-581.

[39] Wyss-Coray T, Mucke L: Inflammation in neurodegenerative disease--a double-edged sword. Neuron 2002;35:419-432.

[40] Stoll G, Jander S, Schroeter M: Detrimental and beneficial effects of injury-induced inflammation and cytokine expression in the nervous system. Adv Exp Med Biol 2002;513:87-113.

[41] Das S, Basu A: Inflammation: A new candidate in modulating adult neurogenesis. J Neurosci Res 2008;86:1199-1208.

[42] Ekdahl CT: Microglial activation - tuning and pruning adult neurogenesis. Front Pharmacol 2012;3:41.

[43] Ekdahl CT, Kokaia Z, Lindvall O: Brain inflammation and adult neurogenesis: The dual role of microglia. Neuroscience 2009;158:1021-1029.

[44] Hanisch UK, Kettenmann H: Microglia: Active sensor and versatile effector cells in the normal and pathologic brain. Nat Neurosci 2007;10:1387-1394.

[45] Walton NM, Sutter BM, Laywell ED, Levkoff LH, Kearns SM, Marshall GP, 2nd, Scheffler B, Steindler DA: Microglia instruct subventricular zone neurogenesis. Glia 2006;54:815-825.

[46] Gonzalez-Perez O, Jauregui-Huerta F, Galvez-Contreras AY: Immune system modulates the function of adult neural stem cells. Curr Immunol Rev 2010;6:167-173.

[47] Gonzalez-Perez O, Quinones-Hinojosa A, Garcia-Verdugo JM: Immunological control of adult neural stem cells. J Stem Cells 2010;5:23-31.

[48] Ekdahl CT, Claasen JH, Bonde S, Kokaia Z, Lindvall O: Inflammation is detrimental for neurogenesis in adult brain. Proc Natl Acad Sci U S A 2003;100:13632-13637.

[49] Monje ML, Toda H, Palmer TD: Inflammatory blockade restores adult hippocampal neurogenesis. Science 2003;302:1760-1765.

[50] Liu YP, Lin HI, Tzeng SF: Tumor necrosis factor-alpha and interleukin-18 modulate neuronal cell fate in embryonic neural progenitor culture. Brain Res 2005;1054:152-158.

[51] Cacci E, Ajmone-Cat MA, Anelli T, Biagioni S, Minghetti L: In vitro neuronal and glial differentiation from embryonic or adult neural precursor cells are differently affected by chronic or acute activation of microglia. Glia 2008;56:412-425.

[52] Ben-Hur T, Ben-Menachem O, Furer V, Einstein O, Mizrachi-Kol R, Grigoriadis N: Effects of proinflammatory cytokines on the growth, fate, and motility of multipotential neural precursor cells. Mol Cell Neurosci 2003;24:623-631.

[53] Cacci E, Claasen JH, Kokaia Z: Microglia-derived tumor necrosis factor-alpha exaggerates death of newborn hippocampal progenitor cells in vitro. J Neurosci Res 2005;80:789-797.

[54] Iosif RE, Ekdahl CT, Ahlenius H, Pronk CJ, Bonde S, Kokaia Z, Jacobsen SE, Lindvall O: Tumor necrosis factor receptor 1 is a negative regulator of progenitor proliferation in adult hippocampal neurogenesis. J Neurosci 2006;26:9703-9712.

[55] Koo JW, Duman RS: Il-1beta is an essential mediator of the antineurogenic and anhedonic effects of stress. Proc Natl Acad Sci U S A 2008;105:751-756.

[56] Vallieres L, Campbell IL, Gage FH, Sawchenko PE: Reduced hippocampal neurogenesis in adult transgenic mice with chronic astrocytic production of interleukin-6. J Neurosci 2002;22:486-492.

[57] Liu BF, Gao EJ, Zeng XZ, Ji M, Cai Q, Lu Q, Yang H, Xu QY: Proliferation of neural precursors in the subventricular zone after chemical lesions of the nigrostriatal pathway in rat brain. Brain Res 2006;1106:30-39.

[58] Liu Z, Fan Y, Won SJ, Neumann M, Hu D, Zhou L, Weinstein PR, Liu J: Chronic treatment with minocycline preserves adult new neurons and reduces functional impairment after focal cerebral ischemia. Stroke 2007;38:146-152.

[59] Rueger MA, Muesken S, Walberer M, Jantzen SU, Schnakenburg K, Backes H, Graf R, Neumaier B, Hoehn M, Fink GR, Schroeter M: Effects of minocycline on endogenous neural stem cells after experimental stroke. Neuroscience 2012;215:174-183.

[60] Simard AR, Rivest S: Role of inflammation in the neurobiology of stem cells. Neuroreport 2004;15:2305-2310.

[61] Hoehn BD, Palmer TD, Steinberg GK: Neurogenesis in rats after focal cerebral ischemia is enhanced by indomethacin. Stroke 2005;36:2718-2724.

[62] Arvidsson A, Collin T, Kirik D, Kokaia Z, Lindvall O: Neuronal replacement from endogenous precursors in the adult brain after stroke. Nat Med 2002;8:963-970.

[63] Nakatomi H, Kuriu T, Okabe S, Yamamoto S, Hatano O, Kawahara N, Tamura A, Kirino T, Nakafuku M: Regeneration of hippocampal pyramidal neurons after ischemic brain injury by recruitment of endogenous neural progenitors. Cell 2002;110:429-441.

[64] Butovsky O, Ziv Y, Schwartz A, Landa G, Talpalar AE, Pluchino S, Martino G, Schwartz M: Microglia activated by il-4 or ifn-gamma differentially induce neurogenesis and oligodendrogenesis from adult stem/progenitor cells. Mol Cell Neurosci 2006;31:149-160.

[65] Aarum J, Sandberg K, Haeberlein SL, Persson MA: Migration and differentiation of neural precursor cells can be directed by microglia. Proc Natl Acad Sci U S A 2003;100:15983-15988.

[66] Morgan SC, Taylor DL, Pocock JM: Microglia release activators of neuronal proliferation mediated by activation of mitogen-activated protein kinase, phosphatidylinositol-3-kinase/akt and delta-notch signalling cascades. J Neurochem 2004;90:89-101.

[67] Nakanishi M, Niidome T, Matsuda S, Akaike A, Kihara T, Sugimoto H: Microglia-derived interleukin-6 and leukaemia inhibitory factor promote astrocytic differentiation of neural stem/progenitor cells. Eur J Neurosci 2007;25:649-658.

[68] Jakubs K, Bonde S, Iosif RE, Ekdahl CT, Kokaia Z, Kokaia M, Lindvall O: Inflammation regulates functional integration of neurons born in adult brain. J Neurosci 2008;28:12477-12488.

[69] Thored P, Arvidsson A, Cacci E, Ahlenius H, Kallur T, Darsalia V, Ekdahl CT, Kokaia Z, Lindvall O: Persistent production of neurons from adult brain stem cells during recovery after stroke. Stem Cells 2006;24:739-747.

[70] Kokaia Z, Thored P, Arvidsson A, Lindvall O: Regulation of stroke-induced neurogenesis in adult brain--recent scientific progress. Cereb Cortex 2006;16 Suppl 1:i162-167.

[71] Bonde S, Ekdahl CT, Lindvall O: Long-term neuronal replacement in adult rat hippocampus after status epilepticus despite chronic inflammation. Eur J Neurosci 2006;23:965-974.

[72] Thored P, Heldmann U, Gomes-Leal W, Gisler R, Darsalia V, Taneera J, Nygren JM, Jacobsen SE, Ekdahl CT, Kokaia Z, Lindvall O: Long-term accumulation of microglia with proneurogenic phenotype concomitant with persistent neurogenesis in adult subventricular zone after stroke. Glia 2009;57:835-849.

[73] Imitola J, Raddassi K, Park KI, Mueller FJ, Nieto M, Teng YD, Frenkel D, Li J, Sidman RL, Walsh CA, Snyder EY, Khoury SJ: Directed migration of neural stem cells to sites of cns injury by the stromal cell-derived factor 1alpha/cxc chemokine receptor 4 pathway. Proc Natl Acad Sci U S A 2004;101:18117-18122.

[74] Wang X, Fu S, Wang Y, Yu P, Hu J, Gu W, Xu XM, Lu P: Interleukin-1beta mediates proliferation and differentiation of multipotent neural precursor cells through the activation of sapk/jnk pathway. Mol Cell Neurosci 2007;36:343-354.

[75] Green HF, Treacy E, Keohane AK, Sullivan AM, O'Keeffe GW, Nolan YM: A role for interleukin-1beta in determining the lineage fate of embryonic rat hippocampal neural precursor cells. Mol Cell Neurosci 2012;49:311-321.

[76] Wong G, Goldshmit Y, Turnley AM: Interferon-gamma but not tnf alpha promotes neuronal differentiation and neurite outgrowth of murine adult neural stem cells. Exp Neurol 2004;187:171-177.

[77] Song JH, Wang CX, Song DK, Wang P, Shuaib A, Hao C: Interferon gamma induces neurite outgrowth by up-regulation of p35 neuron-specific cyclin-dependent kinase 5 activator via activation of erk1/2 pathway. J Biol Chem 2005;280:12896-12901.

[78] Baron R, Nemirovsky A, Harpaz I, Cohen H, Owens T, Monsonego A: Ifn-gamma enhances neurogenesis in wild-type mice and in a mouse model of alzheimer's disease. Faseb J 2008;22:2843-2852.

[79] Widera D, Mikenberg I, Elvers M, Kaltschmidt C, Kaltschmidt B: Tumor necrosis factor alpha triggers proliferation of adult neural stem cells via ikk/nf-kappab signaling. BMC Neurosci 2006;7:64.

[80] Wu JP, Kuo JS, Liu YL, Tzeng SF: Tumor necrosis factor-alpha modulates the proliferation of neural progenitors in the subventricular/ventricular zone of adult rat brain. Neurosci Lett 2000;292:203-206.

[81] Iosif RE, Ahlenius H, Ekdahl CT, Darsalia V, Thored P, Jovinge S, Kokaia Z, Lindvall O: Suppression of stroke-induced progenitor proliferation in adult subventricular

zone by tumor necrosis factor receptor 1. J Cereb Blood Flow Metab 2008;28:1574-1587.

[82] Heldmann U, Thored P, Claasen JH, Arvidsson A, Kokaia Z, Lindvall O: Tnf-alpha antibody infusion impairs survival of stroke-generated neuroblasts in adult rat brain. Exp Neurol 2005;196:204-208.

[83] Bernardino L, Agasse F, Silva B, Ferreira R, Grade S, Malva JO: Tumor necrosis factor-alpha modulates survival, proliferation, and neuronal differentiation in neonatal subventricular zone cell cultures. Stem Cells 2008;26:2361-2371.

[84] Peng H, Kolb R, Kennedy JE, Zheng J: Differential expression of cxcl12 and cxcr4 during human fetal neural progenitor cell differentiation. J Neuroimmune Pharmacol 2007;2:251-258.

[85] Pacher P, Beckman JS, Liaudet L: Nitric oxide and peroxynitrite in health and disease. Physiol Rev 2007;87:315-424.

[86] Bredt DS: Endogenous nitric oxide synthesis: Biological functions and pathophysiology. Free Radic Res 1999;31:577-596.

[87] Dawson TM, Snyder SH: Gases as biological messengers: Nitric oxide and carbon monoxide in the brain. J Neurosci 1994;14:5147-5159.

[88] Alderton WK, Cooper CE, Knowles RG: Nitric oxide synthases: Structure, function and inhibition. Biochem J 2001;357:593-615.

[89] Garthwaite J, Boulton CL: Nitric oxide signaling in the central nervous system. Annu Rev Physiol 1995;57:683-706.

[90] Prast H, Philippu A: Nitric oxide as modulator of neuronal function. Prog Neurobiol 2001;64:51-68.

[91] Garthwaite J, Charles SL, Chess-Williams R: Endothelium-derived relaxing factor release on activation of nmda receptors suggests role as intercellular messenger in the brain. Nature 1988;336:385-388.

[92] Brown GC, Bolanos JP, Heales SJ, Clark JB: Nitric oxide produced by activated astrocytes rapidly and reversibly inhibits cellular respiration. Neurosci Lett 1995;193:201-204.

[93] Dawson VL, Dawson TM: Nitric oxide neurotoxicity. J Chem Neuroanat 1996;10:179-190.

[94] Simmons ML, Murphy S: Induction of nitric oxide synthase in glial cells. J Neurochem 1992;59:897-905.

[95] Murphy S: Production of nitric oxide by glial cells: Regulation and potential roles in the cns. Glia 2000;29:1-13.

[96] Heneka MT, Feinstein DL: Expression and function of inducible nitric oxide synthase in neurons. J Neuroimmunol 2001;114:8-18.

[97] Serou MJ, DeCoster MA, Bazan NG: Interleukin-1 beta activates expression of cyclooxygenase-2 and inducible nitric oxide synthase in primary hippocampal neuronal culture: Platelet-activating factor as a preferential mediator of cyclooxygenase-2 expression. J Neurosci Res 1999;58:593-598.

[98] Chao CC, Hu S, Molitor TW, Shaskan EG, Peterson PK: Activated microglia mediate neuronal cell injury via a nitric oxide mechanism. J Immunol 1992;149:2736-2741.

[99] Galea E, Feinstein DL, Reis DJ: Induction of calcium-independent nitric oxide synthase activity in primary rat glial cultures. Proc Natl Acad Sci U S A 1992;89:10945-10949.

[100] Murphy S, Simmons ML, Agullo L, Garcia A, Feinstein DL, Galea E, Reis DJ, Minc-Golomb D, Schwartz JP: Synthesis of nitric oxide in cns glial cells. Trends Neurosci 1993;16:323-328.

[101] Nathan C, Xie QW: Nitric oxide synthases: Roles, tolls, and controls. Cell 1994;78:915-918.

[102] Schlossmann J, Hofmann F: Cgmp-dependent protein kinases in drug discovery. Drug Discov Today 2005;10:627-634.

[103] Hanafy KA, Krumenacker JS, Murad F: No, nitrotyrosine, and cyclic gmp in signal transduction. Med Sci Monit 2001;7:801-819.

[104] Lee SY, Andoh T, Murphy DL, Chiueh CC: 17beta-estradiol activates ici 182,780-sensitive estrogen receptors and cyclic gmp-dependent thioredoxin expression for neuroprotection. Faseb J 2003;17:947-948.

[105] Cunningham GA, McClenaghan NH, Flatt PR, Newsholme P: L-alanine induces changes in metabolic and signal transduction gene expression in a clonal rat pancreatic beta-cell line and protects from pro-inflammatory cytokine-induced apoptosis. Clin Sci (Lond) 2005;109:447-455.

[106] Calabrese V, Cornelius C, Rizzarelli E, Owen JB, Dinkova-Kostova AT, Butterfield DA: Nitric oxide in cell survival: A janus molecule. Antioxid Redox Signal 2009;11:2717-2739.

[107] Matarredona ER, Murillo-Carretero M, Moreno-Lopez B, Estrada C: Nitric oxide synthesis inhibition increases proliferation of neural precursors isolated from the postnatal mouse subventricular zone. Brain Res 2004;995:274-284.

[108] Park C, Kang M, Kim-Kwon Y, Kim J, Ahn H, Huh Y: Inhibition of neuronal nitric oxide synthase increases adrenalectomy-induced granule cell death in the rat dentate gyrus. Brain Res 2002;933:81-84.

[109] Goldman SA, Chen Z: Perivascular instruction of cell genesis and fate in the adult brain. Nat Neurosci 2011;14:1382-1389.

[110] Moreno-Lopez B, Noval JA, Gonzalez-Bonet LG, Estrada C: Morphological bases for a role of nitric oxide in adult neurogenesis. Brain Res 2000;869:244-250.

[111] Reif A, Schmitt A, Fritzen S, Chourbaji S, Bartsch C, Urani A, Wycislo M, Mossner R, Sommer C, Gass P, Lesch KP: Differential effect of endothelial nitric oxide synthase (nos-iii) on the regulation of adult neurogenesis and behaviour. Eur J Neurosci 2004;20:885-895.

[112] Luo CX, Jin X, Cao CC, Zhu MM, Wang B, Chang L, Zhou QG, Wu HY, Zhu DY: Bi-directional regulation of neurogenesis by neuronal nitric oxide synthase derived from neurons and neural stem cells. Stem Cells 2010;28:2041-2052.

[113] Vieira HL, Alves PM, Vercelli A: Modulation of neuronal stem cell differentiation by hypoxia and reactive oxygen species. Prog Neurobiol 2011;93:444-455.

[114] Charles N, Ozawa T, Squatrito M, Bleau AM, Brennan CW, Hambardzumyan D, Holland EC: Perivascular nitric oxide activates notch signaling and promotes stem-like character in pdgf-induced glioma cells. Cell Stem Cell 2010;6:141-152.

[115] Eyler CE, Wu Q, Yan K, MacSwords JM, Chandler-Militello D, Misuraca KL, Lathia JD, Forrester MT, Lee J, Stamler JS, Goldman SA, Bredel M, McLendon RE, Sloan AE, Hjelmeland AB, Rich JN: Glioma stem cell proliferation and tumor growth are promoted by nitric oxide synthase-2. Cell 2011;146:53-66.

[116] Vercelli A, Repici M, Biasiol S, Jhaveri S: Maturation of nadph-d activity in the rat's barrel-field cortex and its relationship to cytochrome oxidase activity. Exp Neurol 1999;156:294-315.

[117] Bredt DS, Snyder SH: Transient nitric oxide synthase neurons in embryonic cerebral cortical plate, sensory ganglia, and olfactory epithelium. Neuron 1994;13:301-313.

[118] Bredt DS, Glatt CE, Hwang PM, Fotuhi M, Dawson TM, Snyder SH: Nitric oxide synthase protein and mrna are discretely localized in neuronal populations of the mammalian cns together with nadph diaphorase. Neuron 1991;7:615-624.

[119] Garbossa D, Fontanella M, Tomasi S, Ducati A, Vercelli A: Differential distribution of nadph-diaphorase histochemistry in human cerebral cortex. Brain Res 2005;1034:1-10.

[120] Chen J, Tu Y, Moon C, Matarazzo V, Palmer AM, Ronnett GV: The localization of neuronal nitric oxide synthase may influence its role in neuronal precursor proliferation and synaptic maintenance. Dev Biol 2004;269:165-182.

[121] Arancio O, Kiebler M, Lee CJ, Lev-Ram V, Tsien RY, Kandel ER, Hawkins RD: Nitric oxide acts directly in the presynaptic neuron to produce long-term potentiation in cultured hippocampal neurons. Cell 1996;87:1025-1035.

[122] Roskams AJ, Bredt DS, Dawson TM, Ronnett GV: Nitric oxide mediates the formation of synaptic connections in developing and regenerating olfactory receptor neurons. Neuron 1994;13:289-299.

[123] Weruaga E, Brinon JG, Porteros A, Arevalo R, Aijon J, Alonso JR: Expression of neuronal nitric oxide synthase/nadph-diaphorase during olfactory deafferentation and regeneration. Eur J Neurosci 2000;12:1177-1193.

[124] Santacana M, Uttenthal LO, Bentura ML, Fernandez AP, Serrano J, Martinez de Velasco J, Alonso D, Martinez-Murillo R, Rodrigo J: Expression of neuronal nitric oxide synthase during embryonic development of the rat cerebral cortex. Brain Res Dev Brain Res 1998;111:205-222.

[125] Soygüder Z, Karadağ H, Nazli M: Neuronal nitric oxide synthase immunoreactivity in ependymal cells during early postnatal development. Journal of Chemical Neuroanatomy 2004;27:3-6.

[126] Riquelme PA, Drapeau E, Doetsch F: Brain micro-ecologies: Neural stem cell niches in the adult mammalian brain. Philos Trans R Soc Lond B Biol Sci 2008;363:123-137.

[127] Peunova N, Enikolopov G: Nitric oxide triggers a switch to growth arrest during differentiation of neuronal cells. Nature 1995;375:68-73.

[128] Poluha W, Schonhoff CM, Harrington KS, Lachyankar MB, Crosbie NE, Bulseco DA, Ross AH: A novel, nerve growth factor-activated pathway involving nitric oxide, p53, and p21waf1 regulates neuronal differentiation of pc12 cells. J Biol Chem 1997;272:24002-24007.

[129] Cheng A, Wang S, Cai J, Rao MS, Mattson MP: Nitric oxide acts in a positive feedback loop with bdnf to regulate neural progenitor cell proliferation and differentiation in the mammalian brain. Dev Biol 2003;258:319-333.

[130] Kokaia Z, Lindvall O: Neurogenesis after ischaemic brain insults. Curr Opin Neurobiol 2003;13:127-132.

[131] Ekdahl CT, Mohapel P, Elmer E, Lindvall O: Caspase inhibitors increase short-term survival of progenitor-cell progeny in the adult rat dentate gyrus following status epilepticus. Eur J Neurosci 2001;14:937-945.

[132] Kim YM, Talanian RV, Billiar TR: Nitric oxide inhibits apoptosis by preventing increases in caspase-3-like activity via two distinct mechanisms. J Biol Chem 1997;272:31138-31148.

[133] Holscher C: Nitric oxide, the enigmatic neuronal messenger: Its role in synaptic plasticity. Trends Neurosci 1997;20:298-303.

[134] Torroglosa A, Murillo-Carretero M, Romero-Grimaldi C, Matarredona ER, Campos-Caro A, Estrada C: Nitric oxide decreases subventricular zone stem cell proliferation

by inhibition of epidermal growth factor receptor and phosphoinositide-3-kinase/akt pathway. Stem Cells 2007;25:88-97.

[135] Sun Y, Jin K, Childs JT, Xie L, Mao XO, Greenberg DA: Neuronal nitric oxide synthase and ischemia-induced neurogenesis. J Cereb Blood Flow Metab 2005;25:485-492.

[136] Estrada C, Murillo-Carretero M: Nitric oxide and adult neurogenesis in health and disease. Neuroscientist 2005;11:294-307.

[137] Corsani L, Bizzoco E, Pedata F, Gianfriddo M, Faussone-Pellegrini MS, Vannucchi MG: Inducible nitric oxide synthase appears and is co-expressed with the neuronal isoform in interneurons of the rat hippocampus after transient ischemia induced by middle cerebral artery occlusion. Exp Neurol 2008;211:433-440.

[138] Cardenas A, Moro MA, Hurtado O, Leza JC, Lorenzo P, Castrillo A, Bodelon OG, Bosca L, Lizasoain I: Implication of glutamate in the expression of inducible nitric oxide synthase after oxygen and glucose deprivation in rat forebrain slices. J Neurochem 2000;74:2041-2048.

[139] Jander S, Schroeter M, Stoll G: Role of nmda receptor signaling in the regulation of inflammatory gene expression after focal brain ischemia. J Neuroimmunol 2000;109:181-187.

[140] Jiang W, Xiao L, Wang JC, Huang YG, Zhang X: Effects of nitric oxide on dentate gyrus cell proliferation after seizures induced by pentylenetrazol in the adult rat brain. Neurosci Lett 2004;367:344-348.

[141] Parent JM: Adult neurogenesis in the intact and epileptic dentate gyrus. Prog Brain Res 2007;163:529-540.

[142] Covacu R, Danilov AI, Rasmussen BS, Hallen K, Moe MC, Lobell A, Johansson CB, Svensson MA, Olsson T, Brundin L: Nitric oxide exposure diverts neural stem cell fate from neurogenesis towards astrogliogenesis. Stem Cells 2006;24:2792-2800.

[143] Cardenas A, Moro MA, Hurtado O, Leza JC, Lizasoain I: Dual role of nitric oxide in adult neurogenesis. Brain Res Brain Res Rev 2005;50:1-6.

[144] Matarredona ER, Murillo-Carretero M, Moreno-Lopez B, Estrada C: Role of nitric oxide in subventricular zone neurogenesis. Brain Res Brain Res Rev 2005;49:355-366.

[145] Alvaro AR, Martins J, Araujo IM, Rosmaninho-Salgado J, Ambrosio AF, Cavadas C: Neuropeptide y stimulates retinal neural cell proliferation--involvement of nitric oxide. J Neurochem 2008;105:2501-2510.

[146] Zhang RL, Zhang ZG, Zhang L, Chopp M: Proliferation and differentiation of progenitor cells in the cortex and the subventricular zone in the adult rat after focal cerebral ischemia. Neuroscience 2001;105:33-41.

[147] Park C, Shin KS, Ryu JH, Kang K, Kim J, Ahn H, Huh Y: The inhibition of nitric oxide synthase enhances psa-ncam expression and creb phosphorylation in the rat hippocampus. Neuroreport 2004;15:231-234.

[148] Chen J, Zacharek A, Zhang C, Jiang H, Li Y, Roberts C, Lu M, Kapke A, Chopp M: Endothelial nitric oxide synthase regulates brain-derived neurotrophic factor expression and neurogenesis after stroke in mice. J Neurosci 2005;25:2366-2375.

[149] Arvidsson A, Kokaia Z, Lindvall O: N-methyl-d-aspartate receptor-mediated increase of neurogenesis in adult rat dentate gyrus following stroke. Eur J Neurosci 2001;14:10-18.

[150] Shen L, Zhang J: Nmda receptor and inos are involved in the effects of ginsenoside rg1 on hippocampal neurogenesis in ischemic gerbils. Neurol Res 2007;29:270-273.

[151] Luo CX, Zhu XJ, Zhang AX, Wang W, Yang XM, Liu SH, Han X, Sun J, Zhang SG, Lu Y, Zhu DY: Blockade of l-type voltage-gated ca channel inhibits ischemia-induced neurogenesis by down-regulating inos expression in adult mouse. J Neurochem 2005;94:1077-1086.

[152] Megson IL, Webb DJ: Nitric oxide donor drugs: Current status and future trends. Expert Opin Investig Drugs 2002;11:587-601.

[153] Feelisch M, Kelm M: Biotransformation of organic nitrates to nitric oxide by vascular smooth muscle and endothelial cells. Biochem Biophys Res Commun 1991;180:286-293.

[154] Scheiman JM, Fendrick AM: Summing the risk of nsaid therapy. The Lancet 2007;369:1580-1581.

[155] Aalykke C, Lauritsen K: Epidemiology of nsaid-related gastroduodenal mucosal injury. Best practice & research Clinical gastroenterology 2001;15:705-722.

[156] Harirforoosh S, Jamali F: Renal adverse effects of nonsteroidal anti-inflammatory drugs. Expert Opinion on Drug Safety 2009;8:669-681.

[157] Martin GR, Wallace JL: Gastrointestinal inflammation: A central component of mucosal defense and repair. Experimental Biology and Medicine 2006;231:130-137.

[158] Stanek A, Gadowska-Cicha A, Gawron K, Wielkoszynski T, Adamek B, Cieslar G, Wiczkowski A, Sieron A: Role of nitric oxide in physiology and pathology of the gastrointestinal tract. Mini Rev Med Chem 2008;8:1549-1560.

[159] Wallace JL, Reuter B, Cicala C, McKnight W, Grisham M, Cirino G: A diclofenac derivative without ulcerogenic properties. European Journal of Pharmacology 1994;257:249-255.

[160] Wallace JL, Reuter B, Cicala C, McKnight W, Grisham MB, Cirino G: Novel nonsteroidal anti-inflammatory drug derivatives with markedly reduced ulcerogenic properties in the rat. Gastroenterology 1994;107:173-179.

[161] Fiorucci S, Antonelli E, Santucci L, Morelli O, Miglietti M, Federici B, Mannucci R, Soldato PD, Morelli A: Gastrointestinal safety of nitric oxide–derived aspirin is related to inhibition of ice-like cysteine proteases in rats. Gastroenterology 1999;116:1089-1106.

[162] Fiorucci S: No-releasing nsaids are caspase inhibitors. Trends in Immunology 2001;22:232-235.

[163] Fiorucci, Santucci, Federici, Antonelli, Distrutti, Morelli, Renzo, Coata, Cirino, Soldato: Nitric oxide-releasing nsaids inhibit interleukin-1β converting enzyme-like cysteine proteases and protect endothelial cells from apoptosis induced by tnfα. Alimentary Pharmacology & Therapeutics 1999;13:421-435.

[164] Dimmeler S, Haendeler J, Nehls M, Zeiher AM: Suppression of apoptosis by nitric oxide via inhibition of interleukin-1β–converting enzyme (ice)-like and cysteine protease protein (cpp)-32–like proteases. The Journal of Experimental Medicine 1997;185:601-608.

[165] Ajmone-Cat MA, Cacci E, Minghetti L: Non steroidal anti-inflammatory drugs and neurogenesis in the adult mammalian brain. Curr Pharm Des 2008;14:1435-1442.

[166] Wallace JL, Vergnolle N, Muscará MN, Asfaha S, Chapman K, McKnight W, Soldato PD, Morelli A, Fiorucci S: Enhanced anti-inflammatory effects of a nitric oxide–releasing derivative of mesalamine in rats. Gastroenterology 1999;117:557-566.

[167] Griffin WS, Sheng JG, Royston MC, Gentleman SM, McKenzie JE, Graham DI, Roberts GW, Mrak RE: Glial-neuronal interactions in alzheimer's disease: The potential role of a 'cytokine cycle' in disease progression. Brain Pathol 1998;8:65-72.

[168] Shimohama S, Tanino H, Fujimoto S: Changes in caspase expression in alzheimer's disease: Comparison with development and aging. Biochemical and Biophysical Research Communications 1999;256:381-384.

[169] Broe GA, Grayson DA, Creasey HM, Waite LM, Casey BJ, Bennett HP, Brooks WS, Halliday GM: Anti-inflammatory drugs protect against alzheimer disease at low doses. Arch Neurol 2000;57:1586-1591.

[170] Andersen K, Launer LJ, Ott A, Hoes AW, Breteler MM, Hofman A: Do nonsteroidal anti-inflammatory drugs decrease the risk for alzheimer's disease? The rotterdam study. Neurology 1995;45:1441-1445.

[171] Blasko I, Apochal A, Boeck G, Hartmann T, Grubeck-Loebenstein B, Ransmayr G: Ibuprofen decreases cytokine-induced amyloid beta production in neuronal cells. Neurobiology of Disease 2001;8:1094-1101.

[172] in 't Veld BA, Ruitenberg A, Hofman A, Launer LJ, van Duijn CM, Stijnen T, Breteler MMB, Stricker BHC: Nonsteroidal antiinflammatory drugs and the risk of alzheimer's disease. New England Journal of Medicine 2001;345:1515-1521.

[173] Hüll M, Lieb K, Fiebich BL: Anti-inflammatory drugs: A hope for alzheimer's disease? Expert Opinion on Investigational Drugs 2000;9:671-683.

[174] Hoehn BD, Palmer TD, Steinberg GK: Neurogenesis in rats after focal cerebral ischemia is enhanced by indomethacin. Stroke 2005;36:2718-2724.

[175] Parepally JM, Mandula H, Smith QR: Brain uptake of nonsteroidal anti-inflammatory drugs: Ibuprofen, flurbiprofen, and indomethacin. Pharm Res 2006;23:873-881.

[176] Hauss-Wegrzyniak B, Willard LB, Del Soldato P, Pepeu G, Wenk GL: Peripheral administration of novel anti-inflammatories can attenuate the effects of chronic inflammation within the cns. Brain Research 1999;815:36-43.

[177] Jantzen PT, Connor KE, DiCarlo G, Wenk GL, Wallace JL, Rojiani AM, Coppola D, Morgan D, Gordon MN: Microglial activation and β-amyloid deposit reduction caused by a nitric oxide-releasing nonsteroidal anti-inflammatory drug in amyloid precursor protein plus presenilin-1 transgenic mice. The Journal of Neuroscience 2002;22:2246-2254.

[178] Hauss-Wegrzyniak B, Vraniak P, Wenk GL: The effects of a novel nsaid on chronic neuroinflammation are age dependent. Neurobiol Aging 1999;20:305-313.

[179] Ajmone-Cat MA, Nicolini A, Minghetti L: Differential effects of the nonsteroidal anti-inflammatory drug flurbiprofen and its nitric oxide-releasing derivative, nitroflurbiprofen, on prostaglandin e2, interleukin-1β, and nitric oxide synthesis by activated microglia. Journal of Neuroscience Research 2001;66:715-722.

[180] Jurado S, Sanchez-Prieto J, Torres M: Expression of cgmp-dependent protein kinases (i and ii) and neuronal nitric oxide synthase in the developing rat cerebellum. Brain Res Bull 2005;65:111-115.

[181] Ignarro LJ: Signal transduction mechanisms involving nitric oxide. Biochem Pharmacol 1991;41:485-490.

[182] Murad F: Regulation of cytosolic guanylyl cyclase by nitric oxide: The no-cyclic gmp signal transduction system. Adv Pharmacol 1994;26:19-33.

[183] Madhusoodanan KS, Murad F: No-cgmp signaling and regenerative medicine involving stem cells. Neurochemical Research 2007;32:681-694.

[184] Lugnier C: Cyclic nucleotide phosphodiesterase (pde) superfamily: A new target for the development of specific therapeutic agents. Pharmacol Ther 2006;109:366-398.

[185] Krumenacker JS, Hanafy KA, Murad F: Regulation of nitric oxide and soluble guanylyl cyclase. Brain Res Bull 2004;62:505-515.

[186] Lincoln TM, Hall CL, Park CR, Corbin JD: Guanosine 3':5'-cyclic monophosphate binding proteins in rat tissues. Proc Natl Acad Sci U S A 1976;73:2559-2563.

[187] Coquil JF, Brunelle G, Guedon J: Occurrence of the methylisobutylxanthine-stimulated cyclic gmp binding protein in various rat tissues. Biochem Biophys Res Commun 1985;127:226-231.

[188] Wallis RM, Corbin JD, Francis SH, Ellis P: Tissue distribution of phosphodiesterase families and the effects of sildenafil on tissue cyclic nucleotides, platelet function, and the contractile responses of trabeculae carneae and aortic rings in vitro. Am J Cardiol 1999;83:3C-12C.

[189] Shimizu-Albergine M, Rybalkin SD, Rybalkina IG, Feil R, Wolfsgruber W, Hofmann F, Beavo JA: Individual cerebellar purkinje cells express different cgmp phosphodiesterases (pdes): In vivo phosphorylation of cgmp-specific pde (pde5) as an indicator of cgmp-dependent protein kinase (pkg) activation. J Neurosci 2003;23:6452-6459.

[190] Bender AT, Beavo JA: Specific localized expression of cgmp pdes in purkinje neurons and macrophages. Neurochem Int 2004;45:853-857.

[191] Wang L, Gang Zhang Z, Lan Zhang R, Chopp M: Activation of the pi3-k/akt pathway mediates cgmp enhanced-neurogenesis in the adult progenitor cells derived from the subventricular zone. J Cereb Blood Flow Metab 2005;25:1150-1158.

[192] Ballard SA, Gingell CJ, Tang K, Turner LA, Price ME, Naylor AM: Effects of sildenafil on the relaxation of human corpus cavernosum tissue in vitro and on the activities of cyclic nucleotide phosphodiesterase isozymes. J Urol 1998;159:2164-2171.

[193] Mochida H, Takagi M, Inoue H, Noto T, Yano K, Fujishige K, Sasaki T, Yuasa K, Kotera J, Omori K, Kikkawa K: Enzymological and pharmacological profile of t-0156, a potent and selective phosphodiesterase type 5 inhibitor. Eur J Pharmacol 2002;456:91-98.

[194] Chalimoniuk M, Strosznajder JB: Aging modulates nitric oxide synthesis and cgmp levels in hippocampus and cerebellum. Effects of amyloid beta peptide. Mol Chem Neuropathol 1998;35:77-95.

[195] Chen Q, Nakajima A, Choi SH, Xiong X, Sisodia SS, Tang Y-P: Adult neurogenesis is functionally associated with ad-like neurodegeneration. Neurobiology of Disease 2008;29:316-326.

[196] Reneerkens OAH, Rutten K, Akkerman S, Blokland A, Shaffer CL, Menniti FS, Steinbusch HWM, Prickaerts J: Phosphodiesterase type 5 (pde5) inhibition improves object recognition memory: Indications for central and peripheral mechanisms. Neurobiology of Learning and Memory 2012;97:370-379.

[197] Devan BD, Bowker JL, Duffy KB, Bharati IS, Jimenez M, Sierra-Mercado D, Jr., Nelson CM, Spangler EL, Ingram DK: Phosphodiesterase inhibition by sildenafil citrate attenuates a maze learning impairment in rats induced by nitric oxide synthase inhibition. Psychopharmacology (Berl) 2006;183:439-445.

[198] Devan BD, Sierra-Mercado D, Jr., Jimenez M, Bowker JL, Duffy KB, Spangler EL, Ingram DK: Phosphodiesterase inhibition by sildenafil citrate attenuates the learning

impairment induced by blockade of cholinergic muscarinic receptors in rats. Pharmacol Biochem Behav 2004;79:691-699.

[199] Prickaerts J, van Staveren WC, Sik A, Markerink-van Ittersum M, Niewohner U, van der Staay FJ, Blokland A, de Vente J: Effects of two selective phosphodiesterase type 5 inhibitors, sildenafil and vardenafil, on object recognition memory and hippocampal cyclic gmp levels in the rat. Neuroscience 2002;113:351-361.

[200] Singh N, Parle M: Sildenafil improves acquisition and retention of memory in mice. Indian J Physiol Pharmacol 2003;47:318-324.

[201] Puzzo D, Staniszewski A, Deng SX, Privitera L, Leznik E, Liu S, Zhang H, Feng Y, Palmeri A, Landry DW, Arancio O: Phosphodiesterase 5 inhibition improves synaptic function, memory, and amyloid-β load in an alzheimer's disease mouse model. The Journal of Neuroscience 2009;29:8075-8086.

[202] Frost MC, Reynolds MM, Meyerhoff ME: Polymers incorporating nitric oxide releasing/generating substances for improved biocompatibility of blood-contacting medical devices. Biomaterials 2005;26:1685-1693.

[203] Saraiva J, Marotta-Oliveira S, Cicillini S, Eloy Jde O, Marchett J: Nanocarriers for nitric oxide delivery. Journal of Drug Delivery 2011

[204] Liebenberg N, Harvey B, Brand L, Wegener G, Brink C: Chronic treatment with the phosphodiesterase type 5 inhibitors sildenafil and tadalafil display anxiolytic effects in flinders sensitive line rats. Metabolic Brain Disease 2012:1-4.

[205] Zhang RL, Zhang Z, Zhang L, Wang Y, Zhang C, Chopp M: Delayed treatment with sildenafil enhances neurogenesis and improves functional recovery in aged rats after focal cerebral ischemia. J Neurosci Res 2006;83:1213-1219.

[206] Zhang L, Zhang RL, Wang Y, Zhang C, Zhang ZG, Meng H, Chopp M: Functional recovery in aged and young rats after embolic stroke: Treatment with a phosphodiesterase type 5 inhibitor. Stroke 2005;36:847-852.

[207] Zhang R, Wang Y, Zhang L, Zhang Z, Tsang W, Lu M, Chopp M: Sildenafil (viagra) induces neurogenesis and promotes functional recovery after stroke in rats. Stroke 2002;33:2675-2680.

[208] Zhang L, Zhang Z, Zhang RL, Cui Y, LaPointe MC, Silver B, Chopp M: Tadalafil, a long-acting type 5 phosphodiesterase isoenzyme inhibitor, improves neurological functional recovery in a rat model of embolic stroke. Brain Res 2006;1118:192-198.

[209] Scatena R, Bottoni P, Pontoglio A, Giardina B: Pharmacological modulation of nitric oxide release: New pharmacological perspectives, potential benefits and risks. Curr Med Chem 2010;17:61-73.

[210] Uthayathas S, Karuppagounder SS, Thrash BM, Parameshwaran K, Suppiramaniam V, Dhanasekaran M: Versatile effects of sildenafil: Recent pharmacological applications. Pharmacol Rep 2007;59:150-163.

Chapter 8

A Vascular Perspective on Neurogenesis

Joshua S. Goldberg and Karen K. Hirschi

Additional information is available at the end of the chapter

http://dx.doi.org/10.5772/54980

1. Introduction

The vasculature has been identified as a prominent feature across several stem cell niches, suggesting a crucial role in their regulation and maintenance. While a critical component of every organ and tissue, it has adopted specific features for specialized microenvironments. Most notably, the subventricular (SVZ) and subgranular (SGZ) zones of the adult brain harbor unique vascular plexi that are finely tuned to support neural stem cell (NSC) function and behavior. Whether it is through direct contact with, and paracrine signaling from, endothelial and mural cells that comprise blood vessels, or systemically via distribution of soluble factors from the circulation, the vasculature serves as a multifaceted stem cell niche regulator. As emerging evidence continues to emphasize the importance of vascular and nervous system interdependency, it is clear that the vascular compartment in the neural stem cell niche is uniquely poised to coordinate responses of both systems to ensure proper maintenance and regeneration, as needed.

2. Vascular composition and function in the brain

2.1. Brain vascular endothelium

The vasculature is a critical component of every organ and tissue, and has the remarkable ability to integrate systemic signals and directly regulate the local microenvironment. In general, the vasculature provides nutrients and protection; however, it has adopted specialized features for specialized microenvironments. Accordingly, not only does the composition of blood vessels vary (e.g. smooth muscle cell and pericyte coverage, perivascular cell recruitment, extracellular matrix (ECM) deposition), but heterogeneity among the endothelium itself is recognized. Indeed, this endothelial cell (EC) heterogeneity may be at the heart of their vast regulatory potential, allowing control of multiple processes. These include, but are not limited to, angiogenesis, microvascular permeability, vessel wall tone, coagulation and anticoagulation, blood cell generation and trafficking, inflammation, and microenvironment regulation [1-3]. From a functional standpoint, the endothelium displays an incredible division of labor, where a spectrum of responses, both to internal and external stimuli, is carried out. Thus, heterogeneity among the vascular endothelium is a core property bestowing vast regulatory potential [4].

Within the brain, capillaries are tightly integrated within the neural parenchyma. As arterioles traverse deeper into the brain, they become progressively smaller and lose portions of their smooth muscle layer, and are thus termed cerebral capillaries [5]. These capillaries are tubes of EC that are variably surrounded by pericytes or pericyte processes, astrocytes, neurons, and ECM. This minimal composition of capillaries allows for a unique interface that facilitates communication with the underlying tissue environment. Distribution of cerebral capillaries within the brain is relatively heterogeneous, due to regional differences in blood flow and metabolic demand. Owing to their thin walls and slow rate of blood flow, capillaries are engineered to minimize diffusion path length and optimize diffusion time [2]. Surprisingly, the average luminal diameter is ~4.8 μm [6], a length that is somewhat smaller than the diameter of an erythrocyte, requiring red blood cells to deform slightly as they progress through these vessels.

The brain endothelium is characterized by unique features that allow it to selectively control permeability between blood and the central nervous system, which manifests as the blood-brain-barrier (BBB). Specifically, this endothelium is discontinuous and nonfenestrated, with few caveolae at the luminal surface and large numbers of mitochondria [7]. The barrier function is mediated by both a physical barrier, owing to high expression of tight interendothelial junctions, and a highly selective transport system. Interestingly, the basal lamina in capillary beds is common with that of perivascular astrocytic endfeet and pericytes, allowing direct contact of neural cells with the underlying endothelium [8]. These capillary EC are ~0.1 μm thick, giving them a cell volume of only ~20μL/cm^3, cumulatively amounting to just 0.2% of the volume of the entire brain [9]. As the demand for energy must be matched by nutrient supply, the remarkable thinness and surface area of these EC allows for quick, selective, and efficient transport across endothelial membranes.

3. Development and vascularization of the brain

3.1. Embryonic brain development

The initial steps of central nervous system (CNS) development occur prior to gastrulation, beginning with neural plate induction from ectoderm [10]. The neural plate is then patterned along its anterioposterior (AP) and dorsoventral (DV) axes in a dose-dependent fashion, where gradients of secreted morphogens specify distinct neural fates by inducing expression of region-specific transcription factors. It has been reported that fibroblast growth factor (FGF), retinioic acid, and secreted Wnt family members determine AP polarity, while bone morphogenetic proteins (BMPs) and members of the Hedgehog family control mediolateral polarity [11-18]. As the neural tube fuses from the neural plate, the neuroepithelium begins to undergo a complex series of morphological transformations, and begins expressing proteins such as vimentin and nestin, thus marking the first appearance of radial glia in the cerebral cortex [19]. The appearance of projection neurons, originating from the neuroepithelium between E8.5 and E10, is followed closely by the onset of neurogenesis at E11 [20].

In the early stages of embryonic neurogenesis (E11-E13), the first mitotic cortical neurons leave the VZ to form the preplate via interkinetic nuclear migration, independent of radial fibers, creating an intermediate zone (IZ) where postmitotic neurons accumulate to commence differentiation [21]. Subsequently generated neurons continue to leave the VZ and enter the preplate to form the CP (E13-18), further subdividing this region into the subplate and marginal zone (MZ), where the latter becomes lamina I, the most superficial layer of the brain [22]. At this same time, SVZ progenitors generated in the VZ divide and expand the progenitor pool. Excluding layer I, subsequent development is said to occur in an "inside-out" manner, where earlier-born neurons reside in the deeper layers (V, VI), and later-born neurons occupy the more superficial layers (II,III) [23].

Between E11 and E18, neurons proceed radially from the ventricular zone (VZ) to the CP, while interneurons originating from the ganglionic eminence migrate tangentially, traveling perpendicular to radial fibers and parallel to the pial surface [24]. As these neurons reach their final destination, migration ceases, detachment from radial glia occurs, and differentiation begins. Cell lineage studies have revealed that proliferative progenitors of the neural epithelium are for the most part multipotent up until their final mitosis [25-30]. However, committed progenitors appear to be an exception, as their existence in secondary proliferative zones, such as the SVZ and other regions in the adult, have been documented to give rise to various neuronal subtypes, astrocytes, and glia [31, 32].

3.2. Vascularization of the brain

The brain, in general, has a specialized vasculature relative to other organs, and there are specialized microenvironments within the brain that exhibit distinct characteristics and functions. For example, it has been proposed that a unique vascular plexus exists in neurogenic regions of the brain, both during embryonic and adult neurogenesis [33-35], where EC-NSC interactions aid in stem cell maintenance while promoting cell division and NSC expansion [36, 37]. How then, does the vasculature become specialized to fulfill such distinct roles, even within the same tissue?

During early stages of embryonic brain development, the perivascular neural plexus (PVNP) forms around the neural tube at E8.5-E10, from anterior to posterior, yet does not invade the neural tissue until later in development [38]. During E10-E11, the periventricular vascular network advances into the dorsal telencephalon, and by E11 forms a lattice shaped plexus. However, a distinct vascular plexus of periventricular vessels appears in the ventral telencephalon at E9, and by E13 EC invasion into the ventricular zone (VZ) and subventricular zone (SVZ) has generated radially oriented capillaries that extend towards the cortical plate (CP), eventually joining the pial vasculature [39, 40].

Recently, the identification of distinct vascular origins within the developing brain [41] suggests that specialized features of vascular beds of adult germinal regions may begin during embryonic development and persist into adolescence. Previously, the long-standing model of CNS angiogenesis suggested that pial vessels, originating from the perineural plexus surrounding the neural tube, passively sprout into the brain parenchyma and extend radial branches toward the ventricles, where the neurogenic VZ and SVZ are established. Upon arrival to the periventricular area, these pial vessels were thought to form new branches, reverse direction to grow towards the pial surface, and ultimately branch into plexuses [42-44]. However, recent studies suggest that pial and periventricular vessels not only have distinct origins, but develop along independent schedules. In fact, periventricular vessels in the ventral telencephalon are thought to originate from a basal vessel, most likely arising from pharangeal arch arteries [42, 45], situated on the floor of the telencephalic vesicle within the basal ganglia primoridum. As early as E9, pial vessels are observed to encircle the telencephalon, while a spatially distinct population of periventricular vessels is restricted to the ventral telencephalon. From E9-E10, the basal vessel matures to produce periventricular branches in a ventral-to-dorsal and lateral-medial direction, eventually giving rise to a vascular lattice in the dorsal telencephalon. As narrow branches from the periventricular and pial networks fuse, the first arterial-venous communication is thought to occur, as early reports suggest venous sinuses and arterial networks develop from pial and periventricular vessels, respectively [41, 45]. At E15, the first tangential vessels to the pial surface emerge in the intermediate zone, and by E16, these vessels appear in the presumptive rostral migratory stream (RMS). By E18, extensive vascular remodeling has taken place, and the ventricular plexus loses much of its definition. However, upon reaching postnatal ages and adulthood, blood vessels begin to align themselves longitudinally and parallel to each other in the direction of the RMS, presenting a more homogeneous structure [35, 40].

Interestingly, the periventricular vascular network is present in the telencephalon prior to the formation of neuronal networks and before the appearance of radial units and striosome-matrix compartments in the dorsal and ventral telencephalon, respectively. Thus, the periventricular network is temporally and spatially poised to influence neural maturation, as well as guide tangential migration in the developing brain [41]. A similar vascular niche for NSC has been reported to exist in the adult SVZ and SGZ, and may have been established early during embryonic brain development. This suggests that the vasculature may be critical in promoting and regulating neural development.

3.3. Establishment of neurogenic regions of the brain

During brain development, three different NSPC types make their appearance in a tightly coordinated spatiotemporal manner, seeding the brain with committed progenitors that differentiate into the various cell types of the mature brain. The first of these to appear are pseudostratified epithelial cells termed radial glia, regarded as the *bona fide* NSC in the embryonic VZ [46, 47]. Morphological studies have identified two processes emanating from their cell bodies, suggesting an inherent bipolar nature. A short and thick apical process directed towards the ventricle is thought to anchor radial glia, while a longer basal radial fiber projects towards the basement membrane of the pia mater, acting as a scaffold for prospective neuronal migration [21, 48]. These radial fibers are often observed to contact blood vessels and exhibit multiple branched endfeet at the pial surface [49]. Interestingly, their apical regions are typically folded and contain a single cilium [50, 51], reminiscent of the proposed location, structural morphology, and vascular contacts of adult NSC in the SVZ.

During the early stages of cortical development, the cerebral cortex is composed almost exclusively of proliferative radial glia dividing at the ventricular surface in the VZ [52]. As proliferating radial glia progress through the cell cycle, they undergo interkinetic nuclear migration, where the nucleus migrates away from the ventricle during G1 phase, and enters S phase at the top of the VZ. Upon return through the VZ to the ventricular surface, they proceed through G2 phase and M-phase, respectively [19, 52-54]. A switch from symmetric self-renewing to asymmetric neurogenic divisions occurs as development proceeds, leading to pairs of daughter cells with distinct progenitor or early neuronal fates; symmetric divisions have also occasionally been observed to produce early neurons or intermediate progenitor cells (IPC) [19, 23, 49]. However, during peak neurogenesis, radial glia give rise to one radial glial cell, and either one post-mitotic neuron or a neuronally committed IPC [49, 55]. Similarly in adult neurogenesis, adult NSC asymmetrically divide to generate transit-amplifying cells that produce committed progenitors. In both cases, regulation of the symmetry of cell division is critical, and ultimately controls cerebral cortical size during brain development [52].

At the onset of neurogenesis, radial glia progeny migrate away from the ventricle and begin to establish the first layers of the developing brain, separate from the VZ. IPC establish the SVZ as a distinct proliferative region, while young cortical neurons migrate to a superficial position to establish the cortical plate [48]. These migrating cortical neurons are intimately associated with the long pial fiber of radial glia, utilizing it to traverse relatively long distances to the overlying cortex in a process termed radial migration. Once telophase is complete and radial glia have entered M-phase, the apical plasma membrane becomes unequally segregated into the two daughter cells. Interestingly, the apical daughter inherits a larger portion of the membrane while the basal daughter receives a smaller proportion in addition to the radial fiber, indicating the latter assumes the stem cell radial glia fate [46, 47, 49, 56]. However, this is not absolute, as instances of basal daughters becoming post-mitotic and apical daughters remaining proliferative have been reported. Instead, it has been suggested that fate decisions involving asymmetric division may also depend on developmental stage [47, 48]. Thus, the function of radial glia is two-fold, wherein they generate and guide migration of their own daughter cells [52, 57].

After the VZ reaches its maximal size during midstage cortical neurogenesis, the VZ begins to shrink while the SVZ begins to expand [57]. Derived from radial glia, IPC are the first cell types to initially seed the SVZ [48]. While some observations describe the distribution of IPC throughout the upper VZ and lower intermediate zones [49, 58], they are predominantly concentrated in the SVZ, where they almost exclusively divide symmetrically to generate postmitotic daughter neurons [59-62]. In contrast to radial glia, IPC are multipolar, extending and retracting multiple processes [19, 49]. Additionally, they do not appear to sustain contact with either the ventricular or pial surfaces, and are in fact defined by their lack of prominent apical or basal processes and a basal location relative to the apical surface [52]. While their contact with neighboring blood vessels has not been confirmed, their appearance in the cortex seems to follow that of blood vessel invasion in the cortical wall [19]. Furthermore, behavioral differences between radial glia and IPC have been noted. IP cells progress through the cell cycle away from the ventricle, and do not undergo interkinetic nuclear migration, thus, allowing differentiation from radial glia based on spatial location during mitosis [19, 63].

At later stages of cortical development, the SVZ progenitor pool continues to expand via IPC symmetric divisions. Further aiding in expansion, progenitors from the ventral telencephalon may even migrate dorsally to contribute to the SVZ progenitor pool [19]. Upon completion of cortical neurogenesis, radial glia transition into astrocytes and exit the VZ, leaving a single layer of ependymal cells lining the ventricle [57]. Consequently, proliferative IPC become the predominant component of the cortical progenitor pool, and eventually comprise the majority of mitotic progenitors as embryonic neurogenesis nears completion. Interestingly, while only a single layer of VZ-derived ependymal cells remains postnatally, IPC are present in large numbers in the postnatal SVZ, and persist into adulthood [19]. Aside from generating cortical neurons, postnatal and adult progenitors have been demonstrated to generate neurons destined for the olfactory bulb [49, 58, 64], and possibly all excitatory neurons of the upper cortical layers [58]. Thus, IPC in the SVZ play a vital role in cortical neurogenesis during embryonic development as well as in the adult.

Interestingly, a novel progenitor type termed the outer SVZ (OSVZ) progenitor has recently been identified and appears to exist in all mammals, albeit to varying extents [52]. These progenitors have a modified radial morphology, but are exclusively localized to the SVZ. OSVZ cells are enriched in mammals with larger cerebral cortices, and their appearance during mid-gestation seems to coincide with the onset of neurogenesis [52, 65-69]. OSVZ progenitors are peculiar in that they possess characteristics reminiscent of both radial glia and IPC. They display radial morphology and express radial glial markers paired box protein-6 (Pax6), phospho-vimentin, glial fibrillary acidic protein (GFAP), and brain lipid-binding protein (BLBP), and also display random cleavage planes, where both proliferative self-renewing symmetrical divisions and asymmetric divisions producing OSVZ daughter and progenitor have been reported. However, an apical process is absent in these cells while their basal process is retained throughout mitosis [65-67].

3.4. Vascular cues during embryonic neurogenesis

It has been suggested that the developing cortical vasculature within the SVZ promotes IPC expansion during neurogenesis by providing a suitable microenvironment for IPC accumulation and division [39]. It is noteworthy that brain EC share similar molecular profiles with their neighboring NSC. For example, ventral and dorsal EC, as well as NSC, express Dlx1/5 and Nkx2.1, and Pax6, respectively, while pial EC are negative for all three [41]. This strongly suggests that mechanisms of patterning during early angiogenesis and neurogenesis in the brain are shared.

A strong association between NSC and blood vessels exists during embryonic and adult neurogenesis, especially in regard to cell cycle regulation [33-35]. The filopodia of endothelial tip cells extend towards the ventricular surface where radial glia divide, and even interact with pial fibers of radial glia in the hindbrain [39]. Additionally, dividing cells in the embryonic SVZ reside statistically closer to blood vessels than predicted by chance, and recent studies report a synchronization of SVZ cell division with the formation of ventricular vascular plexuses [40]. This vascular relationship is apparent in the emerging RMS as well, where dividing cells have been reported to associate with blood vessels at E16, E18, and P4 [35]. On the other hand, progenitor migration in the RMS does not seem to rely on the vasculature, as the vast majority of neuroblast neurites in P4 RMS have little or no association with blood vessels [35]. This is in contrast to radial migration outside the RMS, where postmitotic doublecortin (DCX) and glial fibrillary acidic protein (GFAP)-positive cells associate with blood vessels during migration into superficial cortical layers [40]. Interestingly, IPC have been suggested to maintain a stronger interaction with blood vessels, as T-brain gene-2 (Tbr2)-eGFP progenitors in M-phase reside closer to blood vessels when compared to total phosphohistone H3-positive progenitors [40]. Furthermore, these dividing IPC are often found at vessel branch points, which have previously been observed to be sites of glial tumor mitosis and subsequent migration [70].

Mounting evidence suggests that Tbr2 progenitors are temporally and spatially correlated with the appearance of cortical vasculature, and even follow and mimic the pattern of nascent blood vessels. Similarly, the positions of IPC during mitosis, migration and differentiation are all correlated with EC development in the SVZ. Even detection of Tbr2-positive cells correlates with the appearance of vascularization, as Tbr2 cell density is highest in the vascularized lateral regions as compared to the nearly avascular medial regions in the dorsal cortex of E12 embryos [39]. Moreover, ectopic overexpression of vascular endothelial growth factor (VEGF)-A causes IPC to follow a pattern of aberrant vascular growth. Interestingly, leading EC tip cells have been observed to associate with some Tbr2-positive IPC in M-phase, suggesting a functional interaction during division. These data collectively suggest the SVZ vasculature serves as a niche for mitotic IPC [39], and provides instructive and permissive cues for stem and progenitor cell expansion and tissue invasion [71].

3.5. Parallels between embryonic neurogenesis and adult neurogenesis

Similarities between embryonic and adult NSC at the cellular level and across their extracellular microenvironments have been reported, and selective labeling of radial glia has demonstrated a direct link between these cells, indicating that NSC are most likely contained within the neuroepithelial-radial glia-astrocyte lineage [72, 73]. Furthermore, reports indicate adult SVZ NSC retain specialized characteristics of radial glia. However, the molecular characteristics that confer progenitor potential onto astroglial cells and distinguish them from those with normal support function remain largely unknown [53].

From adult NSC, also referred to as Type B cells in the SVZ, an apical process at times intercalates between ependymal cells lining the lateral ventricular surface, potentially serving to both anchor and present NSC to circulating factors in the cerebrospinal fluid (CSF). Embryonic radial glia also share this apical process, and most likely contain a similar profile of specialized apical junctions at the site of this primary cilium [53]. Similarly, the longer basal process that radial glia extend towards the surface of the brain during embryonic development is also shared by adult NSC of the SVZ. In the adult SVZ, this basal process projects radially or tangentially, depending on location, eventually terminating in specialized endfeet on the surface of blood vessels [33, 74, 75]. These vascular contacts may be analogous to those of radial glia during development, as branch contacts with the overlying vasculature also occur [53]. This suggests that adult SVZ NSC share core properties with the embryonic radial glia from which they are derived from, allowing them to retain progenitor function throughout life.

These similarities are also observed in adult NSC of the SGZ in the hippocampal dentate gyrus. Early anatomical studies suggest that radial glia in the dentate neurepithelium transition to the different astrocyte populations of the dentate gyrus, including radial astrocytes [76, 77]. While experimental evidence linking radial glia to adult SGZ radial astrocytes is lacking [53], it is possible that this derivation occurs, and further studies will be needed to clarify this lineage relationship. However, the primary cilium of radial glia is present on SGZ progenitors and adult radial astrocytes, and is essential for progenitor proliferation and generation of postnatal radial astrocytes, thus establishing its requirement for neurogenesis. From a signaling standpoint, the primary cilium serves as an integration site for signaling via pathways such as Shh. Interestingly, this cilium seems to be specific to the radial astrocytic NSC pool in the hippocampus. Non-stem cell astrocytes are not affected by lack of primary cilium or Shh signaling, suggesting a unique requirement among these NSC [78, 79].

As in the adult SVZ [80], location seems to dictate specificity, where radial glia in the dorsal telencephalon generate only pyramidal excitatory neurons, while those located in the ventral telencephalon give rise to nonpyramidal inhibitory interneurons [81]. Studies from several mouse models demonstrate that neurons can migrate into the cortical plate (CP) radially or tangentially [82-84]. This type of migration is mirrored in the adult SVZ, where neuroblast progenitors migrate tangentially through the rostral migratory stream towards the olfactory bulb, destined to become inhibitory interneurons. Features of interkinetic nuclear migration are also shared by radial glia and adult NSC. While mitotic cells are found only adjacent to the lumen of the neural tube, nuclei of cells in S-phase are found in the outer half of neural epithelium [19, 54]. This correlation of cell cycle with spatial location occurs in the adult SVZ as well, where basally located blood vessels are proposed to exert growth control over proximal NSC by providing a proliferation-inducing microenvironment. This is in contrast to NSC located apically, either adjacent to or within the ependymal layer, which are immunoreactive for mitotic markers [33, 34, 75, 85].

These findings highlight basic properties that are common to embryonic radial glia and adult SVZ and SGZ NSC. Given the evidence, it is highly likely that a microenvironment similar to the one which supports embryonic neurogenesis persists throughout development and is maintained in the adult neural stem cell niche.

4. The adult NSC niche

4.1. Cellular architecture of the adult SGZ and SVZ

As previously mentioned, two prominent germinal regions of the adult brain have been identified to function as stem cell or neurogenic niches, allowing for continuous generation of new neurons. The SVZ represents the largest neurogenic stem cell region within the adult brain. It resides within a narrow region of the lateral ventricular wall, roughly four to five cells in diameter [86]. Progenitors generated from this region migrate through the RMS towards the olfactory bulb, where differentiation into at least five interneuron subtypes has been reported. In fact, it is estimated that 30,000–60,000 new neurons are generated in the rodent olfactory bulb per day [87, 88]. Differentiation into oligodendrocytes of the corpus callosal white matter also occurs, albeit to a lesser extent [89, 90]. A second neurogenic region, the SGZ, is located between the hilus and the granule cell layer of the dentate gyrus within the hippocampal formation. In contrast to SVZ progenitors, granule neurons born from this region migrate short distances to the granule cell layer, where differentiation commences [91]. Whether *bona fide* NSC exist in the adult mammalian hippocampus is currently under investigation, as *in vivo* lineage tracing assays suggest that separate progenitors responsible for neurogenesis and gliogenesis exist in the SGZ [92-94].

A group of distinct cell types in the adult SVZ help maintain this specialized niche microenvironment: putative NSC (type B cells), transit-amplifying cells (type C cells), neuroblasts (type A cells), ependymal cells, and specialized vascular endothelium [34, 95]. There is no definitive marker of NSC, and researchers rely on combinations of overlapping markers, as well as spatial location within the niche to identify NSC. Accordingly, NSC are usually identified by their apical location, superficial to the ependymal layer, and slow cell cycle time of ~ 28 days [96]; however, their expression of Sox 2 and 9, GFAP, and CD133/prominin-1 are not exclusive [33, 34, 74]. The presumptive lineage progression from stem cell to more differentiated progenitor is as follows: NSC generate transit amplifying cells that differentiate into migrating neuroblast progenitors.

Non-dividing ependymal cells are multiciliated, and form a single layer lining the ventricle surface, acting as a physical barrier separating the brain parenchyma from the cerebrospinal fluid (CSF) [97]. While ependymal cell cilia contribute to CSF flow, they have also been reported to affect the migration of young neurons by creating gradients of Slit chemorepellents that guide anterior neuroblast migration [98]. An *en face* view of the lateral ventricle wall reveals a planar organization, commonly referred to as "pinwheel organization", where the apical process of NSC is surrounded by a mosaic of ependymal cells [33, 74, 99]. Through studies mapping numbers of ventricle-contacting NSC along the ventricular surface, "hot spots," or areas of stem cell activation, have been revealed [100-102]. Currently a topic of debate is whether ependymal cells can function as multipotent NSC. Previous studies have demonstrated that CD133/prominin-1 positive ependymal cells are in fact multipotent, and during ischemia become active to generate neuroblasts and astrocytes [103, 104]. However, a more recent study using split-Cre technology demonstrated a subset of CD133/prominin-1 positive cells within the ependymal layer are immunoreactive for GFAP, suggesting that these double positive radial-like cells are NSC, not ependymal cells [98, 105].

Two astrocytic populations have been proposed to reside in the SVZ [100]. Type B NSC astrocytes reside underneath the ependymal layer, while non-stem cell astrocytes are more superficial and differ in morphology [33, 34, 74, 99]. NSC are closely associated with ependymal cells, and at times extend a short, apical, non-motile primary cilium that innervates between the ependyma to directly contact the CSF within the ventricle [33, 74, 99]. While NSC are relatively quiescent, transit-amplifying cells are highly proliferative, and remain localized to the SVZ [95, 100]. Neuroblasts, on the other hand, migrate through astrocytic tubes in the RMS to the olfactory bulb, where interneuron differentiation occurs [106, 107]. Interestingly, experiments using viral targeting and genetic lineage tracing in neonatal and adult mice have revealed that specific subtypes of interneurons in the olfactory bulb are derived from specific locations within dorsal, medial, and ventral portions of the adult SVZ [108-112]. Interestingly, while the vascular beds of the SVZ and SGZ both support adult neurogenesis, the SVZ vasculature is somewhat unique. Differences in permeability, stability, and perivascular cell coverage are thought to account for these differences. NSC and transit-amplifying cells both display an intimate relationship with SVZ blood vessels, as 3-dimensional niche modeling indicates closer proximity and increased vascular contact relative to other SVZ cells. Interestingly, these vascular associations are further exaggerated in niche regeneration models [33, 34]. Additionally, NSC extend a long basal process that terminates on blood vessels in the form of specialized endfeet, potentially serving to integrate vascular cues [33, 74, 99].

4.2. NSC-Vascular EC associations within the SVZ and SGZ

NSC are not randomly distributed throughout the brain; rather, they are concentrated around blood vessels, allowing constant access to circulating signaling molecules and nutrient metabolites [113, 114]. The SVZ and SGZ both present functional neurogenic environments, maintaining neural stem and progenitor cells (NSPC) in poised and undifferentiated states. Regulatory processes within the SVZ niche can be controlled via secreted neurotrophic and angiogenic factors, such as Wnt, Shh, and TGF-β [115]. For example, circulating complement factors have been shown to promote basal and ischemia-induced neurogenesis, and components of complement signaling are present on transit-amplifying cells and neural progenitors *in vivo* [116]. The vascular-derived factors, stromal cell derived factor (SDF)-1 and angiopoietin (Ang)-1, promote neuroblast proliferation and survival [117], and when expressed on EC, serve as "molecular migratory scaffolds" [118, 119] to damaged areas post stroke [120]. EC themselves have even been shown to regulate NSC self-renewal [37, 86, 101, 121]. Interaction with the vascular endothelium may in fact be a vital component of the niche, as radiation-induced disruption of endothelial cell–SGZ precursor cell interaction results in a loss of neurogenic potential, as is the case after NSC transplantation into an irradiated host. [122].

Within the SGZ, nestin-expressing radial astrocytes are localized to areas near blood vessels [91], and there exists an anatomical relationship between proliferating neural progenitors and EC in the hippocampus [101, 123, 124]. In contrast to the SVZ, where angiogenic sprouting and division of EC are absent [34], surges of EC division are said to be spatially and temporally related to clusters of neurogenesis in the SGZ [101]. In the hippocampus, angiogenesis and neurogenesis are coupled, as suggested by high levels of VEGF and VEGFR2 [101], and the shared responsiveness to similar growth factors, e.g., neurotrophins, neuropilins, semaphorins and ephrins [125-127]. In fact, similar bidirectional communication occurs within the higher vocal center (HVC) of the songbird brain, where increases in angiogenesis are said to be coupled to testosterone-induced upregulation of VEGF and VEGFR2 in neurons and astrocytes, respectively. The newly generated capillaries produce BDNF (brain-derived neurotrophic factor) that subsequently promotes the recruitment and migration of newly born neurons [121]. Similarly, exercise-induced angiogenesis in the hippocampus is met with increased expression of NGF (nerve growth factor) and BDNF [126], leading to robust increases in neurogenesis [128-130].

In contrast to other areas of the brain, where the BBB is strictly maintained by EC tight junctions, pericyte coverage, and astrocyte endfeet, a modified BBB has been proposed to exist in the SVZ. The lack of astrocyte endfeet and endothelial cell tight junctions, as revealed by aquaporin-4 and zonula occludens-1 immunostaining, respectively, demonstrate major structural differences in the SVZ vascular endothelium. Under homeostatic conditions, the majority of $BrdU^+$ label-retaining NSC and transit amplifying cells reside significantly closer, and frequently make direct contact, to the vasculature; furthermore, after antimitotic cytosine-β-D-arabinofuranoside (Ara-C) treatment to ablate rapidly proliferating cells and induce NSC-mediated repopulation, these vascular associations are increased [33, 34]. At times, transit-amplifying cells can be seen contacting the vasculature at sites lacking astrocyte endfeet and pericyte coverage, suggesting that sites along the vessel are primed for intercellular communication. In fact, fluorescent tracer experiments have proposed that differences in the ultrastructural composition of SVZ blood vessels may be responsible for the detection of sodium fluorescein in the SVZ after perfusion into the blood; however, access to the SVZ from the cerebral spinal fluid cannot be dismissed as an entry point. Integrin-$\alpha 6\beta 1$ partially mediates the adhesion between NSPC and blood vessels through the binding of laminins that are highly concentrated around SVZ blood vessels. *In vitro* and *in vivo* blocking experiments using an integrin-blocking antibody (GoH3) have demonstrated a crucial role for this interaction in the attachment, spreading, and proliferation of NSPC [33].

The vascular environment in the RMS has also been suggested to be somewhat specialized, where migrating neuroblasts en route to the olfactory bulb are found closely apposed to blood vessels. Interestingly, blood vessels in this region are parallel and aligned with the direction of the RMS, and the density of vessels is significantly higher when compared to equally cell-dense areas of the brain [131]. It has been reported that over 80% of RMS vessels are lined with migrating neuroblasts [132], and degradation of ECM through vascular EC secretion of matrix metalloproteinases (MMPs) opens a path for their migration [133]. This observation has prompted some to suggest that increased vessel density is a consequence of greater metabolic demand by migrating progenitors.

5. Vascular regulation of adult neurogenesis

5.1. EC regulation of NSPC

Through cytokines and secreted factors, direct contact *in vivo*, or within the confines of the coculture system, EC exert their influence over NSC to regulate fate specification, differentiation, quiescence and proliferation (Figure 1). Early experiments established a crude role for EC regulation of NSC, where increases in neurite outgrowth and maturation, and enhanced migration were observed in cocultures of SVZ explants with EC [134]. NSC are reported to respond to pro-angiogenic factors [135-137] that promote NSPC proliferation, neurogenesis, synaptogenesis, axonal growth, and neuroprotection [138]. Studies in tumor and stroke models have also uncovered neural regulatory roles of EC. EC can protect stem cells and tumor cells from radiation damage [139, 140], and in preclinical models where NSPC isolated from stroke boundary are cocultured with cerebral EC, significant increases in neural progenitor cell proliferation, neuronal differentiation, and capillary tube formation are observed [141]. Even cotransplantation of EC with NSPC increases survival and proliferation as compared to transplantation of neural precursors alone [142]. Similarly, coculture of adult NSC with EC results in self-renewal and symmetric neural cell division, leading to enhanced neurogenesis through an increase in nestin+ precursor number [37]. Interestingly, when stroke-activated rat brain EC are cocultured with SVZ cells, progenitor proliferation and neuron number are increased by 28% and 46%, respectively, when compared to coculture with normal EC. This suggests that activated EC are more potent in promoting neurogenesis, potentially through modulation of Sox2 and Hes6 levels in SVZ cells [143], although further investigation is required to identify the mechanisms involved.

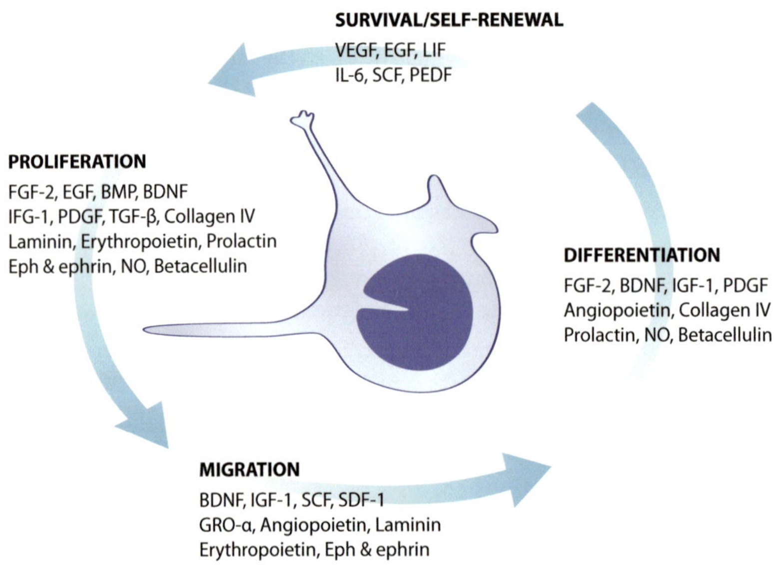

Influence of niche effectors on adult neural stem cell behavior.*Referring to adult suvbentricular and subrangularzones. BDNF: Brain-derived neurotrophicfactor;GRO-oz Growth-related oncogene alpha; LIF: Leukemiainhibitoryfactor;VEGF: Vascularendothelial growth factor; FGF-2: Fibroblastgrowthfactor-2; EGF: Epidermal growth factor; BMP: Bone morphogenetic protein; PDGF: Platelet-derived growth factor; SCF: Stem cell factor; PEDF: Pigmentepithelial-derived factor; SDF-l: Stromal cell-derived factor-l;TGF[3: Transforming growth factor beta; NO: Nitric oxide; CXCR4: Chemokine receptor type 4; EGFR: Epidermalgrowthfactorreceptor;CSPG: Chondroitin sulfate proteoglycan

Figure 1. Regulatory effects on adult neural stem and progenitors

http://dx.doi.org/10.5772/54980

Mock treatment with serum-rich endothelial growth media induces NSC differentiation into neurons and astrocytes [31], indicating that EC-mediated regulation of NSPC self-renewal and differentiation may be mediated through the release of certain growth factors, including PEDF or VEGF [144, 145]. Cytokine expression profiles of human umbilical vein and cerebral microvascular EC reveal that a large number of chemokines, growth factors, adhesion molecules and ECM proteins are expressed by these cells [146]. Levels of these signaling molecules varied under stimulating and nonstimulating conditions as well as by EC type, highlighting the diverse signaling potential that exists even among endothelial subtypes. Studies of adult neurogenic niche regulation have identified a number of growth factors and secreted molecules, although the origin of some remains unknown (Figure 2).

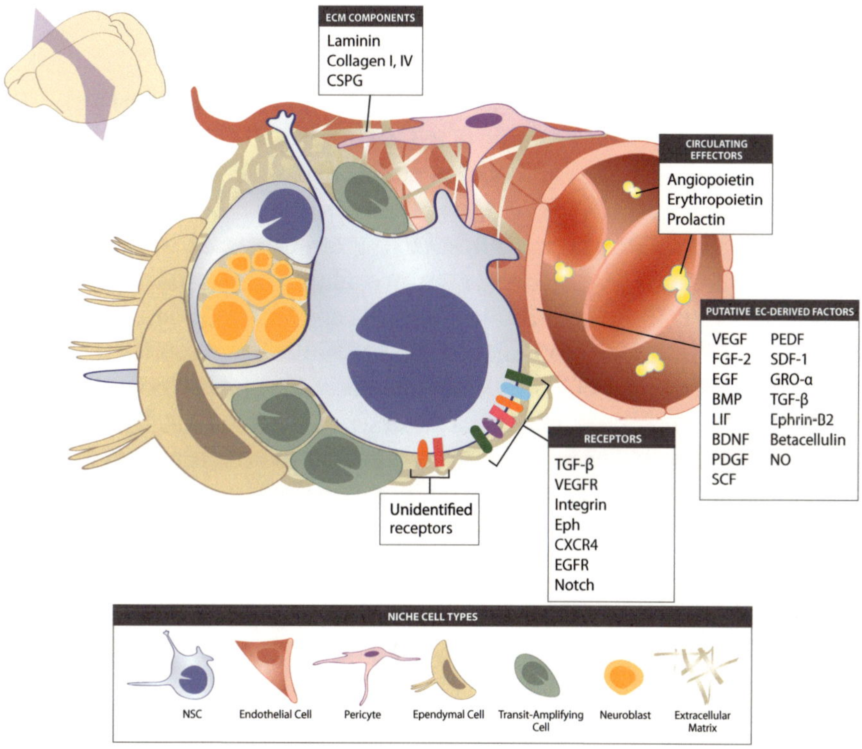

Sources of vascular – derived adult NSC niche effectors horiginate from the endothelium, circulation, ECM deposition, and perivascular cell types. *Referring to the adult suvbentricular zone BDNF: Brain-derived neurotrophic factor; GRO-a: Growth-related oncogene alpha; LIF: Leukemia inhibitory factor; VEGF: Vascular endothelial growth factor; FGF-2: Fibroblast growth factor-2; EGF: Epidermal growth factor; BMP: Bone morphogenetic protein; PDGF: Platelet-derived growth factor; SCF: Stem cell factor; PEDF: Pigment epithelial-derived factor ; SDF-1: Stromal cell-derived factor-1; TGF_: Transforming growth factor beta; NO: Nitric oxide; CXCR4: Chemokine receptor type 4; EGFR: Epidermal growth factor receptor; CSPG: Chondroitin sulfate proteoglycan

Figure 2. Putative vascular – derived regulators of the adult neural stem cell niche*

The vascular-derived molecules shown to locally regulate the adult NSC niche include leukemia inhibitory factor (LIF), brain-derived neurotrophic factor (BDNF), VEGF, platelet-derived growth factor (PDGF), pigment epithelial-derived factor (PEDF), betacellulin (BTC), and laminins and integrins [33, 121, 147]. However, there are additional factors reported to influence NSPC behavior which may be derived from the NSC niche vasculature, although this has not yet been demonstrated, including fibroblast growth factor 2 (FGF-2), epidermal growth factor (EGF), interleukin-6 (IL-6), stem cell factor (SCF), insulin growth factor-1 (IGF-1), transforming growth factor-β (TGF-β), bone morphogenic proteins (BMP), SDF-1/CXCR4, collagen IV, Eph/ephrins, angiopoietin, nitric oxide (NO), erythropoietin and prolactins. We review advances made toward understanding the cellular and molecular role of these factors since last reviewed [148].

5.2. Endothelium-derived niche effectors

5.2.1. VEGFs

VEGF signaling is a complex signaling hierarchy involving several isoforms (e.g., VEF-A, -B, -C, -D) that result from alternative splicing of the *VEGF* gene. VEGFs are highly implicated in NSPC survival, proliferation, and neuroblast migration and maturation [149-154]. Along with the recent identification of VEGFR3/Flt4 expression in the adult SVZ, all corresponding receptors, including VEGFR1/Flt1 and VEGFR2/Flk1, are expressed within the NSPC pool [149, 154, 155]. While VEGFR1 negatively regulates adult olfactory neurogenesis and RMS migration by altering VEGF-A bioavailability, VEGFR2 and VEGFR3 both appear to positively regulate neurogenesis [133, 155]; in addition, VEGFR2 has also been reported to affect vascular proliferation [154]. This ability of VEGF family members to regulate both neurogenesis and angiogenesis may be important in clinical settings of intracerebral hemorrhage, where transplantation of human NSPC overexpressing VEGF have been shown to increase micro-vessel density and promote NSPC engraftment in sites of tissue damage [156, 157].

In VEGFR1 signaling-deficient (Flt-1 $TK^{-/-}$) mice, the increased levels of VEGF-A and subsequent phospohorylation of VEGFR2 in NPSC are thought to account for the altered RMS migration, demonstrating a critical role for VEGF-A in this process [154]. Reported to be required for hippocampal neurogenesis in the adult rat [145], EC, ependymal cells and the choroid plexus secrete VEGF at neurogenic sites, which serves as a survival factor to stimulate NSPC self-renewal. Neurospheres, as well as reactive astrocytes, have been shown to express VEGF-A [158, 159], and infusion into the lateral ventricle after cerebral ischemia acts as a trophic survival factor for NSPC and increases neurogenesis, most likely through the VEGFR2/Flk-1 receptor [37, 150, 152]. Similarly, other studies suggest that *in vitro* VEGF stimulation increases the number of BrdU-labeled precursors, which is attenuated in the presence of SU1598, a Flk-1 receptor tyrosine kinase inhibitor, further supporting mediation through VEGFR2/Flk-1 [123]. Although VEGF-A is reported to have a direct role in signaling during development [101, 123, 150], evidence also supports an indirect role when it is secreted by ependymal cells, through the stimulated release of BDNF from EC [121, 152].

However, in experiments comparing the numbers of primary Ki67^{+} adult neural precursors in Nestin^{cre}Flk1$^{+/-}$ and Nestin^{cre}Flk$^{-/-}$ short-term cultures, it was found that VEGF-A signaling does not appear to affect the proliferation of these cells, and individual neurospheres that proliferate clonally from Flk1$^{+/+}$ and Flk1$^{-/-}$ mice are of comparable size and cell number [160]. Similar studies demonstrate VEGF-A secreted from cerebral endothelial cells promotes migration of oligodendrocyte precursor cells (OPCs), but not proliferation, as treatment with a Flk-1 neutralizing antibody only affected OPC propagation [161]. Therefore, VEGF-A signaling may exert control over NSCs via the regulation of survival; this potential mechanism should be further explored, especially given that an internal autocrine role for VEGF-A in HSC survival has been demonstrated[162].

More recently, a direct requirement for VEGFR3 in neurogenesis has been established, and Vegfr3::YFP reporter mice demonstrate expression in NSC [155]. Interestingly, coexpression with VEGF-C along the walls of the lateral ventricle is also observed. Accordingly, an increase in neurogenesis is said to occur from VEGFR3^{+} NSC after VEGF-C stimulation, deletion of VEGFR3 in neural cells and SVZ astrocytes, as well as VEGFR3 inhibition via blocking antibodies, all lead to a reduction in neurogenesis. *In vitro*, VEGF-C treatment also increases BrdU incorporation in YFP^{+}EGFR^{+} NSC [155].

5.2.2. BDNF & IGF

BDNF is secreted by EC and induces the differentiation of astrocyte precursors [147, 163], and *in vivo* has been shown to influence proliferation and differentiation of NSPC in adult neurogenic regions [121, 134]. As mentioned previously, *in vivo* experiments suggest that VEGF-induced secretion of BDNF from higher vocal center (HVC) capillary vasculature in the songbird brain results in newly born neuron recruitment. Interestingly, BDNF secretion in this region is quite high, as canary brain EC secrete an average of 1 ng BDNF/10^{6} cells/24 h [121, 164]; a study of adult-derived human brain EC revealed a comparable amount of BDNF secretion [147]. *In vitro*, BDNF release from EC supports SVZ-derived neuron outgrowth, survival, and migration [147]. Although subependymal astrocytes also secrete BDNF, it may be sequestered at the cell surface; this is partly mediated by the truncated gp95 extracellular domain of TrkB, a high affinity receptor for BDNF, which prevents its release into the surrounding space [147]. This has been proposed to be a mechanism whereby regions of NSC expansion exclude BDNF, limiting its availability only to those areas supporting differentiation and maturation. Interestingly, while NSC and transit-amplifying cells express the low-affinity neurotrophin receptor p75, expression of TrkB is only found on lineage-restricted neuroblasts. Additionally, it has been suggested that BDNF acts in a positive feedback loop to reduce proliferation and increase neuroblast differentiation through the release of NO by NSPC [165, 166]. Thus, endothelial-derived BDNF appears to serve chemoattraction and survival roles for neuronal progenitors [167].

Studies of exercise-induced neurogenic cognitive enhancement in the dentate gyrus have linked BDNF with IGF-1 [168]. Exercise stimulates uptake of IGF-1 from the bloodstream in the hippocampus, leading to an increase in the number of BrdU^{+} hippocampal neurons as well as upregulation of BDNF mRNA and protein levels [169, 170]. The neurogenic effect of IGF-1 may be mediated in part through estrogen signaling, as estrogen antagonists reduce neurogenesis within the dentate gyrus [171]. From a clinical standpoint, IGF-1 may be involved in neurodegenerative disease progression, such as Alzheimer's and stroke, where levels of circulating IGF-1 are altered [170].

5.2.3. PDGF

PDGF signaling has been shown to affect stem cell properties and lineage bias [98, 172]. Vascular EC secrete PDGF-B as a disulfide-linked homodimer (PDGF-BB), and via a specific positively charged C-terminal retention motif, it interacts with heparin sulfate proteoglycans within the ECM to aid in localized retention [173-176]. Specifically within the CNS, it regulates oligodendrocyte precursor cell number. Interestingly, PDGF-B is implicated in brain tumor formation, where activation of its signaling pathway is present in more than 80% oligodendrogliomas and 50–100% of astrocytomas[177]. Thus, identifying which cells respond to PDGF is crucial to elucidate mechanisms involved in these brain cancers.

In the SVZ, putative NSC and most $GFAP^+$ cells have been shown to express PDGFRα, and become activated in the presence of PDGF-AA [178]. Accordingly, PDGF is reported to have mitogenic and differentiation actions on neural progenitor cells [179-181], and synergy with bFGF has been reported to enhance neurosphere generation [178]. After intracerebroventricular infusion of PDGF-AA, astrocyte-derived periventricular hyperplasias are formed, and increases in oligodendrogenesis are observed at the expense of olfactory bulb neurogenesis. While EGF infusion elicits a similar proliferative response in the SVZ, staining for PDGFRα and EGFR reveals expression in distinct populations, suggesting that they label stem cells and transit-amplifying progenitors, respectively. Conversely, conditional ablation of PDGRα in the SVZ decreases oligodendrogenesis while having little effect on neurogenesis [178]. Thus, PDGF signaling may play a role in maintaining the balance between neurogenesis and oligodendrogenesis.

5.2.4. SCF

SCF, also known as Kit ligand, has been reported to be expressed by a variety of cell types including vascular EC [182, 183]. Previous reports indicate that within the CNS, SCF/Kit-ligand signaling influences oligodendrocyte precursors prior to differentiation towards a myelinated phenotype. Although Kit belongs to the same class of tyrosine kinase receptors as PDGF receptors, their effects on NSPC are different. In $nestin^+$ NSCs isolated from embryonic rat cortex, more than 93% express SCF. More recent studies demonstrate that SCF acts as a chemoattractant and survival factor for NSPCs during early stages of differentiation while having no effect on proliferation or differentiation [184-186].

5.2.5. PEDF

PEDF is secreted by a variety of cell types, and can interact with the ECM, most notably collagen-I [187-189]. Being the first soluble factor shown to selectively activate type B NSC, PEDF seems to contribute to stem cell maintenance within the neurogenic niche. In the adult mouse brain, expression is restricted to endothelial and ependymal cells, suggesting that PEDF is in fact a niche-derived signal. Accordingly, Western blot analysis on conditioned media from cultures indicate that PEDF is specifically secreted by endothelial and ependymal cells [144, 190, 191]. Aside from acting as a brake on cell cycle progression by promoting NSC self-renewal without affecting proliferation [192], recent evidence suggests an additional role in renewing symmetric divisions. Interestingly, PEDF has been implicated in regulating certain aspects of Notch signaling by modulating the NFκB pathway. The role of Notch signaling in NSC maintenance is well characterized [193], and NSC treated with PEDF upregulate Notch effectors Hes1 and Hes5, as well as the Sry-related HMG box-transcription factor Sox2 [194]. In cells with low levels of Notch signaling, PEDF enhances Notch-dependent transcription by relieving repression of Notch-responsive promoters by the transcriptional co-repressor N-CoR, thereby potentiating symmetric cell division [195]. Additionally, BrdU-labeled mice treated with PEDF display an increase in the number of $BrdU^+GFAP^+$ cells, and injection with a C-terminal blocking peptide to PEDF reveals no significant change in the number of $BrdU^+GFAP^+$ cells compared with vehicle-injected controls. Taken together, these data suggest that PEDF may not be a survival factor for NSC, and may instead serve to activate NSC by stimulating self-renewal [194].

5.2.6. Nitric Oxide (NO)

A variety of mechanisms have been proposed for NO regulation of NSPC, perhaps accounting for conflicting studies suggesting opposing roles on NSPC proliferation. Early reports demonstrated a role for NO in the repression of adult neurogenesis, as exposure to NOS inhibitors L-NAME and 7-NI increased neurogenesis in the dentate gyrus and SVZ [165, 166, 196, 197]. However, its effects seem to depend on the signaling pathway involved. Potentially through NO-induced S-nitrosylation of the EGFR [198], NO inhibits PI3K/Akt signaling to suppress NSPC proliferation, both in culture and *in vivo* [199]. However, bypassing the EGFR induces proliferation through activation of p21Ras, leading to an increase in activation levels of c-Myc, p90RSK and Elk-1, and subsequent reduction in $p27^{Kip1}$ [200]. BDNF may be involved in this process as well, as its stimulatory effect on neuronal differentiation is blocked by L-NAME. Interestingly, NSC express and release NO, suggesting a feedback mechanism whereby NSC-produced NO induces production of BDNF from the vascular bed [201]. EC also produce NO via eNOS, and a decrease in SVZ cell proliferation and migration post-stroke is observed in eNOS-deficient mice. Interestingly, BDNF levels are also reduced in $eNOS^{-/-}$ ischemic mice, and BDNF treatment rescues the decrease in neurosphere formation, proliferation, and neurite outgrowth in cultured $eNOS^{-/-}$ neurospheres [202].

5.2.7. Vascular ECM: Laminins, collagens, fractones

The ECM is an integral component of the NSC niche, regulating signaling by providing, storing, and compartmentalizing growth factors and cytokines indispensable for proliferation, differentiation and adhesion. Within the SVZ, a unique basal lamina, rich in laminins, collagen-1 and collagen IV, extends from perivascular cells as 'fractones' [203]. Each fractone consists of a base, attached to the perivascular cell, a stem that crosses the SVZ, and bulbs that terminate just underneath the ependymal layer [204]. The branched configuration of fractones has been suggested to enable sequestration and subsequent presentation of growth factors and other signaling molecules to stem cells and progenitors to regulate their proliferation, activation, and differentiation within the niche [205].

Other important ECM molecules secreted by niche cells are laminin and fibronectin, both of which have been implicated in neural stem cell growth, differentiation, and migration. In addition to promoting neuroepithelial proliferation and differentiation during development, they also function as permissive substrates, supporting the migration of cerebellar neural precursors *in vitro* and neural progenitors through the RMS *in vivo* [206]. Most recently, the importance of laminin–integrin interactions within the SVZ support a role in migration, spreading and proliferation of NSPC [37]. Several *in vitro* studies have highlighted a critical role for β1-integrin in mediating multiple effects of ECM on NSC in a temporally and spatially controlled manner. For example, genetic ablation of β1-integrin results in reduced neural progenitor proliferation, increased cell death, and impairment of cell migration on different ECM substrates [207]. In β1-integrin-deficient neurospheres, β1-integrin signaling is not required for NSC maintenance, and instead seems to cooperate with growth factor signaling to regulate progenitor number [208]. *In vivo*, the role of laminins in migration and recruitment are critical, as injection of intact laminin and peptide infusion mimicking the E8 domain of the laminin α chain dramatically redirect neuroblast migration towards the site of administration. Interestingly, inhibiting the $\alpha6$ or β1 subunits with antibodies also recapitulates the migratory defect without causing neuroblast death [209].

Collagen IV and chondroitin sulfate proteoglycans (CSPG) are also present in the microenvironment, and have been demonstrated to exert control over proliferation, leading to differentiation. While collagen IV inhibits proliferation of rat NSPCs and promotes differentiation into neurons [210], treatment of neurospheres or telencephalic ventricles with enzymes degrading CSPG glycosminoglycans leads to a reduction in cell proliferation and self-renewal of radial glia; interestingly, the increase in astrocyte formation is at the expense of neuronal differentiation [211]. Additionally, sulfation of chondroitin sulfate polymers *in vitro* modulates the activities and effects of various growth and morphogenetic factors that control NSC proliferation, maintenance, and differentiation [212].

5.3. Other putative endothelial-derived niche effectors

5.3.1. FGF-2

FGF-2 (aka (b)FGF) is detected in the endothelium of tumor capillaries *in vivo*, as well as at sites of vessel branching within the basal lamina of capillaries. *In vitro* studies suggest significant amounts of FGF-2 can also localize to the ECM in cell culture. Normally found to be extracellular, FGF-2 is reported to modulate cell function in an autocrine manner, and depending on the molecular weight isoform, may or may not be secreted; secreted FGF acts through intracellular signaling mechanisms [213]. While EC can secrete this potent angiogenic factor [214] to regulate proliferation, migration and differentiation, type B NSC respond to [113, 215-218] and express the corresponding receptor [178, 219, 220]. Within the CNS, FGF-2 has been shown to affect neurogenesis and proliferation of cortical progenitors [220-222]. Interestingly, FGF-2 infusion increases adult SVZ proliferation while decreasing the number of newly born neurons, suggesting that FGF-2 serves to maintain SVZ self-renewal [223]. In fact, *Fgf-2* knockout mice display a decrease in olfactory bulb size, presumably owing to attentuated output from neurogenic regions. Although FGF-2 can promote NSC proliferation, it does not act alone to maintain self-renewal, and must work with other factors to accomplish this [37]. In addition to inducing VEGF expression in EC, FGF-2 can prime neural precursor responsiveness towards EGF [218].

5.3.2. EGF & Betacellulin

While the specific cell type(s) expressing and secreting EGF remains unidentified within the adult NSC niche, several reports suggest an EC orgin. Affymetrix microarray analysis has revealed that human dermal microvascular EC express EGF, and that this expression is further upregulated in coculture with head and neck squamous cell carcinoma cells [224]. Similarly, an antibody-based human cytokine array has demonstrated that EGF is expressed and secreted by dermal microvascular endothelial cells with or without VEGF stimulation, suggesting basal expression of EGF within some EC [146]. Within the SVZ, receptors for EGF are predominantly expressed by the type C transit-amplifying cells apposed to capillaries. Furthermore, EGFR expression can be further induced by SDF-1 and PEDF [34, 85, 225, 226]. Intraventricular infusion of EGF increases the number of type B NSC contacting the ventricle [219], and leads to transit-amplifying cell proliferation while arresting neuroblast production. *In vitro,* EGF stimulates neurosphere generation from transit-amplifying cells, and is said to cause reversion to a more 'stem-like' phenotype [219]. Curiously, transit-amplifying cells with elevated EGFR signaling also show non-cell autonomous defects in Notch signaling, leading to elevated Numb levels in the stem compartment [227].

Recently, another member of the EGF family, BTC, has been shown to play a critical role in SVZ regulation. mRNA transcripts for *BTC* are detectable in EC, and immunofluorescent analysis reveals protein expression in EC of microcapillaries and in the choroid plexus, with the latter demonstrating greater expression. After intraventricular infusion, NSC and neuroblast compartments are expanded, promoting neurogenesis both in the olfactory bulb and the dentate gyrus. Defects in neuroblast regeneration are observed post cytosine-β-arabinofuranoside (Ara-C) infusion in *Btc*-null mice in comparison to wild-type littermates. Although related to EGF, its effects in the SVZ are slightly different, and it has been suggested that its ability to act on distinct receptors expressed on NSC and neuroblasts, EGFR and Erb4 respectively, may account for these differential effects.

5.3.3. BMPs & LIF & IL-6

In addition to high levels of BMP2 and BMP4 production in astroglia within the SVZ, it has been demonstrated that brain EC can act as potential sources of BMP. mRNA transcripts for *BMP2* and *BMP4* were found in the bEnd.3 endothelial cell line, as well as in primary brain EC. Furthermore, BMP4 protein was also detected in these brain EC [228]. Shown to counteract neurogenesis *in vitro* and *in vivo* [229-231], BMP signaling increases astrocyte formation, possibly through activation of transcriptional regulators of Smads to control cell-cycle exit. Indeed, when embryonic and adult NSPC are cocultured with brain EC, the canonical BMP/Smad pathway becomes activated to reduce proliferation and induce NSPC cell-cycle exit in the presence of EGF and FGF-2 [228]. LIF and IL-6 belong to the cohort of endothelial-secreted factors that promote self-renewal of adult NSC [232], and when synergize with BMP factors to promote self-renewal of embryonic stem cells through activation of gp130-mediated STAT signaling, which induces astrogenesis [232, 233].

5.3.4. SDF-1 & growth-related oncogene-α & angiopoietin

A chemokine previously shown to direct migration of leukocytes during inflammation, SDF-1/CXCL12 signaling via its CXCR4 receptor also provides migratory cues for NSPC recruitment from the lateral ventricle to the nascent dentate gyrus during CNS development; interestingly, SDF-1/CXCR4 expression by EC and neurons persists in the adult dentate gyrus. In the SVZ, neural cells express CXCR4 while ependymal cells and vascular EC express SDF-1 [226]. Neuroblasts expressing CXCR4 migrate towards and are attracted to activated EC of cerebral vessels that secrete SDF-1α [234-238]. While neurospheres express CXCR4, human cerebral EC have been shown to secrete growth-related oncogene-α, also a ligand for CXCR4 [239]. More recently, evidence of progenitor homing to SVZ EC in a SDF1/CXCR4-dependent manner has been demonstrated, where SDF1 upregulates EGFR and α6-integrin in activated NSC and transit amplifying cells, thereby enhancing activation state and the binding to laminin on resident vessels. SDF1 was also shown to increase the motility of migrating neuroblasts towards the olfactory bulb [34, 226].

Recent reports also suggest shared receptor/cytokine signaling between NSC and the vasculature concerning growth related CXCR4/oncogene-α and Ang-1/Tie2 [240, 241]. Ang-1 can be expressed by EC, as well as mural cells, and seems to be upregulated poststroke. While also having a general neuroprotective effect on the nervous system, it has been shown to directly regulate stem cell differentiation and migration through the Tie2 and CXCR4 receptors [235, 237, 242-244].

5.3.5. TGF-β1

Latent TGF-β1 is secreted by EC, pericytes, glia and neurons. Reported to induce VEGF expression by vascular EC and gliomas [245, 246], TGF-β1 serves as an important neurogenic growth factor. Produced in a latent form in mesenchymal and epithelial cell types, EC and mural cells have also been shown to produce a latent form of TGF-β1 which can be activated in endothelial cell-mural cell cocultures [247, 248]. Because *Tgf-β1* knockout mouse models demonstrate a reduced potential for neuron survival [249], and transgenic mouse models overexpressing *Tgf-β1* under control of the GFAP promoter show a reduction in NSCP proliferation [250], it is believed that TGF-β1 has no impact on NSP identity or on differentiation. Rather, it is believed to affect proliferative potential, as demonstrated by an arrest in the G0/G1 phase of the cell cycle [251]. In cell culture, NSC and progenitors express TGFβRI, II and III, and TGF-β1 decreases the expansion of these cells in a dose-dependent manner [252].

5.3.6. Ephs and ephrins

Belonging to a family of receptor tyrosine kinases and associated transmembrane ligands, Ephrins and Eph receptors have established roles in vascular development. However, more recent data suggest both a role for Eph receptor and ephrin ligand interaction within the CNS, and this interaction can occur between endothelial and nonvascular tissues. For example, during development, the close proximity of EphB3/4 on intersomitic vessels with ephrin-B1/B2 of somites seems to imply bidirectional communication [253]. Within the SVZ, EphA7 seems to localize to ependymal cells and astrocytes. Interestingly, the cells immunoreactive for EphA7 also express nestin, a marker associated with NSC. Additionally, ephrin-B2/3 is localized to SVZ astrocytes. By contrast, ephrin-A2 is predominantly expressed on transit-amplifying cells and neuroblasts [254, 255]. Although ephrin-B2 is not cell type exclusive within the SVZ, it has been shown to selectively mark arterial endothelium in the adult, as well as surrounding smooth muscle cells and pericytes [256]. While EphA7 and ephrin-A2 negatively regulate NSPC proliferation, EphB1–2/EphA4 and ephrin-B2/3 direct neuroblast migration and directly or indirectly regulate NSPC proliferation. Interestingly, infusion of antibody-clustered ephrin-B2-Fc or EphB2-Fc into the lateral ventricle increases SVZ proliferation, suggesting that B-class ephrins and Ephs may promote proliferation [254, 255, 257]. A correlation between EphB2 and Notch signaling has also been proposed, wherein EphB2 acts downstream of Notch to maintain ependymal identity under homeostatic conditions, while regulating conversion to astrocytes after injury [258].

5.4. Circulating effectors:

5.4.1. Erythropoietin & prolactin

Systemic transport of erythropoietins and prolactins via blood circulation has been demonstrated to have effects within the CNS. Prolactin, in cooperation with TGF-α, promotes SVZ proliferation and neuronal differentiation. It has been proposed that prolactin serves as an important contributor to the increase in neurogenesis during pregnancy [259]; however, the responsiveness to prolactin within the dentate gyrus is negligible [259, 260]. Although erythropoietin synthesis can be activated in astrocytes and neurons [261-263], it is also possible that circulating erythropoietin, from the kidneys, can cross the BBB to exert neuroprotective effects. Significant amounts of the erythropoietin receptor are localized to the surface of EC and within caveoli [264, 265], and systemic administration of erythropoietin has been shown to penetrate the BBB as an intact molecule [266]. These observations certainly suggest that erythropoietin can reach the brain; however, the precise mechanism mediating this transport is unknown. Erythropoietin has been shown to stimulate NSPC production and prevent apoptosis during embryonic development. Additionally, it serves as a paracrine neuroprotective mediator of ischemia in the brain [267], and erythropoietin-activated EC promote the migration of neuroblasts through the secretion of MMP-2 and 9. [268]. Thus, further investigation of the penetrance and potential function of erythropoietin in the adult NSC niche is warranted.

6. Conclusions

Its remarkable ability to penetrate throughout the entire body to regulate and respond to distinct microenvironments simultaneously has truly earned the vasculature the term of 'master regulator'. It plays a crucial role in embryonic and adult neurogenesis, where its secretion and/or systemic circulation of growth factors, serves to regulate the growth and behavior of stem and progenitor cells. Although a handful of signaling molecules have been identified thus far, it is likely that there are many more unidentified effectors that influence NSC behavior. From a neuro-regenerative perspective, identifying factors responsible for modulating specific stem cell behaviors is crucial, whether the goal is preventing further tissue loss or potentiating endogenous repair mechanisms. Towards this goal, stem cell-based therapies offer the intriguing possibility of accomplishing both. Therefore, understanding the intrinsic and extrinsic mechanisms responsible for modulating NSPC behavior will be critical for the development of more targeted therapies. As mounting evidence points to a strong interdependent relationship between neurogenesis and the vasculature, therapies aimed at targeting both compartments hold great promise.

Acknowledgements

This work was supported by NIH R01-HL077675 and R01-HL096360 to KKH.

Author details

Joshua S. Goldberg and Karen K. Hirschi*

*Address all correspondence to: karen.hirschi@yale.edu

Yale Cardiovascular Research Center and Yale Stem Cell Center, Yale University School of Medicine, New Haven, USA

References

[1] Aitsebaomo, J, et al. Brothers and sisters: molecular insights into arterial-venous heterogeneity. Cir Res, (2008)., 929-939.

[2] Aird, W. C. Phenotypic heterogeneity of the endothelium: I. Structure, function, and mechanisms. Circ Res, (2007)., 158.

[3] Aird, W. C. Phenotypic heterogeneity of the endothelium: II. Representative vascular beds. Cir Res, (2007)., 174-190.

[4] Aird, W. C. Endothelium in health and disease. Pharmacol Rep, (2008)., 139-143.

[5] Girouard, H. and I. C., Neurovascular coupling in the normal brain and in hypertension, stroke, and Alzheimer disease. J Appl Genet, (2006)., 328-335.

[6] Wiederhold, K. H, Bielser, S. U, Jr, W, Veteau, M. J, & Hunziker, O. Three-dimensional reconstruction of brain capillaries from frozen serial sections. Microvasc Res, (1976)., 175-180.

[7] Atkins, G. B, Jain, M. K, & Hamik, A. Endothelial differentiation: molecular mechanisms of specification and heterogeneity. Arterioscler Thromb Vasc Biol, (2011)., 1476-1484.

[8] Wolburg, H, Noell, M. A, Wolburg-buchholz, S, & Fallier-becker, K. P., Brain endothelial cells and the glio-vascular complex. Cell Tissue Res, (2009)., 75-96.

[9] Bär, T. The vascular system of the cerebral cortex. Adv Anat Embryol Cell Biol, (1980). I-VI):, 1-62.

[10] Kiecker, C, & Lumsden, A. Recent advances in neural development. F1000 Biol Rep, (2009)., 1.

[11] Ciani, L, & Salinas, P. C. WNTs in the vertebrate nervous system: from patterning to neuronal connectivity. Nat Rev Neurosci, (2005)., 351-362.

[12] De Robertis, E. M, et al. The establishment of Spemann's organizer and patterning of the vertebrate embryo. Nat Rev Genet, (2000)., 171-181.

[13] Maden, M. Retinoic acid in the development, regeneration and maintenance of the nervous system. Nat Rev Neurosci, (2007)., 755-765.

[14] Mason, I. Initiation to end point: the multiple roles of fibroblast growth factors in neural development. Nat Rev Neurosci, (2007)., 583-596.

[15] Niehrs, C. Regionally specific induction by the Spemann-Mangold organizer. Nat Rev Genet, (2004)., 425-434.

[16] Rhinn, M, Picker, A, & Brand, M. Global and local mechanisms of forebrain and midbrain patterning. Curr Opin Neurobiol, (2006)., 5-12.

[17] Stern, C. D, et al. Head-tail patterning of the vertebrate embryo: one, two or many unresolved problems? Int J Dev Biol, (2006)., 3-15.

[18] Wilson, S. W, & Houart, C. Early steps in the development of the forebrain. Dev Cell, (2004)., 167-181.

[19] Noctor, S. C, Martinez-cerdeno, V, & Kriegstein, A. R. Contribution of intermediate progenitor cells to cortical histogenesis. Arch Neurol, (2007)., 639-642.

[20] Anthony, T. E, et al. Radial glia serve as neuronal progenitors in all regions of the central nervous system. Neuron, (2004)., 881-890.

[21] Nadarajah, B, Brunstrom, G. J, Wong, J. E, & Pearlman, R. O. AL., Two modes of radial migration in early development of the cerebral cortex. Nat Neurosci, (2001). , 143-150.

[22] Parnavelas, J. G. The origin of cortical neurons. Braz J Med Biol Res, (2002). , 1423-1429.

[23] Haubensak, W, Attardo, D. W, & Huttner, A. WB., Neurons arise in the basal neuroepithelium of the early mammalian telencephalon: a major site of neurogenesis. Proc Natl Acad Sci U S A., (2004). , 3196-3201.

[24] Marín, O, & Long, R. J. L. , A. remarkable journey: tangential migration in the telencephalon. Nat Rev Neurosci, (2001). , 780-790.

[25] Galileo, D. S, et al. Neurons and glia arise from a common progenitor in chicken optic tectum: demonstration with two retroviruses and cell type-specific antibodies. Proc Natl Acad Sci U S A., (1990). , 458-462.

[26] Golden, J. A. and C. CL., Clones in the chick diencephalon contain multiple cell types and siblings are widely dispersed. Development, (1996). , 65-78.

[27] Gray, G. E, Glover, M. J, & Sanes, J. C. JR., Radial arrangement of clonally related cells in the chicken optic tectum: lineage analysis with a recombinant retrovirus. Proc Natl Acad Sci U S A., (1988). , 7356-7360.

[28] Leber, S. M, Breedlove, S. J, & Lineage, S. M. arrangement, and death of clonally related motoneurons in chick spinal cord. J Neurosci, (1990). , 2451-2462.

[29] Turner, D. L. and C. CL., A common progenitor for neurons and glia persists in rat retina late in development. Nature, (1987). , 131-136.

[30] Walsh, C. and C. CL., Widespread dispersion of neuronal clones across functional regions of the cerebral cortex. Science, (1992). , 434-440.

[31] Luskin, M. B, & Pearlman, S. J. AL, Cell lineage in the cerebral cortex of the mouse studied in vivo and in vitro with a recombinant retrovirus. Neuron, (1988). , 635-647.

[32] Price, J. and T. L., Cell lineage in the rat cerebral cortex: a study using retroviral-mediated gene transfer. Development, (1988). , 473-482.

[33] Shen, Q, Wang, K. E, Lin, Y, Chuang, G, Goderie, S. M, Roysam, S. K, & Temple, B. S., Adult SVZ stem cells lie in a vascular niche: a quantitative analysis of niche cell-cell interactions. Cell Stem Cell, (2008). , 289-300.

[34] Tavazoie, M, Van Der Veken, S. -V. V, Louissaint, L, Colonna, M, Zaidi, L, Garcia-verdugo, B, & Doetsch, J. M. F., A specialized vascular niche for adult neural stem cells. Cell Stem Cell, (2008). , 279-288.

[35] Nie, K, Molnar, Z, & Szele, F. G. Proliferation but not migration is associated with blood vessels during development of the rostral migratory stream. Dev Neurosci, (2010). , 163-172.

[36] Daneman, R, et al. The mouse blood-brain barrier transcriptome: a new resource for understanding the development and function of brain endothelial cells. PLoS One, (2010). , e13741.

[37] Shen, Q, Goderie, J. L, Karanth, S. K, Sun, N, Abramova, Y, Vincent, N, Pumiglia, P, & Temple, K. S., Endothelial cells stimulate self-renewal and expand neurogenesis of neural stem cells. 304, (2004).

[38] Hogan, K. A, et al. The neural tube patterns vessels developmentally using the VEGF signaling pathway. Development, (2004). , 1503-1513.

[39] Javaherian, A, & Kriegstein, A. A stem cell niche for intermediate progenitor cells of the embryonic cortex. Cereb Cortex, (2009). Suppl 1: , i70-i77.

[40] Stubbs, D, et al. Neurovascular congruence during cerebral cortical development. Cereb Cortex, (2009). Suppl 1: , i32-i41.

[41] Vasudevan, A, et al. Compartment-specific transcription factors orchestrate angiogenesis gradients in the embryonic brain. Nat Neurosci, (2008). , 429-439.

[42] Kurz, H. Physiology of angiogenesis. J Neurooncol, (2000). , 17-35.

[43] Plate, K. H. Mechanisms of angiogenesis in the brain. J Neuropathol Exp Neurol, (1999). , 313-320.

[44] Greenberg, D. A, & Jin, K. From angiogenesis to neuropathology. Nature, (2005). , 954-959.

[45] Hiruma, T, Nakajima, Y, & Nakamura, H. Development of pharyngeal arch arteries in early mouse embryo. J Anat, (2002). , 15-29.

[46] Noctor, S. C, et al. Neurons derived from radial glial cells establish radial units in neocortex. Nature, (2001). , 714-720.

[47] Miyata, T, et al. Asymmetric inheritance of radial glial fibers by cortical neurons. Neuron, (2001). , 727-741.

[48] Tabata, H, Yoshinaga, S, & Nakajima, K. Cytoarchitecture of mouse and human subventricular zone in developing cerebral neocortex. Exp Brain Res. 216(2): , 161-168.

[49] Noctor, S. C, et al. Cortical neurons arise in symmetric and asymmetric division zones and migrate through specific phases. Nat Neurosci, (2004). , 136-144.

[50] Aaku-saraste, E, Hellwig, A, & Huttner, W. B. Loss of occludin and functional tight junctions, but not ZO-1, during neural tube closure--remodeling of the neuroepithelium prior to neurogenesis. Dev Biol, (1996). , 664-679.

[51] Bancroft, M, & Bellairs, R. Differentiation of the neural plate and neural tube in the young chick embryo. A study by scanning and transmission electron microscopy. Anat Embryol, (1975). , 309-335.

[52] Lehtinen, M. K, & Walsh, C. A. Neurogenesis at the brain-cerebrospinal fluid interface. Annu Rev Cell Dev Biol. 27: , 653-679.

[53] Kriegstein, A, & Alvarez-buylla, A. The glial nature of embryonic and adult neural stem cells. Annu Rev Neurosci, (2009). , 149-184.

[54] Taverna, E, & Huttner, W. B. Neural progenitor nuclei IN motion. Neuron, (2010). , 906-914.

[55] Takahashi, T, Nowakowski, R. S, & Caviness, V. S. Jr., The leaving or Q fraction of the murine cerebral proliferative epithelium: a general model of neocortical neuronogenesis. J Neurosci, (1996). , 6183-6196.

[56] Kosodo, Y, et al. Asymmetric distribution of the apical plasma membrane during neurogenic divisions of mammalian neuroepithelial cells. EMBO J, (2004). , 2314-2324.

[57] Rakic, P. A century of progress in corticoneurogenesis: from silver impregnation to genetic engineering. Cereb Cortex, (2006). Suppl 1: , i3-17.

[58] Tarabykin, V, et al. Cortical upper layer neurons derive from the subventricular zone as indicated by Svet1 gene expression. Development, (2001). , 1983-1993.

[59] Kriegstein, A, Noctor, S, & Martinez-cerdeno, V. Patterns of neural stem and progenitor cell division may underlie evolutionary cortical expansion. Nat Rev Neurosci, (2006). , 883-890.

[60] Martinez-cerdeno, V, Noctor, S. C, & Kriegstein, A. R. The role of intermediate progenitor cells in the evolutionary expansion of the cerebral cortex. Cereb Cortex, (2006). Suppl 1: , i152-i161.

[61] Noctor, S. C, Martinez-cerdeno, V, & Kriegstein, A. R. Distinct behaviors of neural stem and progenitor cells underlie cortical neurogenesis. J Comp Neurol, (2008). , 28-44.

[62] Pontious, A, et al. Role of intermediate progenitor cells in cerebral cortex development. Dev Neurosci, (2008). , 24-32.

[63] Takahashi, T, Nowakowski, R. S, & Caviness, V. S. Jr., Early ontogeny of the secondary proliferative population of the embryonic murine cerebral wall. J Neurosci, (1995). , 6058-6068.

[64] Lois, C, & Alvarez-buylla, A. Proliferating subventricular zone cells in the adult mammalian forebrain can differentiate into neurons and glia. Proc Natl Acad Sci U S A, (1993). , 2074-2077.

[65] Fietz, S. A, et al. OSVZ progenitors of human and ferret neocortex are epithelial-like and expand by integrin signaling. Nat Neurosci, (2010). , 690-699.

[66] Hansen, D. V, et al. Neurogenic radial glia in the outer subventricular zone of human neocortex. Nature, (2010). , 554-561.

[67] Reillo, I, et al. A role for intermediate radial glia in the tangential expansion of the mammalian cerebral cortex. Cereb Cortex, (2011). , 1674-1694.

[68] Shitamukai, A, Konno, D, & Matsuzaki, F. Oblique radial glial divisions in the developing mouse neocortex induce self-renewing progenitors outside the germinal zone that resemble primate outer subventricular zone progenitors. J Neurosci, (2011). , 3683-3695.

[69] Wang, X, et al. A new subtype of progenitor cell in the mouse embryonic neocortex. Nat Neurosci, (2011). , 555-561.

[70] Farin, A, et al. Transplanted glioma cells migrate and proliferate on host brain vasculature: a dynamic analysis. Glia, (2006). , 799-808.

[71] Goldman, S. A, & Chen, Z. Perivascular instruction of cell genesis and fate in the adult brain. Nat Neurosci, (2011). , 1382-1389.

[72] Alvarez-buylla, A, Garcia-verdugo, J. M, & Tramontin, A. D. A unified hypothesis on the lineage of neural stem cells. Nat Rev Neurosci, (2001). , 287-293.

[73] Merkle, F. T, et al. Radial glia give rise to adult neural stem cells in the subventricular zone. Proc Natl Acad Sci U S A, (2004). , 17528-17532.

[74] Mirzadeh, Z, et al. Neural stem cells confer unique pinwheel architecture to the ventricular surface in neurogenic regions of the adult brain. Cell Stem Cell, (2008). , 265-278.

[75] Shen, Q, et al. Adult SVZ stem cells lie in a vascular niche: a quantitative analysis of niche cell-cell interactions. Cell Stem Cell, (2008). , 289-300.

[76] Altman, J, & Bayer, S. A. Mosaic organization of the hippocampal neuroepithelium and the multiple germinal sources of dentate granule cells. J Comp Neurol, (1990). , 325-342.

[77] Eckenhoff, M. F, & Rakic, P. Radial organization of the hippocampal dentate gyrus: a Golgi, ultrastructural, and immunocytochemical analysis in the developing rhesus monkey. J Comp Neurol, (1984). , 1-21.

[78] Han, Y. G, et al. Hedgehog signaling and primary cilia are required for the formation of adult neural stem cells. Nat Neurosci, (2008). , 277-284.

[79] Breunig, J. J, et al. Primary cilia regulate hippocampal neurogenesis by mediating sonic hedgehog signaling. Proc Natl Acad Sci U S A, (2008). , 13127-13132.

[80] Merkle, F. T, Mirzadeh, Z, & Alvarez-buylla, A. Mosaic organization of neural stem cells in the adult brain. Science, (2007). , 381-384.

[81] Takahashi, T, Nowakowski, C. V. J, & Interkinetic, R. S. and migratory behavior of a cohort of neocortical neurons arising in the early embryonic murine cerebral wall. J Neurosci, (1996). , 5762-5776.

[82] Nadarajah, B, Jones, E. W, & Parnavelas, A. M. JG., Differential expression of connexins during neocortical development and neuronal circuit formation. J Neurosci, (1997). , 3096-3111.

[83] Bittman, K. S. and LoTurco JJ, Differential regulation of connexin 26 and 43 in murine neocortical precursors. Cereb Cortex, (1999). , 188-195.

[84] Bittman, K, et al. Connexin expression in homotypic and heterotypic cell coupling in the developing cerebral cortex. J Comp Neurol, (2002). , 201-212.

[85] Tavazoie, M, et al. A specialized vascular niche for adult neural stem cells. Cell Stem Cell, (2008). , 279-288.

[86] Luo, J, Daniels, L. J, Notti, S. B, & Conover, R. Q. JC., The aging neurogenic subventricular zone. Aging Cell, (2006). , 139-152.

[87] Lois, C, & Long-distance, A. -B. A. neuronal migration in the adult mammalian brain. Science, (1994). , 1145-1148.

[88] Biebl, M, et al. Analysis of neurogenesis and programmed cell death reveals a self-renewing capacity in the adult rat brain. Neurosci Lett, (2000). , 17-20.

[89] Luskin, M. B. Restricted proliferation and migration of postnatally generated neurons derived from the forebrain subventricular zone. Neuron, (1993). , 173-189.

[90] Menn, B, Garcia-verdugo, Y. C, Gonzalez-perez, J. M, Rowitch, O, & Alvarez-buylla, D. A., Origin of oligodendrocytes in the subventricular zone of the adult brain. J Neurosci, (2006). , 7907-7918.

[91] Seri, B, García-verdugo, C. -M. L, Mcewen, J. M, & Alvarez-buylla, B. S. A., Cell types, lineage, and architecture of the germinal zone in the adult dentate gyrus. J Comp Neurol, (2004). , 359-378.

[92] Riquelme, P. A, & Drapeau, D. F. E, Brain micro-ecologies: neural stem cell niches in the adult mammalian brain. Philos Trans R Soc Lond B Biol Sci, (2008). , 123-137.

[93] Seaberg, R. M. and v.d.K. D., Adult rodent neurogenic regions: the ventricular subependyma contains neural stem cells, but the dentate gyrus contains restricted progenitors. J Neurosci, (2002). , 1784-1793.

[94] Clarke, L. and v.d.K. D., The Adult Mouse Dentate Gyrus Contains Populations of Committed Progenitor Cells that are Distinct from Subependymal Zone Neural Stem Cells. Stem Cells, (2011). , 1448-1458.

[95] Doetsch, F, & García-verdugo, A. -B. A. JM, Cellular composition and three-dimensional organization of the subventricular germinal zone in the adult mammalian brain. J Neurosci, (1997). , 5046-5061.

[96] Morshead, C. M, Reynolds, C. C, Mcburney, B. A, Staines, M. W, Morassutti, W. A, Weiss, D, & Van Der Kooy, S. D., Neural stem cells in the adult mammalian forebrain: a relatively quiescent subpopulation of subependymal cells. Neuron, (1994). , 1071-1082.

[97] De Pina-benabou, M. H, Srinivas, S. D, & Scemes, M. E., Calmodulin kinase pathway mediates the K+-induced increase in Gap junctional communication between mouse spinal cord astrocytes. J Neurosci, (2001). , 6635-6643.

[98] Ihrie, R. A, & Lake-front, A. -B. A. property: a unique germinal niche by the lateral ventricles of the adult brain. Neuron, (2011). , 674-686.

[99] Mirzadeh, Z, Merkle, S. -N. M, Garcia-verdugo, F. T, & Alvarez-buylla, J. M. A., Neural stem cells confer unique pinwheel architecture to the ventricular surface in neurogenic regions of the adult brain. Cell stem Cell, (2008). , 265-278.

[100] Doetsch, F. A niche for adult neural stem cells. Curr Opin Gen Dev, (2003). , 543-550.

[101] Palmer, T. D, & Willhoite, G. F. AR, Vascular niche for adult hippocampal neurogenesis. J Comp Neurol, (2000). , 479-494.

[102] Lie, D. C, Song, C. S, Ming, H, & Gage, G. L. FH., Neurogenesis in the adult brain: new strategies for central nervous system diseases. Annu Rev Pharmacol Toxicol, (2004). , 399-421.

[103] Coskun, V, Wu, B. B, Tsao, H, Kim, S, Zhao, K, Biancotti, J, Hutnick, J. C, Krueger, L, Jr, R. C, Fan, G, De Vellis, J, & Sun, Y. E. CD133+ neural stem cells in the ependyma of mammalian postnatal forebrain. Proc Natl Acad Sci U S A., (2008). , 1026-1031.

[104] Meletis, K, Barnabé-heider, C. M, Evergren, F, Tomilin, E, Shupliakov, N, & Frisén, O. J., Spinal cord injury reveals multilineage differentiation of ependymal cells. PLoS Biol, (2008). , e182.

[105] Beckervordersandforth, R, et al. In vivo fate mapping and expression analysis reveals molecular hallmarks of prospectively isolated adult neural stem cells. Cell Stem Cell, (2010). , 744-758.

[106] Luskin, M. B, Zigova, S. B, & Stewart, T. RR., Neuronal progenitor cells derived from the anterior subventricular zone of the neonatal rat forebrain continue to proliferate in vitro and express a neuronal phenotype. Nol Cell Neurosci, (1997). , 351-366.

[107] García-verdugo, J. M, et al. Architecture and cell types of the adult subventricular zone: in search of the stem cells. J Neurobiol, (1998). , 234-248.

[108] Kelsch, W, Mosley, L. C, & Lois, C. P. C., Distinct mammalian precursors are committed to generate neurons with defined dendritic projection patterns. PLoS Biol, (2007)., e300.

[109] Kohwi, M, Petryniak, L. J, Ekker, M. A, Obata, M, Yanagawa, K, Rubenstein, Y, & Alvarez-buylla, J. L. A., A subpopulation of olfactory bulb GABAergic interneurons is derived from Emx1- and Dlx5/6-expressing progenitors. J Neurosci, (2007). , 6878-6891.

[110] Merkle, F. T, & Mirzadeh, A. -B. A. Z, Mosaic organization of neural stem cells in the adult brain. Science, (2007)., 381-384.

[111] Ventura, R. E. and G. JE., Dorsal radial glia generate olfactory bulb interneurons in the postnatal murine brain. J Neurosci, (2007)., 4297-4302.

[112] Young, K. M, Fogarty, K. N, & Richardson, M. WD., Subventricular zone stem cells are heterogeneous with respect to their embryonic origins and neurogenic fates in the adult olfactory bulb. J Neurosci, (2007)., 8286-8296.

[113] Gage, F. H, et al. Survival and differentiation of adult neuronal progenitor cells transplanted to the adult brain. Proc Natl Acad Sci U S A., (1995)., 11879-11883.

[114] Kuhn, H. G, & Dickinson-anson, G. F. H, Neurogenesis in the dentate gyrus of the adult rat: age-related decrease of neuronal progenitor proliferation. J Neurosci, (1996)., 2027-2033.

[115] Temple, S. The development of neural stem cells. Nature, (2001)., 112-117.

[116] Rahpeymai, Y, Hietala, W. U, Fotheringham, M. A, Davies, A, Nilsson, I, Zwirner, A. K, Wetsel, J, Gerard, R. A, Pekny, C, Pekna, M, & Complement, M. a novel factor in basal and ischemia-induced neurogenesis. EMBOJ, (2006)., 1364-1374.

[117] Shingo, T, Sorokan, S. T, & Weiss, S. T. S., Erythropoietin regulates the in vitro and in vivo production of neuronal progenitors by mammalian forebrain neural stem cells. J Neurosci, (2001)., 9733-9743.

[118] Thored, P, Wood, A. A, Cammenga, J, Kokaia, J, Lindvall, Z, & Long-term, O. neuroblast migration along blood vessels in an area with transient angiogenesis and increased vascularization after stroke. Stroke, (2007)., 3032-3039.

[119] Yamashita, T, Ninomiya, H. A. P, García-verdugo, M, Sunabori, J. M, Sakaguchi, T, Adachi, M, Kojima, K, Hirota, T, Kawase, Y, Araki, T, Abe, N, Okano, K, & Sawamoto, H. K., Subventricular zone-derived neuroblasts migrate and differentiate into mature neurons in the post-stroke adult striatum. J Neurosci, (2006)., 6627-6636.

[120] Ohab, J. J, Fleming, B. A, & Carmichael, S. ST., A neurovascular niche for neurogenesis after stroke. J Neurosci, (2006)., 13007-13016.

[121] Louissaint A Jr and LC. Rao S, Goldman SA., Coordinated interaction of neurogenesis and angiogenesis in the adult songbird brain. Neuron, (2002)., 945-960.

[122] Monje, M. L, Mizumatsu, F. J, & Palmer, S. TD., Irradiation induces neural precursor-cell dysfunction. Nat Med, (2002)., 955-962.

[123] Jin, K, Zhu, S. Y, Mao, Y, Xie, X. O, & Greenberg, L. DA., Vascular endothelial growth factor (VEGF) stimulates neurogenesis in vitro and in vivo. Proc Natl Acad Sci U S A., (2002)., 11946-11950.

[124] Wurmser, A. E, Palmer, G. F, & Neuroscience, T. D. Cellular interactions in the stem cell niche. Science, (2004)., 1253-1255.

[125] Shima, D. T. and M. C., Vascular developmental biology: getting nervous. Curr Opin Genet Dev, (2000)., 536-542.

[126] Ding, Y, Li, L. X, Ding, J, Lai, Y. H, Rafols, Q, Phillis, J. A, Clark, J. W, & Diaz, J. C. FG., Exercise pre-conditioning reduces brain damage in ischemic rats that may be associated with regional angiogenesis and cellular overexpression of neurotrophin. Neuroscience, (2004)., 583-591.

[127] Vasconcellos, J. P, Melo, S. R, Bressanim, M. B, Costa, N. C, & Costa, F. F. VP., A novel mutation in the GJA1 gene in a family with oculodentodigital dysplasia. Arch Ophthalmol, (2005)., 1422-1426.

[128] Barnea, A, & Nottebohm, F. Seasonal recruitment of hippocampal neurons in adult free-ranging black-capped chickadees. Proc Natl Acad Sci U S A., (1994). , 11217-112121.

[129] Kempermann, G, & Kuhn, G. F. HG, More hippocampal neurons in adult mice living in an enriched environment. Nature, (1997)., 493-495.

[130] Kempermann, G, Kuhn, G. F, & Experience-induced, H. G. neurogenesis in the senescent dentate gyrus. J Neurosci, (1998)., 3206-3212.

[131] Whitman, M. C, et al. Blood vessels form a migratory scaffold in the rostral migratory stream. J Comp Neurol, (2009)., 94-104.

[132] Bovetti, S, et al. Blood vessels form a scaffold for neuroblast migration in the adult olfactory bulb. J Neurosci, (2007)., 5976-5980.

[133] Yang, X. T, Bi, Y. Y, & Feng, D. F. From the vascular microenvironment to neurogenesis. Brain Res Bull, (2011)., 1-7.

[134] Leventhal, C, Rafii, R. D, Shahar, S, & Goldman, A. SA., Endothelial trophic support of neuronal production and recruitment from the adult mammalian subependyma. Mol Cell Neurosci, (1999)., 450-464.

[135] Wolswijk, G, Riddle, N. M, & Platelet-derived, P. N. growth factor is mitogenic for O-2Aadult progenitor cells. Glia, (1991)., 495-503.

[136] Barres, B. A, et al. Multiple extracellular signals are required for long-term oligodendrocyte survival. Development, (1993)., 283-295.

[137] Cameron, H. A, Hazel, T. G, & Mckay, R. D. Regulation of neurogenesis by growth factors and neurotransmitters. J Neurobiol, (1998). , 287-306.

[138] Park, J. A, Choi, K. S, & Kim, K. S. KW., Coordinated interaction of the vascular and nervous systems: from molecule- to cell-based approaches. Biochem Biophys Res Commun, (2003). , 247-253.

[139] Paris, F, Fuks, K. A, Capodieci, Z, Juan, P, Ehleiter, G, Haimovitz-friedman, D, Cordon-cardo, A, & Kolesnick, C. R., Endothelial apoptosis as the primary lesion initiating intestinal radiation damage in mice. Science, (2001). , 293-297.

[140] Garcia-barros, M, et al. Tumor response to radiotherapy regulated by endothelial cell apoptosis. Science, (2003). , 1155-1159.

[141] Guzman, R, Bliss, D. L. A. A, Moseley, T, Palmer, M, & Steinberg, T. G., Neural progenitor cells transplanted into the uninjured brain undergo targeted migration after stroke onset. J Neurosci Res, (2008). , 873-882.

[142] Nakagomi, N, et al. Endothelial cells support survival, proliferation, and neuronal differentiation of transplanted adult ischemia-induced neural stem/progenitor cells after cerebral infarction. Stem Cells, (2009). , 2185-2195.

[143] Teng, H, Zhang, W. L, Zhang, Z. G, Zhang, R. L, Morris, L, Gregg, D, Wu, S. R, Jiang, Z, Lu, A, Zlokovic, M, & Chopp, B. V. M., Coupling of angiogenesis and neurogenesis in cultured endothelial cells and neural progenitor cells after stroke. J Cereb Blood Flow Metab, (2008). , 764-771.

[144] Ramírez-castillejo, C, Sánchez-sánchez, A. -A. C, Ferrón, F, Aroca-aguilar, S. R, Sánchez, J. D, Mira, P, Escribano, H, & Fariñas, J. I., Pigment epithelium-derived factor is a niche signal for neural stem cell renewal. Nat Neurosci, (2006). , 331-339.

[145] Cao, L, et al. VEGF links hippocampal activity with neurogenesis, learning and memory. Nat Genet, (2004). , 827-835.

[146] Schmidt, N. O, Koeder, M. M, Mueller, D, Aboody, F. J, Kim, K. S, Black, S. U, Carroll, P. M, Westphal, R. S, & Lamszus, M. K., Vascular endothelial growth factor-stimulated cerebral microvascular endothelial cells mediate the recruitment of neural stem cells to the neurovascular niche. Brain Res, (2009). , 24-37.

[147] Mi, H, & Haeberle, B. B. H, Induction of astrocyte differentiation by endothelial cells. J Neurosci, (2001). , 1538-1547.

[148] Goldberg, J. S, & Hirschi, K. K. Diverse roles of the vasculature within the neural stem cell niche. Regen Med, (2009). , 879-897.

[149] Zhang, H, et al. VEGF is a chemoattractant for FGF-2-stimulated neural progenitors. J Cell Biol, (2003). , 1375-1384.

[150] Schänzer, A, Wachs, W. D, Acker, F. P, Cooper-kuhn, T, Beck, C, Winkler, H, Aigner, J, Plate, L, & Kuhn, K. H. HG., Direct stimulation of adult neural stem cells in vitro

and neurogenesis in vivo by vascular endothelial growth factor. Brain Pathol, (2004). , 237-248.

[151] Meng, H, et al. Biphasic effects of exogenous VEGF on VEGF expression of adult neural progenitors. Neurosci Lett, (2006). , 97-101.

[152] Wada, T, Haigh, E. M, Hitoshi, J. J, Chaddah, S, Rossant, R, Nagy, J, & Van Der Kooy, A. D., Vascular endothelial growth factor directly inhibits primitive neural stem cell survival but promotes definitive neural stem cell survival. J Neurosci, (2006). , 6803-6812.

[153] Mani, N, et al. Vascular endothelial growth factor enhances migration of astroglial cells in subventricular zone neurosphere cultures. J Neurosci Res, (2010). , 248-257.

[154] Wittko, I. M, et al. VEGFR-1 regulates adult olfactory bulb neurogenesis and migration of neural progenitors in the rostral migratory stream in vivo. J Neurosci, (2009). , 8704-8714.

[155] Calvo, C. F, et al. Vascular endothelial growth factor receptor 3 directly regulates murine neurogenesis. Genes Dev, (2011). , 831-844.

[156] Lee, H. J, et al. Human neural stem cells over-expressing VEGF provide neuroprotection, angiogenesis and functional recovery in mouse stroke model. PLoS One, (2007). , e156.

[157] Sun, Y, et al. VEGF-induced neuroprotection, neurogenesis, and angiogenesis after focal cerebral ischemia. J Clin Invest, (2003). , 1843-1851.

[158] Maurer, M. H, Tripps, F. R. J, & Kuschinsky, W. K. W., Expression of vascular endothelial growth factor and its receptors in rat neural stem cells. Neurosci Lett, (2003). , 165-168.

[159] Chow, J, Ogunshola, F. S, Li, O, Ment, Y, Madri, L. R, & Astrocyte-derived, J. A. VEGF mediates survival and tube stabilization of hypoxic brain microvascular endothelial cells in vitro. Brain Res Dev Res, (2001). , 123-132.

[160] Lucitti, J. L, Jones, H. C, Chen, E. A, Fraser, J, & Dickinson, S. E. ME., Vascular remodeling of the mouse yolk sac requires hemodynamic force. Development, (2007). , 3317-3326.

[161] Hayakawa, K, et al. Cerebral endothelial derived vascular endothelial growth factor promotes the migration but not the proliferation of oligodendrocyte precursor cells in vitro. Neurosci Lett, (2012). , 42-46.

[162] Gerber, H. P, et al. VEGF regulates haematopoietic stem cell survival by an internal autocrine loop mechanism. Nature, (2002). , 954-958.

[163] Shah, N. M, & Groves, A. D. AK, Alternative neural crest cell fates are instructively promoted by TGFbeta superfamily members. Cell, (1996). , 331-343.

[164] Nottebohm, F. The neural basis of birdsong. PLoS Biol, (2005)., e164.

[165] Cheng, A, Wang, C. J, Rao, S, & Mattson, M. S. MP., Nitric oxide acts in a positive feedback loop with BDNF to regulate neural progenitor cell proliferation and differentiation in the mammalian brain. Dev Biol, (2003)., 319-333.

[166] Packer, M. A, Stasiv, B. A, Chmielnicki, Y, Grinberg, E, Westphal, A, Goldman, H, & Enikolopov, S. A. G., Nitric oxide negatively regulates mammalian adult neurogenesis. Proc Natl Acad Sci U S A., (2003)., 9566-9571.

[167] Snapyan, M, et al. Vasculature guides migrating neuronal precursors in the adult mammalian forebrain via brain-derived neurotrophic factor signaling. J Neurosci, (2009)., 4172-4188.

[168] Vaynman, S, & Ying, G. -P. F. Z, Interplay between brain-derived neurotrophic factor and signal transduction modulators in the regulation of the effects of exercise on synaptic-plasticity. Neuroscience, (2003)., 647-657.

[169] Ding, Q, Vaynman, A. M, Ying, S, Gomez-pinilla, Z, & Insulin-like, F. growth factor I interfaces with brain-derived neurotrophic factor-mediated synaptic plasticity to modulate aspects of exercise-induced cognitive function. Neuroscience, (2006). , 823-833.

[170] Busiguina, S, et al. Neurodegeneration is associated to changes in serum insulin-like growth factors. Neurobiol Dis, (2000). Pt B):, 657-665.

[171] Perez-martin, M, Azcoitia, T. J, Sierra, I, & Garcia-segura, A. LM., An antagonist of estrogen receptors blocks the induction of adult neurogenesis by insulin-like growth factor-I in the dentate gyrus of adult female rat. Eur J Neurosci, (2003)., 923-930.

[172] Ihrie, R. A, & Alvarez-buylla, A. Lake-front property: a unique germinal niche by the lateral ventricles of the adult brain. Neuron, (2011)., 674-686.

[173] Armulik, A, et al. Pericytes regulate the blood-brain barrier. Nature, (2010)., 557-561.

[174] Bjarnegard, M, et al. Endothelium-specific ablation of PDGFB leads to pericyte loss and glomerular, cardiac and placental abnormalities. Development, (2004). , 1847-1857.

[175] Enge, M, et al. Endothelium-specific platelet-derived growth factor-B ablation mimics diabetic retinopathy. EMBO J, (2002)., 4307-4316.

[176] Andrae, J, Gallini, R, & Betsholtz, C. Role of platelet-derived growth factors in physiology and medicine. Genes Dev, (2008)., 1276-1312.

[177] Dai, C, et al. PDGF autocrine stimulation dedifferentiates cultured astrocytes and induces oligodendrogliomas and oligoastrocytomas from neural progenitors and astrocytes in vivo. Genes Dev, (2001)., 1913-1925.

[178] Jackson, EL, Garcia-Verdugo, G.-P.S, Roy, JM, Quinones-Hinojosa, M, & Vanden, A. , Alvarez-Buylla A., PDGFR alpha-positive B cells are neural stem cells in the adult

SVZ that form glioma-like growths in response to increased PDGF signaling. Neuron, 2006. 51(2): p. 187-199.

[179] Gensburger, C, Labourdette, G, & Sensenbrenner, M. Brain basic fibroblast growth factor stimulates the proliferation of rat neuronal precursor cells in vitro. FEBS Lett, (1987). , 1-5.

[180] Richards, L. J, & Kilpatrick, B. P. TJ, De novo generation of neuronal cells from the adult mouse brain. Proc Natl Acad Sci U S A., (1992). , 8591-8595.

[181] Johe, K. K, Hazel, M. T, Dugich-djordjevic, T. G, & Mckay, M. M. RD., Single factors direct the differentiation of stem cells from the fetal and adult central nervous system. Genes Dev, (1996). , 3129-3140.

[182] Broudy, V. C, et al. Human umbilical vein endothelial cells display high-affinity c-kit receptors and produce a soluble form of the c-kit receptor. Blood, (1994). , 2145-2152.

[183] Ashman, L. K. The biology of stem cell factor and its receptor C-kit. In J Biochem Cell Biol, (1999). , 1037-10351.

[184] Jin, K, Mao, S. Y, Xie, X. O, & Greenberg, L. DA., Stem cell factor stimulates neurogenesis in vitro and in vivo. J Clin Invest, (2002). , 311-319.

[185] Erlandsson, A, & Larsson, F. -N. K. J, Stem cell factor is a chemoattractant and a survival factor for CNS stem cells. Exp Cell Res, (2004). , 201-210.

[186] Sun, Y, et al. Vascular endothelial growth factor-B (VEGFB) stimulates neurogenesis: evidence from knockout mice and growth factor administration. Dev Biol, (2006). , 329-335.

[187] Aparicio, S, et al. Expression of angiogenesis factors in human umbilical vein endothelial cells and their regulation by PEDF. Biochem Biophys Res Commun, (2005). , 387-394.

[188] Kozaki, K, et al. Isolation, purification, and characterization of a collagen-associated serpin, caspin, produced by murine colon adenocarcinoma cells. J Biol Chem, (1998). , 15125-15130.

[189] Meyer, C, Notari, L, & Becerra, S. P. Mapping the type I collagen-binding site on pigment epithelium-derived factor. Implications for its antiangiogenic activity. J Biol Chem, (2002). , 45400-45407.

[190] Sugita, Y, Becerra, C. G, & Schwartz, S. P. JP., Pigment epithelium-derived factor (PEDF) has direct effects on the metabolism and proliferation of microglia and indirect effects on astrocytes. J Neurosci Res, (1997). , 710-718.

[191] Pignolo, R. J, Francis, R. M, & Cristofalo, M. K. VJ., Putative role for EPC-1/PEDF in the G0 growth arrest of human diploid fibroblasts. J Cell Physiol, (2003). , 12-20.

[192] Pumiglia, K, & Pedf, T. S. bridging neurovascular interactions in the stem cell niche. Nat Neurosci, (2006). , 299-300.

[193] Ables, J. L, et al. Not(ch) just development: Notch signalling in the adult brain. Nat Rev Neurosci, (2011). , 269-283.

[194] Ramirez-castillejo, C, et al. Pigment epithelium-derived factor is a niche signal for neural stem cell renewal. Nat Neurosci, (2006). , 331-339.

[195] Andreu-agullo, C, et al. Vascular niche factor PEDF modulates Notch-dependent stemness in the adult subependymal zone. Nat Neurosci, (2009). , 1514-1523.

[196] Park, C, et al. Inhibition of neuronal nitric oxide synthase enhances cell proliferation in the dentate gyrus of the adrenalectomized rat. Neurosci Lett, (2001). , 9-12.

[197] Moreno-lopez, B, et al. Nitric oxide is a physiological inhibitor of neurogenesis in the adult mouse subventricular zone and olfactory bulb. J Neurosci, (2004). , 85-95.

[198] Murillo-carretero, M, et al. S-Nitrosylation of the epidermal growth factor receptor: a regulatory mechanism of receptor tyrosine kinase activity. Free Radic Biol Med, (2009). , 471-479.

[199] Torroglosa, A, et al. Nitric oxide decreases subventricular zone stem cell proliferation by inhibition of epidermal growth factor receptor and phosphoinositide-3-kinase/Akt pathway. Stem Cells, (2007). , 88-97.

[200] Carreira, B. P, et al. Nitric oxide stimulates the proliferation of neural stem cells bypassing the epidermal growth factor receptor. Stem Cells, (2010). , 1219-1230.

[201] Li, Q, et al. Modeling the neurovascular niche: VEGF- and BDNF-mediated cross-talk between neural stem cells and endothelial cells: an in vitro study. J Neurosci Res, (2006). , 1656-1668.

[202] Chen, J, et al. Endothelial nitric oxide synthase regulates brain-derived neurotrophic factor expression and neurogenesis after stroke in mice. J Neurosci, (2005). , 2366-2375.

[203] Kerever, A, Schnack, V. D, Ichikawa, J, Moon, N, Arikawa-hirasawa, C, Efird, E, & Mercier, J. T. F., Novel extracellular matrix structures in the neural stem cell niche capture the neurogenic factor fibroblast growth factor 2 from the extracellular milieu. Stem cells, (2007). , 2146-2157.

[204] Mercier, F, Kitasako, H. G, & Fractones, J. T. and other basal laminae in the hypothalamus. J Comp Neurol, (2003). , 324-340.

[205] Mercier, F, & Kitasako, H. G. JT, Anatomy of the brain neurogenic zones revisited: fractones and the fibroblast/macrophage network. J Comp Neurol, (2002). , 170-188.

[206] Tsang, K. Y, Cheung, C. D, & Cheah, M. C. KS., The developmental roles of the extracellular matrix: beyond structure to regulation. Cell Tissue Res, (2010). , 93-110.

[207] Tsang, K. Y, Cheung, C. D, & Cheah, M. C. KS., The developmental roles of the extracellular matrix: beyond structure to regulation. Cell Tisue Res, (2010)., 93-110.

[208] Leone, D. P, et al. Regulation of neural progenitor proliferation and survival by beta1 integrins. J Cell Sci, (2005). Pt 12):, 2589-2599.

[209] Emsley, J. G, & Hagg, T. alpha6beta1 integrin directs migration of neuronal precursors in adult mouse forebrain. Exp Neurol, (2003)., 273-285.

[210] Ali, S. A, Pappas, I. S, & Parnavelas, J. G. Collagen type IV promotes the differentiation of neuronal progenitors and inhibits astroglial differentiation in cortical cell cultures. Brain Res Dev Brain Res, (1998)., 31-38.

[211] Brückner, G, et al. Region and lamina-specific distribution of extracellular matrix proteoglycans, hyaluronan and tenascin-R in the mouse hippocampal formation. J Chem Nauroanat, (2003)., 37-50.

[212] Gates, M. A, et al. Cell and molecular analysis of the developing and adult mouse subventricular zone of the cerebral hemispheres. J Comp Neurol, (1995)., 249-266.

[213] Seghezzi, G, Patel, R. C, Gualandris, S, Pintucci, A, Robbins, G, Shapiro, E. S, Galloway, R. L, Rifkin, A. C, & Mignatti, D. B. P., Fibroblast growth factor-2 (FGF-2) induces vascular endothelial growth factor (VEGF) expression in the endothelial cells of forming capillaries: an autocrine mechanism contributing to angiogenesis. J Cell Biol, (1998)., 1659-1673.

[214] Biro, S, et al. Expression and subcellular distribution of basic fibroblast growth factor are regulated during migration of endothelial cells. Circ Res, (1994)., 485-494.

[215] Kilpatrick, T. J. and B. PF., Cloned multipotential precursors from the mouse cerebrum require FGF-2, whereas glial restricted precursors are stimulated with either FGF-2 or EGF. J Neurosci, (1995). Pt 1):, 3653-3651.

[216] Gritti, A, Parati, C. L, Frolichsthal, E. A, Galli, P, Wanke, R, Faravelli, E, Morassutti, L, Roisen, D. J, Nickel, F, & Vescovi, D. D. AL., Multipotential stem cells from the adult mouse brain proliferate and self-renew in response to basic fibroblast growth factor. J Neurosci, (1996)., 1091-1100.

[217] Palmer, T. D, & Takahashi, G. F. J, The adult rat hippocampus contains primordial neural stem cells. Mol Cell Neurosci, (1997)., 389-404.

[218] Ciccolini, F. and S. CN., Fibroblast growth factor 2 (FGF-2) promotes acquisition of epidermal growth factor (EGF) responsiveness in mouse striatal precursor cells: identification of neural precursors responding to both EGF and FGF-2. J Neurosci, (1998)., 7869-7880.

[219] Doetsch, F, Petreanu, C. I, Garcia-verdugo, L, & Alvarez-buylla, J. M. A., EGF converts transit-amplifying neurogenic precursors in the adult brain into multipotent stem cells. Neuron, (2002)., 1021-1034.

[220] Vaccarino, F. M, Schwartz, R. R, Nilsen, M. L, Rhee, J, Zhou, J, Doetschman, M, Coffin, T, Wyland, J. D, & Hung, J. J. YT., Changes in cerebral cortex size are governed by fibroblast growth factor during embryogenesis Nat Neurosci, (1999). , 848.

[221] Raballo, R, Rhee, L. -C. R, Leckman, J, Schwartz, J. F, & Vaccarino, M. L. FM., Basic fibroblast growth factor (Fgf2) is necessary for cell proliferation and neurogenesis in the developing cerebral cortex. J Neurosci, (2000). , 5012-5023.

[222] Alzheimer, C, & Werner, S. Fibroblast growth factors and neuroprotection. Adv Exp Med Biol, (2002). , 3335-3351.

[223] Zheng, W, Nowakowski, R. S, & Vaccarino, F. M. Fibroblast growth factor 2 is required for maintaining the neural stem cell pool in the mouse brain subventricular zone. Dev Neurosci, (2004). , 181-196.

[224] Neiva, K. G, Zhang, M. M, Warner, Z, Karl, K. A, & Nör, E. JE., Cross talk initiated by endothelial cells enhances migration and inhibits anoikis of squamous cell carcinoma cells through STAT3/Akt/ERK signaling. Neoplasia, (2009). , 583-593.

[225] Gomez-gaviro, M. V, et al. The Vascular Stem Cell Niche. J Cardiovasc Transl Res, (2012).

[226] Kokovay, E, et al. Adult SVZ lineage cells home to and leave the vascular niche via differential responses to SDF1/CXCR4 signaling. Cell Stem Cell, (2010). , 163-173.

[227] Aguirre, A, Rubio, M. E, & Gallo, V. Notch and EGFR pathway interaction regulates neural stem cell number and self-renewal. Nature, (2010). , 323-327.

[228] Mathieu, C, Sii-felice, F. P, Etienne, K, Haton, O, Mabondzo, C, Boussin, A, & Mouthon, F. D. MA., Endothelial cell-derived bone morphogenetic proteins control proliferation of neural stem/progenitor cells. Mol Cell Neurosci, (2008). , 569-577.

[229] Lim, D. A, Tramontin, T. J, Herrera, A. D, García-verdugo, D. G, & Alvarez-buylla, J. M. A., Noggin antagonizes BMP signaling to create a niche for adult neurogenesis. Neuron, (2000). , 713-726.

[230] Gross, R. E, Mehler, M. P, Zang, M. F, Santschi, Z, & Kessler, L. JA., Bone morphogenetic proteins promote astroglial lineage commitment by mammalian subventricular zone progenitor cells. Neuron, (1996). , 595-606.

[231] Chen, H. L. and P. DM., Concise review: bone morphogenetic protein pleiotropism in neural stem cells and their derivatives--alternative pathways, convergent signals. Stem Cells, (2007). , 63-68.

[232] Bauer, S, & Patterson, P. H. Leukemia inhibitory factor promotes neural stem cell self-renewal in the adult brain. J Neurosci, (2006). , 12089-12099.

[233] Ying, Q. L, Nichols, C. I, & Smith, J. A., BMP induction of Id proteins suppresses differentiation and sustains embryonic stem cell self-renewal in collaboration with STAT3. Cell, (2003). , 281-292.

[234] Hill, W. D, Hess, M. -S. A, Carothers, D. C, Zheng, J. J, Hale, J, Maeda, D, Fagan, M, Carroll, S. C, Conway, J. E, & Sdf-1, S. J. CXCL12) is upregulated in the ischemic penumbra following stroke: association with bone marrow cell homing to injury. J Neuropathol Exp Neurol, (2004). , 84-96.

[235] Imitola, J, Raddassi, P. K, Mueller, K, Nieto, F. J, Teng, M, Frenkel, Y. D, Li, D, Sidman, J, Walsh, R. L, Snyder, C. A, & Khoury, E. Y. SJ., Directed migration of neural stem cells to sites of CNS injury by the stromal cell-derived factor 1alpha/CXC chemokine receptor 4 pathway. Proc Natl Acad Sci U S A., (2004). , 18117-18122.

[236] Tran, P. B, Ren, V. T, & Miller, D. RJ., Chemokine receptors are expressed widely by embryonic and adult neural progenitor cells. J neurosci Res, (2004). , 20-34.

[237] Robin, A. M, Zhang, W. L, Zhang, Z. G, Katakowski, R. L, Zhang, M, Wang, L, Zhang, Y, & Chopp, C. M., Stromal cell-derived factor 1alpha mediates neural progenitor cell motility after focal cerebral ischemia. J Cereb Blood Flow Metab, (2006). , 125-134.

[238] Thored, P, Arvidsson, C. E, Ahlenius, A, Kallur, H, Darsalia, T, Ekdahl, V, Kokaia, C. T, & Lindvall, Z. O., Persistent production of neurons from adult brain stem cells during recovery after stroke. Stem Cells, (2006). , 739-747.

[239] Kallmann, B. A, Wagner, H. V, Buttmann, S, Bayas, M, Tonn, A, & Rieckmann, J. C. P., Characteristic gene expression profile of primary human cerebral endothelial cells. FASEB J, (2002). , 589-591.

[240] Carmeliet, P. and T.-L. M., Common mechanisms of nerve and blood vessel wiring. Nature, (2005). , 193-200.

[241] Ward, N. L. and L. JC., The neurovascular unit and its growth factors: coordinated response in the vascular and nervous systems. Neurol Res, (2004). , 870-883.

[242] Jones, N, Iljin, D. D, & Alitalo, K. K., Tie receptors: new modulators of angiogenic and lymphangiogenic responses. Nat Rev Mol Cell Biol, (2001). , 257-267.

[243] Stumm, R. K, Rummel, J. V, Culmsee, J, Pfeiffer, C, Krieglstein, M, Höllt, J, & Schulz, V. S., A dual role for the SDF-1/CXCR4 chemokine receptor system in adult brain: isoform-selective regulation of SDF-1 expression modulates CXCR4-dependent neuronal plasticity and cerebral leukocyte recruitment after focal ischemia. J Neurosci, (2002). , 5865-5878.

[244] Arai, F, et al. Tie2/angiopoietin-1 signaling regulates hematopoietic stem cell quiescence in the bone marrow niche. Cell, (2004). , 149-161.

[245] Koochekpour, S, & Merzak, P. G. A, ascular endothelial growth factor production is stimulated by gangliosides and TGF-beta isoforms in human glioma cells in vitro. Cancer Lett, (1996). , 209-215.

[246] Breier, G, et al. Transforming growth factor-beta and Ras regulate the VEGF/VEGF-receptor system during tumor angiogenesis. Int J Cancer, (2002)., 142-148.

[247] Hirschi, K. K, Lai, B. N, Dean, L, Schwartz, D. A, & Zimmer, R. J. WE., Transforming growth factor-beta induction of smooth muscle cell phenotpye requires transcriptional and post-transcriptional control of serum response factor. J Biol Chem., (2002)., 6287-6295.

[248] Antonelli-orlidge, A, et al. An activated form of transforming growth factor beta is produced by cocultures of endothelial cells and pericytes. Proc Natl Acad Sci U S A., (1989)., 4544-4548.

[249] Sato, Y, Tsuboi, L. R, Moses, R, & Rifkin, H. DB., Characterization of the activation of latent TGF-beta by co-cultures of endothelial cells and pericytes or smooth muscle cells: a self-regulating system. J Cell Biol, (1990)., 757-763.

[250] Brionne, T. C, et al. Loss of TGF-beta 1 leads to increased neuronal cell death and microgliosis in mouse brain. Neuron, (2003)., 1133-1145.

[251] Buckwalter, M. S, et al. Chronically increased transforming growth factor-beta1 strongly inhibits hippocampal neurogenesis in aged mice. Am J Pathol, (2006)., 154-164.

[252] Wachs, F. P, et al. Transforming growth factor-beta1 is a negative modulator of adult neurogenesis. J Neuropathol Exp Neurol, (2006)., 358-370.

[253] Adams, R. H, & Klein, R. Eph receptors and ephrin ligands. essential mediators of vascular development. Trends Cardiovas Med, (2000)., 183-188.

[254] Holmberg, J, Armulik, S. K, Edoff, A, Spalding, K, Momma, K, Cassidy, S, Flanagan, R, Frisén, J. G, & Ephrin-a, J. reverse signaling negatively regulates neural progenitor proliferation and neurogenesis. Genes Dev, (2005)., 462-471.

[255] Conover, J. C, Doetsch, G. -V. J, Gale, F, Yancopoulos, N. W, & Alvarez-buylla, G. D. A., Disruption of Eph/ephrin signaling affects migration and proliferation in the adult subventricular zone. Nat Neurosci, (2000)., 1091-1097.

[256] Gale, N. W, Baluk, P. L, Kwan, P, Holash, M, Dechiara, J, Mcdonald, T. M, Yancopoulos, D. M, & Ephrin-b, G. D. selectively marks arterial vessels and neovascularization sites in the adult, with expression in both endothelial and smooth-muscle cells. Dev Biol, (2001)., 151-160.

[257] Alvarez-buylla, A, & Lim, D. A. For the long run: maintaining germinal niches in the adult brain. Neuron, (2004)., 683-686.

[258] Nomura, T, et al. EphB signaling controls lineage plasticity of adult neural stem cell niche cells. Cell Stem Cell, (2010)., 730-743.

[259] Shingo, T, Gregg, E. E, Fujikawa, C, Hassam, H, Geary, R, Cross, C, Weiss, J. C, & Pregnancy-stimulated, S. neurogenesis in the adult female forebrain mediated by prolactin. Science, (2003). , 117120.

[260] Abrous, D. N, & Koehl, M. and Le Moal M, Adult neurogenesis: from precursors to network and physiology. Physiol Rev., (2005). , 523-569.

[261] Masuda, S, Chikuma, S. R, & Insulin-like, M. growth factors and insulin stimulate erythropoietin production in primary cultured astrocytes. Brain Res, (1997). , 63-70.

[262] Johns, L, Sinclair, D. J, & Hypoxia, A. J. hypoglycemia-induced amino acid release is decreased in vitro by preconditioning. Biochem Biophys Res Commun, (2000). , 134-136.

[263] Blondeau, N, et al. K(ATP) channel openers, adenosine agonists and epileptic preconditioning are stress signals inducing hippocampal neuroprotection. Neuroscience, (2000). , 465-474.

[264] Brines, M. L, et al. Erythropoietin crosses the blood-brain barrier to protect against experimental brain injury. Proc Natl Acad Sci U S A., (2000). , 10526-10531.

[265] Eid, T, Brines, C. A, Spencer, M. L, Kim, D. D, Schweitzer, J. H, Ottersen, J. S, & De Lanerolle, O. P. NC., Increased expression of erythropoietin receptor on blood vessels in the human epileptogenic hippocampus with sclerosis. J Neuropathol Exp Neurol, (2004). , 73-83.

[266] Hassouna, I, et al. Erythropoietin augments survival of glioma cells after radiation and temozolomide. Int J Radiat Oncol Biol Phys, (2008). , 927-934.

[267] Ruscher, K, Freyer, K. M, Isaev, D, Megow, N, Sawitzki, D, Priller, B, Dirnagl, J, & Meisel, U. A., Erythropoietin is a paracrine mediator of ischemic tolerance in the brain: evidence from an in vitro model. J Neurosci, (2002). , 10291-10301.

[268] Calabrese, C, et al. A perivascular niche for brain tumor stem cells. Cancer Cell, (2007). , 69-82.

Chapter 9

Parenchymal Neuro-Glio-Genesis Versus Germinal Layer-Derived Neurogenesis: Two Faces of Central Nervous System Structural Plasticity

Luca Bonfanti, Giovanna Ponti, Federico Luzzati,
Paola Crociara, Roberta Parolisi and
Maria Armentano

Additional information is available at the end of the chapter

http://dx.doi.org/10.5772/56100

1. Introduction

The discovery of neural stem cells (NSCs) at the beginning of the nineties led many people to consider definitively broken the dogma of a static central nervous system (CNS) made up of non-renewable elements [1-3]. In parallel, the occurrence and characterization of adult neurogenesis in the olfactory bulb and hippocampus [3-5] triggered new hopes for brain repair. Twenty years after, the dream of regenerative medicine applied to brain/spinal cord injuries and neurodegenerative diseases is still very far [6,7]. As a matter of fact, adult neurogenesis in mammals occurs mainly within two restricted areas known as 'neurogenic sites' [3,8]: the forebrain subventricular zone (SVZ); reviewed in [9] and the hippocampal dentate gyrus (subgranular zone, SGZ); reviewed in [10]. As a direct consequence of such topographical localization, most of the CNS parenchyma out of the two 'classic' neurogenic sites remains substantially a non-renewable tissue. Actually, most of the traumatic/vascular injuries and neurodegenerative diseases do occur in 'non-neurogenic' regions and no efficacious therapies capable of restoring CNS structure and functions through cell replacement are at present available. Thus, two decades after the discovery of NSCs and the reaching of a satisfactory characterization of adult neurogenic sites, a gap remains between the occurrence of stem/progenitor cells in the CNS of adult mammals and their effective capability to serve in brain repair. Several aspects do converge in explaining this gap [11], partially accounting for the heterogeneity of CNS structural plasticity in mammals (summarized in Table 1)

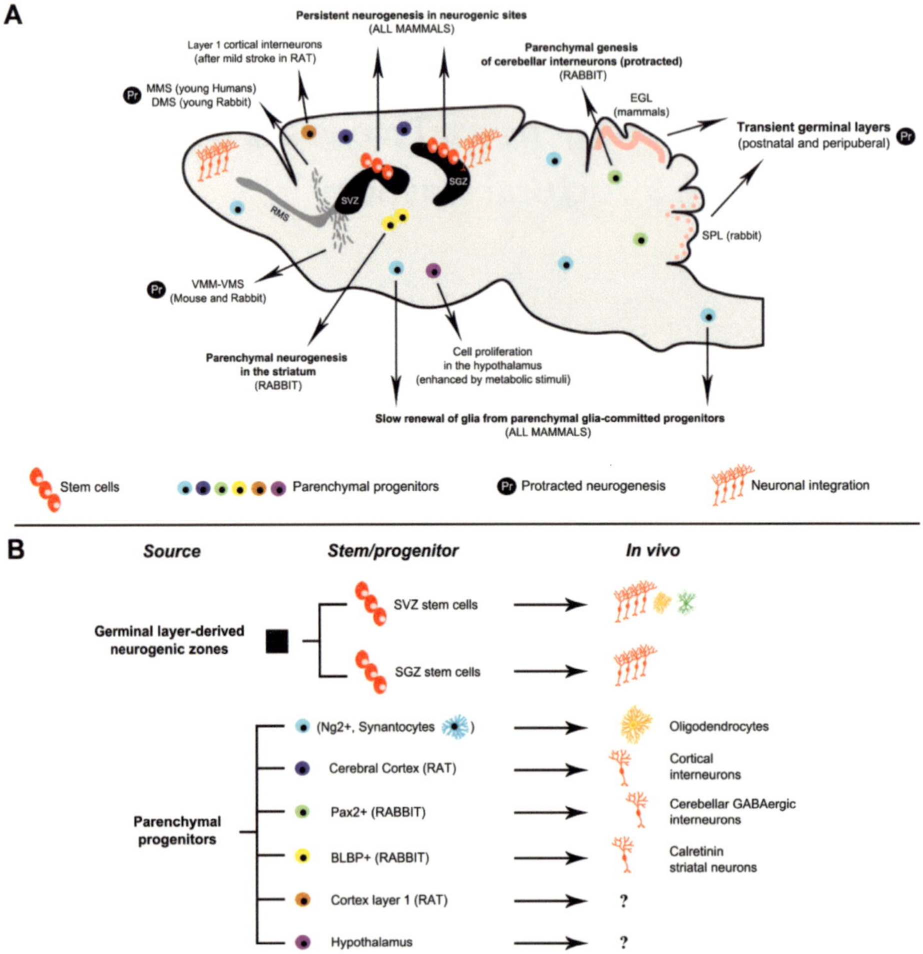

Figure 1. Heterogeneity of postnatal / adult neurogenic processes in different mammals by considering different aspects and mammalian species. B, Schematic summary of the main sources (progenitor cells) of adult mammalian neurogenesis, its outcome in vivo / in culture system, and its possible activation after lesion. In the case of many parenchymal regions, some of these steps are still obscure. BLBP, brain lipid-binding protein; EGL, external germinal layer; GABA, c-aminobutyric acid; Ng2, nerve / glial antigen 2 proteoglycan; NPY, Neuropeptide Y; SGZ, subgranular zone; SPL, subpial layer; SVZ, subventricular zone; VMM, ventral migratory mass; VMS, ventrocaudal migratory stream; MMS, medial migratory stream; DMS, dorsal migratory stream. Adapted from Ref [30].

In this chapter the neurogenic/gliogenic potential of the mammalian brain parenchyma *in vivo* will be analyzed with particular reference to variables involved in its heterogeneity (e.g., animal species, age, CNS regions; see Figure 1 and Table 1). In particular, these variables do determine the tissue environment in which stem/progenitor cells are immersed, what seems to be extremely important for their activity and outcome. In addition, the origin and nature of stem/progenitor cells would also contribute to their neurogenic/gliogenic potential. It is now well known that cells may have a broader potential than they normally exhibit in vivo when

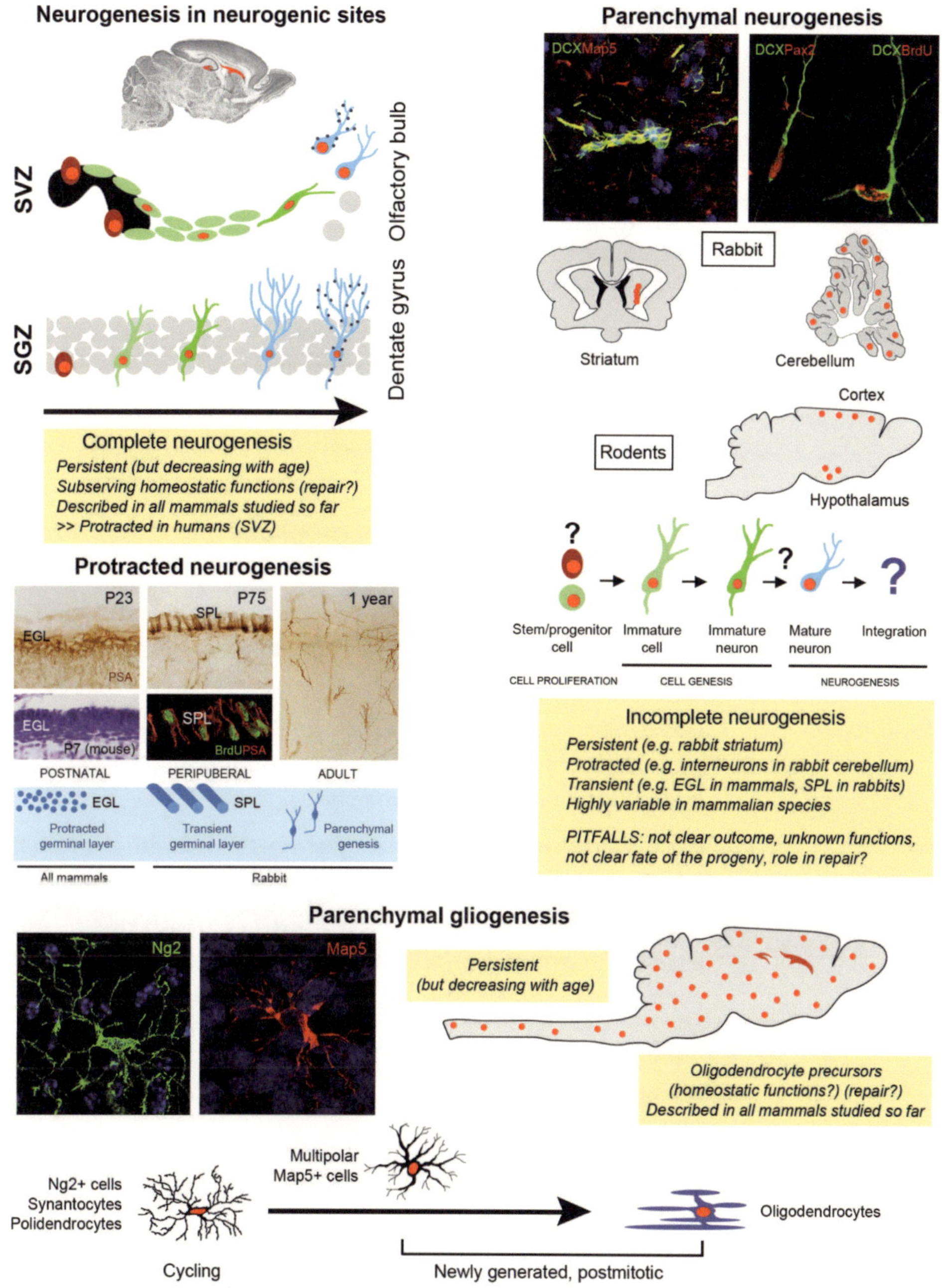

Figure 2. Schematic summary of the features and location of different neurogenic/gliogenic processes occurring spontaneously in the CNS of postnatal and adult mammals. Red dots indicate newlyborn cells. SVZ, subventricular zone; SGZ, subgranular zone; EGL, external germinal layer; SPL, subpial layer (rabbit); PSA, PSA-NCAM; Map5, microtubule-associated protein 1B; P23, postnatal day 23. Question marks indicate lack of knowledge about the origin, late differentiative steps, and final integration of newly generated parenchymal neurons. Adapted from Ref [32].

exposed to a different environment, either in vitro or in vivo [29]. Hence, in order to avoid one of the most common misunderstandings, namely the confusion between occurrence of *de novo* cell proliferation in the CNS tissue and existence of true gliogenic/neurogenic processes, here the attention will be focused on the outcome(s) of the newly generated progeny [30].

A. Variables affecting the nature and features of adult neurogenesis	
Animal species (animal world)	*General plasticity and persistent neurogenesis are usually reduced across phylogeny; in parallel, the reparative/regenerative potential is also reduced*
Animal species (mammals)	*Unlike previous belief and current bias, remarkable differences in the location and extension of adult neurogenesis do exist among mammals*
Age	*Some neurogenic processes are extensions of delayed developmental programs (postnatal/protracted neurogenesis) whereas others persist throughout life (persistent neurogenesis). All neurogenic processes are progressively reduced with age*
Microenvironment (niche)	*A well defined neural stem cell niche sustains neurogenesis in neurogenic sites (SVZ, SGZ), whereas a niche has not been characterized in parenchymal neurogenesis*
Origin of stem/progenitor cells	*Neurogenic sites (SVZ, SGZ) directly derive from persistence and modification of pre-existing, embryonic germinal layers, whereas for parenchymal cell genesis such direct link is not clear*
Location in the CNS	*Location either within a germinal layer-derived niche or in the parenchyma redirects to the two previous points; in parenchymal neurogenesis many variations are linked to local cues of the different CNS regions involved*
Function	*In physiology: linked to the different ecological niches of the animals (present in all animals)*
	In repair: linked to the species; in invertebrates and non-mammalian vertebrates the physiological function is associated with function in repair, whereas in birds and mammals it is only linked to physiology/homeostasis of specific systems

B. Main differences between cell genesis in adult neurogenic sites and in the parenchyma		
	Neurogenic sites	***Parenchyma***
Location	*Restricted*	*Widespread*
Primary progenitor cells	*Stem cells*	*Progenitors*
Microenvironment	*Stem cell niche*	*Mature neuropil*
Origin	*Germinal layer-derived*	*No direct link with germinal layers*
Fate (progeny)	*Mainly neurons (some astrocytes and oligodendrocytes)*	*Mainly glial cells (some neurons)*
Fate (process)	*Complete*	*Incomplete*

Table 1. Heterogeneity of adult neuro-glio-genesis

Since developmental changes also account for loss of CNS reparative/regenerative capacities and neuro-glio-genic potential, a paragraph will be devoted to the progenitor cell developmental origin. Then, a brief summary of comparative adult neurogenesis will be given. Evolutionary explanations can provide an understanding of the logic followed (or not) by neurogenic processes through phylogeny, also accounting for the failure in mammalian CNS repair/regeneration and scarce usefulness of adult neurogenesis as a possible solution for brain repair [31,32].

2. Developmental origin of adult neurogenic/gliogenic processes

What makes it possible the remarkable neurogenesis occurring in neurogenic sites is their direct origin from embryonic germinal layers which retain stem/progenitor cells along with the 'niche' environment allowing their activity [10,33]. The SVZ and SGZ actually are remnants of their embryonic counterpart, from which they maintain several cellular and molecular aspects [9] in parallel with an adaptation to the changing anatomy of the postnatal and adult brain [34,35].

During development, the CNS originates from the neuroepithelium, pseudostratified epithelial cells that maintain contact with both the ventricular and pial surfaces. As brain thickness increases, neuroepitheial cells transform into radial glia [33,36]. Beside their classic role as scaffolding for migrating neurons during embryogenesis and their subsequent transformation into parenchymal astrocytes of the mature CNS [37,38], radial glia cells behave as stem cells, leading to the genesis of astrocytes, neurons [39,40], and, to a lesser extent, oligodendrocytes [41]. Thus radial glial cells not only serve as progenitors for many neurons and glial cells soon after birth, but also give rise to adult SVZ stem cells that continue to produce neurons throughout adult life [41]. The origin of astrocytes that function as neural progenitors in the adult hippocampus has not been determined experimentally. A connection to radial glial cells, has been suggested even in the hippocampal SGZ [42,43]. The relationship of adult NSCs to their developmental precursors offers clues to the unique characteristics that distinguish these germinal astrocytes from other astroglial cells in the brain parenchyma [33]. Indeed, parenchymal astrocytes lose very early their stem cell potential (around postnatal day 10 in mice [44]), although they can still proliferate in the severe gliosis induced after lesion [45], and resume multipotentiality *in vitro* [46].

On the other hand, gliogenesis persists throughout the CNS in the form of parenchymal cell genesis capable of creating new oligodendrocytes and, to a lesser extent, astrocytes, throughout life [12,15]. Most of this gliogenic activity is attributed to synantocytes/polydendrocytes (Ng2+ cells; see below) which are widespread in the CNS tissue and whose origin is still partially obscure. Oligodendrocytes originate from migratory and mitotic embryonic precursors which progressively mature into postmitotic myelin-producing cells. The sequential expression of developmental markers defines distinct phenotypic stages in the oligodendrocyte lineage, characterized by proliferative capacities, migratory abilities and changes in morphology. Most knowledge on this issue comes from studies on the rodent embryonic spinal cord. The first oligodendrocyte-committed cell appears at embryonic day 12 (E12) in two columns in the ventral ventricular zone of the motor neuron progenitor domain [47], which is defined by the expression of Olig2 [48]. The embryonic oligodendrocyte precursors are identified by their expression of platelet-derived growth factor alpha receptor (PDGFRα) [49]. The appearence of the oligodendrocyte lineage-associated markers Olig2 (essential for oligodendrocyte specification and differentiation) and PDGFRα (which permits the expansion of the original precursor population) is dependent on the concentrations of Sonic hedgehog (Shh) [50,51]. One or two days after their appearance, PDGFRα+ cells exit the ventricular zone and expand by local proliferation and migration first in the ventral spinal cord region and then dorsally [52]. Finally, they occupy the entire parenchyma by the time of birth [49]. A dorsal source of oligodendrocyte precursors was also shown to contribute to oligodendrogenesis in the spinal cord and hindbrain [53,54]. Fate mapping experiments revealed a double source of oligodendrocyte precursors in the forebrain: cells expressing oligodendrocyte lineage markers, such as Olig1, Olig2, Sox10 and PDGFRα, first appear ventrally, in the neuroepithelium of the medial ganglionic eminence, and then migrate laterally and dorsally into all parts of the developing forebrain by E16 to birth [55]. However, several studies have provided evidence for a dorsal and later source of oligodendrocyte precursors in the lateral and/or caudal ganglionic eminence(s), which constitute a second wave of cells invading the cortex only by E18 [54,56]. Nevertheless, adult oligodendrocyte derive only by dorsal precursors, since medial ganglionic eminence-derived precursors were demonstrated to completely disappear after birth [56]. On the whole, it is thought that a unique oligodendrocyte population can derive from progenitor domains defined by different signaling molecules, in contrast to what has been established for neuronal specification during embryonic development, where different parts of the ventricular zone generate distinct types of neurons. In the rodent CNS, once PDGFRα+ cells have left the ventricular zone, they start to be termed 'oligodendrocyte progenitor cells' and acquire their most typical marker: an integral membrane chondroitin sulphate proteoglycan named Ng2 (nerve/glial antigen 2). Ng2 expression becomes detectable only at E14 [57], thus, from E17 to adulthood all PDGFRα+ cells are Ng2+, and, conversely, all the parenchymal (non-vascular) Ng2+ cells are PDGFRα+ [57,58]. Early embryonic Ng2+/PDGFαR+ OPCs are small, undifferentiated, proliferative and motile cells [59]. During embryogenesis, their morphology changes rapidly from a simple oval or polygonal cell body with few unbranched processes to a more differentiated and branched shape with a smaller cell soma [57,60].

Coming back to adult neurogenesis, non mammalian vertebrates including fish, amphibians, and reptiles harbor a more widespread genesis of neurons in the parenchyma. Such processes, due to their location, are apparently independent from the primitive germinal layers. Nevertheless, recent studies which analysed in more detail the origin of adult neurogenesis in fish show that all neurogenic processes likely originate from remnants of the germinal layers; reviewed in [61]. Teleost proliferation zones reflect a general proliferation pattern along the ventricular walls of the brain, distinctly localized in all its subdivisions along the rostrocaudal axis. Between 12 and 16 distinct proliferation zones have been recognized in different teleost species [61]. Hence, across different animal classes, most stem cell populations retain contact to the ventricular system, and they appear as neuroepithelial cells, radial glial or astroglial cell types. The different shapes of these progenitors have been suggested to be a secondary consequence of the architecture of the developing parenchyma overlying the ventricular stem cell zone of the embryo [9]. This common pattern across animal species, along with data reported above on the origin of cycling glial progenitors in mammals, indirectly suggests that adult parenchymal neuro-glio-genesis ultimately derives from embryonic germinal layers, yet being able to persist independently in some cases.

3. Comparative adult neurogenesis and brain repair

Unlike mammals, other classes of vertebrates including fish, amphibians, and reptiles, harbor a more widespread adult neurogenesis in the parenchyma. In these animals, stem and progenitor cells, in addition to their role in physiological plasticity, also participate in brain repair and regeneration. Failure in mammalian brain repair after traumatic, vascular, and neurodegenerative injuries is due to: i) a strong reduction in the extension of neurogenic regions within the whole CNS; ii) a substantial lack of CNS reparative/regenerative capacity; iii) the fact that adult neurogenic sites subserve specific physiological functions rather than brain repair; for review, see [11,62,63]. It is important to note that although the occurrence of good neurogenic potentials would generally favor brain repair (at least by making available stem/progenitor cells) there is not a direct, linear relationship between occurrence of stem/progenitor cells and repair/regeneration, the latter processes strongly depending on the tissue environment and/or tissue reactions; for selected examples of neurogenesis and regeneration see [64].

Neurogenic processes are detectable in wide regions of the CNS in invertebrates and non-mammalian vertebrates [61,65,66], whereas in mammals they are restricted to two privileged areas (neurogenic sites) and the remaining CNS is largely made up of non-renewable tissue [30,67,68]. The state of substantial 'general plasticity' and cell renewal existing in the oldest living metazoans, so that all cell types, including neurons, are balanced in their production and loss [69,70], is progressively reduced in vertebrates, although fish and amphibians still maintain remarkable regenerative capacities [71,72]. Then, in birds and mammals a transition between regeneration permissive and non-permissive stages occurs soon after birth, and highly-restricted spots of adult neurogenesis subserve homeostatic functions in specific neural circuits [73,74]. The decrease in neurogenic abilities occurs in parallel with topographical/numerical restriction of germinal layer-derived stem cell niches, whereas the decrease in regenerative abilities occurs in parallel with other aspects: the impossibility to re-access to embryonic developmental programs during adulthood [75], the lack of differentiated cells capable of dedifferentiation [76], the development of a strong immune surveillance [77] and the consequent tissue reactions, most of which detrimental (reviewed in [11,64]). In some cases, the stem cells found in the CNS of non-mammalian vertebrates are deployed for postnatal development of parts of the brain until the final structure is reached. In other cases, postnatal neurogenesis continues into adulthood leading to a net increase of the number of neurons with age. Finally, in other cases, stem cells fuel neuronal turnover. An example is the protracted development of the cerebellar granular layer in mammals, which in adult teleosts actually becomes a persistent neurogenesis, where the granular layer continuously grows and no definite adult cerebellar size is reached [61].

In addition, when considering mammals, the failure in CNS repair is a result of evolutionary constraints in which the injured tissue would not favor a strategy of regeneration but rather one of minimizing further damage (e.g., gliotic reaction [78]). Hence, as a consequence of multiple, converging aspects, CNS regenerative capacity in mammals could have reached a point of non-return, in parallel with the persistence of some neurogenic processes which remain mainly focused on physiological functions (e.g., cell renewal/addition in selective neural circuits linked to learning/memory tasks [73,74]).

An increased consciousness that the scarce reparative capacity of the mammalian CNS depends on multiple aspects should indicate that it is very unlike the finding of a single molecular factor or pharmacological treatment capable of eliciting repair/regeneration. Comparative results from vertebrate species of different classes have demonstrated that adult neurogenesis is widespread among vertebrates but is employed by different species in different functional contexts [74,79,80], and a growing number of reports show a remarkable heterogeneity even among mammals [17-19]. This variability concerns both the organization/extension/function of the two neurogenic sites and many examples of parenchymal neurogenesis; reviewed in [30] (see below). This fact, along with our still incomplete knowledge of adult neurogenesis in humans (especially within the parenchyma), partially hampers the reaching of well established 'common rules' which might be used in the translation of experimental preclinical data to human medicine. Thus, dealing with mammalian CNS structural plasticity, high levels of heterogeneity involving different 'types' of neurogenic processes should be taken into account.

4. Heterogeneity of cell genesis in the mammalian CNS

We now know that 'classic' neurogenic sites are consistently present in all mammals studied, although with some differences, particularly when the outcome(s) of the neurogenic process are involved [30]. The occurrence of a rostral migratory stream which is active throughout life in rodents but temporally restricted to the postnatal period in humans [81] is a prototypical example of variability among mammals. Indeed, in humans this neurogenic process seems to fall in a delayed developmental process rather than adult neurogenesis (see below).

In addition to neurogenic sites, studies carried out during the last two decades revealed the presence of local, parenchymal progenitors which retain some proliferative capacity in most of the mature mammalian CNS [12,14,15,17-19,82] (Figure 1). This fact suggests that structural plasticity involving *de novo* cell genesis in the CNS could be more widespread than previously thought. As a consequence of the increasing number of reports investigating adult neurogenesis in mammals, our perception of this biological process has gained new perspectives and nuances; for deeper analysis see [30,66,83,84]. What was previously thought as "the genesis of new neurons in restricted brain areas endowed with NSCs", can now be intended as a highly heterogeneous phenomenon (summarized in Figures 1 and 2), whose heterogeneity depends on several variables (see Table 1). The main elements of heterogeneity can be summarized as follows: i) the location of progenitors (gathered within restricted neurogenic sites or widely spread out in the parenchyma); ii) the nature of the progenitors (*bona fide* NSCs versus different types of progenitors); iii) the genetic and molecular features of the progenitors (cell lineage: neuronal-like versus glial-like; identification of differentiative stages dependent on the available markers); iv) the existence or not of well characterized neurogenic niches (absence of niches or occurrence of atypical/non-identified niches in the parenchyma?); v) the extension in time after birth (protracted, transient persistent neurogenesis); vi) the ultimate fate of the progeny in terms of cell lineage (neuronal versus glial; astrocytic versus oligodendrocytic); vii) the ultimate fate of the progeny in terms of cell integration into circuits (complete versus incomplete neurogenesis); viii) the spontaneous occurrence of the process versus its injury-induced appearance. This latter point could be considered a further step beyond the so-called 'constitutive' neurogenesis, namely the spontaneous, continuous genesis of new neurons as part of a physiologic, homeostatic process [85].

Due to the multifaceted aspects of the above mentioned processes, some problems of terminology can also be raised (see Refs. [30,32]). A common misunderstanding consists of a different use of the word 'neurogenesis', which can be intended either as 'genesis of neurons' or as 'genesis of neural cells', i.e. neurons and glia. Embryonic neurogenesis, namely the process of building up the whole CNS, involves both neuro- and glio-genesis, occurring in largely overlapping and strictly intermingled phases, whereas neurogenesis and gliogenesis can occur separately in the adult. The landscape is even more complex, since research on adult neurogenesis brought developmental neuroscience within the mature brain, and the intermix of structurally plastic changes involving cell genesis/differentiation with the fully assembled adult tissue is accompanied by a previously unexpected intermix of cell lineages (e.g., newly formed neuroblasts arising from astrocytic-like stem cells *in vivo*). For this reason, in this review article, when not speaking of well characterized cell lineages, the notion of 'cell genesis' instead of 'neurogenesis' will be used, since in most 'neurogenic' processes different cell types can be considered among the progenitors, and different progenies can be generated. Hence, apart from detailed knowledge gathered around the activity of SVZ and SGZ neurogenic sites, many aspects of parenchymal cell genesis remain obscure and/or unesplored, as a consequence of the heterogeneity depicted above. In the last few years, parenchymal neuro-glio-genesis was among the most studied, yet less known, issues, due to the widespread location of the progenitor cells and to the substantial lack of markers which specifically identify their real origin as well as the stage-specific steps of their differentiation. As a consequence, the presence/absence of neurogenic processes within different CNS parenchymal regions in different mammalian species is still quite controversial and debatable. In most cases, parenchymal cell genesis occurs at low levels, at the limit of technical detection. Furthermore, in some cases it is very difficult to show its final outcome(s), most of the parenchymal neurogenesis appearing 'incomplete' as to the final differentiation/integration of the progeny [30] (Figure 2). Finally, to correctly classify both germinal layer-derived and parenchymal neurogenesis some other aspects should be taken into account, such as the temporal extension of 'protracted'/'transient'developmental neurogenic processes with respect to a 'constitutive'/'persistent' neurogenesis [30]. A further aspect is that of lesion-induced neuro-glio-genesis, namely the genesis of new cells as a consequence of different types of CNS injury [18,25,26,86] or altered homeostasis [87]. This is an important point since many lines of research in the field of neural repair directed to manipulate stem cells in the perspective of intracerebral transplantation did not produced substantial therapeutic innovations. As an alternative, another approach might be that of stimulating/modulating the endogenous sources of cell progenitors present both in germinal layer-derived stem cell niches (SVZ and hippocampus) and in the parenchyma.

5. Parenchymal neurogenesis

Spontaneous (constitutive) parenchymal neurogenesis can be considered as a very rare phenomenon in mammals, and its regional location has been shown to be dependent on the animal species, age, and physiological/pathological states [30]. Different examples of neurogenesis occurring outside the two neurogenic sites have been described in rodents [17,82], rabbits [18,19] and monkeys [22,88]. Remarkable differences can be observed between closely related orders (e.g., rodents and lagomorphs [18,19]), between species (e.g., rat and mouse [17,23,89,90]), and even different strains [91,92].

Most parenchymal neurogenesis described in adult rodents seems to occur spontaneously at very low levels, rather being elicited/enhanced after specific physiological or pathological conditions [17,82,86,87] (see below). Dayer and colleagues [17] showed the occurrence of new neurons in the deep layers of the rat cerebral cortex. By labelling newlyborn cells with multiple intra-peritoneal injections of BrdU and using markers of both immature and mature neurons to characterize the new cells through a detailed confocal analysis at different survival times, they demonstrated genesis of new GABAergic interneurons in both neocortex and striatum. At 4-5 weeks survival time, the 0.4 +/- 0.13% of the BrdU+ cells were mature NeuN+ neurons in the neocortex. Morphologic and phenotypic analyses assert these cells belong to different categories of cortical interneurons. Interestingly, although several BrdU+/DCX+/Tuc4+ neuroblasts were identified close to the SVZ periventricular region, the great majority of cortical BrdU+ cells were positive for Ng2. From these data the Authors suggested that adult cortical newborn interneurons might originate from *in situ* progenitors. Other examples of spontaneous parenchymal neurogenesis have been described in lagomorphs. In rabbits, newly generated neurons are spontaneously produced in other regions of the adult brain starting from local, parenchymal progenitors. In the caudate nucleus, newly formed neuroblasts form longitudinally-arranged, doublecortin (DCX) and PSA-NCAM immunoreactive striatal chains similar to the SVZ chains [18]. These neuroblasts are generated from clusters of proliferating cells which express the astroglial marker brain lipid binding protein (BLBP), and about 1/6 of surviving cells differentiate into calretinin striatal interneurons. Always in rabbits, in sharp contrast with our common knowledge concerning the CNS of other mammals studied so far, a remarkable genesis of cells is detectable in the peripuberal, and to a lesser extent, adult cerebellar cortex [19]. Systemically-administered BrdU detected at different post-injection survival times (up to two months) reveals newly generated PSA-NCAM+/DCX+/Pax2+ interneurons of neuroepithelial origin homogeneously distributed in the cerebellar cortex. Thus, in the striatal and cerebellar parenchyma of lagomorphs new neurons are generated independently from the remnants of germinal layers, yet their final outcome and their role in the adult neural circuits remains obscure; reviewed in [30].

The heterogeneity in parenchymal neurogenesis adds to that described for neurogenic processes occurring in adult neurogenic sites, which have been related to adaptation to ecological pressures [80]. At present, this is one of the most satisfactory functional explanations for adult neurogenesis in the entire phylogenetic tree, along with multiple, genetically-determined variables spanning from the brain anatomy/developmental history to the animal lifespan [93]. This range of possibilities can also be increased by non-genetic variables, such as experience-dependent cues [79,80].

Among the unsolved issues of parenchymal neurogenesis are the numerous reports which have not been confirmed by further studies or by other laboratories [22,23,26,94-96], along with a series of data which have been denied in studies trying to reproduce the same results [24,97-99]. Without entering in the scientific and technical discussion about these controversies, it is evident that we still not grasp the real limits of parenchymal neurogenesis and that further studies are required before finally accept or deny the existence of some neurogenic processes.

A case placed in between the spontaneous and experimentally-induced neurogenesis, is that of the hypothalamus. Several publications based on experiments carried out on rodents have been reporting data on this brain region as a new site for adult constitutive neurogenesis in mammals (for review see [100]). Under physiological conditions, both in rats [101] and mice [102,103], proliferative activity does occur in the ependymal layer of the third ventricle and within the surrounding parenchyma. In rats, Xu and collaborators using electron microscopy and immunohistochemistry showed that tanycytes lining the 3rd ventricle proliferate and express molecules usually found in glial, stem-like progenitor cells, such as BLBP and nestin. The presence of putative neural progenitors was further supported by the isolation of cells able to give rise to neurospheres from the hypothalamus. One month after BrdU injection, proliferating cells, some of which expressing Hu protein, were detected in the surrounding parenchyma. Similar results were obtained in mice [102], yet in both rodent species no clear evidence has supported constitutive and complete hypothalamic adult neurogenesis under physiological conditions. A significant increase in hypothalamic proliferating cells can be obtained by performing i.v. delivery of BrdU (350% more positive nuclei, in comparison to i.p. treated animals), nevertheless, in spite of such cell proliferation the level of neurogenesis in the intact hypothalamus seems to be arrested at a very premature stage. On the other hand, growth factor infusion [82,101,104] or certain experimental conditions/models, such as prolonged heat exposure [105] and the mutant mice investigated by Pierce and Xu (2010), seem to increase neurogenesis in the hypothalamus. Intracerebroventricular infusion of insulin growth factor I in rats [104] triggered an intense proliferation along the 3rd periventricular area and in the parenchyma of the caudal hypothalamus. As concerns the genesis of new neurons, after i.v. treatment with bFGF in rats [101], and CNTF in mice [82], it was shown that proliferation induced by growth factors can be followed by genesis of newborn neurons. Detailed morphological and molecular analyses of the 3rd periventricular region of these animals showed interesting architectural similarities with the SVZ neurogenic niche (e.g., proliferating astroglial cells contacting the ventricle by an apical process bearing a single cilium), with tanycytes as primary proliferating elements lining the 3rd ventricle [104]. Yet, additional studies are necessary to clearly demonstrate/confirm whether hypothalamic newborn neurons generated after physiological/pathological stimulation actually become part of the pre-existing circuits playing a role in energy-balance mechanisms.

Taking into account the multifaceted aspects dealing with parenchymal neurogenesis, difficulties encountered in such type of research are not only technical. They are also linked to the occurrence of processes placed in the middle between two well characterized extremes of structural plasticity, such as synaptic plasticity, and 'complete' adult neurogenesis. In a recent review article [30] five levels have been dissected in the neurogenic processes in order to critically evaluate/compare different parenchymal neurogenic events (see also Figure 2). The subsequent steps span from cell division to possible integration of specified/differentiated elements into the CNS tissue, and according to this view, only when any of the five steps are filled the neurogenic process should be classified as 'complete'. As a result, all the parenchymal neurogenic processes described until now can actually be considered as incomplete. This could explain why many claims of neurogenic processes were subsequently refuted because not sustained by experimental evidence. The piriform cortex is one of those regions in which results reported by different researchers are quite controversial; see for example [88,106-108]. Since long time, this cortical region is known to harbor a population of neurons immunoreactive for PSA-NCAM and DCX [108-110], which are two markers highly expressed in newly generated neurons but also present in non newly generated cells [110]. Indeed, deeper investigations have shown that the piriform cortex contains a population of immature, non-newly generated neurons which display very few (or no) synapses and are frequently ensheathed by glial lamellae [108]. These cells, by remaining in an immature state for indeterminate time, can represent a 'reservoir' of neurons that could possibly be recruited into the preexisting neural circuits although not generated *ex novo* [111].

In conclusion, alternative and multiple forms of plasticity involving neurons can overlap within the so-called non-neurogenic tissue, affecting preexisting cells/circuits and increasing the complexity of the whole picture of brain structural remodeling.

6. Lesion-induced (reactive) neurogenesis

Brain lesions have been shown to stimulate neurogenesis in normally non-neurogenic regions such as the neocortex and the striatum. In the neocortex these responses are limited to specific conditions such as targeted apoptosis or mild ischemia [23,86,112,113]. By contrast, several lesion paradigms, associated to both strong or mild degeneration and inflammation, have been shown to induce neurogenesis in the striatum [25,28,114]. It is unknown if lesioned neocortex and striatum have distinct needs for immature neurons or if the neocortical tissue response is more detrimental for neurogenesis. This fundamental point reveals our very poor knowledge of lesion-induced neurogenesis. Indeed, despite an intense research, we have only little information regarding the nature, fate and potential of the progenitors stimulated by brain lesions, the mechanisms that trigger their activation and eventually their functional role.

Initial studies in both cortex and striatum reported that a tiny fraction of lesion-induced neurons may differentiate into projection neurons, suggesting that endogenous neuronal progenitors may have the potential to replace degenerated neurons [23,25,115]. However, these results have not been confirmed by others [28,116]. Moreover, it is now clear that most of the lesion-induced neurons have a transient existence and, at least in the striatum, they do not express markers of projection neurons nor transcription factors involved in their specification [28,116]. Several attempts have been made to increase the survival of these cells, with little success [117]. An intriguing possibility to be explored is that lesion-induced neuroblasts occurring in multiple forms of brain injury are committed to transient neuronal types, which contribute to restorative rather than replacement mechanisms [28,63]. This idea is further supported by data showing that a transient existence often characterizes also cortical and striatal neurons generated in normal conditions [18,89].

Neuronal progenitors in the SVZ and SGZ have been shown to respond to injuries by strongly increasing their proliferation, in the SVZ, also migrating towards damaged regions [25,115]. In parallel, recent reports have showed that in the degenerated neocortex and striatum, new neurons can also be produced locally from parenchymal neuronal progenitors [28,83,113]. In the neocortex, Ohira et al. [86] showed that mild ischemia might stimulate the generation of newborn GABAergic interneurons from progenitors residing in cortical layer I. These cells were not quiescent in normal conditions as they expressed the endogeneous marker of cell proliferation Ki67 and they could be labeled with retroviral vectors. Ohira and coworkers could not define the exact nature of the parenchymal progenitors, which, intriguingly, are very close to the leptomeninges, from which neuronal progenitors have been recently isolated [118,119].

More specific lineage tracing study will be necessary to confirm the real origin of neural progenitors activated after lesion. Lineage tracing has shown that reactive astrocytes isolated from the adult neocortex can give rise to neurospheres *in vitro* [46,120]. To date, the only *in vivo* evidence that neocortical astrocytes can be neurogenic has been obtained in early post-natal mice after hypoxia/Ischemia [113]. A recent study showed that even if the neural stem cells derived from adult neocortical astrocytes maintain the capacity for self-renew when transplanted in the SVZ, they were still unable to produce neurons [121]. This observation casted some doubts over the actual role of these cells as neuronal progenitors *in vivo*. Nonetheless, this result may only indicate that the neurogenic potential of cortical and SVZ progenitors rely on distinct factors.

Another example of the *in vivo* genesis of new neurons within the lesioned brain parenchyma has been obtained in the striatum in a mice model of progressive striatal degeneration, the Creb1^{Camkcre}Crem$^{-/-}$ mutant mice (CBCM) [28,122]. In this model the SVZ acts as a source of postmitotic neuroblasts that enter the striatum from a specific subcallosal migratory stream, as individual elements. Luzzati and coworkers [28] showed that the striatum of CBCM mice contains also tightly clustered neuroblasts which originate locally from parenchymal proliferating progenitors. These cells showed features of intermediate neuronal progenitors of the SVZ and SGZ such as clustering and co-expression of glial (Sox2, Sox9, BLBP) and neuronal markers (Dlx, Sp8, DCX), and the expression of the EGFr [123-126]. This study clearly shows that the mature parenchyma can be permissive to neuronal genesis, although Luzzati and coworkers could not trace the origin of the observed striatal parenchymal neuronal progenitors. Nonetheless, two possibilities can be considered: i) striatal neuronal progenitors could derive from the displacement of primary/intermediate progenitors from the SVZ; ii) they could represent local cells becoming neurogenic in response to neurodegeneration.

Together, these data suggests that specific degenerative conditions can stimulate the production of new neurons not only in the neurogenic niches but also in the mature brain parenchyma. This tissue has been classically considered non-permissive for neuronal progenitors, an idea mainly derived from the observation that SVZ and SGZ neural stem cells differentiate only into glial cells when transplanted into the brain parenchyma (for review, see [127]). In light on the accumulating evidence for parenchymal neurogenesis, the classical concept that the mature brain parenchyma is not permissive for the genesis of new neurons should be restricted to SVZ and SGZ progenitors. Yet, future studies should better analyze whether factors modulating the lesion-induced parenchymal neurogenic potential may differ from those acting on 'classic' neurogenic site progenitors.

7. Parenchymal gliogenesis

In the past, neurogenesis and gliogenesis had always been kept separate, the latter being considered less important than the former. In recent years, adult gliogenesis has been re-evaluated as many populations of progenitor cells with glial-like features and proliferative capacity have been shown to exist in the mature mammalian CNS [13,15]. Actually, parenchymal cell genesis in the so-called non-neurogenic regions is mainly gliogenic. In most regions of the CNS, parenchymal progenitors assure a slow process of 'constitutive' gliogenesis leading to renewal of oligodendrocytes and, to a lesser extent, astrocytes [12,15,128]. In rodents, the major population of cycling progenitors located outside the germinal niches are Ng2+ cells morphologically, antigenically, functionally distinct from mature astrocytes, oligodendrocytes and microglia [12-15]. These cells are also called 'polydendrocytes' to highlight their stellate morphology and lineal relationship to oligodendrocytes [15], 'synantocytes' [14] for their contiguity to neurons, or 'oligodendrocyte progenitor cells' (OPCs) because found able of generating myelinating oligodendrocytes [12,129,130]. Nevertheless, many polydendrocytes remain as a resident cell population of Ng2-expressing cells in the mature white and grey matter after oligodendrocytes are generated. Thus it is widely accepted they represent the fourth CNS major glial population [15], representing 2-9% of total cells [13]. In the last decade, Ng2+ cells have generated a lot of interest among neuroscientists, because they show a series of features quite unusual in OPCs. These include: i) an almost uniform distribution in both grey and white matter areas; ii) a stellate morphology; iii) an intimate association with neurons from which they receive synapses [13,14]; iv) proliferative capacity in the adult brain [13,131, 132], and v) a potential for giving rise to astrocytes and neurons that may be recruited to areas of lesion in the context of brain injury or pathology [128]. At present, it is generally accepted that polydendrocytes are OPCs, even if the demonstration that polydendrocytes differentiate into mature myelinating oligodendrocytes *in vivo* is challenging, because Ng2 expression is lost before the terminal differentiation of these cells and the appearance of mature oligodendrocyte antigens. Some observations provide circumstantial evidences for the oligodendroglial fate of polydendrocytes *in vivo*. For instance, they co-express the PDGFRα, and during the first postnatal week, in the corpus callosum and cortex, they start expressing the immature oligodendrocyte antigen O4 [133]. Polydendrocytes also express the basic helix-loop-helix (bHLH) transcription factors Olig1 and Olig2, which are required for oligodendrocyte specification and differentiation [132,134] as well as Sox9 and Sox10 transcription factors. Moreover, pulse-chase labelling of proliferating cells using 5-bromo-2′-deoxyuridine (BrdU) revealed that the number of BrdU+Ng2+ cells decreases while that of BrdU+ oligodendrocytes increases over time [12,135]. Cell grafting experiments have shown that polydendrocytes give rise to myelinating cells when they are transplanted into an environment free of endogenous myelinating cells [136]. Recently, more direct evidence for the oligodendroglial fate of polydendrocytes was obtained from cell fate-mapping experiments using transgenic mice that express Cre recombinase (Cre) in Ng2-expressing cells or that express inducible Cre (CreeR), under the regulation of the Cspg4, PDGFRα or Olig2 genes, which enable determination of the fate of polydendrocytes at a given time during development [95,137,138]. These studies showed that oligodendrocytes continue to be generated in the mature brain.

Early cell-culture studies showed that OPCs purified from rat optic nerves differentiate not only into oligodendrocytes but also into process-bearing 'type-2 astrocytes' in the presence of serum factors, which led to the concept of bipotential oligodendrocyte type-2-astrocyte (O-2A) progenitor cells [139]. There are now controversial observations suggesting that bipotentiality of polydendrocytes might be real or an *in vitro* artifact [136,140,141], and most likely these cells are inherently capable of differentiating into astrocytes but are prevented from fulfilling their astroglial fate in the normal *in vivo* environment [128].

On the whole, while all of these studies consistently support the oligodendrocyte lineage of the Ng2+ cells, the genesis of astrocytes from Ng2+ cells is confirmed only during postnatal ages. All these different and sometimes controversial results may be explained by some methodological/technical differences, but may also reflect heterogeneity in progenitor cell populations/subpopulations (mostly not yet identified), which is far to be elucidated [98]. In this context, we have recently identified a population of multipolar glial cells immunoreactive for the microtubule associated protein 5 (Map5) [142], which share features but also differences with Ng2+ progenitor cells [19]. These multipolar, Map5+ cells are newly generated, parenchymal elements of the oligodendroglial lineage, which represent a stage-specific population of polydendrocytes (Crociara et al., in preparation; Figure 2).

8. Conclusion and future perspectives

The CNS of mammals, in spite of having lost most of its regenerative/repair capacity with respect to other phyla, is endowed with remarkable plasticity. This property is heterogeneously distributed in different regions and can manifest in different ways. A better knowledge of the various forms of spontaneous and lesion-induced structural plasticity, of their mutual relationships and of the relative underlying mechanisms is fundamental in order to figure out new efficacious therapeutic perspectives for brain repair. During the last two decades, the discovery of neural stem cells and the studies on adult neurogenesis have opened the intriguing possibility of cell replacement-aimed therapeutic strategies. Under pressure of this perspective, studies on CNS stem cells and progenitors have increased exponentially, sometimes leading to excessive emphasis about theoretical correlations between neuro-glio-genic processes and brain repair. In this context, focusing on the 'real' neurogenic/gliogenic potential of the mammalian CNS should avoid to turn an exciting biological discovery into a therapeutic illusion. Indeed, the approach of regenerative medicine applied to the CNS is still hampered by overwhelming problems concerning the final integration of both transplanted and endogenously-induced cells [6]. The reason of this failure might be mostly due to evolutionary constraints [78], and to the fact that cell renewal typical of adult constitutive neurogenesis is primarily involved in tissue homeostasis of highly restricted regions, being hardly useful in response to external injury and neurodegenerative brain damage affecting the parenchyma [11,62]. On the other hand, the parenchymal cell genesis might represent a new plastic potential to be explored within wide regions of the CNS, including those areas affected by different neurodegenerative diseases and traumatic injuries. With respect to classic SVZ and SGZ neurogenesis, parenchymal neuro-glio-genesis does constitute an alternative source of progenitors, although with different outcomes [30]. Indeed, a vast number of reports currently published in this domain, although accurate and carried out with multiple technical approaches, do suggest that in most cases newly formed elements barely survive and do not fully integrate. In addition, the extreme heterogeneity of parenchymal neuro-glio-genesis makes the brain parenchyma a harsh territory, in which many questions remain unanswered and new ones are opened (see Box 1). For instance, beside a deep knowledge on the cell cycle and early cell lineage in neurogenic sites (see for example [143,144]), such information is starting to be gathered only in specific perenchymal regions and or situations [145,146]. Hence, further studies of parenchymal stem/progenitor cells, on their origin and their different fates and outcomes, should grant new challenges in the multifaceted field of CNS structural plasticity and repair.

- Which is the real extension of parenchymal cell genesis in the CNS of different mammals and in humans?
- Do parenchymal progenitors divide asymmetrically?
- Which are the real stemness properties of different parenchymal progenitors?
- Which stem/progenitor cells do contribute to postnatal neurogenesis but become depleted as their progeny differentiates, and which continue to replenish the stem/progenitor cell reservoir?
- Which is the origin of the different types of parenchymal progenitors?
- What is(are) the ultimate fate(s) of parenchymal neuro-glio-genesis?
- Which are the specific stimuli that can trigger quiescent parenchymal progenitor cell division and differentiation?
- Can the fate of parenchymal progenitors be altered by microenvironmental cues or it is predetermined? To which extent these changes do depend on regional localization?
- Can distinct parenchymal stem/progenitor cells be forced to produce unusual progeny if needed?
- Which are the factors leading to the progressive decrease of neurogenic and gliogenic activity with increasing age, both in neurogenic sites and parenchyma?

Box 1. Some open questions

http://dx.doi.org/10.5772/56100

Author details

Luca Bonfanti*, Giovanna Ponti, Federico Luzzati, Paola Crociara, Roberta Parolisi and Maria Armentano

*Address all correspondence to: luca.bonfanti@unito.it

Neuroscience Institute Cavalieri Ottolenghi (NICO), University of Turin, Italy

References

[1] B. A. Reynolds, S. Weiss, "Generation of neurons and astrocytes from isolated cells of the adult mammalian central nervous system," Science, vol. 255, pp. 1707-1710, 1992.

[2] C. G. Gross, "Neurogenesis in the adult brain: death of a dogma," Nature Review Neurosciences, vol. 1, pp. 67-73, 2000.

[3] F. H. Gage, "Mammalian neural stem cells," Science,,vol. 287, pp. 1433-1438, 2000.

[4] C. Lois, A. Alvarez-Buylla, "Long-distance neuronal migration in the adult mammalian brain," Science, vol. 264, pp. 1145-1148, 1994.

[5] F. H. Gage, G. Kempermann, T. D. Palmer, D. A. Peterson, J. Ray, "Multipotent progenitor cells in the adult dentate gyrus," Journal of Neurobiology, vol. 36, pp. 249-266, 1998.

[6] E. Arenas, "Towards stem cell replacement therapies for Parkinson's disease", Biochemical and Biophysical Research Communications, vol. 396, pp. 152-156, 2010.

[7] O. Lindvall, Z. Kokaia, "Stem cells in human neurodegenerative disorders--time for clinical translation?" Journal of Clinical Investigation, vol. 120, pp. 29-40, 2010.

[8] L. Bonfanti, G. Ponti, "Adult mammalian neurogenesis and the New Zealand white rabbit," Veterinary Journal, vol. 175, pp. 310-331, 2008.

[9] Kriegstein, A. Alvarez-Buylla, "The glial nature of embryonic and adult neural stem cells," Annual Review of Neuroscience, vol. 32, pp. 149-184, 2009.

[10] Zhao, E. M., Teng, R. G. Jr Summers, G. L. Ming, F. H. Gage, "Distinct morphological stages of dentate granule neuron maturation in the adult mouse hippocampus," Journal of Neuroscience, vol. 26, pp. 3-11, 2006.

[11] L. Bonfanti, "From hydra regeneration to human brain structural plasticity: a long trip through narrowing roads," The Scientific Word Journal, vol. 11, pp. 1270-1299, 2011.

[12] P. J., Horner, A. E., Power, G. Kempermann, H. G. Kuhn, T. D. Palmer, J. Winkler, L. J. Thal, F. H. Gage, "Proliferation and differentiation of progenitor cells throughout the intact adult rat spinal cord," Journal of Neuroscience, vol. 20, pp. 2218-2228, 2000.

[13] M. R. Dawson, A. Polito, J. M. Levine, R. Reynolds, "NG2-expressing glial progenitor cells: an abundant and widespread population of cycling cells in the adult rat CNS," Molecular and Cellular Neuroscience, vol. 24, pp. 476-488, 2003.

[14] M. Butt, N. Hamilton, P. Hubbard, M. Pugh, M. Ibrahim, "Synantocytes: the fifth element," Journal of Anatomy, vol. 207, pp. 695-706, 2005.

[15] Nishiyama, M. Komitova, R. Suzuki, X. Zhu, "Polydendrocytes (NG2 cells): multifunctional cells with lineage plasticity," Nature Reviews Neuroscience, vol. 10, pp. 9-22, 2009.

[16] J. Trotter, K. Karram, A. Nishiyama, "NG2 cells: Properties, progeny and origin," Brain Research Review, vol. 63, pp. 72-82, 2010.

[17] Dayer, K. Cleaver, T. Abouantoun, H. Cameron, "New GABAergic interneurons in the adult neocortex and striatum are generated from different precursors," Journal of Cell Biology, vol. 168, pp. 415-427, 2005.

[18] F. Luzzati, S. De Marchis, A. Fasolo, P. Peretto, "Neurogenesis in the caudate nucleus of the adult rabbit," Journal of Neuroscience, vol. 26, pp. 609-621, 2006.

[19] G. Ponti, P. Peretto, L. Bonfanti, "Genesis of neuronal and glial progenitors in the cerebellar cortex of peripuberal and adult rabbits," PLoS One, vol. 3, e2366, 2008.

[20] T.D. Palmer, E. A. Markakis, A. R. Willhoite, F. Safar, F. H. Gage, "Fibroblast growth factor-2 activates a latent neurogenic program in neural stem cells from diverse regions of the adult CNS," Journal of Neuroscience, vol. 19, pp. 8487–8497, 1999.

[21] S. Belachew, R. Chittajallu, A. A. Aguirre, X. Yuan, M. Kirby, S. Anderson, V. Gallo, "Postnatal NG2 proteoglycan-expressing progenitor cells are intrinsically multipotent and generate functional neurons," Journal of Cell Biology, vol. 161, pp. 169-186, 2003.

[22] E. Gould, A. J. Reeves, M. S. Graziano, C. G. Gross, "Neurogenesis in the neocortex of adult primates" Science, vol. 286, pp. 548-552, 1999.

[23] S. S. Magavi, B. R. Leavitt, J. D. Macklis, "Induction of neurogenesis in the neocortex of adult mice. Nature, vol. 405, pp. 951-955, 2000.

[24] D. R. Kornack, P. Rakic, "Cell proliferation without neurogenesis in adult primate neocortex," Science, vol. 294, pp. 2127-2130, 2001.

[25] Arvidsson, T. Collin, D. Kirik, Z. Kokaia, O. Lindvall, "Neuronal replacement from endogenous precursors in the adult brain after stroke," Nature Medicine, vol. 8, pp. 963-970, 2002.

[26] H. Nakatomi, T. Kuriu, S. Okabe, S. Yamamoto, O. Hatano, N. Kawahara, A. Tamura, T. Kirino, M. Nakafuku, "Regeneration of hippocampal pyramidal neurons after ischemic brain injury by recruitment of endogenous neural progenitors," Cell, vol. 110, pp. 429-441, 2002.

[27] P. Thored, A. Arvidsson, E. Cacci, H. Ahlenius, T. Kallur, V. Darsalia, C.T. Ekdahl, Z. Kokaia, O. Lindvall, "Persistent production of neurons from adult brain stem cells during recovery after stroke," Stem Cells, vol. 24, pp. 739-747, 2006.

[28] F. Luzzati, S. De Marchis, R. Parlato, S. Gribaudo, G. Schütz, A. Fasolo, P. Peretto, "New striatal neurons in a mouse model of progressive striatal degeneration are generated in both the subventricular zone and the striatal parenchyma," PLoS One, vol. 6, e25088, 2011.

[29] J. Ninkovic, M. Gotz, Fate specification in the adult brain: lessons from eliciting neurogenesis from glial cells, Bioessays, 2013, in press.

[30] L. Bonfanti, P. Peretto, "Adult neurogenesis in mammals - A theme with many variations," European Journal of Neuroscience, vol. 34, pp. 930-950, 2011.

[31] L. Bonfanti, F. Rossi, G. K. Zupanc, "Towards a comparative understanding of adult neurogenesis," European Journal of Neuroscience, vol. 34, pp. 845-846, 2011.

[32] L. Bonfanti, The (real) neurogenic/gliogenic potential of the postnatal and adult brain parenchyma, ISRN Neuroscience, vol 2013, pp. 1-14, 2013.

[33] R. A. Ihrie, A. Alvarez-Buylla, "Lake-front property: A unique germinal niche by the lateral ventricles of the adult brain," Neuron, vol. 70, pp. 674-686, 2011.

[34] P. Peretto, C. Giachino, P. Aimar, A. Fasolo, L. Bonfanti, "Chain formation and glial tube assembly in the shift from neonatal to adult subventricular zone of the rodent forebrain," Journal of Comparative Neurology, vol. 487, pp. 407-427, 2005.

[35] L. Bonfanti, P. Peretto, "Radial glial origin of the adult neural stem cells in the subventricular zone," Progress in Neurobiology, vol. 83, pp. 24-36, 2007.

[36] W. Haubensak, A. Attardo, W. Denk, W. B. Huttner, "Neurons arise in the basal neuroepithelium of the early mammalian telencephalon: a major site of neurogenesis," Proceedings of the National Academy of Sciences U.S.A, vol. 101, pp. 3196-3201, 2004.

[37] T. Voigt, "Development of glial cells in the cerebral wall of ferrets: direct tracing of their transformation from radial glia into astrocytes," Journal of Comparative Neurology, vol. 289, pp. 74-88, 1989.

[38] J. P. Misson, T. Takahashi, V. S. Caviness, "Ontogeny of radial and other astroglial cells in murine cerebral cortex," Glia, vol. 4, pp. 138-148, 1991.

[39] P. Malatesta, E. Hartfuss, M. Gotz, "Isolation of radial glial cells by fluorescent-activated cell sorting reveals a neuronal lineage," Development, vol. 127, pp. 5253-5263, 2000.

[40] S. C. Noctor, A. C. Flint, T. A. Weissman, R. S. Dammerman, A. R. Kriegstain, "Neurons derived from radial glial cells establish radial units in neocortex," Nature, vol. 409, pp. 714-720, 2001.

[41] F. T. Merkle, A. D. Tramontin, J. M. Garcìa-Verdugo, A. Alvarez-Buylla, "Radial glia give rise to adult neural stem cells in the subventricular zone" Proceedings of the National Academy of Sciences U.S.A, vol. 101, pp. 17528-17532, 2004.

[42] M. F. Eckenhoff, P. Rakic, "Radial organization of the hippocampal dentate gyrus: a Golgi, ultrastructural, and immunocytochemical analysis in the developing rhesus monkey" Journal of Comparative Neurology, vol. 223, pp. 1-21, 1984.

[43] Seri, J. M. Garcìa-Verdugo, L. Collaudo-Morente, B. S. McEwen, A. Alvarez-Buylla, "Cell types, lineage, and architecture of the germinal zone in the adult dentate gyrus," Journal of Comparative Neurology, vol. 478, pp. 359-378, 2004.

[44] E. D. Laywell, P. Rakic, V. G. Kukekov, E. C. Holland, D. A. Steindler, "Identification of a multipotent astrocytic stem cell in the immature and adult mouse brain," Proceedings of the National Academy of Science U.S.A., vol. 97, pp. 13883-13888, 2000.

[45] M. V. Sofroniew, "Molecular dissection of reactive astrogliosis and glial scar formation," Trends in Neuroscience, vol. 32, pp. 638-647, 2009.

[46] Buffo, I. Rite, P. Tripathi, A. Lepier, D. Colak, A. P. Horn, T. Mori, M. Gotz, "Origin and progeny of reactive gliosis: A source of multipotent cells in the injured brain," Proceedings of the National Academy of Science U.S.A., vol. 105, pp. 3581-3586, 2008.

[47] T. M. Jessell, "Neuronal specification in the spinal cord: inductive signals and transcriptional codes," Nature Reviews Genetics, vol. 1, pp. 20-9, 2000.

[48] H. Takebayashi, S. Yoshida, M. Sugimori, H. Kosako, R. Kominami, M. Nakafuku, Y. Nabeshima, "Dynamic expression of basic helix-loop-helix Olig family members: implication of Olig2 in neuron and oligodendrocyte differentiation and identification of a new member, Olig3," Mechanisms of Development, vol. 99, pp. 143-148, 2000.

[49] N. P. Pringle, W. D. Richardson, "A singularity of PDGF alpha-receptor expression in the dorsoventral axis of the neural tube may define the origin of the oligodendrocyte lineage," Development, vol. 117, pp. 525-533, 1993.

[50] Poncet, C. Soula, F. Trousse, P. Kan, E. Hirsinger, O. Pourquie, A. M. Duprat, P. Cochard, "Induction of oligodendrocyte progenitors in the trunk neural tube by ventralizing signals: effects of notochord and floor plate grafts, and of sonic hedgehog", Mechanisms of Development, vol. 60, pp. 13-32, 1996.

[51] N. P. Pringle, W. P. Yu, S. Guthrie, H. Roelink, A. Lumsden, A. C. Peterson, W. D. Richardson, "Determination of neuroepithelial cell fate: induction of the oligoden-

drocyte lineage by ventral midline cells and sonic hedgehog," Developmental Biology, Vol. 177, pp. 30-42, 1996.

[52] B. C. Warf, J. Fok-Seang, R. H. Miller, "Evidence for the ventral origin of oligodendrocyte precursors in the rat spinal cord. Journal of Neuroscience, Vol. 11, pp. 2477-2488, 1991.

[53] Vallstedt, J. M. Klos, J. Ericson, "Multiple dorsoventral origins of oligodendrocyte generation in the spinal cord and hindbrain" Neuron, vol. 45, pp. 55-67, 2005.

[54] W. D. Richardson, N. Kessaris, N. Pringle, "Oligodendrocyte wars. Nature Reviews Neuroscience, vol. 7, pp. 11-8, 2006.

[55] N. Tekki-Kessaris, R. Woodruff, A. C. Hall, W. Gaffield, S. Kimura, C. D. Stiles, D. H. Rowitch, W. D. Richardson, "Hedgehog-dependent oligodendrocyte lineage specification in the telencephalon," Development, vol. 128, pp. 2545-2554, 2001.

[56] N. Kessaris, M. Fogarty, P. Iannarelli, M. Grist, M. Wegner, W. D. Richardson, "Competing waves of oligodendrocytes in the forebrain and postnatal elimination of an embryonic lineage," Nature Neuroscience, vol. 9, pp. 173-179, 2006.

[57] Nishiyama, X. H. Lin, N. Giese, C.H. Heldin, W.B. Stallcup, "Co-localization of NG2 proteoglycan and PDGF alpha-receptor on O2A progenitor cells in the developing rat brain," Journal of Neuroscience Research, vol. 43, pp. 299-314, 1996.

[58] M. Diers-Fenger, F. Kirchhoff, H. Kettenmann, J. M. Levine, J. Trotter, "AN2/NG2 protein-expressing glial progenitor cells in the murine CNS: isolation, differentiation, and association with radial glia," Glia, vol. 34, pp. 213-228, 2001.

[59] M. Berry, P. Hubbard, A. M. Butt, "Cytology and lineage of NG2-positive glia," Journal of Neurocytology, vol. 31, pp. 457-467, 2002.

[60] J. M. Levine, F. Stincone, Y. S. Lee, "Development and differentiation of glial precursor cells in the rat cerebellum," Glia, vol. 7, pp. 307-321, 1993.

[61] H. Grandel, M. Brand, "Comparative aspects of adult neural stem cell activity in vertebrates," Development Genes and Evolution, in press, 2012.

[62] Y. Kozorovitskiy, E. Gould, "Adult neurogenesis: a mechanism for brain repair?," Journal of Clinical Experimental Neuropsychology, vol. 25, pp. 721-732, 2003.

[63] G. Martino, S. Pluchino, L. Bonfanti, M. Schwartz, "Brain regeneration in physiology and pathology: The immune signature driving therapeutic plasticity of neural stem cells," Physiological Reviews, vol. 91, pp. 1281-1304, 2011.

[64] P. Ferretti, "Is there a relationship between adult neurogenesis and neuron generation following injury across evolution?" European Journal of Neuroscience, vol. 34, pp. 951-962, 2011.

[65] G. K. Zupanc, "Neurogenesis and neuronal regeneration in the adult fish brain," Journal of Comparative Physiology A, vol. 192, pp. 649-670, 2006.

[66] B. W. Lindsey, V. Tropepe, "A comparative framework for understanding the biological principles of adult neurogenesis," Progress in Neurobiology, vol. 80, pp. 281-307, 2006.

[67] U. S., Sohur, J. G. Emsley, B. D., Mitchell, J. D. Macklis, "Adult neurogenesis and cellular brain repair with neural progenitors, precursors and stem cells," Philosophical Transactions of the Royal Society B-Biological Sciences, vol. 361, pp. 1477-1497, 2006.

[68] G. Ponti, P. Crociara, M. Armentano, L. Bonfanti, "Adult neurogenesis without germinal layers: the "atypical" cerebellum of rabbits," Archives Italiennes de Biologie, vol. 148, pp. 147-158, 2010.

[69] O. Koizumi, H. R. Bode, "Plasticity in the nervous system of adult hydra. III. Conversion of neurons to expression of a vasopressin-like immunoreactivity depends on axial location," Journal of Neuroscience, vol. 11, pp. 2011-2020, 1991.

[70] Y. Umesono, K. Agata, "Evolution and regeneration of the planarian central nervous system," Development Growth and Differentiation, vol. 51, pp. 185-195, 2009.

[71] R. F. Sirbulescu, G. K. Zupanc, "Spinal cord repair in regeneration-competent vertebrates: Adult teleost fish as a model system," Brain Research Review, vol. 67, pp. 73-93, 2011.

[72] T. Endo, J. Yoshino, K. Kado, S. Tochinai, "Brain regeneration in anuran amphibians," Development Growth and Differentiation, vol. 49, pp. 121-129, 2007.

[73] P. M. Lledo, M. Alonso, M.S. Grubb, "Adult neurogenesis and functional plasticity in neuronal circuits," Nature Reviews Neuroscience, vol. 7, pp. 179-193, 2006.

[74] G. Kempermann, "New neurons for 'survival of the fittest'," Nature Reviews Neuroscience, vol. 13, pp. 727-736, 2012.

[75] K. Whalley, S. Gogel, S. Lange, P. Ferretti, "Changes in progenitor populations and ongoing neurogenesis in the regenerating chick spinal cord," Developmental Biology, vol. 332, pp. 234-245, 2009.

[76] Jopling, C., Sleep, E., Raya, M., Marti, M., Raya, A., and Belmonte, J.C. (2010) Zebrafish heart regeneration occurs by cardiomyocyte dedifferentiation and proliferation. Nature 464, 606–609.

[77] L. Mescher, A. W. Neff, "Limb regeneration in amphibians: immunological considerations," ScientificWorldJournal, vol. 6, pp. 1-11, 2006.

[78] Z. M. Weil, G. J. Norman, A. C. DeVries, R. J. Nelson, "The injured nervous system: a Darwinian perspective," Progress in Neurobiology, vol. 86, pp. 48-59, 2008.

[79] K. M. Johnson, R. Boonstra, J. M. Wojtowicz, "Hippocampal neurogenesis in food-storing red squirrels: the impact of age and spatial behaviour," Genes Brain and Behaviour, vol. 9, pp. 583-591, 2010.

[80] J. M. Barker, R. Boonstra, J. M. Wojtowicz, "From pattern to pourpose: how comparative studies contribute to understanding the function of adult neurogenesis," European Journal of Neuroscience, vol. 34, pp. 963-977, 2011.

[81] N. Sanai, T. Nguyen, R. A. Ihrie, Z. Mirzadeh, H.H. Tsai, M. Wong, N. Gupta, M.S. Berger, E. Huang, J. M. Garcia-Verdugo, D.H. Rowitch, A. Alvarez-Buylla, "Corridors of migrating neurons in the human brain and their decline during infancy," Nature, vol. 478, pp. 382-386, 2011.

[82] M. V. Kokoeva, H. Yin, J. S. Flier, "Neurogenesis in the hypothalamus of adult mice: potential role in energy balance," Science, vol. 310, pp. 679-683, 2005.

[83] G. H. Kuhn, K. Blomgren, "Developmental dysregulation of adult neurogenesis," European Journal of Neuroscience, vol. 33, pp. 1115-1122, 2011.

[84] M. Feliciano, A. Bordey, "Newborn cortical neurons: only for neonates?" Trends in Neurosciences, in press, 2012.

[85] J. G. Emsley, B. D. Mitchell, G. Kempermann, J. D. Macklis, "Adult neurogenesis and repair of the adult CNS with neural progenitors, precursors, and stem cells," Progress in Neurobiology, vol. 75, pp. 321-341, 2005.

[86] K. Ohira, T. Furuta, H. Hioki, K. C. Nakamura, E. Kuramoto, Y. Tanaka, N. Funatsu, K. Shimizu, T. Oishi, M. Hayashi, T. Miyakawa, T. Kaneko, S. Nakamura, "Ischemia-induced neurogenesis of neocortical layer 1 progenitor cells," Nature Neuroscience, vol. 13, pp. 173-179, 2009.

[87] A. Pierce, A. W. Xu, "De novo neurogenesis in adult hypothalamus as a compensatory mechanism to regulate energy balance," Journal of Neuroscience, vol. 30, pp. 723-730, 2010.

[88] P. J. Bernier, A. Bédard, J. Vinet, M. Lévesque, A. Parent, "Newly generated neurons in the amygdala and adjoing cortex of adult primates," Proceedings of the National Academy of Science U.S.A., vol. 99, pp. 11464-11469, 2002.

[89] Gould, N. Vail, M. Wagers, C.G. Gross, "Adult-generated hippocampal and neocortical neurons in macaques have a transient existence," Proceedings of the National Academy of Science U.S.A., vol. 98, pp. 10910-10917, 2001.

[90] J. S. Snyder, J. S. Choe, M. A. Clifford, S. I. Jeurling, P. Hurley, A. Brown, J. F. Kamhi, H. A. Cameron, "Adult-born hippocampal neurons are more numerous, faster-maturing and more involved in behavior in rats than in mice," Journal of Neuroscience, vol. 29, pp. 14484-14495, 2010.

[91] Kempermann, E. J. Chesler, L. Lu, R. W. Williams, F. H. Gage, "Natural variation and genetic covariance in adult hippocampal neurogenesis," Proceedings of the National Academy of Science U.S.A., vol. 103, pp. 780-785, 2006.

[92] P. J. Clark, R. A. Kohman, D. S. Miller, T. K. Bhattacharya, W. J. Brzezinska, J. S. Rhodes, "Genetic influences on exercise-induced adult hippocampal neurogenesis across 12 divergent mouse strains," Genes Brain and Behaviour, vol. 10, pp. 345-353, 2011.

[93] Amrein, K. Isler, H. P. Lipp, "Comparing adult hippocampal neurogenesis in mammalian species and orders: influence of chronological age and life history stage," European Journal of Neuroscience, vol. 34, pp. 978–987, 2011.

[94] M. Zhao, S. Momma, K. Delfani, M. Carlen, R. M. Cassidy, C. B. Johansson, H. Brismar, O. Shupliakov, J. Frisen, A. M. Janson, "Evidence for neurogenesis in the adult mammalian substantia nigra," Proceedings of the National Academy of Science U.S.A., vol. 100, pp. 7925-7930, 2003.

[95] L. E. Rivers, K. M. Young, M. Rizzi, F. Jamen, K. Psachoulia, A. Wade, N. Kessaris, W. D. Richardson, "PDGFRA/NG2 glia generate myelinating oligodendrocytes and piriform projection neurons in adult mice," Nature Neuroscience, vol. 11, pp. 1392-1401, 2008.

[96] Guo, Y. Maeda, J. Ma, J. Xu, M. Horiuchi, L. Miers, F. Vaccarino, D. Pleasure, "Pyramidal neurons are generated from oligodendroglial progenitor cells in adult piriform cortex," Journal of Neuroscience, vol. 30, pp. 12036-12049, 2010.

[97] S. H. Kang, M. Fukaya, J. K. Yang, J. D. Rothstein, D. E. Bergles, "NG2+ CNS glial progenitors remain committed to the oligodendrocyte lineage in postnatal life and following neurodegeneration," Neuron, vol. 68, pp. 668-681, 2010.

[98] W. D. Richardson, K. M. Young, R. B. Tripathi, I. McKenzie, "NG2-glia as multipotent neural stem cells: fact or fantasy?" Neuron, vol. 70, pp. 661-673, 2011.

[99] Frielingsdorf, K. Schwarz, P. Brundin, P. Mohapel, "No evidence for new dopaminergic neurons in the adult mammalian substantia nigra," Proceedings of the National Academy of Science U.S.A., vol. 101, pp. 10177-10182, 2004.

[100] M. Migaud, M. Batailler, S. Segura, A. Duittoz, I. Franceschini, D. Pillon, "Emerging new sites for adult neurogenesis in the mammalian brain: a comparative study between the hypothalamus and the classical neurogenic zones," European Journal of Neuroscience, vol. 32, pp. 2042-2052, 2010.

[101] Y. Xu, N. Tamamaki, T. Noda, K. Kimura, Y. Itokazu, N. Matsumoto, M. Dezawa, C. Ide, "Neurogenesis in the ependymal layer of the adult rat 3rd ventricle," Experimental Neurology, vol. 192, pp. 251-264, 2005.

[102] M. V. Kokoeva, H. Yin, J. S. Flier, "Evidence for constitutive neural cell proliferation in the adult murine hypothalamus," Journal of Comparative Neurology, vol. 505, pp. 209-220, 2007.

[103] L. Bennett, M. Yang, G. Enikolopov, L. Iacovitti, "Circumventricular organs: a novel site of neural stem cells in the adult brain," Molecular and Cellular Neuroscience, vol. 41, pp. 337-347, 2009.

[104] M. Pérez-Martín, M. Cifuentes, J. M. Grondona, M. D. López-Avalos, U. Gómez-Pinedo, J. M. García-Verdugo, P. Fernández-Llebrez, "IGF-I stimulates neurogenesis in the hypothalamus of adult rats," European Journal of Neuroscience, vol. 31, pp. 1533-1548, 2010.

[105] K. Matsuzaki, M. Katakura, T. Hara, G. Li, M. Hashimoto, O. Shido, "Proliferation of neuronal progenitor cells and neuronal differentiation in the hypothalamus are enhanced in heat-acclimated rats," Pflugers Archiv-European Journal of Physiology, vol. 458, pp. 661-673, 2009.

[106] Pekcec, W. Löscher, H. Potschka, "Neurogenesis in the adult rat piriform cortex. Neuroreport, vol. 17, pp. 571-574, 2006.

[107] L. A. Shapiro, K. L. Ng, Q. Y. Zhou, C. E. Ribak, "Olfactory enrichment enhances the survival of newly born cortical neurons in adult mice," Neuroreport, vol. 18, pp. 981-985, 2007.

[108] M. A. Gomez-Climent, E. Castillo-Gomez, E. Varea, R. Guirado, J. M. Blasco-Ibanez, C. Crespo, F. J. Martinez-Guijarro, J. Nacher, "A population of prenatally generated cells in the rat paleocortex maintains an immature neuronal phenotype into adulthood," Cerebral Cortex, vol. 18, pp. 2229-2240, 2008.

[109] T. Seki, Y. Arai, "Expression of highly polysialylated NCAM in the neocortex and piriform cortex of the developing and the adult rat," Anatomy and Embryology (Berl.), vol. 184, pp. 395-401, 1991.

[110] L. Bonfanti, "PSA-NCAM in mammalian structural plasticity and neurogenesis," Progress in Neurobiology, vol. 80, pp. 129-164, 2006.

[111] L. Bonfanti, J. Nacher, "New scenarios for neuronal structural plasticity in non-neurogenic brain parenchyma: the case of cortical layer II immature neurons," Progress in Neurobiology, vol. 98, pp. 1-15, 2012.

[112] M. Vessal, C. Darian-Smith, "Adult neurogenesis occurs in primate sensorimotor cortex following cervical dorsal rhizotomy," Journal of Neuroscience, vol. 30, pp. 8613-8623, 2010.

[113] Bi, B, Salmaso N, Komitova M, Simonini MV, Silbereis J, Cheng E, Kim J, Luft S, Ment LR, Horvath TL, Schwartz ML, Vaccarino FM, Cortical glial fibrillary acidic protein-positive cells generate neurons after perinatal hypoxic injury. Journal of Neuroscience 31, 9205-9221, 2011.

[114] Tattersfield A, Croon R, Liu Y, Kells A, Faull R, Connor B. Neurogenesis in the striatum of the quinolinic acid lesion model of Huntington's disease. Neuroscience 127, 319-332, 2004.

[115] Parent J, Vexler Z, Gong C, Derugin N, Ferriero D. Rat forebrain neurogenesis and striatal neuron replacement after focal stroke. Annals of Neurology, 52, 802-813, 2002.

[116] Liu F, You Y, Li X, Ma T, Nie Y, Wei B, Li T, Lin H, Yang Z, Brain injury does not alter the intrinsic differentiation potential of adult neuroblasts. Journal of Neuroscience, 29, 5075-5087.

[117] Hoehn B, Palmer T, Steinberg G. Neurogenesis in rats after focal cerebral ischemia is enhanced by indomethacin. Stroke 36, 2718-2724, 2005.

[118] Nakagomi T, Molnár Z, Nakano-Doi A, Taguchi A, Saino O, Kubo S, et al. Ischemia-induced neural stem/progenitor cells in the pia mater following cortical infarction. Stem cells and development 20, 2037-2051, 2011.

[119] Decimo I, Bifari F, Rodriguez F, Malpeli G, Dolci S, Lavarini V, et al. Nestin- and doublecortin-positive cells reside in adult spinal cord meninges and participate in injury-induced parenchymal reaction. Stem cells 29, 2062-2076, 2011.

[120] Shimada IS, Peterson BM, Spees JL Isolation of locally derived stem/progenitor cells from the peri-infarct area that do not migrate from the lateral ventricle after cortical stroke. Stroke 41, 3552-3560, 2010.

[121] Shimada I, LeComte M, Granger J, Quinlan N, Spees J. Self-renewal and differentiation of reactive astrocyte-derived neural stem/progenitor cells isolated from the cortical peri-infarct area after stroke. The Journal of neuroscience: the official journal of the Society for Neuroscience 32, 7926-7940, 2012.

[122] T. Mantamadiotis, T. Lemberger, S. C. Bleckmann, H. Kern, O. Kretz, A. Martin Villalba, F. Tronche, C. Kellendonk, D. Gau, J. Kapfhammer, C. Otto, W. Schmid, G. Schütz, "Disruption of CREB function in brain leads to neurodegeneration," Nature Genetics, vol. 31, pp. 47-54, 2002.

[123] Suh H, Consiglio A, Ray J, Sawai T, D'Amour KA, Gage FH, In vivo fate analysis reveals the multipotent and self-renewal capacities of Sox2+ neural stem cells in the adult hippocampus. Cell Stem Cell, 1, 515-528, 2007.

[124] Cheng LC, Pastrana E, Tavazoie M, Doetsch F miR-124 regulates adult neurogenesis in the subventricular zone stem cell niche. Nature Neuroscience 12, 399-408, 2009.

[125] Steiner B, Klempin F, Wang L, Kott M, Kettenmann H, Kempermann G, Type-2 cells as link between glial and neuronal lineage in adult hippocampal neurogenesis. Glia 54, 805-814, 2006.

[126] Doetsch F, Petreanu L, Caille I, Garcia-Verdugo JM, Alvarez-Buylla A EGF converts transit-amplifying neurogenic precursors in the adult brain into multipotent stem cells. Neuron 36, 1021-1034, 2002.

[127] Burns T, Verfaillie C, Low W. Stem cells for ischemic brain injury: a critical review. Journal of Comparative Neurology 515, 125-144, 2009.

[128] E. Boda, A. Buffo, "Glial cells in non-germinal territories: insights into their stem/progenitor properties in the intact and injured nervous tissue," Archives Italiennes de Biologie, vol. 148, pp. 119-136, 2010.

[129] W. B. Stallcup, L. Beasley, "Bipotential glial precursor cells of the optic nerve express the NG2 proteoglycan," Journal of Neuroscience, vol. 7, pp. 2737-2744, 1987.

[130] R. Reynolds, M. Dawson, D. Papadopoulos, A. Polito, I. C. Di Bello, D. Pham-Dinh, J. Levine, "The response of NG2-expressing oligodendrocyte progenitors to demyelination in MOG-EAE and MS," Journal of Neurocytology, vol. 31, pp. 523-536, 2002.

[131] A. Aguirre, R. Chittajallu, S. Belachew, V. Gallo, "NG2-expressing cells in the subventricular zone are type C-like cells and contribute to interneuron generation in the postnatal hippocampus," Journal of Cell Biology, vol. 165, pp. 575-589, 2004.

[132] K. L. Ligon, S. P. Fancy, R. J. Franklin, D. H. Rowitch, "Olig gene function in CNS development and disease," Glia, vol. 54, pp. 1-10, 2006.

[133] R. Reynolds, R. Hardy, "Oligodendroglial progenitors labeled with the O4 antibody persist in the adult rat cerebral cortex in vivo," Journal of Neuroscience Research, vol. 47, pp. 455-470, 1997.

[134] Q. Zhou, S. Wang, D. J. Anderson, "Identification of a novel family of oligodendrocyte lineage-specific basic helix-loop-helix transcription factors," Neuron, vol. 25, pp. 331-343, 2000.

[135] Bu, A. Banki, Q. Wu, A. Nishiyama, "Increased NG2(+) glial cell proliferation and oligodendrocyte generation in the hypomyelinating mutant shiverer," Glia, vol. 48, pp. 51-63, 2004.

[136] R. J. Franklin, S. C. Barnett, "Do olfactory glia have advantages over Schwann cells for CNS repair?" Journal of Neuroscience Research, vol. 50, pp. 665-672, 1997.

[137] L. Dimou, C. Simon, F. Kirchhoff, H. Takebayashi, M. Gotz, "Progeny of Olig2-expressing progenitors in the gray and white matter of the adult mouse cerebral cortex," Journal of Neuroscience, vol. 28, pp. 10434-10442, 2008.

[138] M. Komitova, E. Perfilieva, B. Mattsson, P.S. Eriksson, B. B. Johansson, "Enriched environment after focal cortical ischemia enhances the generation of astroglia and NG2 positive polydendrocytes in adult rat neocortex," Experimental Neurology, vol. 199, pp. 113-121, 2006.

[139] M. C. Raff, R. H. Miller, M. Noble, "A glial progenitor cell that develops in vitro into an astrocyte or an oligodendrocyte depending on culture medium," Nature, vol. 303, pp. 390-396, 1983.

[140] Espinosa de los Monteros, R. Bernard, B. Tiller, P. Rouget, J. de Vellis, "Grafting of fast blue labeled glial cells into neonatal rat brain: differential survival and migration

among cell types," International Journal of Developmental Neuroscience, vol. 11, pp. 625-639, 1993.

[141] K. Groves, S. C. Barnett, R. J. Franklin, A. J. Crang, M. Mayer, W. F. Blakemore, M. Noble, "Repair of demyelinated lesions by transplantation of purified O-2A progenitor cells. Nature, vol. 362, pp. 453-455, 1993.

[142] M. Riederer, "Microtubule-associated protein 1B, a growth and phosphorilated scaffold protein," Brain Research Bulletin, vol. 71, pp. 541-558, 2007.

[143] Encinas JM, Michurina TV, Peunova N, Park JH, Tordo J, Peterson DA, Fishell G, Koulakov A, Enikolopov G, Division-coupled astrocytic differentiation and age-related depletion of neural stem cells in the adult hippocampus, Cell Stem Cell, vol. 8, pp. 566-579, 2011.

[144] G. Ponti, K. Obernier, C. Guinto, L. Jose, L. Bonfanti, A. Alvarez-Buylla, The cell cycle and lineage progression of neural progenitors in the ventricular-subventricular zones of adult mice. Proceedings of the National Academy of Science U.S.A., in press, 2013.

[145] Simon, M. Götz, L. Dimou, "Progenitors in the adult cerebral cortex: cell cycle properties and regulation by physiological stimuli and injury," Glia, vol. 59, pp. 869-881, 2011.

[146] S. Sugiarto, A. I. Persson, E. G. Munoz et al., "Asymmetricdefective oligodendrocyte progenitors are glioma precursors," Cancer Cell, vol. 20, pp. 320–340, 2011.

Section 3

Neural Stem Cells and Regenerative Medicine

Chapter 10

A Survey of the Molecular Basis for the Generation of Functional Dopaminergic Neurons from Pluripotent Stem Cells: Insights from Regenerative Biology and Regenerative Medicine

Kaneyasu Nishimura, Yoshihisa Kitamura,
Kiyokazu Agata and Jun Takahashi

Additional information is available at the end of the chapter

http://dx.doi.org/10.5772/55442

1. Introduction

Animals that possess regenerative abilities are widespread in the animal kingdom [1]. Hydra, planarian, zebrafish, newt and axolotl are known prominent species, and the cellular aspects of the stem cell system for regeneration are well elucidated [2]. However, few animals can be used to investigate the molecular basis of neuronal regeneration, in spite of the presence of prominent regenerative animals, as mentioned above. Planarians, for instance, can regenerate a functional brain after amputation in a few days, even from non-brain tissue [3,4]. Newts can regenerate several tissues and organs (*i.e.,* lens, limbs, jaws, hearts and tails) with recovery of function and physiology after injury or tissue removal [5-8]. These animals achieve regeneration of missing nervous system utilizing stem cells. However, it is difficult to regenerate nervous system in mammalians, including human beings, although these animals possess neural stem cells. Therefore, regenerative animals provide unique opportunities to investigate the generation and utilization of stem cells to repair lost or injured tissue in non-regenerative animals. On the other hand, the successful derivation of neural cells from human embryonic stem cells (ESCs) [9] and induced pluripotent stem cells (iPSCs) [10,11] under *in vitro* conditions provides a new experimental strategy for clinical translation. In other words, although human beings lack regenerative abilities, the new clinical strategy of "regenerative medicine," including cell-transplantation therapy, has been developed to recover lost neural functions by using stem cells. This research field has become a greatly advancing scientific field worldwide.

In this chapter, we focus on the molecular systems of generation of functional dopaminergic (DA) neurons *in vivo* and/or *in vitro* in regenerative and non-regenerative animals. The first topic investigates how regenerative animals recruit new DA neurons from stem cells after injury. The second topic explores how to generate DA neurons from mammalian ESCs and iPSCs under *in vitro* conditions. The third topic evaluates clinical applications for human neural disease, especially Parkinson's disease.

2. DA neuronal regeneration in regenerative animals

Freshwater planarians, *Plathelminthes*, have a primitive central nervous system (CNS) that is composed of a well-organized brain and a pair of ventral nerve cords [12]. Large-scale expression analyses have revealed that many neural genes involved in vertebrate brain development and function are also expressed in distinct domains of the planarian CNS [13,14]. These results indicate that the planarian CNS is functionally regionalized according to a discrete expression of neural-specific genes.

We recently discovered that the planarian CNS contains dopamine, serotonin, γ-aminobutyric acid (GABA) and acetylcholine, which are known to be present in mammalians, and that the planarian nervous system constitutes particular neural networks and functions [15-19]. Since planarians possess pluripotent stem cells throughout their entire bodies, their CNS can be completely regenerated along with recovery of morphology and function after amputation [20].

Recently, two different processes of DA neuroregeneration in the planarian *Dugesia japonica* have been described. One involves DA neuroregeneration accompanied by brain regeneration after artificial amputation [15,21]. The other involves DA neuronal regeneration after selective degeneration of DA neurons by the DA neurotoxin 6-hydroxydopamine (6-OHDA) [22]. Although both of these processes are achieved by manipulating pluripotent stem cells, the systems of the processes are different between these regenerative processes [23]. In this section, we focus on these two processes of DA neuroregeneration in planarians, one of prominent regenerative animals.

2.1. Brain regeneration and DA neuroregeneration in planarians

Planarians can regenerate a functional brain within 7-10 days after amputation. This regenerative process can be divided into at least five steps: (1) anterior blastema formation, (2) brain rudiment formation, (3) pattern formation, (4) neural network formation and (5) functional recovery. Each step is defined by sequential gene expression alterations that are similar to those observed in mammalian brain development [20].

2.1.1. The early stage of brain regeneration after amputation

The first step of head regeneration after amputation is wound closure, which involves adhesion of the dorsal and ventral tissues, thereby inducing activation of *noggin-like gene A* (*DjnlgA*) at the edge of the amputated site [24]. Subsequently, a blastema (a mass of cells derived from pluripotent stem cells) is formed by mitogen-activated protein (MAP) kinase activation 24 hours after amputation [25,26]. It has been revealed that blastema cells are supplied from the post-blastema region via mitosis from G2-phase pluripotent stem cells. Activation of c-Jun-N-terminal kinase (JNK) after amputation induces G2/M transition and supplies blastema cells. Subsequently, activation of extracellular signal-related kinase (ERK) signalling is required for blastema cells to exit the undifferentiated state and enter the differentiation state in order to form the brain rudiment [27]. Similarly, activation of ERK signalling facilitates exit from self-renewal and regulates differentiation signals, as in mouse ESCs [28]. These findings indicate that planarian stem cells and mouse ESCs may possess a similar molecular basis for cell fate determination.

After the formation of the brain rudiment, the wnt and bone morphogenic protein (BMP) signaling pathways regulate brain polarity along the anterior-posterior axis [29,30] and the dorso-ventral axis [31,32], respectively. Therefore, pluripotent stem cells are regulated by various signals in spacio and temporal manners to form the brain. Similarly, wnt and BMP signaling pathways also regulate polarity of neural tube formation in mammalian early development [33,34].

2.1.2. The stem cell system of DA neuroregeneration in the head regeneration process

During head regeneration, DA neurons begin to appear in future brain regions starting three days after amputation. On day 5, the number of new DA neurons increases and axons start to extend. On day 7, brain regeneration is complete along with complete reconstruction of DA neurons (Figure 1A) [15]. Other neurotransmitter-synthesizing neurons, such as GABAergic, octopaminergic (OA) and cholinergic neurons are also regenerated in a similar manner from pluripotent stem cells during brain regeneration [17-19]. Recently, we revealed that the numbers of each type of different brain neurons are maintained in a constant ratio that is dependent on body size in intact planarians [21]. For instance, the ratio of DA neurons to OA neurons is 2:1 in intact planarian brains. Interestingly, in the early stage of brain regeneration, the ratio is larger than 2:1; however, it is gradually restored to 2:1 during brain regeneration. The ratio among different neuronal cell types fluctuates in the early stages of regeneration and is gradually restored to the original ratio. These data suggested that non-cell-autonomous mechanisms utilized to adjust the ratio among different types of brain neurons [21].

2.2. DA neuronal regeneration after selective DA lesions

We recently established an experimental model to investigate cell-type specific regeneration following selective degeneration by 6-hydroxydopamine (6-OHDA), a DA neurotoxin [22]. According to our observations, DA neurons are completely degenerated within 24 hours after 6-OHDA-administration. Newly generated DA neurons begin to appear in the brain four days after 6-OHDA-induced lesions. Thereafter, the number and the axons of DA neurons gradually recover over a period of several days. Finally, DA neurons are completely recovered within 14 days after 6-OHDA-induced lesions (Figure 1B). Bromodeoxyuridine (BrdU)-pulse and chase experiments indicate that newly generated DA neurons are derived from proliferative stem cells that enter the S-phase in the trunk region and migrate to the head region from the trunk region without entering the M-phase (*i.e.*, they remain in the G2-phase) and then give rise to DA neurons in the head region (Figure 1C). In addition, histological analyses support the hypothesis that pluripotent stem cells may directly give rise to differentiated DA neurons in planarians (Figure 1D, E). This observation suggests that G2-phase proliferating stem cells can respond to degeneration of DA neurons and are committed to DA neurons in planarians [22].

Although most vertebrates show low regenerative capabilities, newts have powerful regenerative abilities among adult vertebrates. Surprisingly, adult newts regenerate brain tissue after partial brainectomy in spite of having a complex brain structure [35,36]. Recent reports indicate that newts maintain the neurogenic potential to repair lost midbrain DA neurons, even in the adult state [37]. Although proliferative abilities are essentially quiescent in the midbrain of adult newts, quiescent ependymoglia cells are activated by the degeneration of midbrain DA neurons. Additionally, activated ependymoglia cells start to proliferate and differentiate to DA neurons to repair lost DA neurons. This neurogenic potential is activated under conditions of injury-responsive cell-replacement and not under homeostatic conditions [38,39]. These reports indicated that these regenerative animals possess unique stem cells system to regenerative missing DA neurons.

3. Strategies for generation of DA neurons in non-regenerative animals

Regenerative animals easily regenerate lost brain and neural tissues by maturating stem cells. Regenerative animals provide us unique ideas to generate neural tissue from mammalian pluripotent stem cells such as ESCs and iPSCs.

Although the capacity for brain formation is present during the developmental stage among animals, including human beings, it is difficult to regenerate missing neurons and brain tissue in the adult state. It has been reported that neurogenesis homeostatically occurs in the restricted regions of the adult mammalian brain such as hippocampus and subventricular zone [40,41]. However, it remains controversial whether dopaminergic neurogenesis/neuroregeneration occurs in the adult mammalian midbrain [42,43] and whether neurogenic potential is "lost" or "quiescent" in the adult mammalian brain. Regardless, this potential is not adequate to recover missing neurons and brain tissue in the adult mammalian brain. If human beings had an adequate regenerative potential, some types of neural disorders and brain injuries might be self-curable. However, it is difficult to self-repair neurodegenerative disorders. Therefore, it is strongly expected that neuronal differentiation techniques will contribute for therapeutic applications, such as cell-transplantation therapy using ESCs/iPSCs.

Parkinson's disease (PD) is a candidate disease for the expected application of cell-transplantation therapy. PD is an intractable neurodegenerative disorder that arises from the progressive death of DA neurons in the substantia nigra pars compacta. Although human beings lack adequate abilities to regenerate DA neurons, techniques to generate midbrain DA neurons from ESCs/iPSCs *in vitro* have been developed and are being improved for use in cell-transplantation therapy (Figure 2).

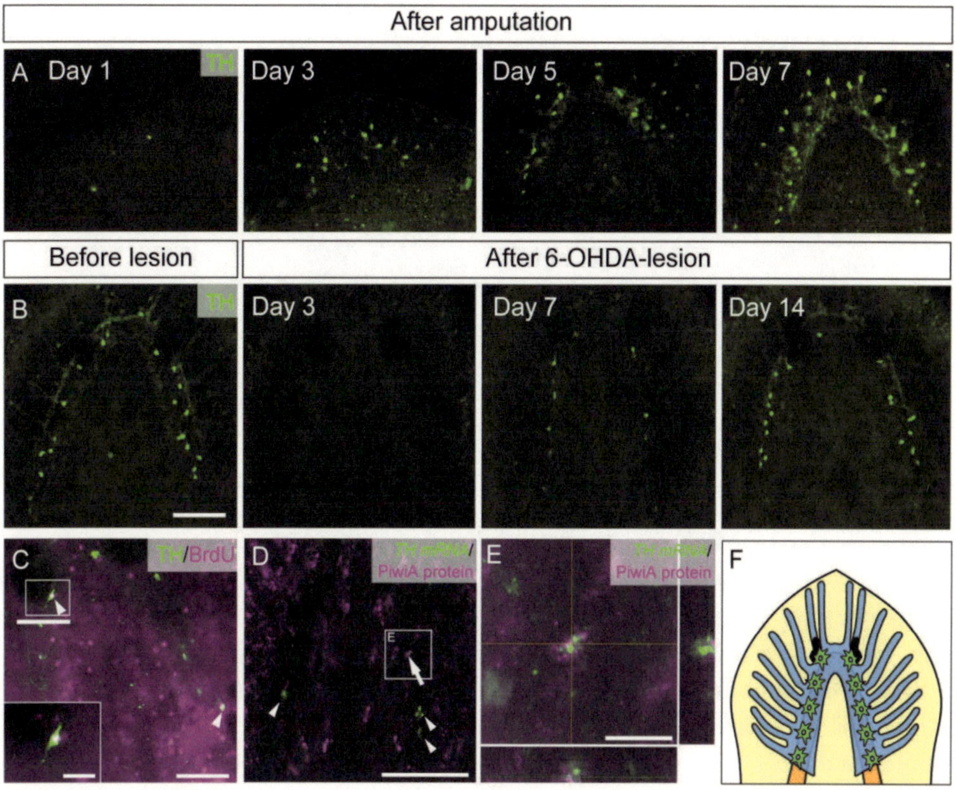

Figure 1. **A**) The regeneration process of DA neurons after decapitation. Immunofluorescence of brain DA neurons one day, three days, five days and seven days after decapitation. (**B**) The process of DA neuronal regeneration in the brain after 6-OHDA-induced-lesions. Immunofluorescence of brain DA neurons in an intact brain three days, seven days and 14 days after 6-OHDA-administration. (**C**) BrdU-incorporation is detected in newly generated DA neurons five days after 6-OHDA-administration. Double-immunofluorescence of tyrosine hydroxylase (TH; green) and BrdU (magenta). The arrowheads indicate BrdU/TH-double positive neurons. (**D, E**) Double-fluorescence of *TH mRNA* (green) and PiwiA proteins (a marker for pluripotent stem cells; magenta) seven days after 6-OHDA-administration. Scale bars: 200 μm (**A-D**), 50 μm (high magnification image in C) and 50 μm (**E**). (**F**) A schematic drawing of the planarian brain nervous system and an image of the distribution of DA neurons. The blue color represents the brain. The orange color represents ventral nerve cords (VNCs). The green cells are DA neurons.

3.1. Induction of midbrain DA neurons *in vitro* from ESCs/iPSCs

In 2000, it was reported that stromal cell-derived inducing activity (SDIA) strongly promotes neural induction in mouse ESCs co-cultured with mouse PA6 stromal cells under serum-free conditions without growth factors [44]. SDIA-induced neurons contain high amounts of DA neurons and are integrated into the 6-OHDA-lesioned mouse striatum after transplantation. Additionally, it has been proven that functional DA neurons can be differentiated from primate and human ESCs using SDIA [45,46]. Therefore, the establishment of the SDIA method opened new fields for both basic neuroscience research and therapeutic applications.

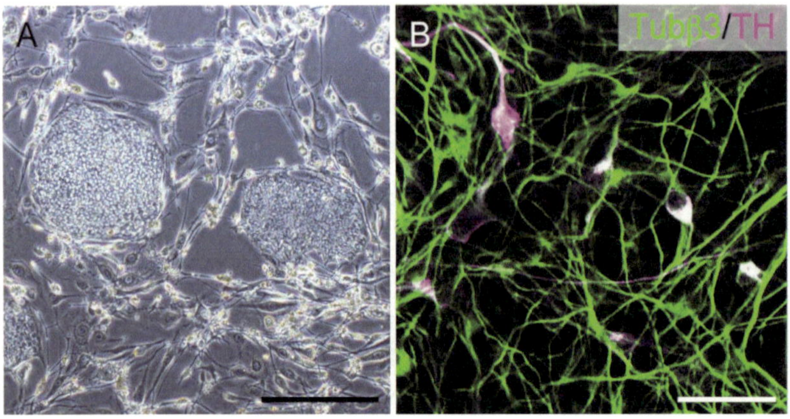

Figure 2. A) A phase contrast image of colonies of undifferentiated human iPSCs on SNL feeder cells. (**B**) DA neurons derived from human iPSCs differentiated using the SFEB method *in vitro*. Double-immunofluorescence of ßIII-tubulin (TubßIII), a neuronal marker (green), and TH, a marker of DA neurons (magenta). Scale bars: 500 μm (**A**) and 50 μm (**B**).

Recently, the molecular mechanisms of mammalian brain development have become better understood. The expression and secretion of patterning factors facilitate neural induction and define anterior-posterior and dorso-ventral patterning of the mammalian brain. For instance, mammalian midbrain development is governed by fibroblast growth factor 8 (FGF8) and Sonic hedgehog (Shh), which are locally expressed at the midbrain-hindbrain boundary and the ventral neural tube, respectively [47]. The floor plate is located along the ventral midline of the neural tube and is known to function as a signaling center during brain development and a source of midbrain DA neurons [48,49]. Midbrain DA neuronal specification is regulated by several transcription factors, including Lmx1a, FoxA2, Nurr1 and Pitx3 [50-53]. Therefore, cellular aspects of brain development provide ideas for improving the differentiation methods of authentic neural identity and subtype specification, such as differentiating DA neurons from ESCs/iPSCs *in vitro*.

Recently, neural lineage commitment has been improved with dual inhibition of SMAD signaling by bone morphogenic protein (BMP) inhibitor (noggin and dorsomorphine) and transforming growth factor-ß (TGF-ß)/activin/nodal inhibitor (SB431542). These combinations promote efficient neural induction of both human ESCs and iPSCs in serum-free, floating cultures of embryoid body-like aggregates (SFEB) and stromal (PA6) feeder co-cultures [54,55]. In addition, GSK3ß inhibitor (CHIR99021) strongly activates wnt signaling and induces the Lmx1a expression in FoxA2-positive floor plate precursors and the neurogenic conversion of human ESC-derived midbrain floor plates towards DA neurons [56]. This differentiation method mimics mammalian brain development.

Recently, human ESCs/iPSCs maintenance and differentiation have been advanced in order to achieve clinical translation, which results in chemically defined conditions and the elimination of animal-derived components and the need for feeder cells [57,58].

http://dx.doi.org/10.5772/55442

4. Towards the clinical application of ESCs/iPSC-derived DA neurons in Parkinson's disease

Clinically, PD is estimated to affect approximately 1% of the population over 65 years of age, and PD patients often exhibit muscle rigidity, tremors, bradykinesia and akinesia. Currently, the primary clinical treatment for PD is dopamine replacement therapy using L-dihydroxy-phenylalanine (L-DOPA) and/or dopamine receptor agonists. Although pharmacotherapy temporarily improves parkinsonian symptoms, the efficacy of pharmacotherapy is gradually lost over long-term treatment, and the wearing-off phenomenon, the on-off phenomenon and drug-induced dyskinesia develop. In addition, the progression of the degeneration of midbrain DA neurons cannot be delayed. Therefore, new strategies such as cell transplantation therapy are expected to recover lost DA neurons.

The first clinical trial of cell transplantation using human fetal ventral midbrain in PD patients was performed in the latter half of the 1980's [59]. Currently, over 400 PD patients have been evaluated in this clinical trial. Some PD patients who underwent grafting have exhibited drastic improvements in movement symptoms. However, strict ethical problems remain regarding the use of human fetal cells for the treatment of human disease. Therefore, efficient methods need to be developed for differentiating DA neurons from ESCs/iPSCs, instead of using human fetal cells, as described above. In addition, the efficiency and safety of ESCs/iPSCs-derived DA neurons should be evaluated *in vivo* experiments using animal models before conducting human trials. In general, DA neurotoxins such as rotenone, 6-OHDA and 1-methyl-4-phenyl-1,2,3,6-tetrahydropyridine (MPTP) are used to create parkinsonian animal models with rodents and non-human primates. In particular, since the pathological symptoms and brain anatomy of monkeys are similar to those of humans, the outcomes of monkey trials strongly contribute to realizing human trials.

4.1. Preclinical trials of human ESCs/iPSC-derived DA neurons for PD

In 2005, it was reported that intra-striatal primate ESC-derived DA neurons survive and function in the putamen in MPTP-lesioned primate parkinsonian models [60]. In that study, positron emission tomography imaging revealed that [^{18}F]-F-DOPA uptake increased in grafted monkeys 14 weeks after transplantation. In addition, the neurological scores of the grafted monkeys improved in comparison with that observed in sham-operated monkeys starting from 10 weeks after transplantation. This study is the first report to indicate the functional efficiency of grafted DA neurons derived from primate ESCs and has opened up the possibility for transplantation therapy using ESC-derived DA neurons.

Meanwhile, human ESC/iPSC-derive DA neurons have the potential to improve motor function in PD model rats after intra-striatal grafting [56,61,62]. However, some problems, such as tumorigenicity, remain in clinical trials. We recently reported that long-term neural maturation (> 28 days) of human ESCs reduces tumorigenicity after grafting in primate parkinsonian models. In addition, motor symptoms are also improved by grafting human ESC-derived DA neurons that have maturated over a long term (42 days). These results suggest that human ESC-derived DA neurons that are differentiated for appropriate terms strongly contribute to both reducing the risk of tumor genesis and improving parkinsonian motor dysfunction [46]. Recently, we first reported that human iPSC-derived DA neurons that are differentiated under feeder-free and serum-free conditions survived in an MPTP-lesioned primate parkinsonian model for six months [63]. This report may support the therapeutic potential of human iPSCs for future clinical trials.

5. Stem cell utilization *in vivo* and *in vitro* for non-regenerative animals: Lessons from regenerative animals

Although the number of regenerative animals used to investigate DA neuronal regeneration is limited, the cellular mechanisms of DA neuronal regeneration are starting to be understood [22,39]. The cellular and molecular mechanisms for DA neuronal regeneration in regenerative animals provide new ideas for generation of DA neurons in non-regenerative animals. Our histological analysis of regeneration indicates that pluripotent stem cells may directly give rise to differentiated DA neurons in the planarian head region (Figure 1D, E) [22]. In addition, we have not yet obtained direct evidence for the presence of neural stem cells in planarians. First, we have not observed proliferating cells in the brain, either during regeneration or in intact brains. Second, the expression of the planarian *musashi* family of genes supports the above hypothesis. *Musashi*, an RNA binding protein, is expressed in neural stem cells in various animals[64]. Although three *musashi*-like genes (*DjmlgA-C*) have been isolated from planarians, the expressions of these genes are detected in the planarian CNS and are not eliminated by X-ray irradiation [65]. These results indicate that planarian *musashi*-like genes are expressed in differentiated cells, not in proliferative stem cells. Based on these observations, we speculate that the neural stem cell system most likely evolved at a later stage of evolution [66].

In the case of brain regeneration, the brain rudiment is formed inside of the blastema. The cells that participate in blastema formation already exist from the proliferative state, and a portion of these cells start to form the brain rudiment [25,26]. Therefore, the commitment of DA neurons may immediately occur after brain rudiment and brain pattern formation.

In the case of DA neuronal regeneration after 6-OHDA-induced lesions, G2-phase stem cells are recruited into DA neurons in the brain. Since the presence of neural stem cells were not clarified in planarians, the key roles for differentiation into DA neurons from pluripotent stem cells would function in G2-phase in the cell cycle. The neurons remaining in the brain after 6-OHDA-induced lesions may play an important role in sensing the loss of DA neurons and recruiting G2-phase stem cells into DA neurons [22]. Recent findings using eukaryote cells indicate that cell fate determination of either self-renewal or differentiation occurs during the G2/M phase in the cell cycle [67]. Therefore, the results of our study concerning the cellular system of DA neuronal regeneration are supported by these observations. Therefore, G2-phase stem cell strongly contributed to DA neuronal regeneration by response to extrinsic environment in planarians. However, we have not yet understand a clear answer regarding which signal molecules contribute to the recruitment of G2-phase stem cells to DA neurons in planarians.

In the case of newts, another regenerative animals, quiescent ependymoglia cells can sense degeneration of DA neurons, and re-enter into the cell cycle to restore lost DA neurons [39]. In this case, DA receptor-expressing surrounding cells (*i.e.*, ependymoglia cells) respond to the degeneration of DA neurons via DA receptor signaling, and contribute to regenerate DA neurons. Therefore, the responsibility of stem cells to extrinsic environment is important for DA neuronal regeneration in both planarians and newts.

The findings from regenerative animals may provide any idea to generate DA neurons from ESCs/iPSCs of non-regenerative animals. Recent strategies for DA neuronal induction from ESCs/iPSCs have based on mimicking midbrain development, that is, the activation of morphogenic factors and transcriptional regulation. In addition, the establishment of 3-dimensional (3D) culture systems strongly contributes to mimicking the complicated organogenesis and lead to the acquisition of sub-regional identities because the cells respond to extrinsic signals [57, 69].

Recently, Kirkeby *et al.* developed an induction method for human ESC-derived DA neurons using dual-SMAD inhibition with embryoid body formation. The gene expression profile and transplantation aspects of human ESC-derived DA neurons can be recaptured in the human fetal midbrain [70], suggesting that 3D cultures are suitable for mimicking brain organogenetic processes in authentic midbrain DA neurons.

In addition, our findings in planarians may contribute to improving strategies of cell-transplantation therapy. That is, how to integrate ESCs/iPSC-derived grafting neurons into the host brain. We found that commitment to DA neurons in stem cells occurs during the G2-phase of the cell cycle. Which state of committed cells can be easily incorporated into lesioned regions and whether the location of commitment activates important factor(s) in the incorporation of committed cells into appropriate positions should be considered. Additionally, both planarians and newts showed drastic behavioral recovery according to DA neuronal regeneration [22,37]. This suggested that regenerated DA neurons contributed to functional recover by integrating to the existing neural circuit. We speculate that this phenomenon may provide to the any ideas how to integrate ESCs/iPSC-derived neurons into existing neural circuit after grafting and contribute to recover the motor function. We will attempt to answer several important questions in our future works. A deeper understanding of the answers to the above questions may provide unique clues regarding not only how to commit DA neurons in vitro, but also how to optimize cell-transplantation therapy in the future.

6. Conclusion

Stem cell research is an interesting field in basic science due to its potential therapeutic applications. It is now extensively studied to investigate that the cellular and molecular mechanisms of DA neuronal regeneration in many regenerative animals *in vivo* have been described over the last several years, although numerous unclear issues remain. Important aspects learned from regenerative animals suggest that the cell sources and stem cell systems of DA neuronal regeneration may reflect several concepts in the achievement of regeneration, even among regenerative animals. Therefore, it is difficult to find a common rule for DA neuronal regeneration among animals. However, this trend in regenerative biology will be more emphasized by discoveries made in future studies of regenerative animals, and will provide hints about more efficient utilization of stem cells towards clinical application. Importantly, studying the diversity of regenerative contexts in different animals may contribute to highlighting diverse concepts and attractive clues for investigating the generation of DA neurons *in vivo* and *in vitro* in regenerative medicine.

Acknowledgements

We thank all members of the laboratory for discussions and support. This work was supported by a Grant-in-Aid for a Grant-in-Aid for Creative Research to K.A. and a Grant-in-Aid for Global COE Program A06.

Author details

Kaneyasu Nishimura[1], Yoshihisa Kitamura[2], Kiyokazu Agata[3] and Jun Takahashi[1]

1 Department of Biological Repair, Institute for Frontier Medical Sciences, and Laboratory of Neuronal Regeneration, Department of Clinical Application, Center for iPS cell Research and Application (CiRA), Kyoto University, Japan

2 Department of Neurobiology, Kyoto Pharmaceutical University, Japan

3 Department of Biophysics, Graduate School of Science, Kyoto University, Japan

References

[1] Agata K, Inoue T. Survey of the differences between regenerative and non-regenerative animals. Development Growth and Differentiation 2012;54(2) 143-152.

[2] Tanaka EM, Reddien PW. The cellular basis for animal regeneration. Developmental Cell 2011;21(1) 172-185.

[3] Agata K, Watanabe K. Molecular and cellular aspects of planarian regeneration. Seminars in Cell and Developmental Biology 1999;10(4) 377-383.

[4] Reddien PW, Sánchez Alvarado A. Fundamentals of planarian regeneration. Annual Review of Cell and Developmental Biology 2004;20 725-757.

[5] Eguchi G. Cellular and molecular background of wolffian lens regeneration. Cell Differentiation and Development 1988;Suppl. 147-158.

[6] Brockes JP, Kumar A. Plasticity and reprogramming of differentiated cells in amphibian regeneration. Nature Review Molecular Cell Biology 2002;3(8) 566-574.

[7] Tanaka EM. Regeneration: if they can do it, why can't we? Cell 2003;113(5) 559-562.

[8] Kurosaka H, Takano-Yamamoto T, Yamashiro T, Agata K. Comparison of molecular and cellular events during lower jaw regeneration of newt (*Cynops pyrrhogaster*) and West African clawed frog (*Xenopus tropicalis*). Developmental Dynamics 2008;237(2) 354-365.

[9] Thomson JA, Itskovitz-Eldor J, Shapiro SS, Waknitz MA, Swiergiel JJ, Marshall VS, Jones JM. Embryonic stem cell lines derived from human blastocysts. Science 1998;282(5391) 1145-1147.

[10] Takahashi K, Tanabe K, Ohnuki M, Narita M, Ichisaka T, Tomoda K, Yamanaka S. Induction of pluripotent stem cells from adult human fibroblasts by defined factors. Cell 2007;131(5) 861-872.

[11] Yu J, Vodyanik MA, Smuga-Otto K, Antosiewicz-Bourget J, Frane JL, Tian S, Nie J, Jonsdottir GA, Ruotti V, Stewart R, Slukvin II, Thomson JA. Induced pluripotent stem cell lines derived from human somatic cells. Science 2007;318(5858) 1917-1920.

[12] Agata K, Soejima Y, Kato K, Kobayashi C, Umesono Y, Watanabe K. Structure of the planarian central nervous system (CNS) revealed by neuronal cell markers. Zoological Science 1998;15(3) 433-440.

[13] Cebrià F, Nakazawa M, Mineta K, Ikeo K, Gojobori T, Agata K. Dissecting planarian central nervous system regeneration by the expression of neural-specific genes. Development Growth and Differentiation 2002;44(2) 135-146.

[14] Mineta K, Nakazawa M, Cebria F, Ikeo K, Agata K, Gojobori T. Origin and evolutionary process of the CNS elucidated by comparative genomics analysis of planarian ESTs. Proceedings of the National Academy of Sciences of the United States of America 2003;100(13) 7666-7671.

[15] Nishimura K, Kitamura Y, Inoue T, Umesono Y, Sano S, Yoshimoto K, Inden M, Takata K, Taniguchi T, Shimohama S, Agata K. Reconstruction of dopaminergic neural network and locomotion function in planarian regenerates. Developmental Neurobiology 2007;67(8) 1059-1078.

[16] Nishimura K, Kitamura Y, Inoue T, Umesono Y, Yoshimoto K, Takeuchi K, Taniguchi T, Agata K. Identification and distribution of tryptophan hydroxylase (TPH)-pos-

itive neurons in the planarian *Dugesia japonica.* Neuroscience Research 2007;59(1) 101-106.

[17] Nishimura K, Kitamura Y, Umesono Y, Takeuchi K, Takata K, Taniguchi T, Agata K. Identification of glutamic acid decarboxylase gene and distribution of GABAergic nervous system in the planarian, *Dugesia japonica.* Neuroscience 2008;153(4) 1103-1114.

[18] Nishimura K, Kitamura Y, Inoue T, Umesono Y, Taniguchi T, Agata K. Characterization of tyramine ß-hydroxylase in planarian *Dugesia japonica*: Cloning and expression. Neurochemistry International 2008;53(6-8) 184-192.

[19] Nishimura K, Kitamura Y, Taniguchi T, Agata K. Analysis of motor function modulated by cholinergic neurons in planarian *Dugesia japonica.* Neuroscience 2010;168(1) 18-30.

[20] Agata K, Umesono Y. Brain regeneration from pluripotent stem cells in planarian. Philosophical Transactions of the Royal Society of London Series B, Biological Science 2008;363(1500) 2071-2078.

[21] Takeda H, Nishimura K, Agata K. Planarians maintain a constant ratio of different cell types during changes in body size by using the stem cell system. Zoological Science 2009;26(12) 805-813.

[22] Nishimura K, Inoue T, Yoshimoto K, Taniguchi T, Kitamura Y, Agata K. Regeneration of dopaminergic neurons after 6-hydroxydopamine-induced lesion in planarian brain. Journal of Neurochemistry 2011;119(6) 1217-1231.

[23] Nishimura K, Kitamura Y, Agata K. Regeneration of brain and dopaminergic neurons utilizing pluripotent stem cells: lessons from planarians. In: Tao S. (Ed.) Neural stem cells and Therapy Rijeka: InTech; 2012. p141-158.

[24] Ogawa K, Ishihara S, Saito Y, Mineta K, Nakazawa M, Ikeo K, Gojobori T, Watanabe K, Agata K. Induction of a *noggin*-like gene by ectopic DV interaction during planarian regeneration. Developmental Biology 2002;250(1) 59-70.

[25] Tasaki J, Shibata N, Sakurai T, Agata K, Umesono Y. Role of c-Jun N-terminal kinase activation in blastema formation during planarian regeneration. Development Growth and Differentiation 2011;53(3) 389-400.

[26] Tasaki J, Shibata N, Nishimura O, Itomi K, Tabata Y, Son F, Suzuki N, Araki R, Abe M, Agata K, Umesono Y. ERK signaling controls blastema cell differentiation during planarian regeneration. Development 2011;138(12) 2417-2427.

[27] Umesono Y, Tasaki J, Nishimura K, Inoue T, Agata K. Regeneration in an evolutionarily primitive brain -the planarian *Dugesia japonica* model-. European Journal of Neuroscience 2011;34(6) 863-869.

[28] Ying QL, Wray J, Nichols J, Batlle-Morera L, Doble B, Woodgett J, Cohen P, Smith A. The ground state of embryonic stem cell self-renewal. Nature 2008;453(7194) 519-523.

[29] Gurley KA, Rink JC, Sánchez Alvarado A. ß-catenin defines head versus tail identity during planarian regeneration and homeostasis. Science 2008;319(5861) 323-327.

[30] Petersen CP, Reddien PW. Smed-*ß*-catenin-1 is required for anteroposterior blastema polarity in planarian regeneration. Science 2008;319(5861) 327-330.

[31] Molina MD, Neto A, Maeso I, Gómez-Skarmeta JL, Saló E, Cebrià F. Noggin and noggin-like genes control dorsoventral axis regeneration in planarians. Current Biology 2011;21(4) 300-305.

[32] Gaviño MA, Reddien PW. A Bmp/Admp regulatory circuit controls maintenance and regeneration of dorsal-ventral polarity in planarians. Current Biology 2011;21(4) 294-299.

[33] Kiecker C, Niehrs C. A morphogen gradient of Wnt/ß-catenin signalling regulates anteroposterior neural patterning in *Xenopus*. Development 2001;128(21):4189-4201.

[34] De Robertis EM, Kuroda H. Dorsal-ventral patterning and neural induction in *Xenopus* embryos. Annual Review of Cell and Developmental Biology 2004;20 285-308.

[35] Endo T, Yoshino J, Kado K, Tochinai S. Brain regeneration in anuran amphibians. Development Growth and Differentiation 2007;49(2) 121-129.

[36] Okamoto M, Ohsawa H, Hayashi T, Owaribe K, Tsonis PA. Regeneration of retinotectal projections after optic tectum removal in adult newts. Molecular Vision 2007;15(13) 2112-2118.

[37] Parish CL, Beljajeva A, Arenas E, Simon A. Midbrain dopaminergic neurogenesis and behavioural recovery in a salamander lesion-induced regeneration model. Development 2007;134(15) 2881-2887.

[38] Berg DA, Kirkham M, Beljajeva A, Knapp D, Habermann B, Ryge J, Tanaka EM, Simon A. Efficient regeneration by activation of neurogenesis in homeostatically quiescent regions of the adult vertebrate brain. Development 2010;137(24) 4127-4134.

[39] Berg DA, Kirkham M, Wang H, Frisén J, Simon A. Dopamine controls neurogenesis in the adult salamander midbrain in homeostasis and during regeneration of dopamine neurons. Cell Stem Cell 2011;8(4) 426-433.

[40] Doetsch F, García-Verdugo JM, Alvarez-Buylla A. Cellular composition and three-dimensional organization of the subventricular germinal zone in the adult mammalian brain. Journal of Neuroscience 1997;17(13) 5046-5061.

[41] Eriksson PS, Perfilieva E, Björk-Eriksson T, Alborn AM, Nordborg C, Peterson DA, Gage FH. Neurogenesis in the adult human hippocampus. Nature Medicine 1998;4(11) 1313-1317.

[42] Zhao M, Momma S, Delfani K, Carlen M, Cassidy RM, Johansson CB, Brismar H, Shupliakov O, Frisén J, Janson AM. Evidence for neurogenesis in the adult mammalian substantia nigra. Proceedings of the National Academy of Sciences of the United States of America 2003;100(13) 7925-7930.

[43] Frielingsdorf H, Schwarz K, Brundin P, Mohapel P. No evidence for new dopaminergic neurons in the adult mammalian substantia nigra. Proceedings of the National Academy of Sciences of the United States of America 2004;101(27) 10177-10182.

[44] Kawasaki H, Mizuseki K, Nishikawa S, Kaneko S, Kuwana Y, Nakanishi S, Nishikawa SI, Sasai Y. Induction of midbrain dopaminergic neurons from ES cells by stromal cell-derived inducing activity. Neuron 2000;28(1) 31-40.

[45] Kawasaki H, Suemori H, Mizuseki K, Watanabe K, Urano F, Ichinose H, Haruta M, Takahashi M, Yoshikawa K, Nishikawa S, Nakatsuji N, Sasai Y. Generation of dopaminergic neurons and pigmented epithelia from primate ES cells by stromal cell-derived inducing activity. Proceedings of the National Academy of Sciences of the United States of America 2002;99(3) 1580-1585.

[46] Doi D, Morizane A, Kikuchi T, Onoe H, Hayashi T, Kawasaki T, Motono M, Sasai Y, Saiki H, Gomi M, Yoshikawa T, Hayashi H, Shinoyama M, Refaat MM, Suemori H, Miyamoto S, Takahashi J. Prolonged maturation culture favors a reduction in the tumorigenicity and the dopaminergic function of human ESC-derived neural cells in a primate model of Parkinson's disease. Stem Cells 2012;30(5) 935-945.

[47] Ye W, Shimamura K, Rubenstein JL, Hynes MA, Rosenthal A. FGF and Shh signals control dopaminergic and serotonergic cell fate in the anterior neural plate. Cell 1998;93(5) 755-766.

[48] Kittappa R, Chang WW, Awatramani RB, McKay RD. The foxa2 gene controls the birth and spontaneous degeneration of dopamine neurons in old age. PLoS Biology 2007;5(12) e325.

[49] Ono Y, Nakatani T, Sakamoto Y, Mizuhara E, Minaki Y, Kumai M, Hamaguchi A, Nishimura M, Inoue Y, Hayashi H, Takahashi J, Imai T. Differences in neurogenic potential in floor plate cells along an anteroposterior location: midbrain dopaminergic neurons originate from mesencephalic floor plate cells. Development 2007;134(17) 3213-3225.

[50] Andersson E, Tryggvason U, Deng Q, Friling S, Alekseenko Z, Robert B, Perlmann T, Ericson J. Identification of intrinsic determinants of midbrain dopamine neurons. Cell 2006;124(2) 393-405.

[51] Ferri AL, Lin W, Mavromatakis YE, Wang JC, Sasaki H, Whitsett JA, Ang SL. Foxa1 and Foxa2 regulate multiple phases of midbrain dopaminergic neuron development in a dosage-dependent manner. Development 2007;134(15) 2761-2769.

[52] Nunes I, Tovmasian LT, Silva RM, Burke RE, Goff SP. Pitx3 is required for development of substantia nigra dopaminergic neurons. Proceedings of the National Academy of Sciences of the United States of America 2003;100(7) 4245-4250.

[53] Saucedo-Cardenas O, Quintana-Hau JD, Le WD, Smidt MP, Cox JJ, De Mayo F, Burbach JP, Conneely OM. Nurr1 is essential for the induction of the dopaminergic phenotype and the survival of ventral mesencephalic late dopaminergic precursor neurons. Proceedings of the National Academy of Sciences of the United States of America 1998;95(7) 4013-4018.

[54] Chambers SM, Fasano CA, Papapetrou EP, Tomishima M, Sadelain M, Studer L. Highly efficient neural conversion of human ES and iPS cells by dual inhibition of SMAD signaling. Nature Biotechnology 2009;27(3) 275-280.

[55] Morizane A, Doi D, Kikuchi T, Nishimura K, Takahashi J. Small-molecule inhibitors of bone morphogenic protein and activin/nodal signals promote highly efficient neural induction from human pluripotent stem cells. Journal of Neuroscience Research 2011;89(2) 117-126.

[56] Kriks S, Shim JW, Piao J, Ganat YM, Wakeman DR, Xie Z, Carrillo-Reid L, Auyeung G, Antonacci C, Buch A, Yang L, Beal MF, Surmeier DJ, Kordower JH, Tabar V, Studer L. Dopamine neurons derived from human ES cells efficiently engraft in animal models of Parkinson's disease. Nature 2011;480(7378) 547-551.

[57] Watanabe K, Kamiya D, Nishiyama A, Katayama T, Nozaki S, Kawasaki H, Watanabe Y, Mizuseki K, Sasai Y. Directed differentiation of telencephalic precursors from embryonic stem cells. Nature Neuroscience 2005;8(3) 288-296.

[58] Chen G, Gulbranson DR, Hou Z, Bolin JM, Ruotti V, Probasco MD, Smuga-Otto K, Howden SE, Diol NR, Propson NE, Wagner R, Lee GO, Antosiewicz-Bourget J, Teng JM, Thomson JA. Chemically defined conditions for human iPSC derivation and culture. Nature Methods 2011;8(5):424-429.

[59] Brundin P, Strecker RE, Lindvall O, Isacson O, Nilsson OG, Barbin G, Prochiantz A, Forni C, Nieoullon A, Widner H, Gage FH, Björklund A. Intracerebral grafting of dopamine neurons. Experimental basis for clinical trials in patients with Parkinson's disease. Annals of the New York Academy of Sciences 1987;495 473-496.

[60] Takagi Y, Takahashi J, Saiki H, Morizane A, Hayashi T, Kishi Y, Fukuda H, Okamoto Y, Koyanagi M, Ideguchi M, Hayashi H, Imazato T, Kawasaki H, Suemori H, Omachi S, Iida H, Itoh N, Nakatsuji N, Sasai Y, Hashimoto N. Dopaminergic neurons generated from monkey embryonic stem cells function in a Parkinson primate model. The Journal of Clinical Investigation 2005;115(1) 102-109.

[61] Brederlau A, Correia AS, Anisimov SV, Elmi M, Paul G, Roybon L, Morizane A, Bergquist F, Riebe I, Nannmark U, Carta M, Hanse E, Takahashi J, Sasai Y, Funa K, Brundin P, Eriksson PS, Li JY. Transplantation of human embryonic stem cell-de-

rived cells to a rat model of Parkinson's disease: effect of *in vitro* differentiation on graft survival and teratoma formation. Stem Cells 2006;24(6) 1433-1440.

[62] Hargus G, Cooper O, Deleidi M, Levy A, Lee K, Marlow E, Yow A, Soldner F, Hockemeyer D, Hallett PJ, Osborn T, Jaenisch R, Isacson O. Differentiated Parkinson patient-derived induced pluripotent stem cells grow in the adult rodent brain and reduce motor asymmetry in Parkinsonian rats. Proceedings of the National Academy of Sciences of the United States of America 2010;107(36) 15921-15926.

[63] Kikuchi T, Morizane A, Doi D, Onoe H, Hayashi T, Kawasaki T, Saiki H, Miyamoto S, Takahashi J. Survival of human induced pluripotent stem cell-derived midbrain dopaminergic neurons in the brain of a primate model of Parkinson's disease. Journal of Parkinson's Disease 2011;1 395-412.

[64] Okano H, Imai T, Okabe M. *Musashi*: a translational regulator of cell fate. Journal of Cell Science 2002;115(Pt 7) 1355-1359.

[65] Higuchi S, Hayashi T, Tarui H, Nishimura O, Nishimura K, Shibata N, Sakamoto H, Agata K. Expression and functional analysis of *musashi*-like genes in planarian CNS regeneration. Mechanisms of Development 2008;125(7) 631-645.

[66] Agata K, Nakajima E, Funayama N, Shibata N, Saito Y, Umesono Y. Two different evolutionary origins of stem cell systems and their molecular basis. Seminars in Cell and Developmental Biology 2006;17(4) 503-509.

[67] Sharif J, Endoh M, Koseki H. Epigenetic memory meets G2/M: to remember or to forget? Developmental Cell 2011;20(1) 5-6.

[68] Watanabe K, Kamiya D, Nishiyama A, Katayama T, Nozaki S, Kawasaki H, Watanabe Y, Mizuseki K, Sasai Y. Directed differentiation of telencephalic precursors from embryonic stem cells. Nature Neuroscience 2005;8(3) 288-296.

[69] Eiraku M, Watanabe K, Matsuo-Takasaki M, Kawada M, Yonemura S, Matsumura M, Wataya T, Nishiyama A, Muguruma K, Sasai Y. Self-organized formation of polarized cortical tissues from ESCs and its active manipulation by extrinsic signals. Cell Stem Cell 2008;3(5) 519-532.

[70] Kirkeby A, Grealish S, Wolf DA, Nelander J, Wood J, Lundblad M, Lindvall O, Parmar M. Generation of regionally specified neural progenitors and functional neurons from human embryonic stem cells under defined conditions. Cell Reports 2012 28;1(6) 703-714.

Systemic Neural Stem Cell-Based Therapeutic Interventions for Inflammatory CNS Disorders

Matteo Donegà, Elena Giusto, Chiara Cossetti and Stefano Pluchino

Additional information is available at the end of the chapter

http://dx.doi.org/10.5772/55426

1. Introduction

Regenerative processes occurring under physiological (*maintenance*) [1-3] and pathological (*reparative*) [4-6] conditions are a fundamental part of life, and vary greatly among different species, individuals, and tissues. Despite the central nervous system (CNS) has been considered for years as a perennial tissue, it has recently become clear that both physiological and reparative regeneration occur also within the CNS to sustain tissue homeostasis and repair. Importantly, the proliferation and differentiation of endogenous neural stem cells (NSCs) residing within the healthy CNS, or surviving injury, are considered crucial in sustaining these events. However, these processes are not robust enough to promote a functional and stable recovery of the nervous system architecture. Thus, the development of cell-based therapies designed to promote functional (direct *vs.* indirect) neural cell replacement was anticipated [7]. Nevertheless, most of the experimental cell therapies with neural lineage-committed progenitors have failed to foster substantial repair in disease models where the anatomical and functional damage is widespread and an inflamed and/or degenerative microenvironment coexists. Conversely, the systemic injection of *in vitro* expanded neural stem/precursor cells (NPCs) – both as neurospheres as well as plastic-adherent monolayers - has provided a remarkable amelioration of the clinico-pathological features of rodents affected by experimental inflammatory CNS disorders that include experimental autoimmune encephalomyelitis (EAE), cerebral ischemic/haemorrhagic stroke, spinal cord injury (SCI) and traumatic brain injury (TBI). This has been shown to be dependent on the capacity of transplanted NPCs to engage multiple mechanisms of action within specific microenvironments *in vivo* [8]. Among a wide range of potential therapeutic actions – and in addition to the expected cell replacement – this phenomenon may also occur via several *bystander effects*. These effects are heterogeneous and likely exerted by undifferentiated NPCs releasing immune regulatory and neuroprotective molecules within specific microenvironments in response to local stimuli elicited by inflammatory cells (*therapeutic plasticity*). The molecular and cellular mechanism(s) that sustain the multifaceted therapeutic plasticity of NPCs remain far from being fully characterized [9].

The transplantation of undifferentiated exogenous NPCs very efficiently protects the CNS from experimental chronic degeneration induced by inflammation both in small rodents (mice and rats) [10-14] as well as in non-human primates [15]. Specific homing of systemically injected NPCs is shown, so far, in experimental models of multiple sclerosis (MS), ischemic/ haemorrhagic stroke, SCI and TBI, and epilepsy. *In vitro* and *in vivo* data provide extensive evidence of the molecular mechanisms behind the ability of NPCs to cross the blood-brain barrier (BBB) and specifically accumulate at the sites of inflammation/tissue damage [16-18]. After entering the CNS using constitutively functional cell adhesion molecules and inflammatory chemokine receptors, systemically injected NPCs accumulate at the level of perivascular CNS areas, where they establish *atypical ectopic perivascular niches* [16, 19]. In these areas, a much likely active cell-to-cell communication takes place between transplanted NPCs and the different cells of the *atypical niche*. As consequence of this, transplanted NPCs survive while displaying undifferentiated features, and promote neuroprotection by releasing immune modulatory molecules and neurotrophic factors *in situ*. Further evidence exists about an additional peripheral immune-modulatory effect exerted by NPCs [20, 21]. Systemically injected NPCs, in fact, enter also peripheral organs (e.g. draining lymph nodes and spleen) were they accumulate at the boundaries of blood vessels and interact closely with lymphocytes and professional antigen presenting cells (APCs), impairing their maturation and functional activation [15, 22, 23].

NPC-based therapies have been therefore considered a plausible alternative strategy for the treatment of neurological inflammatory disorders. However, some urgent and still unclear questions have to be solved prior to straightforwardly translate most of these exciting experimental observations into clinical medicines, such as: (i) the ideal stem cell source, whether it has to be derived from pluripotent or multipotent sources; (ii) the ideal route of cell administration, whether it has to be focal or systemic; (iii) the optimal time point for cell administration, depending on the disease characteristics; (iv) the ideal balance between differentiation and persistence of stem cells into the targeted tissue and (v) the ideal mechanism of tissue repair to foster, whether it has to be cell replacement or tissue protection/healing. Further, while some encouraging efforts are being devoted towards the development of guidelines and establishment of explorative phase I clinical trials, still one of the major constraints to the easy translation into human medicines is represented by the immunogenicity of allogeneic stem cells, and the modest expandability of somatic human NPCs *in vitro*. Within this scenario, the emerging figure of induced pluripotent stem (iPS) cells [24], induced neuronal (iN) cells [25] and/or induced neural stem cells (iNSCs) [26] holds a new exciting promise.

In this chapter we will describe the most recent evidence of the remarkable therapeutic plasticity of transplanted NPCs, when injected systemically in inflammation-driven CNS degeneration experimental models. We will first focus on the evidence that inspired the modern stem cell experimental therapies and then elaborate on the mechanisms regulating the cross talk between somatic NPCs and the dysfunctional microenvironment, both at the outer and inner endothelial sides, and their clinico-pathological impact. Finally, we will discuss the rationale of the most recent explorative trials that are bringing neural stem cell therapies into the clinic.

2. Adult neural stem cells

2.1. A change in the dogma

Stem cells (SCs) possess the unique ability to self-renew and differentiate into different cell types in the body. Their contribution is essential during embryonic and early post-natal life, where they regulate morphogenesis and development by properly balancing proliferation and differentiation. Though their number is destined to decrease with time, their presence in adult organisms is still required to ensure *homeostasis* and *repair*. While the regenerating properties of some tissues (e.g., the skin) and organs (e.g., the liver) are undisputed, the brain with its unique organization and complexity was considered for long time an exception. In fact, the dogmatic concept '*no new neurons after birth*' (1913) expressed by Santiago Ramon y Cajal, sustaining the limitation of neurogenesis to prenatal life, has been resonating for decades within the scientific community, finally becoming an established belief. NSCs were thought to be present within the brain only during the developmental stage. It was only in the late 60's, thanks to the availability of new techniques and advanced tools of investigation, that the picture of the brain as an immutable organ started to be reviewed. Altman and colleagues, using (^{3}H)-thymidine pulses and autoradiographs, first demonstrated the presence of proliferating neurons in different regions of the post-natal brain in rats [1, 2, 27]. However, the turning point was marked later in 1983 when Goldman and Nottebohm at Rockfeller University (USA) described newly generated neurons at the level of the hyperstriatum ventrale, pars caudalis (HVc), of the ventricular zone in intact adult female canaries [3] Subsequently, numerous pioneering experiments contributed in demonstrating that specific regions of the mammalian CNS undergo a continuous, though moderate, level of neurogenesis throughout adult life [28].

2.2. Adult neurogenesis in physiological conditions

Today it is widely accepted that in the adult mammalian brain, newly generated cells derive from NSCs residing in two regions [29], the ventricular-subventricular zone (V-SVZ) of the forebrain lateral ventricles [2, 27, 30] and the subgranular zone (SGZ) of the dentate gyrus (DG) of the hippocampus [1, 31, 32] (Figure 1). Because of the peculiar cellular organization and exclusive microenvironment, these neurogenic regions are commonly referred to as *germinal-like niches* [33, 34]. Although different, these two areas share an extremely organized and specialized microenvironment where NSCs can strategically interact with a rich vascular plexus [35, 36], while communicating with their progeny and neighbouring NSCs as well as with differentiated neural cells trough specialized structures (e.g. primary cilium, basal and apical processes). Altogether, these cellular components provide a unique milieu of extracellular matrix proteins and growth factors other than electrical stimuli, which define the dynamic characteristic of the adult brain stem cell *niches*. In here, a strictly regulated balance between proliferation and differentiation of NSCs ensure the maintenance of a constant, though quantitatively modest, pool of progenitor cells throughout lifetime [37].

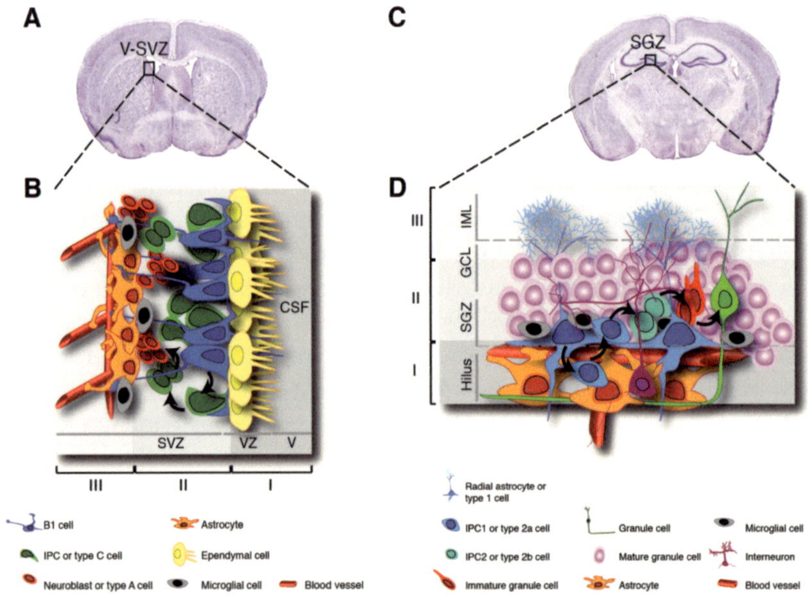

Figure 1. Schematic representations of the adult V-SVZ and SGZ neurogenic compartments.A and **C**, coronal sections of the adult mouse brain showing the localization of the V-SVZ and SGZ of the hippocampus. **B** and **D**, cytoarchitecture of the V-SVZ (**B**), and of the SGZ of the DG of the hippocampus (**D**) in the adult mammalian brain. **B**, Composition of the B1 cell domain into the V-SVZ. NSCs or type B1 cells (blue) extend from the proximal domain (domain I, dark grey) to the distal domain (domain III, light grey). At the level of the ventricles, B1 cells contact the CSF with their primary cilium extruding in the centre of a rosette of multi-ciliated ependymal cells (yellow), forming the typical pinwheel-like structures on the ventricular surface. Here, NSCs can sense different signals circulating into the CSF. In the distal domain, type B1 cells contact the blood vessels (red) with their specialized end-foot terminations. In the intermediate domain (or domain II) type B1 cells give rise to IPCs (or type C cells, green), which are transit-amplifying cells generating neuroblasts (or type A cells, red). In this domain they are also in contact with their progeny, neighbouring cells and neuronal terminations. **D**, Composition of the RA domain at the level of the DG of the SGZ. RAs (or type 1 cells, blue) extend from the hilus of the hippocampus (domain I, dark gray) to the IML (distal domain or domain III, light gray). At the level of domain I, RAs sense the hilus microenvironment with their primary cilium and contact other RAs, IPCs and blood vessels (red). RAs extend, trough their main shaft, into the distal domain where their arborisations receive signals from glial cells and neuronal terminations. RAs give rise to IPCs that mature (trough blue IPC1 or type 2a cells, and light green IPC2 or type 2b cells) and differentiate into immature granule cells (IGC, red). During their maturation, IPCs move from the proximal domain to the intermediate domain (or domain II, composed by SGZ and GCL), where RAs receive signals from the progeny, neighbouring NSCs, interneurons (purple) and microglia (grey). Finally IGC differentiate into mature GC (green), which extend their axons into the hilus and arbores dendrites into the distal domain. Only few new-born neurons survive and become a long-lasting GC (pink).

2.2.1. Defining the cellular composition of the V-SVZ

The V-SVZ is situated in proximity of the lateral ventricles and contains slow-cycling SCs with astroglial properties that express glial-fibrillary acidic protein (GFAP), called type B1 cells. These cells give rise to intermediate progenitor cells (IPCs) or type C cells, which lose GFAP immunoreactivity and acquire the expression of the distal-less homeobox (Dlx)-2. These cells finally give origin to a pool of neuroblasts (type A cells) expressing the polysialylated form of neural cell adhesion molecule (PSA-NCAM) and the early neuronal marker doublecourtin (DCX) [38]. Within rodent's brain these neuroblasts form chains of migration along the rostral migratory stream (RMS) to reach the olfactory bulb (OB), where they terminally differentiate into at least six different subtypes of OB interneurons, depending on their origin along the axes of the V-SVZ [39-41]. The V-SVZ niche (Figure 1 A-B) can be divided in three differently organized domains where self-renewing B1 cells receive different signals: *proximal* (or apical, I), *intermediate* (II) and *distal* (or basal, III) [37]. Type B1 cells retain the typical apical-basal bi-polarity of their embryonic predecessors (radial glia) [42] extending their processes along the three different domains and spanning the cerebrospinal fluid (CSF) and the blood stream. In the proximal domain (composed by VZ and part of the SVZ) Type B1 cells are enclosed within a cluster of ependymal (type E) cells, which sense the CSF by means of motile cilia and create an appropriate gradient of molecules within the VZ [43]. Type B1 cells are therefore physically separated from the ventricles. Nevertheless, their contact with the CSF is still made possible by a single apical primary cilium extruding in the centre of a rosette of type E cells. Typically, these apical end-foot terminations cluster together to finally arise in the middle of a layer of E cells forming a characteristic *pinwheel structure* resembling the embryonic forebrain germinal zone [36, 42]. Recently, it has been shown that the expression of the adhesion and signalling molecule vascular cell adhesion molecule (VCAM)-1 is critical for the correct positioning of these protrusions and the preservation of this complex structure [44]. The small apical surface of B1 cells gives them the chance to sense the CSF which contains soluble factors, such as insulin-like growth factor (IGF)-2, bone morphogenetic proteins (BMPs) and Noggin, Wnts, Sonic hedgehog (Shh) and retinoic acid, able to modulate NSCs behaviour [45]. At the same time a long basal process from the opposite pole (distal domain), bridges B1 cells to the surrounding vascular plexus that runs in the parenchymal side of the V-SVZ. Here, with a specialized end-foot termination, type B1 cells contact endothelial cells (ECs) of the blood vessels, thus being influenced from soluble factors released from ECs and/or possibly by molecules produced far away from the niche and released in the blood stream. The intermediate domain (composed by the SVZ) contains B1 cell progeny, such as IPCs and neuroblasts, which participate in the maintenance of the niche equilibrium perhaps trough mechanisms of direct feedback on NSCs providing information about the number of new neurons already generated. This balance, seems to be regulated on one side by canonical Notch signalling trough ligands released or expressed by both IPCs and neighbouring B1 cells [46, 47] and on the other side by neurotransmitters [e.g. gamma-aminobutyric acid (GABA)] secreted by neuroblasts [48]. Importantly, while many studies have focussed on the role of the microenvironment on the functionality of NSCs [49], much less is known about the role that NSCs themselves exert on the definition of the niche. Recently it has been shown that NSCs in the germinal niches do secrete a multitude of factors, among which some with immune modulatory potentials that may influence the behaviour of the surrounding cells, including microglia [50]. In parallel to the rodent CNS, the lateral wall of the lateral ventricles (and the hippocampus) of the human brain contains NSCs that generate new neurons throughout adult life [51-53]. A total of four layers have been observed forming the human lateral ventricular wall, which comprise a monolayer of ependymal cells, a hypocellular gap, a ribbon of astrocytes, and a transitional zone into the brain parenchyma [52, 54]. Unlike the rodent and non-human primate brain [55], SVZ astrocytes of the human brain are separated from the ependyma by a hypocellular gap [52]. The presence of prominent neurogenesis in the V-SVZ as well as of a RMS of migrating neuroblasts in the human brain has been, however, intensively debated (for a preview, see [56]). Initially it was reported the existence of a ribbon of astrocytes in the adult human V-SVZ that function as multipotent NSCs in culture although, only few proliferating cells and no evidence of chains of migratory (β-III tubulin positive) immature neurons were observed [52]. In contrast, a later report evidenced a robust cell proliferation in adult human V-SVZ and the presence of a RMS of neuroblasts along a lateral ventricular extension that connects the lateral ventricle to the OB [51]. Finally, two recent studies have provided evidence of a small ventricular lumen connecting the lateral ventricles to the OB that is observed only in the foetal [57], but not adult, human brain [55, 58]. Interestingly, the absence of this ventricular extension has been confirmed even in the postnatal infant human brain [58], whereas a new medial migratory stream (MMS) targeting the prefrontal cortex has been observed. Altogether these findings suggest a dynamic evolution in human SVZ neurogenesis throughout life; with the infant human SVZ, RMS and MMS activity, undergoing a progressive extinction at ages older than 18 months post-natal [58].

2.2.2. Defining the cellular composition of the SGZ

The second putative progenitor cell compartment is located in the SGZ of the DG of the hippocampus (Figure 1C-D), namely the region of the brain involved in learning and memory [1, 31, 32]. In this area, NSCs residing at the interface of the hilus and dentate gyrus are called type-1 progenitors or radial astrocytes (RAs) [59] and they mainly correspond to astroglial cells [60]. They mature in dentate granule cells and migrate towards the granule cell layer (GCL) to finally integrate into hippocampal circuitry [59]. RAs, unlike B1 cells of the V-SVZ, are found deeper into the brain parenchyma, surrounded by neurons, neighbouring RAs and other glial cells but without any chance to contact the CSF [37]. However, B1 cells and RAs share some key features: they both express astroglial markers, have ultrastructural characteristics of astrocytes [41] and possess long processes reaching different compartments of the niche far away from where the cell bodies reside [37]. RAs function as the primary precursors for the generation of new dentate granule neurons, either directly or via the generation of IPC1 (type 2a cells) and IPC2 (type 2b cells) [61]. Similarly to the V-SVZ, also the SGZ can be subdivided in a proximal, intermediate and distal domain along which RAs, with their polarized structure (apical-basal), span from the hilus interface (proximal domain) to the inner molecular layer (IML, distal domain) [37]. The proximal domain contains the primary cilium (important for Sonic hedgehog (Shh) signalling), which sense the hilus microenvironment, and lateral processes contacting other RAs and IPCs and, importantly, blood vessels. Here ECs release vascular endothelial growth factor (VEGF), IGF and brain-derived neurotrophic factor (BDNF) responsible for the regulation of the balance between proliferation and differentiation. RAs have their cell bodies in the SGZ and extend their main shaft along the GCL, which compose the intermediate domain. In this area astrocytes receive inputs from their progeny, including immature and mature granule neurons, IPCs and different neuronal and glial (e.g. microglia) cell types. Type 2a cells expressing Achaete-scute homolog (Ascl)-1 (also known as Mash-1) - a transcription factor important for neuronal commitment - are likely to originate in the proximal domain and then rapidly migrate into the intermediate one, where they divide once before differentiating into type 2b cells that will express DCX [62]. Similarly to V-SVZ, feedback mechanisms from the progeny, such as canonical and non-canonical Notch signalling are responsible for the quiescence of RAs or their transition to IPCs [63, 64]. In the IML, RAs terminate with an elaborate and branched structure contacting glial cells, neuronal processes and synapses. Although the contacts taking place in this area are still not completely understood, it seems probable that the GABAergic and glutamatergic inputs coming from interneurons and mossy cells, are important for the regulation of NSCs [65]. These astrocyte-like cells of CNS germinal areas work as real pacemakers of adult neurogenesis, as they receive internal and external inputs from their main shaft as well as from the end-foot of their radial processes that contact ECs in the V-SVZ [42], or embedded into the molecular layer in the SGZ [41]. However, despite the relatively high rate of neurogenesis, only a minority of new born cells eventually survive, mature and integrate within the existing circuitries at the level of the GCL of the hippocampus [66]. In parallel, postnatal SGZ neurogenesis in the human brain has been demonstrated to occur across the lifespan [55]. Although the role of new born neurons generating in the SGZ is not yet fully understood, increasing evidence suggest a possible role in learning and memory function [55].

3. CNS inflammation effects on endogenous adult NSC niches

3.1. Switching from an immune-privileged to an immune-specialized state

Protection and homeostasis are fundamental keystones for the proper maintenance of the CNS. Hence, brain and spinal cord must be kept under an extreme security state to ensure their fully functionality and, ultimately, the survival of an organism. However, in the past the CNS has been often regarded as an immune *privileged* site, where immune cells were not supposed to enter and interact with cells of the nervous system. This common belief was strongly supported by observations showing lack of lymphatic vessels, absence of parenchymal APCs, low expression of constitutive major histocompatibility complex (MHC) class I and II molecules within the brain parenchyma, as well as poor rejection of transplanted allo- or xeno-graft. In the last decades this historical concept has been extensively revised, and there is now convincing evidence that the CNS is instead an immune *specialized* site, where a complex regimen of immune surveillance does occur under physiological as well as pathological conditions and is essential to guarantee its optimal functionality [67]. It is now clear that cells of both the innate (microglia and monocyte-derived macrophages) and the adaptive (mainly $CD4^+$ cells) immune system are present within the brain parenchyma and exert beneficial effects on adult brain plasticity and neurogenesis, as well as on the spontaneous attempt of the CNS to self-repair following an injury [68].

3.2. Effects of inflammation on neurogenesis

Studies conducted over the last decade have extensively proved that the immune and nervous systems interact by engaging an active bidirectional crosstalk. Indeed, the expression of receptors able to recognize inflammatory mediators released by activated immune cells allows endogenous progenitor cells to increase their proliferation rate and specifically home to the site of inflammation after a trauma. As a consequence, both acute [69, 70] and chronic CNS inflammation [6, 71] has been shown to perturb the anatomical architecture and functional activity of adult germinal niches.

Work on EAE mice, the most widely accepted model of MS, has shown that chronic CNS inflammation in myelin oligodendrocyte glycoprotein $(MOG)_{35-55}$-immunized mice causes a transient decrease in the proliferation rate of both C and B1 type cells and a contemporary increased accumulation of neuroblasts within the V-SVZ [6]. This effect, observed during the peak of the disease, was attributed to cell non-autonomous factors, such as pro-inflammatory (Th1) cytokines [e.g. interferon (IFN)- and its intracellular effector Stat-1]. However, these data contrast with other studies showing how inflammatory demyelination in $MOG^{+/-}$ mice immunized with purified mouse myelin increased proliferation and mobilization of neural progenitor cells from the V-SVZ of adult mice. Surprisingly, while new born cells generated at the level of V-SVZ commonly intended to differentiate into neurons, in response to EAE, these cells were able to generate astrocytes and oligodendrocytes as well, thus suggesting that inflammation can diverge (at least partially) their intrinsic nature [4]. Increased proliferation, measured in terms of BrdU-positive cells, has been found also at the level of the hippocampus both during the acute and chronic phases of the disease in MOG35-55 immunized mice. Similarly to the observed accumulation of neuroblasts in the V-SVZ, autoimmune inflammation leads to increased numbers of immature DCX-positive cells in the DG of the hippocampus [72]. Even though some alterations in the Notch, Wnt/β-catenin, Shh, and BDNF signalling pathways have been observed, their real contribution to the deregulation of hippocampal neurogenesis in the course of chronic autoimmune neuroinflammation needs to be further confirmed [72]. In addition, magnetic resonance imaging (MRI) techniques revealed structural alterations in the hippocampus, evidencing marked hippocampal atrophy [73], which may correlate with deficits in attention, information processing capacity and long-term memory observed in the majority of MS patients. Enhanced proliferation during the acute phase of the disease has been observed in proteolipid protein $(PLP)_{139-151}$-induced relapsing EAE in SJL mice. However, during both the relapsing and chronic phase of the disease, the number of SVZ progenitors cells decreased, without changes in the ultrastructural features of the type B, C or A cells, but accompanied by an impaired maturation of oligodendrocyte progenitor cells (OPCs). This suggests that the chronic activation of glial cells (namely microglia and astrocytes) might be deleterious for the repair potential of endogenous brain stem/progenitor cells. Indeed, minocycline-induced inactivation of microglia during the chronic phase in relapsing-remitting EAE mice was associated with an improvement in the number of proliferating Sox2/Bromodeoxyuridine $(BrdU)^{+}$ neural stem cells [74]. Finally, models of targeted focal EAE, obtained by stereotactic injection of cytokines [e.g. tumor necrosis factor (TNF)-α and INF-γ] in rodents pre-immunized with a sub-clinical amount of myelin peptides, allowed to better analyse the time course effect of auto-immune inflammation in the neurogenic areas. In this experimental model a decreased proliferation in the proximity of the V-SVZ was observed at 3 days, followed by an increase at 7 days after the injection of the cytokines, suggesting a regenerative attempt at the level of the V-SVZ area. Interestingly, the concomitant death of neuroblasts, the decreased type C cell proliferation, and the reduction of type A migrating cells, during the initial phase, might explain the impaired long-term olfactory memory observed by means of behavioural analysis [75]. Altogether, these findings suggest the existence of a compensatory mechanism of the injured brain in its attempt to counteract neuronal injury and disturbed conductivity resulting from T cell attack to the myelin sheaths wrapping the axons, which is among the most accepted causes of EAE and MS [76].

In agreement with what described in animal models, SVZ activation and expansion have been found at the level of periventricular active and chronic active lesions in MS patients, thus suggesting that the repetitive exposure to inflammatory insults does not completely exhaust the proliferative potential of the SVZ [77]. V-SVZ from post-mortem brains shows an altered balance between neurogenesis and gliogenesis, likely related to these inflammation effects within the neurogenic niche of MS patients [78]. Interestingly, the majority of MS patients show deficits in attention, information processing capacity and long-term memory, thus suggesting that neuronal damage in MS can result not only in motor and sensory deficits but also cognitive impairment. In support of these MRI techniques revealed structural alterations in the hippocampus, evidencing marked hippocampal atrophy [73].

Acute events, occurring in non-autoimmune diseases such as stroke, have been similarly proved of giving rise to increased proliferation of endogenous NSCs in the V-SVZ. These cells migrate from the neurogenic niche towards the ischemic boundary regions of the striatum and cerebral cortex, where they differentiate into mature striatal neurons [79-81]. During this (injury-reactive) site-specific homing, newly generated neuroblasts form chain-like structures in association with reactive astrocytes and blood vessels in the striatum, a reminiscence of the embryonic migration of type A cells along the RMS [82, 83]. Initially, this potential self-repair mechanism was supposed to happen only during the acute post-stroke phase. However, subsequent studies showed that stroke-induced neurogenesis is an extensive and long-lasting (up to 2 weeks) event, with continuous production of mature striatal neurons for several months after the insult [84]. Unfortunately, the vast majority of migrating new born neurons die within few weeks after the ischemia, and only few damaged cells (about 0.1%) are replaced by newly generated neurons [85]. Similar evidence of stroke-induced neurogenesis has been reported in post-mortem brains, where new born neurons are present in the ischemic penumbra surrounding cerebral cortical infarcts, preferentially localized in the vicinity of blood vessels [80]. The identification of those factors able to influence NSCs proliferation, homing and survival after stroke may have a great therapeutic impact. Several cytokines and growth factors that may be released by injured cells are thought to play a substantial role in promoting the observed neurogenic response after stroke. Among these, ciliary neurotrophic factor (CNTF) [86], transforming growth factor (TGF)-α [87], VEGF [88], fibroblast growth factor (FGF)-2 [89] and erythropoietin (Epo) [90] have been proposed.

Much less is known about the presence of adult neurogenesis after SCI. Most likely, this can be ascribed to a more diffuse scepticism concerning the existence of stem cells within the spinal cord. Indeed, even if the spinal cord is generally considered a non-neurogenic tissue, multipotent precursors can be isolated and propagated *in vitro* [91, 92]. In addition, spinal neurogenesis has been shown to occur to a limited extent in response to several types of trauma [93-95]. In a very recent study, the modulation of neurogenesis in the more canonical niches of the adult brain has been investigated following SCI in rats. Interestingly, $BrdU^+$ positive cells were found to be significantly decreased both at the level of the V-SVZ and the SGZ in subacute [15 days post injury (dpi)] condition. However, while V-SVZ proliferation returns to normal levels at 90 dpi, this does not happen at the hippocampal level. This could be equally explained by either a higher plasticity in the V-SVZ or a higher sensibility of the SGZ [96]. Alterations in adult neurogenesis have been extensively observed in a multitude of models of other neurodegenerative diseases. Rats suffering by pilocarpine-induced temporal lobe epilepsy (TLE), exhibit increased neurogenesis in the V-SVZ [97] as well as in the SGZ [98] after a period of latency and then it lasts for several weeks following prolonged seizures activity. Further, status epilepticus (SE) seems to accelerate the maturation and integration of adult new born DG cells [99]. However, chronic TLE induces a decrease of neurogenesis, as children affected by frequent seizures show decreased numbers of newly generated neurons and proliferating cells [100]. Impaired (cell type specific) proliferation of V-SVZ as well as SGZ progenitors has been observed also in experimental models of Alzheimer's disease (AD) [101]. While it is not clear whether this is reflected in an increased [102] or decreased neurogenesis [103], mainly because of the high number of different models used [104, 105], a recent study suggested that abnormalities at the level of both the neurogenic niches might precede the onset of amyloid deposition and memory impairments [106]. Interestingly, in post-mortem brains of patients with AD, it has been observed an increase of neurogenesis into the SGZ accompanied by depletion in the V-SVZ [107].

3.3. The double face of inflammation

According to what described, it is suggestive that the CNS is able to start a beneficial, though limited, process of self-repair. However, most of the new born cells generated following injury are destined to die within few weeks, maybe due to a failure in their integration or due to the inflammatory milieu. Even if these cells have been shown to fully differentiate into mature neurons [82], the very low rate of occurrence imposes logical concerns regarding the therapeutic value of this regenerative response to brain injury [81, 108]. Because of the rearrangements occurring in the neurogenic niches after an inflammatory event, the immune system has been accredited as one of the major responsible of this failure. This assumption is further corroborated by the observation that the increasing complexity of the immune system over the evolutionary process has been accompanied by a concomitant loss of regenerative capacity [109]. Also, several findings link inflammation to the pathogenesis of neurodegenerative disorders and anti-inflammatory drugs seem to be promising candidates for their treatment. Recent studies suggest that inflammation may indeed have a neuroprotective effect [110]. Nevertheless, the real effect of inflammation in several of these pathologies still needs to be completely clarified.

The early (acute) post-traumatic phase of a neuroinflammatory process involves the action of resident microglial cells. These cells of myeloid origin are usually present in a resting but dynamic state, ready to shift their activity and undergo morphological and functional transformations in response to any kind of brain damage or injury [111, 112]. While on one side the selective ablation of microglia has been shown to exacerbate the ischemic injury in a mouse model of focal cerebral ischemia [113], on the other side mounting evidence indicates that chronic microglial activation may also contribute to the development and progression of neurodegenerative disorders, mainly through the release of pro-inflammatory cytokines such as interleukin (IL)-1, IL-6, and TNF-α [114]. As an example, infiltrating blood-derived macrophages have been shown to exert a beneficial role in an experimental model of SCI, where they contribute to limit the action of activated resident microglia, whose prolonged presence would finally lead to detrimental consequences [115]. Indeed, when the activity of microglia is not properly contained, their action may lead to a prolonged (chronic) inflammation eventually culminating in the formation of fibrotic tissue.

Endogenous T cells are key components of the protective immunity process. As such, physiological trafficking of lymphocytes through the CNS is required to support the essential function of immune surveillance [116]. Cellular composition analysis of the CSF of healthy patients has revealed that up to 80% of the total number of cells is represented by central memory and effector memory T-cells. This atypical composition, which is very different from the one of the blood, also suggests a major role for the CSF in the defence of the CNS [117]. Indeed, the CSF drains the interstitial fluid of the CNS and brings CNS antigens to the cervical lymph nodes, thus supplying for the absence of a proper lymphatic drainage [118, 119]. Following the seminal observation that self-specific T-cells recognizing myelin basic protein were able to protect injured CNS neurons from secondary degeneration in a rat model of optic nerve crash injury [120], several studies further supported the idea that T cell-dependent autoimmunity might promote recovery from CNS injuries [121, 122]. These studies finally culminated with the idea that boosting T-cell response to CNS antigens by means of immunization with CNS myelin-associated self-antigens could have enhanced this therapeutic potential [123-125]. Also, myelin-reactive T-cells possess neuroprotective effects, which may be essentially attributed to their ability to release neurotrophic factors such as BDNF, nerve growth factor (NGF) and CNTF [126]. Importantly though, auto-reactive T-cells showing this protective effect may turn out to be harmful if escaping the control exerted by the immune system, finally resulting into the development of autoimmune diseases such as MS. Therefore, a strict control is required to finely tune the balance between the *good* and the *bad* [76].

To further complicate the scene, several inflammatory mediators, such as TNF-α, TGF-β, IL-1, IL-6, IL-10 and IL-12 may have contrasting effects (e.g neuroprotective *vs.* neurotoxic) depending on the overall context. As an example, the role of IL-6 is crucial for the induction of EAE [127], and its overexpression exacerbates tissue injury in experimental models of SCI [128, 129]. Also, high levels of this inflammatory marker in the blood of patients undergoing inflammatory response after stroke correlates with the disease severity and poor clinical outcome [130]. Accordingly, the use of monoclonal antibody directed towards IL-6 proved to be beneficial in the treatment of acute SCI and MOG_{35-55}-induced EAE [131, 132]. However,

IL-6 knockout mice showed significantly increased of chronic (but not acute) lesion volumes and worse long-term functional outcome after stroke [133]. This may imply the need of a finely tuned regulation, most likely depending on their precise timing and location other than on the specific nature of the disease. Indeed, while some pathologies such as MS, SCI, stroke, TBI and are characterized by an acute inflammatory event followed by secondary neurodegeneration, others, such as epilepsy, Alzheimer's disease (AD), Parkinson's disease (PD), and Huntington's disease (HD) are instead caused by primary neurodegeneration subsequently leading to secondary reactive inflammation [19].

As described in this paragraph, the CNS is able to regulate the proliferation rate within the adult neurogenic niches as an extreme attempt to respond to damages in both primary and secondary inflammatory neurodegenerative diseases. Nevertheless, this process is not robust enough to effectively re-establish the complex functionality of the CNS. Therefore, protocols aiming at pharmacological manipulation of endogenous precursors from germinal niche(s), *in vivo,* might be therapeutically inefficacious in inflammatory CNS disorders. Thank to the development of protocols allowing *in vitro* growth and large scale-up of brain-derived NPCs [134], innovative therapies, for both acute and chronic CNS inflammatory disorders, based on stem cell transplants have been proposed [7]. Transplantation of adult exogenous NPCs represents, in fact, an alternative, and possibly more efficacious, therapeutic approach that might overcome the limited endogenous repair. Motivated by the ambitious expectation to achieve CNS repair (and/or regeneration) via functional neural cell replacement, many different preclinical studies have evidenced a potential benefit of NPC-based treatments in experimental animal models of several neurological diseases [8].

4. Exogenous NPC-based therapies: The systemic administration

4.1. Systemic injection and functional recovery

Following the seminal observation that systemically delivered NPCs were able to target an intracranial tumour in rodent (both mice and rats) model of experimental brain tumours [135], numerous studies started to investigate the validity of this administration route in a variety of CNS disorders. Over the last decade, data have been provided on the feasibility of systemic NPC transplants via either intravenous (i.v.) cell injection into the blood stream, or interacerebroventricular (i.c.v.)/intrathecal (i.t.) into the CSF, in experimental CNS disease models [10-13, 16, 136-138]. Adult somatic mouse NPCs, administered 22 days post immunization (dpi) greatly reduced the functional impairment observed in chronic MOG_{35-55}-immunized EAE mice [11]. Also, rat NPC neurospheres, administered i.c.v. or i.t. in rats affected by acute EAE, attenuate the clinical symptoms when administered at the same day of disease induction (0 dpi) [136]. Intravenous injected mouse NPCs were proved to be efficacious in $PLP_{139-151}$ immunized relapsing EAE mice [16]. Indeed, mice, treated with NPCs at the disease onset or at the time of the first relapse, recovered faster and showed a decrease in the relapse rate compared to controls. At the end of the follow-up (90 dpi) both treatments resulted in a lower relapsing remitting EAE cumulative score [16]. Finally, MOG_{35-55}-immunized chronic EAE mice receiving either mouse neurospheres (i.c.v) or single cell NPCs (i.v.), at 6 and 8 dpi respectively also showed significant clinical amelioration [20, 139].

The therapeutic efficacy of systemically administered NPCs has been later observed in a different pathological context, such as stroke. Mouse somatic NPCs, systemically transplanted 3 days after middle cerebral artery occlusion (MCAo), resulted in a better recovery, significantly improving the neurological severity score starting from 18 days post transplantation (dpt) until the end of the follow-up (30 dpt) [140]. Similar neurological improvements were observed in rats subjected to MCAo and common carotid artery occlusion (CCAo) after i.t. transplantation (7 days after stroke) of rat NPCs [141]. Significant locomotor recovery was also observed after acute systemic NPC transplant in mice suffering from contusion SCI [13].

Similarly to rodent cells, human NPCs have been proved to be therapeutically efficacious. Foetal NPCs administered either i.v. or i.t. at the disease onset, reduced the severity of MOG_{1-125}-induced chronic EAE in common marmosets [15]. Further, human embryonic stem cell (ESC)-derived NPCs have been shown to reduce disease severity of chronic EAE mice [142]. Human immortalized NPCs have been widely enrolled in stroke models. The systemic injection of the HB1.F3 NPC line resulted in neurological improvements of rats treated 1 day after MCAo or intracerebral hemorrhage (ICH) stroke models [12, 137, 138]. The same line resulted therapeutically effective also in quinolinic acid-induced experimental HD in rats, where NPCs administered intravenously at 7 days after disease induction, significantly ameliorated the behavioural outcome [14].

All these evidences showing behavioural recovery upon systemic injection of NPCs in different CNS inflammatory models, led to the investigation of the molecular mechanisms standing behind, since this capacity bears the hope of developing less invasive surgical techniques to implant therapeutic adult human stem cells into patients affected by highly debilitating CNS disorders, such as MS, stroke, SCI, epilepsy, PD, AD and HD.

4.2. Homing capacity: NPCs breach the CNS barriers

Brain and spinal cord are protected by a complex control system, composed by tight barriers shielding the action of a unique troop of immune cells. Indeed, to access the brain and spinal cord parenchyma, circulating cells have to breach through all the barriers that closely seal the CNS from the surrounding environment. Namely, these are the blood-brain barrier (BBB) at the level of parenchymal capillaries and post-capillary venules, the blood-cerebrospinal fluid barrier (BCSFB) at the level of the choroid plexus in the brain ventricles, and the blood-leptomeningeal barrier (BLMB) at the level of the leptomeningeal/subarachnoid space. The main role of these barriers is to maintain the chemical composition of the CNS microenvironment, thus ensuring the proper functionality of neuronal circuits, synaptic transmission and remodelling, angiogenesis, and neurogenesis in the adult brain, while their rupture is involved in many neuroinflammatory disorders [143]. Because of the existence of such barriers, the access of systemically injected NPCs to the CNS parenchyma seems quite unlikely.

Structurally, the main component of the BBB is represented by specialized ECs characterized by the absence of fenestrae, low pinocytotic activity and by the presence of intercellular tight junctions (TJs) [144]. This clutched arrangement prevents the free passage of molecules, while the transport of nutrients into the CNS and the discard of toxic molecules into the circulation is ensured by active mechanisms, thus guarantying a proper neuronal activity [145]. Moreover, the BBB is an essential constituent of the so-called *neurovascular unit*, a boundary zone defined on one side by the *endothelium basement membrane* (located in the abluminal side of the vasculature) and on the other by the *parenchymal basement membrane*, which establishes the ultimate border between the perivascular space and the CNS parenchyma. In post-capillary venules these two membranes lay in close proximity, leaving just a narrow perivascular space in between, which becomes more significant at the level of arteries and veins. In this area, occasional APCs (leptomeningeal mesothelial cells) reside and play a major role in the immunosurveilance program of the CNS. Finally, the inner and outer sides of the parenchymal basement membrane are juxtaposed to the *glia limitans*, whose crossing seems to be crucial for the effective triggering of a neuroinflammation process [146, 147]. The functionality of the BBB in clinical situations, such as those occurring in some neurodegenerative disorders like MS, ischemic stroke, sub-arachnoid haemorrhage, TBI and AD is markedly reduced, leading to an increased permeability and trafficking of immune cells into the CNS parenchyma [148-152].

Most of the knowledge about the mechanisms that allow circulating cells to breach the barrier(s) [117, 153, 154] and move into the CNS parenchyma comes from observations conducted with models of CNS inflammation. Initial studies showed how intravenously injected radioactively labelled encephalitogenic T cells were able to cross the BBB of healthy recipients [153]. It was also shown that, while activation is mandatory for T-cells to cross the endothelial barrier and reach perivascular spaces, antigen specificity is dispensable to further cross the glia limitans and invade the CNS parenchyma after having encountered the appropriate APCs [155].

The extravasation of specific T cells requires a multistep process [156]. The first step, known as *capture* (in non-inflamed endothelia) or *tethering* (in inflamed endothelia) *and rolling* is represented by an initial, transient contact promoted by the specific interaction between members of the selectin and integrin families expressed by the activated endothelium with their respective ligands on circulating immune cells. It has been shown how the recruitment of inflammatory cells across the BBB involves α4-integrin and its ligands of the immunoglobulin (Ig) superfamily, namely vascular cell adhesion molecule (VCAM)-1 and mucosal addressin cell adhesion molecule (MAdCAM)-1 [147]. Upon this initial contact, circulating cells decrease their initial speed and resist the shear stress created by the blood flow, mainly through endothelial intercellular adhesion molecule (ICAM)-1 and VCAM-1, but not ICAM-2 [157]. Elegant studies have consistently shown that the inhibition of the dimeric $\alpha4\beta1$-integrin and its cognate receptor VCAM-1 on the activated endothelium prevented the accumulation of leukocytes in the CNS and the development of EAE [158]. Interestingly, when the inflammatory process is started, $\alpha4\beta1$-integrin is no more dispensable for T-cell capture or rolling [159].

The following step requires the *firm adhesion* and *crawling* of T-cells along the vascular wall. This is orchestrated by some chemokines and chemoattractants, such as stromal cell-derived factor (SDF)-1α/CXCL12 [160], monocyte chemoattractant protein (MCP)-1/CCL2 [161], regulated and normal T cell expressed and secreted (RANTES)/CCL5 [162], EBI1 ligand chemokine (ELC)/CCL19, secondary lymphoid-tissue chemokine (SLC)/CCL21 released by the inflamed endothelia and CCR7 [163]. By binding to their G-coupled receptor (e.g.: C-X-C chemokine receptor -CXCR- type 4 for SDF-1 and CCR2 for MCP-1), these chemokines transmit an inside out signalling to T-cell surface integrins, which undergo dramatic conformational modifications thus increasing their *avidity* (specificity for the ligand). Once engaged in such a firm adhesion, T-cells need to make their way through the endothelium. To this purpose they start probing the vasculature to find the optimal site to breach the wall. Following adhesion to blood vessel walls, leukocytes undergo a series of actin rearrangements that eventually mark their transition to a more flatten and polarized shape [164]. Finally, T-cells cross the border either by paracellular or transcellular diapedesis or by creating pores through the cells - *transcellular diapedesis* -. While the former require the disassembly of the intercellular junction structure, the latter involves the formation of "cell-in-cell" interactions through the arrangement of docking structures or *transmigratory cups* enriched in ICAM-1 and VCAM-1, which partially embrace migrating leukocytes [165].

A very similar sequential process has been shown being recapitulated when systemically injected NPCs specifically home to the site of damage. In fact, NPCs posses the ability to reach the cerebral parenchyma where they eventually induce recovery in animal models of neurodegenerative diseases such as EAE [10, 11] stroke [12, 166], SCI [167, 168], epilepsy [138, 169], HD [14], other than glioblastoma [135, 170]. The first studies showing the extravasation capacity of NPCs [10-12] clearly demonstrated that this capacity was strictly related to the activation of the ECs by an inflammation process occurring within the brain. NPCs administered either i.c.v. or i.v. to healthy animals were, in fact, never observed inside the CNS, while mainly accumulating in peripheral organs, or remaining confined in the ventricles or subarachnoid space. Only after activation of endothelial cells, exogenous NPCs were observed to accumulate into the CNS. Systemically injected NPCs are, in fact, able to follow a gradient of chemoattractants (e.g. pro-inflammatory cytokines and chemokines) released by the inflammatory lesions into the blood stream and CSF. Following these signals, NPCs rapidly reach the source of pro-inflammatory molecules within and interact with the activated endothelial/ependymal cells around inflamed CNS tissues. At this level, NPCs and endothelial cells start an organized sequence of events resembling those described for T cell extravasation that allow the selective entrance and specific *homing* of transplanted cells in multifocal inflammatory CNS areas [16]. Interestingly enough, only small percentages (between 1-5%) of the systemically administered NPCs actually infiltrate and integrate within the CNS [11, 13, 140]. Mouse SVZ-derived adult NPCs transplanted in a subacute model of brain inflammation were shown to adhere to the CD49d counterligand VCAM-1 [16]. Further *in vivo* experiments showed that migration of mouse NPCs towards the site of damage is dependent on the CXCR4-SDF-1α signalling in mouse models of MS and brain tumour [17, 171]. In stroke models, the up-regulation of VCAM-1 on the surface of endothelial cells facilitates the targeting and the subsequent extravasation of VLA-4 expressing NPCs to the site of injury [18]. In line with this, mouse NPCs sorted via FACS for the presence of VLA-4 revealed a more efficient transendothelial migration in a mouse model of stroke after intracarotid injection [18]. More recently it was shown that also the CCR2/CCL2 interaction is substantially involved in the recruitment of systemically delivered NPCs in a mouse model of stroke [172].

In vitro experiments confirmed that mouse NPCs express many functional receptors on theirs surfaces, among which the $\alpha4$ subunit of the integrin VLA-4 [16], the SDF-1α receptor CXCR4 [173] and CD44, a cell-surface glycoprotein that binds to hyaluronic acid (HA) and is expressed also in activated T cells [174, 175]. Interestingly, NPCs led to the formation of transmigratory cups, enriched in multiple adhesion molecules such as ICAM-1 and VCAM-1, on the surface of endothelial cells [175] as previously shown for T lymphocytes diapedesis [176].

Similarly, also immortalized human NPC lines express CD44 [175] and CXCR4 [173]. However, in a recent study, human NPCs were shown to interact with activated ECs through integrins $\alpha2$, $\alpha6$ and $\beta1$ rather than CXCR4 [177]. Further, human NPCs express the receptors CXCR1 and CXCR5, which mediate their *in vitro* migration across a monolayer of human brain ECs in response to IL-8/CXCL8 and B lymphocyte chemoattractant (BLC)/CXCL13, chemokines previously known to favour the trans-endothelial migration of immune cells [178].

All these evidences suggest that systemically injected mouse and human NPCs share the expression of a variety of functional immune-like receptors, such as functional cell adhesion molecules (e.g. CD44 and VLA-4) and inflammatory chemokine receptors (e.g. CCR2, CCR5 and CXCR4), giving them a unique leukocyte-like molecular signature. This characteristic, allowing NPC interaction with activated endothelial and ependymal cells, represents an essential requirement in the therapeutic paradigm of systemic delivery. Therefore, the discovery of the specific homing ability of NPCs across the BBB opened new frontiers for the treatment of CNS diseases, in particular for those diseases characterized by disseminated damage.

4.3. NPC interaction with the dysfunctional CNS microenvironment: The establishment of ectopic niches

Consistent data exists reporting the ability of i.v. injected NPCs to cross the BBB and accumulate into the CNS. Here, exogenous NPCs co-exist with different host components, such as ECs, infiltrating inflammatory cells, activated macrophages/microglia and reactive astrocytes [19]. In this context, the intimate association with ECs, the physical proximity to the vasculature and the enhanced expression of stem cell regulators and growth factors involved both in angiogenesis and neurogenesis has been described to play a major role in defining a molecular architecture reminiscent of prototypical germinal stem cell niches [16]. Within these *atypical ectopic perivascular niches*, in addition to hierarchical (mother-to-daughter) communication, a sophisticated level of cell-to-cell horizontal communication takes place between transplanted NPCs and resident cells. Recent evidences confirm that NPCs are able to communicate with host cells via cellular contacts. For instance, functional gap junction formation has been shown to allow exogenous NPCs to rescue host neurons and their projections in animal models of Purkinje neurodegeneration. Gap junctions permitted the trans-cellular delivery of homeostasis-modulating molecules, as well as directly influenced the coordinated activity of the host network via Ca^{++} waves. Moreover, hypoxic preconditioning of NPCs before their *in vitro* engraftment increased Connexin 43 expression and improved subsequent communication with host cells [179]. Possible mechanisms of communication include also secretion of growth factors, hormones, cytokines, chemokines and small molecular mediators [180], cell-to-cell interactions via tunnelling nanotubes [181] and secretion of circular membrane vesicles [182], other than cell-to-cell contacts [183].

Correlative evidence suggest that, depending on local inflammatory milieu, transplanted NPCs may either remain in the niche while maintaining an undifferentiated state, or move out from the niche, finally acquiring a terminally differentiated phenotype [16]. When systemically injected in chronic EAE mice, syngenic NPCs were found almost exclusively in areas of CNS damage, mainly within the submeningeal space in close proximity to subpial inflammatory foci (after i.t. stem cell injection), or around post-capillary venules (after i.v. stem cell injection) [11]. Ten days after transplantation, relatively few cells were found in the CNS parenchyma and at 30 dpi many of the surviving donor cells were localized deeply within the brain parenchyma and displayed a marked distribution pattern: most of them were confined within areas of demyelination and axonal loss, and only very few cells were found within regions where the myelin architecture was preserved [11]. Similar results were obtained after i.c.v NPC injection at the peak of EAE in rats: cells entered into the brain or spinal cord parenchyma and mostly accumulated at sites of inflamed white matter but not into adjacent grey matter regions. In line with the previous study, after 2 weeks cells had migrated into distant white matter tracts but, on the contrary, most of them had acquired specific markers of the astroglial and oligodendroglial lineages [184]. Mouse NPC transplants in rodents affected by EAE are also associated with significantly reduced glial scar formation [11] and an increased survival and recruitment of endogenous neural cells participating to the naturally occurring brain reparative response upon myelin damage [10, 15, 16, 185, 186].

Human NPCs have shown a higher rate of cell integration after being administered in different animal models. In particular, the HB1.F3 immortalized cell line, i.v. injected in a model of ischemic stroke, were shown to enter the ischemic area and differentiate into neurons and astrocytes, similarly to what observed with focal injected cells [12, 187]. Transplanted cells seemed to adapt their fate accordingly to the region of engraftment, showing the appropriate neuronal and glial markers. NeuN^{+} and NF^{+} cells were identified primarily in the CA1 area of the hippocampus and in the dentate gyrus, mixed with GFAP^{+} cells. Vimentin, GFAP and NF markers showed a progressive expression during the first 2-3 weeks after transplantation, suggesting a step-by-step maturation of the cells [187]. The very same line of cells, injected in a rat model of ICH [137], was observed to infiltrate the brain, survive and migrate towards the peri-hematomal areas. The cells were found mainly differentiating into GFAP^{+} and NeuN^{+} cells. However, the rapid behavioural recovery observed in ICH rats as soon as 2 weeks after transplantation, suggested that the NPC therapeutic effect was mainly related to neuroprotection, rather than to integration into neuronal circuitry [137, 187], since the latter would require longer time to produce clinical ameliorations. A similar trend towards human NPC differentiation has been observed in animal models of SCI, SE and HD. HB1.F3 hNPCs administered in mice subjected to compression SCI, were observed to differentiate into βIII-tubulin^{+} neurons at 21 days after transplantation [167]. GABA-immunoreactive interneurons were, instead, observed originating from HB1.F3 when systemically administered the day after lithium-pilocarpine induction of experimental SE in rats [138]. Further, HB1.F3 cells injected 7 days after unilateral quinolinic acid (QA)-induced model of HD in rats were found to stay confined around blood vessels, mostly in the damaged hemisphere and only partially differentiating in $GFAP^+$ and $NeuN^+$ cells at 3 weeks post transplantation [14].

Despite these evidences showing the ability of exogenous NPCs to survive and differentiate into multiple derivatives according to local cues, the majority of the data provided has substantially failed to show convincing relevant differentiation and integration of transplanted NPCs *in vivo*. It is now quite evident that NPCs (and more generally somatic adult SCs) might protect the CNS trough mechanisms alternative to direct cell replacement, which imply the interaction of NPCs with both resident neural and immune cells [7, 188]. Cell replacement is therefore only one of the multiple ways by which transplanted NPCs can promote tissue repair, and a much more complex therapeutic scenario should be foreseen. The concept of *stem cell therapeutic plasticity* (or *functional multipotency*) has therefore emerged, describing the different way(s) NPCs use to interact with tissue-resident *vs.* infiltrating immune cells, at the level of the inflammatory tissue context in which they are either transplanted or migrate to after transplantation. These bystander effects, are mainly represented by *neuroprotection,* which might occur both through secretion of soluble factors and cell-to-cell contact interactions and *immunomodulation,* intended as the capacity of NPCs to influence the activity of the immune system in the CNS and/or in the periphery, at the level of secondary lymphoid organs [5, 19].

4.3.1. Tissue trophic effects

NPCs may exert their neuroprotective effect by increasing *in situ* bioavailability of several molecules, such as neurotrophins, growth factors and developmental stem cell regulators, thus promoting the survival and function of endogenous glial and neuronal progenitors that escaped the primary insults [19].

Mouse NPCs systemically injected in mice affected by middle cerebral artery occlusion (MCAo) were observed to mostly maintain an undifferentiated phenotype, while accumulating at the boundaries of the lesioned area [140, 141]. Tissue survival was associated with a down regulation of inflammation, glial scar formation and neuronal apoptotic cell death at both mRNA and protein levels [140]. Increased levels of BDNF, NGF and neurotrophin (NT)-3 were found in the CSF of stroke-affected rats after intra-cisterna magna administration of NPCs. In addition, immunohistochemical analysis of the injured brain revealed an increase of MHC class I levels in treated rats [141]. Interestingly, this neuroprotective effect in the ischemic microenvironment seems to start very soon after the systemic administration of cells. In fact, data have been provided showing an increase in the gene expression levels of IGF-1, VEGF, TGF-1β, BDNF and CXCL12/SDF1-α in the NPC-transplanted MCAo brain, as soon as 24 hours after the acute i.c.v. injection [189]. Further, NPCs have been proved to increase *in vivo* vascularisation when administered after stroke, most likely due to their ability to increase the presence of VEGF, FGF, BDNF and chemoattractant factors (such as SDF-1α), which promote angiogenesis and mobilization of endogenous endothelial progenitors [18, 190]

More recently, adult mouse NPCs systemically injected in mice (3 injections few hours after the injury) suffering from acute contusion SCI, showed an undifferentiated morphology (similarly to what observed in EAE) at the level of the damaged CNS. *Ex vivo* RT-PCR analysis showed NPC-driven up regulation of BDNF, NT-3, NGF, leukemia inhibitory factor (LIF) and TNF- α only at 48h after treatments, while no differences were observed neither at 24h or 7 days after transplantation [13]. Similarly to what observed by indirect evidence *in vivo*, real-time PCR gene expression analysis directly revealed high levels of NGF, BDNF and NT-3 and glial cell line-derived neurotrophic factor (GDNF) in the transcripts of cultured rat NPCs [141, 191]. In addition, in line with the observed pro-angiogeneic effect *in vivo* after transplantation of mouse NPCs in stroke models [18, 190], human NPCs were proved to secrete VEGF *in vitro* [192].

All these evidences suggested that the underlying molecular mechanisms by which transplanted NPCs instruct tissue protection effects are partly related to increased *in vivo* bioavailability of major neurotrophins [11, 138, 139, 193] able to modulate the host environment resulting more permissive to regeneration. Neurotrophins exert important roles as mediators in cell cycle regulation, cell survival, and differentiation during development and adulthood. The delivery of diffusible proteins to the CNS has been seen as a possible therapeutic weapon for neurological diseases. However, because the CNS is likely impenetrable for many of these diffusible proteins, NPCs might be envisaged as carrier of neurotrophic factors. To this aim, NPCs have been genetically modified to act as *Trojan horses* to deliver the desired diffusible molecules at the site of injury, thus fostering their innate capacity to secrete neurotrophic and growth factors [194]. Among all the candidate neurotrophic factors to be delivered, GDNF has shown a potent neuroprotective effect on a variety of neuronal inflammatory models, such as stroke and PD [195-197]. However, its effects are generally transient and need consecutive administrations to obtain long-standing results. NPCs over-expressing GDNF can instead provide durable neuroprotective effects, as shown with mouse NPCs, transplanted i.c.v in rats 3 days after MCAo [198]. The exogenous cells resulted in an overall increase of cell survival of endogenous cells after the insult, which in turn was associated to a partial functional recovery. Interestingly, treated rats also displayed a significant increase of the synaptic proteins synaptophysin and post-synaptic density protein (PSD)-95, suggesting an enhanced neuronal function and a possible reconstruction of endogenous neural circuitries after the grafting [198]. Finally, a recent study showed that the intravenous administration HB1.F3 human NPCs transduced with INF-β and cytosine deaminase (CD), was able to interfere with toll-like-receptor (TLR)-4 (up-regulated into the site of injury) suppressing the SCI-induced proliferation of reactive astrocytes and promoting functional recovery [199]. Other neurotrophic factors, such as BDNF and VEGF, have been over-expressed in NPCs and mainly tested upon intraparenchymal injection in models of SCI [200] or ICH [201, 202]. Taken together, these data suggest that the clinical amelioration observed in CNS disease animal models are, at least in part, mediated by a multilayered NPC neurotrophic signature.

4.3.2. Regulation of the immune system

Considerable evidence of the immune modulatory capacity of NPCs has derived from transplantation studies through different routes in the EAE model. As mentioned, transplanted NPCs are consistently found around inflamed blood vessels, in close contact with both endogenous neural cells (e.g. astrocytes and neurons) and CNS-infiltrating blood-borne $CD45^+$ immune cells [185]. Also, i.c.v. administered NPCs were found to attenuate brain inflammation primarily through a reduction of perivascular infiltrates and $CD3^+$ T cells with a concomitant increase of $CD25^+$ and $CD25^+/CD62L^+$ regulatory T cells [136]. Interestingly, i.v. injection of NPCs also protects against chronic neural tissue loss as well as disease-related disability in EAE, via induction of apoptosis of blood-borne CNS-infiltrating encephalitogenic T cells [185]

NPCs have been shown to possess immune modulatory capacity also in models of stroke, where T cells do not have a major role in the disease pathology. Irrespectively of the route of administration (systemically *vs.* intraparechimally), transplanted NPCs migrate towards the site of infarct in MCAo and ICH models [12, 137, 138, 203-207] and once reached the ischemic boundary zone (IBZ), grafted NPCs interact with the inflammatory environment. The sub-acute (*delayed*) i.v. injection of mouse NPCs after MCAo in mice, significantly down-regulates multiple RNA species involved in inflammation, including IFN-γ, TNF-α, IL-1β, IL-6 and leptin receptor [140]. Therefore, NPCs may exert an immune modulatory action, causing a profound down regulation of inflammatory *lymphoid* (T cells) and *myeloid* (macrophages) cells within inflamed brain areas.

While the inhibition of the T cell responses by NPCs is a quite established concept [208], the effect of the interaction between transplanted NPCs and microglia/macrophages is still controversial, mainly because of the non-univocal data regarding the role sustained by professional phagocytes under CNS inflammatory conditions. *In vivo* studies have in fact produced opposite evidences that might underline once more the bimodal action of some immune regulators [209]. NPC transplantation promote the infiltration of $CD11b^+$ myeloid cells into the brain of MCAo mice, thus suggesting that myeloid cell activation might be required for transplanted NPCs to exert part of their neuroprotective action [189]. Indeed, MCAo mice in which $CD11b^+$ microglia have been selectively ablated showed exacerbation of the ischemia-dependent brain injury [113]. However, several studies have showed a significant reduction of microglia/macrophages in the brain of mice, with either ischemic or haemorrhagic stroke, together with improved neuronal survival and locomotor functions after NPC transplantation [22, 140]. Also in this case, NPCs have been engineered in order to increase their immune modulatory capacity. Recently, mouse NPCs were transduced with IL-10, which has been proved to efficiently suppress EAE symptoms and promote survival of neurons and oligodendrocytes [210-212]. Mouse NPCs, transduced with a lentiviral vector encoding IL-10, showed enhanced ability to induce remyelination, neuronal repair and immune suppression after systemic injection in EAE mice compared to control NPCs [213].

In vitro, NPCs are shown to increase the apoptosis of $PLP_{139-151}$-specific Th1 pro-inflammatory (but not Th2 anti-inflammatory) cells through the engagement of death receptors, including FasL, TNF-related apoptosis-inducing ligand (TRAIL), and APO3L, on the surface of NPCs [16]. Mouse and rat NPCs also inhibit T cell activation and proliferation in response to T cell receptor (TCR)-mediated stimuli (e.g., concanavalin A and anti-CD3/anti-CD28) [136, 214]. NPC/T lymphocyte co-culture experiments suggest that part of the anti-proliferative effect of NPCs might depend on the inhibition of IL-2 and IL-6 signalling on T lymphocytes [214]. In addition, NPCs have shown a selective pro-apoptotic effect on Th17 cells *in vitro* via a FasL (CD95L)-dependent mechanism, identifying the axis Fas-Birc3 as an additional survival pathway for NPCs [215]. Mouse C17.2 NPCs also suppress T-cell proliferation, at least in part, by reactive production of the soluble mediators nitric oxide (NO) and prostaglandin E2 (PGE2). High levels of NO and PGE2 are in fact induced in T cells when co-cultured with NPCs. In addition, inducible NO synthase (iNOS) and microsomal type 1 PGES (mPGES-1) are detected in NPCs in co-culture with T-cells, suggesting that NO and PGE2 production in NPCs is induced by exposure to activated T cells [216].

Human NPCs have been proved to suppress the proliferation and alter the cytokine secretion profiles of activated T cells on both xenogeneic antigen-specific T cells derived from EAE induced non-human primates (common marmosets), and allogeneic mitogen-activated T cells. Co-culture of human NPCs with T cells, revealed their immune modulatory capacity through both direct cell-to-cell contacts as well as via the release of soluble mediators into the culture medium [15]. Notably, in contrast to the mouse counterpart, human NPCs have a limited cytotoxicity against T cells *in vitro*, given that FasL is only barely detectable on their surface. However, human NPCs exposed to cytokines express high levels of TNF-α, resulting in a higher cytotoxic potential against monocytes and macrophages [217]. In line with this, immortalized human NPCs were also proved, trough direct *in vitro* experiments, to reduce T cell activation and proliferation. Conditioned media collected from human NSCs (HB1.F3 line) *in vitro*, directly suppress the proliferation of activated human T cells through both induction of apoptosis and cell cycle arrest. Nonetheless, human NPC-released mediators alter the cytokine production pattern of T lymphocytes, increasing the expression of IL-4, IL-10, TNF-α, and IFN-γ and decreasing IL-2, thus affecting the overall activation [200].

4.4. NPC interactions with the dysfunctional non-CNS microenvironment

In parallel to the observed immune modulation and neuroprotection into the CNS, other studies have shown that systemically injected cells may act also outside the injured CNS. Different studies, in fact, have reported the capacity of NPCs to target and synergize with immune cells at the level of secondary lymphoid organs (e.g. draining lymph nodes) and the spleen, resulting in the attenuation of the inflammatory response following EAE, stroke and SCI.

It was initially showed *in vitro* that NPCs strongly inhibited the ability of EAE-derived lymphocytes to produce pro-inflammatory Th1 cytokines in response to MOG_{35-55} stimulation. In addition, specifically activated T cells isolated from EAE mice treated with NPCs, were deficient in their ability to adoptively transfer EAE (to a naïve host), thus suggesting a long-lasting inhibition of encephalitogenicity of T cells [20]. Further data have been provided about a specific and almost exclusive targeting of the peripheral immune system in SJL mice with PLP-induced EAE, in which NPCs had been injected subcutaneously (s.c.) at 3 and 10 dpi [21]. This alternative administration protocol showed a significant clinical improvement in EAE mice despite injected cells were never been consistently found into the inflamed CNS. Sub-cutaneously injected, s.c.-injected cells were mainly found accumulating and persisting (up to 2 months) at the level of the perivascular areas of the draining lymph nodes, where they interacted with resident cells. Similarly to what observed in the CNS parenchyma, NPCs accumulated as focal clusters around blood vessels of the hilum and medullary/paracortical areas. Here they established close interactions with endothelial cells and cell-to-cell contacts with $CD11c^+$ DCs, $F4/80^+$ professional phagocytes and MHC class II^+ immune cells [21]. Further, *ex vivo* analyses of lymph nodes isolated from NPC-treated EAE mice, showed hampered activation and maturation of myeloid DCs. This was associated, according to both *in vivo* and *in vitro* analyses, to the release of BMP-4 and TNF-α or TLR agonists. The BMP-dependent effect is highly specific for immune regulatory NPCs and, in turn, leads to the restraint of encephalitogenic T cell expansion at sites of antigen presentation. In addition to BMP-4, transplanted NPCs modulated the local increase of major stem cell fate determinants, including BMP-7, the extracellular matrix protein tenascin C, Shh and Noggin. The pattern of NPC accumulation, the secretion of extracellular matrix proteins and stem cell regulators, and the lack of expression of neural lineage antigens (e.g. PSA-NCAM, class III β–tubulin, NeuN, NG-2 and GFAP) once more suggest the establishment of perivascular *atypical ectopic niche-like* areas in the peripheral lymph nodes, similarly to what already seen in the CNS [15]. Supported by these experiments, a successive step forward was undertaken to test NPCs therapeutic capacity in a non-human primate model of EAE. Systemically injected foetal human NPCs into MOG_{74-96}-immunized common marmosets delivered at either the clinical onset of the disease or at subclinical occurrence of MRI detectable brain lesions, were found not only at the level of perivascular inflammatory CNS areas but also in secondary lymphoid organs. In parallel to these observations, human NPCs interfere *in vitro* with a number of key functions, such as the differentiation of myeloid precursor cells (MPCs) into immature DC (iDC) and the maturation of iDC to functional mature DCs. A significant impairment of the differentiation of $CD14^+$ MPCs into $CD1a^+$ iDCs has been reported when MPCs were cultured with granulocyte macrophage colony-stimulating factor (GM-CSF) and IL-4 in presence of NPCs [15]. In the same study, NPCs influenced the up regulation of the co-stimulatory molecules CD80, CD86 and MHC-II on LPS-treated DCs, thus impairing their capacity to induce a proliferative allogeneic response in mixed leukocyte reaction *in vitro*. Clinically wise, the i.v. NPC injection resulted more efficacious than the i.t. NPC treatment. This might be related to either the higher number of surviving cells or to an additional peripheral effect. Systemic NPCs were, in fact, subjected to selective capturing into cervical lymph nodes where they persisted up to 3 months while establishing close contacts with blood-borne inflammatory cells [15].

Similarly to what observed in EAE models, i.v. administered human NPCs, in a rat model of ICH, revealed a peripheral therapeutic function in attenuating the inflammatory response to the insult [22]. In line with previous studies, cells were rarely observed into the injured brain while the majority of NPCs were found distributed within the systemic organs. In particular, few NPCs were observed in mesenteric lymph nodes while large numbers were detected in the spleen, especially in the marginal zone area, which is typically enriched in macrophages. Once again NPCs were found in close contact with immune cells and some of them were establishing cell-to-cell contact interactions with $CD11b^+$ spleen macrophages. This result was probably due to the existence of a link between brain and spleen inflammation, called "brain-spleen inflammatory coupling". Remarkably, splenectomy prior to ICH has been shown to

http://dx.doi.org/10.5772/55426

reduce the initial cerebral oedema and inflammatory cell infiltration caused by stroke [22]. NPC accumulation into the spleen, in this case, modulated brain inflammation by reducing the level of major inflammatory mediators in stroke, such as TNF-α, IL-6 and nuclear factor-kappa B (NF-κB), and consequently improved neurologic outcome.

It has been recently shown that mouse NPCs (from fully mismatched C57BL/6 mice) co-transplanted with pancreatic islet under the kidney capsule of Balb/c diabetic mice prevents acute islet allograft rejection. This effect was related to a significant reduction of $CD4^+$ T cells and with a concomitant enrichment of $CD4^+/CD25^+/FoxP3^+$ regulatory T cells in the spleen, inducing active tolerance. These data suggest that the peripheral immune-modulation exerted by NPCs could alleviate the immune reaction leading to organ rejection. Unfortunately this condition appeared strictly associated with the development of NPC-derived tumours mainly sustained by insulin secretion from the co-transplanted islets [218].

Whether most of the immune regulatory effects of systemically injected NPCs act mainly into the CNS or in the periphery is still under debate. Peripheral lymphoid organs have been demonstrated to play an important role in the regulation of the immune responses to myelin antigens in EAE and a very sophisticated modulation of T-cell self-reactivity is known to take place [219-221]. A very recent study proposes a molecular mechanism sustaining NPCs immune modulation capacity in EAE. The preventive (0 dpi) or therapeutic (10 dpi) i.v. administration of NPCs resulted in their accumulation in lymph nodes and spleen, with rare cells observed into the CNS and without any evidence of myelin repair. Nevertheless, treated mice showed partial clinical recovery. Remarkably, the authors achieved the same results even transplanting NPC conditioned-medium or minimally irradiated NPCs (unable to differentiate but capable of secreting cytokines and neurotrophins), evidence sustaining a true peripheral function of NPCs. In particular, the observed clinical amelioration seem to be related to the selective inhibition of encephalitogenic Th17 cell differentiation through secreted factors. LIF has been identified as the key factor responsible for the observed inhibition of Th17 cell differentiation and the authors elucidated the signalling pathway behind this novel mechanism of action, where LIF antagonizes IL-6-induced Th17 cell differentiation through ERK-dependent inhibition of STAT3 phosphorylation [23]. Further studies will be needed to establish the absolute relevance of these pre-clinical data in EAE, where peripheral lymphoid organs play an important role in the regulation of the immune responses to myelin antigens, and their potential for future applications in MS. All the preclinical data describing NPC therapeutic effect upon systemic administration are summarized in Table 1.

Disease Model	**Species**	**Transplant Features**				**Observed Effect(s)**	**Proposed Mechanism(s)**	**Refs**
		Cell type	*Cell no./ animal*	*Route*	*Time*			
			Experimental Autoimmune Encephalomyelitis (EAE)					
Acute EAE	Rat	Rat neurospheres	1.5-2x10^4	i.c.v. or i.t.	Disease peak	Cell differentiatio	None	[10]

Disease Model	Species	Transplant Features				Observed Effect(s)	Proposed Mechanism(s)	Refs
		Cell type	*Cell no./ animal*	*Route*	*Time*			
						n (neuronal and glial and tissue trophism		
Acute EAE	Rat	Rat neurospheres	$2x10^4$	i.c.v.	0 dpi	Immune regulation (central)	None	[136]
Chronic EAE	Mouse	Mouse NPCs	$1x10^6$	i.v. or i.t.	22 dpi	(Low) cell differentiation and tissue trophism	Inhibition of reactive gliosis	[11]
Chronic EAE	Mouse	Mouse neurospheres	$2.5x10^3$	i.c.v.	6 dpi	Immune regulation (local)	Reduction of CNS inflammatory infiltrates, increase of regulatory T cells	[139]
Chronic EAE Passive EAE	Mouse	Mouse NPCs	$1x10^6$	i.v.	8 dpi	Immune regulation (peripheral)	Suppression of encephalitogenic T cells	[20]
Chronic EAE	Mouse	Human ES cell-derived NPCs	$5x10^5$	i.c.v.	7 dpi	Immune regulation (local)	Suppression of encephalitogenic T cells	[142]
Chronic EAE	Mouse	IL-10-transduced mouse NPCs	$1.5x10^6$	i.v. or i.c.v.	10, 22 or 30 dpi	Immune regulation (local, peripheral) and cell differentiation	Induction of T cell apoptosis, promotion of myelin debris clearance	[213]
Chronic EAE	Mouse	Mouse and human ES cell-derived NPCs	$2x10^6$	i.v.	0 or 10 dpi	Immune regulation (peripheral)	LIF-mediated inhibition of Th17 cell differentiation	[23]

http://dx.doi.org/10.5772/55426

Disease Model	Species	Transplant Features				Observed Effect(s)	Proposed Mechanism(s)	Refs
		Cell type	*Cell no./ animal*	*Route*	*Time*			
Chronic EAE	Mouse	Mouse MSC-derived NPCs	$3.5x10^4$-6. $1x10^6$	i.t.	21, 28 and 35 dpi	Tissue trophism	None	[258]
Chronic EAE Passive EAE	Mouse	CCR5-transduced mouse BM-derived NPCs	$1.5x10^6$	i.v.	22 dpi (peak)	Immune regulation	None	[259]
Chronic EAE	Common Marmoset	Human NPCs	$2\text{-}6x10^6$	i.t. or i.v.	Disease onset	Immune regulation (central)	Suppression of encephalitogen ic T cells, imparment of dendritic cell maturation	[15]
Relapsing EAE	Mouse	Mouse NPCs	$1x10^6$	i.v.	Disease onset or first relapse	Immune regulation (central)	Induction of T cell apoptosis	[16]
Relapsing EAE	Mouse	Mouse NPCs	$0.5x10^6$	s.c.	3 and 10 dpi, or 10 dpi only	Immune regulation (peripheral)	BMP-4-dependent hindrance of dendritic cell maturation	[21]
Relapsing EAE	Mouse	Mouse NPCs and Olig2-transduced NPCs	$1.5x10^5$	i.c.v.	Disease onset or first relapse	Immune regulation (central) and Tissue trophism	None	[260]
				Stroke				
MCAo	Rat	Rat NPCs	$1x10^5$	i.t.	2 dpi	Cell differentiatio n (neuronal)	None	[204]
MCAo (10' or 90')	Rat	Human NPCs	$5x10^6$	i.v.	1 dpi	Cell differentiatio n (neuronal, glial)	None	[12, 187]
MCAo (180')	Rat	Rat NPCs	$1x10^5$	i.t.	2 dpi	Tissue trophism	Increased angiogenesis	[190]

Disease Model	Species	Transplant Features				Observed Effect(s)	Proposed Mechanism(s)	Refs
		Cell type	*Cell no./ animal*	*Route*	*Time*			
MCAo (180'	Rat	Human NPCs	$1x10^6$	i.v.	2 dpi	Tissue trophism	None	[261]
CCAo + global hypoxia-ischemia	Mouse	Mouse 17.2 NSCs	$3x10^5$	i.ca.	2 dpi	Tissue trophism	Increased angiogenesis	[18]
MCAo (120')	Rat	GDNF-transduced rat NPCs	$5x10^5$	i.c.v.	3 dpi	Tissue trophism, Cell differentiation (neuronal)	None	[198]
MCAo (45')	Mouse	Mouse NPCs	$1x10^6$	i.v.	3 dpi	Immune regulation (local) and Tissue trophism	Reduction of microglial activation and neuronal death	[140]
MCAo and CCAo	Rat	Rat NPCs	$1.5x10^5$	i.t.	7 dpi	Immune regulation and tissue trophism	Neuroprotection mediated by NGF and modulation of class I MHC expression	[141]
MCAo (90')	Rat	HIF-1α-transduced rat NPCs	$1x10^6$	i.c.v.	1 dpi	Tissue trophism	Promotion of angiogenesis	[227]
MCAo (45')	Mouse	TAT-Hsp70-transduced mouse NPCs	$1x10^6$ or $5x10^5$	i.v. or i.p.c.	Acute	Tissue trophism, reduction of ROS formation and BBB leakage	Neuroprotection and enhanced neurogenesis	[232]
ICH	Rat	Human NPCs	$5x10^6$	i.v.	1 dpi	Tissue trophism, Cell differentiatio	Neuroprotection and integration in endogenous	[137]

Disease Model	Species	Transplant Features				Observed Effect(s)	Proposed Mechanism(s)	Refs
		Cell type	*Cell no./ animal*	*Route*	*Time*			
						n (glial and neuronal)	neuronal circuitries	
				Spinal Cord Injury (SCI)				
Contusion (T8)	Mouse	Mouse NPCs	$1x10^6$ or $1x10^5$	i.v. or i.p.c.	Acute	Tissue trophism	Reduction of apoptosis and modulation of TNF-α expression	[13]
Compression (T8)	Mouse	Human NPCs	$1x10^7$	i.v.	7 dpi	Cell differentiation (neuronal, glial)	None	[167]
Contusion (T12)	Mouse	Mouse NPCs and MOG_{35-55} immunization	$5x10^5$	i.c.v.	7 dpi	Immune regulation (local) and Tissue trophism	T-cell mediated activation of microglia with a protective phenotype	[226]

BBB: blood-brain barrier; BM: bone marrow; BMP-4: bone morphogenetic protein 4; CCAo: common carotid artery occlusion; CCR5: C-C chemokine receptor type 5; dpi: days post immunization/injury; ES cells: embryonic stem cells; GDNF: glial-derived neurotrophic factor; HIF-1α: hypoxia-inducible factor 1α; i.ca.: intracarotid; ICH: intracerebral haemorrhage; i.c.v.: intracerebroventricular; i.p.c.: intraparenchymal (perilesional); i.t.: intrathecal; I.v.: intravenous; LIF: leukemia inhibitory factor; MCAo: middle cerebral artery occlusion; MSC: mesenchymal stem cells; ROS: reactive oxygen species; s.c.: subcutaneous; TAT-Hsp70: TAT-heat shock protein.

Table 1. Neuro-immune interaction following systemic neural stem cell transplantation in experimental disease models.

5. Pros and cons of NPC systemic administration

In parallel to the investigation concerning the principal mechanism(s) sustaining NPC therapeutic efficacy, other questions, such as (i) the ideal administration route, (ii) the amount of cells to be transplanted and (iii) the optimal time point for cell delivery need to be answered. Among the different possible routes of cell administration, intravenous cell delivery represents one of the most attractive because of its technical simplicity and clinical practicability. However, i.v. and i.t. administrations result in lower numbers of cells infiltrating the CNS, compared to local stereotaxic-driven intracerebral injections, a reason why local injections of cells are commonly preferred in clinical trials (see next section) despite the higher invasiveness of the procedure. Even though initially investigated for multifocal disorders (e.g. MS), in order to deliver exogenous cells to all the disseminated inflammatory foci, all the previous experimental data suggest that intravenous or intrathecal administration routes could be desirable even for focal damages, such as those occurring in stroke and spinal cord injury [222]. In experimental animal studies, i.p.c [223-225], i.v. [137, 185] i.a. [222], i.t. [204, 226] and i.c.v. [142, 227] protocols have been tested so far. However, only few comparative studies have been conducted, testing pros and cons of the different administration routes. These studies (mainly in animal models of stroke) evidenced the obvious capacity of intraparenchymal injection to deliver higher numbers of cells *in situ*, compared to i.c.v. and i.v. [228]. By contrast, systemic injections are thought to lead to a wider distribution of cells around the focal lesioned area. This aspect is extremely important if we consider that human stem cells (and in particular hNPCs) are still a limited resource [229]. Intravenously injected NPCs are firstly delivered to peripheral organs, such as lungs, liver, spleen and kidney [16, 230]. This whole-body distribution of exogenous systemic injected NPCs significantly reduces cell homing to the injured brain [222]. To avoid this problem, at least partially, intra-arterial administration could be a valid alternative (possibly coupled with pre-interventional imaging-based planning) to selectively cover an injured volume supplied by several target vessels. Intracarotid injection has already been proved to be functional for delivering stem cells in models of stroke, TBI and SCI, resulting in higher numbers of extravasating cells (20%) compared to i.v. injections [18]. Nevertheless, although the number of cells infiltrating the CNS has been sometimes described as fundamental, or at least proportional to their therapeutic effect [231], others have shown that very low numbers of cells [140] can result into similar outcomes (in term of functional recovery) compared to higher numbers of locally injected cells. This effect may be explained by the fact that cell replacement is unlikely the only mechanism sustaining stem-cell therapeutic potential. Higher starting numbers of cells, in fact, increase the therapeutic potential of intracerebral administered cells, but did not affected the efficacy of the i.v. injected cells. This again suggests that the number of cells is much more important for focal than systemic injections [232].

Importantly, when evaluating the optimal protocol, we should consider the procedure itself, so that the risk should not outweigh the benefits of the treatment. From this point of view, i.e. cell injection might be accompanied by increased mortality during cell delivery, probably due to further ischemia or thrombosis [233, 234]. By contrast, cell transplantation trough the vertebral artery, into patients affected by SCI, showed no adverse effects [235].

Another important unsolved issue for experimental stem cell therapies is the ideal time point of transplantation. As described, the inflammatory activation of the CNS, characterizing MS, stroke, SCI, epilepsy, AD, PD, HD is necessary for the homing of systemically injected cells. Because of the rapid and dynamic changes occurring into the CNS during these inflammatory conditions, the time of transplantation should be evaluated carefully. In fact, cell death, excitotossicity, reactive oxygen species accumulation, inflammatory cell infiltrations and glial scar formation, cause a rapid evolution in the damaged tissue, while creating an hostile microenvironment for the engraftment of exogenous cells. This is important irrespectively to the route of administration. For example, the acute focal transplantation of cells into the ischemic brain or the injured spinal cord reduces the therapeutic efficacy of the cells, which

http://dx.doi.org/10.5772/55426

are subjected to highly inflammatory conditions causing cell death [205, 236]. On the contrary, the sub acute phase (few days after insult, in rodents) of the injury seems to be characterized by better conditions for stem cell survival and a permissive microenvironment for tissue repair/healing [237]. Although higher inflammation generally correlates with higher number of cells infiltrating the CNS, it has been shown that greater numbers of cells accumulated into the spinal cord after i.v. injection at 7 dpi compared to 3 and 10 dpi [167]. However, the optimal time window for cell transplantation is still elusive and depends mainly on the type of pathology and aim of the treatment. While neuroprotection should be addressed in the early stage of the inflammatory disease, just after the initial insult, cell replacement and neuroregeneration should be targeted in a later stage, when the lesion has stabilized. Indeed, administration route, number of cells and time window and seem intimately related and it is not so difficult to envisage a future in which a combination of early i.v. and late i.p.c administration of different stem cell sources will be enrolled for the treatment of so far incurable CNS disorders.

6. Clinical trials

In the last two decades the clinical potential of stem cells in the field of regenerative/restorative medicine has been often matter of debate, mainly because of its inconsistent outcome. As an example, the first attempt to treat a CNS disorder by means of stem cell transplantation took place in the '80s: autologous adrenal medulla cells were intracerebrally transplanted into the striatum of PD patients to provide a local source of catecholamine. The study was proved safe although with minimal beneficial effect. Further, the first intrastriatal grafts of human foetal ventral mesencephalic (neuronal preparations) tissue have provided proof-of-concept that cell therapy can work in patients affected by PD [238]. However, subsequent randomized, double-blind, placebo-controlled trials brought to much more sceptical conclusions because of patients showing functional decline (post transplantation) due to dyskinesias (graft-induced involuntary movements), originated by excessive graft function [239, 240].

Prospectively, many factors can be contended to (partially) justify these patchy results. First, it is now clear that different cell types are needed for different diseases. If on one side PD and amyotrophic lateral sclerosis (ALS) patients will require cells with dopaminergic and motor neuron properties respectively, on the other side, cell replacement in AD patients is much more complicated by the necessity to replace a large variety of cell populations lost in different brain areas. Second, even though initially expected and long-term envisaged, neuronal replacement and circuitry integration of transplanted NPCs have been poorly proved. Third, it as to be considered that pre-clinical animal studies only represent models of human conditions, and, as such, they offer an exceptionally homogeneous platform, where the genetic background, age, and environment are all alike. Clearly, this is not the case with patients. Further, even if multiple models have been established to investigate different aspects of a given disease, none of them can faithfully emulate the human pathology in its complexity [241, 242]. This is particularly challenging considering the rate of progression and lack of validated surrogate disease markers typical of many neurodegenerative disorders. While these aspects are most likely destined to remain unsolved pitfalls, others (including the amount of cells to be transplanted, the manipulation protocols used, the time of transplantation, the route of cell delivery and the statistics adopted to analyse the data) need to be ameliorated through the establishment of common guidelines. In particular, the International Society for Stem Cell Research (ISSCR) composed with a group of international experts (scientists, surgeons, ethicists and patient advocates) "The ISSCR Guidelines for the Clinical Translation of Stem Cells" [243] to trace a roadmap guiding the application of experimental stem cell therapeutics in patients. Importantly, when translating into clinical trials, the choice of the "ideal patient" imposes major scientific and ethical constraints. Indeed, if on one side the treatment of the most chronic/severe patients who were not able to respond to previous treatment lowers the blame for a possible ineffectiveness of a therapy, on the other side the scenario offered by such a compromised tissue may hinder the potential effect of the treatment.

The primary importance of patient's care dramatically impacts also on the choice of the best route of cell delivery. Indeed, if one side the intravenous injection allows for a less invasive procedure, on the other, the number of cells delivered to the site of interest is lower compared to local injections. Further, the intracerebral transplantation has been widely accepted, by both clinicians and patients, after years of clinical applications and technical improvements. These, together with the relatively limited availability of human NPCs explains why most of the clinical trials started so far have nevertheless favoured the adoption of more invasive procedures, such as intraparenchymal/intracerebroventricular ones (see Table 2). However, as discussed, the correlation between the number of cells entering the CNS and their efficacy still need to be confirmed.

Sponsor and place	**Disease**	**Trial phase**	**Patients (no)**	**Age at enrolment (y)**	**Follow up (months)**	**Transplant Features**					**Status**	**Principal Investigator**	**Trial Identifier**	**Outcome and Notes**
						Cell type	*Cell no./ patient*	*Route*	*Time after disease/ injury*	*Immune suppression*				
StemCells, Inc. at University Hospital Balgrist-Uniklinik Zurich, (Switzerland)	Thoracic spinal cord injuries (SCI)	I/II	12	18-60	12	HuCNS-SC* (Foetal, Brain-derived, Allogeneic, single donor)	$2x10^7$	Multiple injections, Single dose, Intramedullar	≥ 3 months	Y (9 months pt)	AR	Armin Curt, MD	NCT01321333	NA

Sponsor and place	Disease	Trial phase	Patients (no)	Age at enrolment (y)	Follow up (months)	Transplant Features					Status	Principal Investigator	Trial Identifier	Outcome and Notes
						Cell type	*Cell no./ patient*	*Route*	*Time after disease/ injury*	*Immune suppression*				
ReNeuron, Ltd. at Glasgow Southern General Hospital, Glasgow (UK)	Stable Ischemic Stroke (PISCES)	I	12	60-85	24	CTX0E303 (Foetal, Brain-derived, c-myc immortalized, Allogeneic, single donor)	2-20x10^6	Single injection, Four Ascending doses, Intracerebral (putamen)	0.5-5 years	NA	AR	Keith Muir, MD	NCT01151124	NA
Neuralstem, Inc. at Emory University, Atlanta (USA)	Amyotrophic Lateral Sclerosis (ALS)	I	18	> 18	48	NSI-566RSC (Foetal, Spinal cord-derived, Allogeneic, single donor)	0.5-1x10^6	Multiple injections, Intraspinal	≥ 1.5 years	Y (≥ 4 months pt)	AnR	Eva Feldman, MD, PhD	NCT01348451	[248, 249]
Azienda Ospedaliera Santa Maria, Terni (Italy)	ALS	I	18	20-75	36	Foetal, Brain-derived, Allogeneic, single donor	2.5x10^5/ injection	Multiple injections, Single dose, Intraspinal	> 6 months	NA	AR	Angelo Vescovi, PhD	NCT01640067	NA
StemCells, Inc. at University of California, San Francisco (USA)	Pelizaeus Merzbacher disease (PMD)	I	4	0.5-5	12	HuCNS-SC*	3x10^8	Multiple injections, Single dose, Intracerebral	NA	Y (9 months pt)	AnR	Stephen Huhn, MD	NCT01005004	[250]

Sponsor and place	**Disease**	**Trial phase**	**Patients (no)**	**Age at enrolment (y)**	**Follow up (months)**	**Transplant Features**					**Status**	**Principal Investigator**	**Trial Identifier**	**Outcome and Notes**
						Cell type	*Cell no./ patient*	*Route*	*Time after disease/ injury*	*Immune suppression*				
StemCells, Inc. at Oregon Health and Science University, Portland (USA)	Neuronal Ceroid Lipofuscinosis (NCL)	I	6	1.5-12	13	HuCNS-SC*	$0.5\text{-}1 \times 10^9$	Multiple injections, Single dose, Intracerebral	NA	Y (12 months pt)	C	Robert Steiner, MD	NCT00337636	[247]
StemCells, Inc. Retina Foundation of the Southwest, Dallas (USA)	Age-related Macular Degeneration (AMD)	I/II	16	> 50	12	HuCNS-SC*	$0.2\text{-}1 \times 10^6$	Single injection, Single dose, Subretinal	NA	Y (3 months pt)	AR	David Birch, PhD	NCT01632527	NA

pt: post-transplant; NA: information not available. AR: Active Recruiting; AnR: Active not Recruiting; C: Completed.

Table 2. Active clinical trials with neural stem/precursor cells.

Other challenging problems that need to be faced when approaching the clinic, are related to safety, product potency, and manufacturing quality of the cell source. Indeed, principles of good tissue practice (GTP) and good manufacturing practice (GMP) are mandatory requirements, especially when dealing with cells of human origin [244].

Last, but not least, some major issues related to the long-term safety of the cellular product need to be solved. It is important to stress how, differently from the classical drug-therapy, a cell-based treatment cannot be discontinued, since once the cells are within the patient they cannot be removed. Therefore, long-term pre-clinical data need to be collected before translating from the bench to the bed-side to avoid the occurrence of dramatic outcomes, such as the one involving a young patient suffering of Ataxia Telangiectasia who developed a donor-derived brain tumour following neural stem cell transplantation [245].

http://dx.doi.org/10.5772/55426

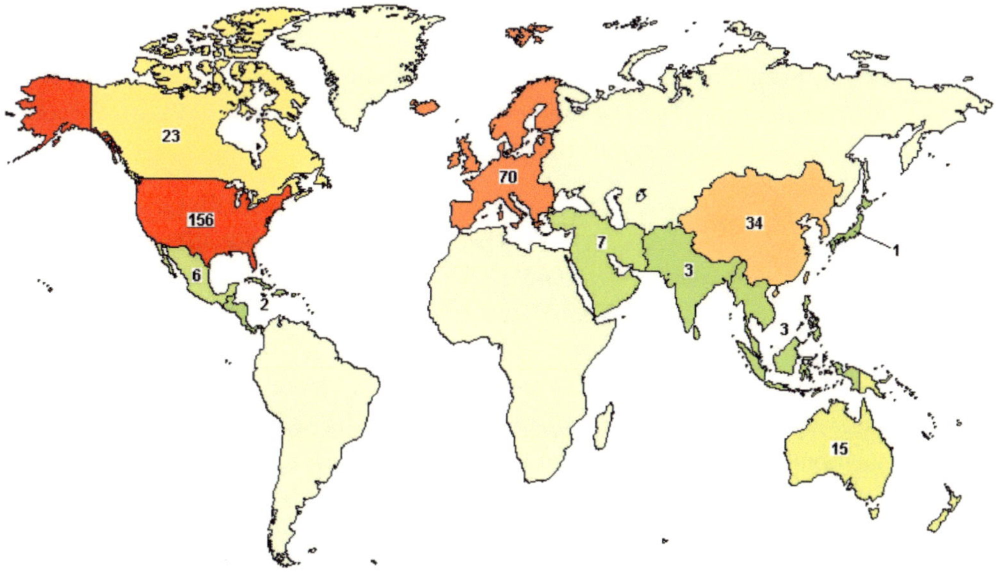

Figure 2. Worldwide open experimental clinical trials involving the use of stem cells for CNS disorders. The map shows the distribution of open stem cell-based clinical trials enrolling patients affected by CNS pathologies and treated with different source of SCs. Figure has been generate by clinicaltrials.gov, using the keywords *stem cells* AND *CNS diseases*; the inclusion criteria were only open studies; while the exclusion criteria where studies with unknown status. Stem cells in the figure include: neural stem/precursor cells, mesenchymal stem cells from the bone marrow and the adipose tissue, hematopoietic stem cells, embryonic stem cells.

Nevertheless, translational research did not (and should not) stop. As a matter of fact, the huge amount of data collected so far led to the development of numerous early stage clinical trials. There are currently 1750 studies employing the use of stem cells in interventional clinical trials, among which 280 are testing NPCs. Within the total number of clinical studies, 277 worldwide open trials involve SCs and patients affected by CNS disorders (Figure 2), such as MS, stroke, SCI, epilepsy, ALS, PD, AD, HD, neuronal ceroid lipofuscinose (NCL), Pelizaeus–Merzbacher disease (PMD), age-related macular degeneration (AMD) [246]. Several phase I and II clinical studies with NPCs (9 in highly debilitating CNS disorders, Table 2) have now been started with the primary aim to verify the safety (mainly in terms of toxicity) and feasibility - rather than the efficacy - of the treatment. Further, clinical studies are often accompanied by non-official secondary end-points usually concerning the potential impact in the clinical outcomes. These explorative trials have certainly a key role in stem cell medicine development, as both phase II dose-escalation studies and the inclusion of non-fatal diseases with larger population bases will definitely be facilitated once human safety will be established.

In May 2006 at Oregon Health and Science University (OHSU, Portland, OR, USA) the first human study involving the transplantation of allogeneic NPCs was started on a fatal rare neurometabolic syndrome, such as the NCL Batten's disease. In this open-label dose-escalating phase I trial, a total of 6 subjects with infantile and late-infantile NCL were transplanted in a single-stage procedure. StemCell, Inc. proprietary, single donor allogeneic free-floating cultured, foetal-derived brain human NPCs (HuCNS- SC®) were directly administered to the cerebral hemispheres and lateral ventricles. Immune suppression was administered for 12 months after transplantation. This study has now been completed with one out of 6 patients died for disease progression, 11 months after treatment. The cell transplantation and combination with prolonged immune suppression were both well tolerated [247].

In September of 2009, NeuralStem, Inc. sponsored a phase I trial in ALS at the Emory University School of Medicine (Atlanta, GA, USA), using proprietary single donor allogeneic, adherent cultured, foetal-derived spinal NPCs (NSI-566RSC). NSI-566 cells were surgically implanted on a total of 12 patients via multiple injections directly into the thoracic spinal cord (either unilateral or bilateral). The clinical assessments demonstrated no evidence of acceleration of disease progression with the planned 18 months post-transplantation follow up [248, 249]. StemCell, Inc. is also sponsoring other two phase I trials with HuCNS-SC® in X chromosome linked connatal leukodistrophy PMD (in which oligodendrocytes cannot myelinate axons) and AMD. With the PMD trial at the University of California, San Francisco (UCSF, San Francisco, CA, USA), HuCNS-SC® were directly delivered through multiple injections into the brain of a total of 4 male patients (clinicaltrials.gov identifier no. NCT01005004). Data regarding this clinical trial has been recently published [250]. The transplantation procedure, the immunosuppression and the cells were well tolerated by all the 4 patients. No adverse effects related to the implant were detected. MRI investigation before and after the transplantation of cells, revealed, after 9 months, a consistent donor cell-derived myelination *in situ*, in three of the patients. However, these data are just published and under intense scientific discussion. With the AMD trial at the Retina Foundation of the Southwest (Dallas, TX, USA), HuCNS-SC® are being delivered directly into the subretinal space of one eye in a single transplant procedure in a total 16 patients. The estimated completion date of this study is March 2014 (clinicaltrials.gov identifier no. NCT01632527).

In June 2012, the Glasgow Southern General Hospital (Glasgow, Scotland) enrolled the first patient (of 12 total) of the dose-escalating Pilot Investigation of Stem Cells in Stroke (PISCES) phase I trial to be transplanted in a single-stage procedure with direct cerebral (intraparenchymal) delivery of Reneuron, Ltd. proprietary single donor allogeneic adherent cultured, c-myc immortalized foetal-derived brain human NPCs (CTX0E03) (clinicaltrials.gov identifier no. NCT01151124).

In March 2011, the University Hospital Balgrist (Zurich, Switzerland) enrolled the first patient (of 12 total) with chronic thoracic (T2–T11) SCI (3 to 12 months after complete and incomplete cord injuries) to be transplanted with HuCNS-SC® in a further StemCell, Inc. sponsored phase I/II clinical trial estimated to be concluded in March 2016. A single dose ($20x10^6$ cells) of HuCNS-SC® has been directly implanted through multiple injections into the thoracic spinal cord, and immune suppression administered for 9 months after transplantation (clinicaltrials.gov identifier no. NCT01321333). In November 2012 started the consequent long-term follow up of the 12 patients subjected to HuCNS-SC® transplantation that will last until March 2018 (clinicaltrials.gov identifier no. NCT01725880).

In June 2012, the Azienda Ospedaliera Santa Maria (Terni, Italy) enrolled the very first of total 18 ALS patients to treat with intraspinally implanted allogeneic free-floating cultured, foetal-derived brain NPCs. (clinicaltrials.gov identifier no. NCT01640067).

Importantly, there are not yet clinical trials with NPCs in MS. However, a consensus paper has recently been produced by a group of experts to define the uniform guidelines on the development of haematopoietic and non-haematopoietic stem cell therapies for MS [9]. All the current clinical trials envolving NPCs for CNS disorders are described in Table 2.

While in this paragraph we offer an overview of the current clinical trials involving solely human NPCs, it has to be said that, in the light of the neuroprotective/immunomodulatory (rather than cell replacement) properties attributed to stem cells, the therapeutic plasticity of cells of non-neural origin are being tested as well. Among these, MSCs are emerging as a good potential candidate, mainly because of their great accessibility and remarkable proliferation. Also, growing evidence suggests that other than giving origin to multiple derivatives of the mesodermal lineage (from which they derive), under particular conditions MSCs seem able to *transdifferentiate* into neuro-ectodermal cells *in vitro* [251-254]. However, this ability to convert from one lineage to another is still highly questionable and opened to different interpretations. Several studies have also proved the ability of MSCs to survive, migrate and eventually bring about functional recovery when transplanted into the CNS of different experimental models of neurological diseases (for a review see [255]). However, the mechanisms yielding to such rescue are unlikely ascribable merely to cell replacement.

Since 2006, the advent of induced pluripotent stem cells (iPSCs, [24]) technology has brought new excitement in medical research and clinical therapy, since these cells provide a valuable alternative without being constrained by ethical issues and immunological incompatibility [256]. Although still under debate about their long-term safety, the methods for iPSC generation, reprogramming and differentiation efficiency, iPSCs represent a break-trough for both study of disease mechanisms and investigation of potential new treatments (for a perspective analysis, see [257]).

The potential impact of this technological platform has been further boosted by the scientific stream emerged from iPSC technology that is the "direct reprogramming" from one somatic lineage to another. In fact, the direct conversion of fibroblasts to functional neurons (iN cells) or iNSCs [25, 26], for example, represents one of the most exciting, ultimate technologies for future application in CNS pathologies. Thanks to these next generation techniques, it will be possible to derive virtually unlimited numbers of specific neural/neuronal population bypassing the pluripotent stage, thus likely eliminating the potential presence of unwanted undifferentiated cells. However, many issues, such as the purity of the cell preparation, the use of virus-based technologies and the proper *in vivo* integration and differentiation still need to be better addressed. Importantly, the availability of such a high number of cells will release the intravenous protocol from one of its major limit, thus casting new light on its clinical potentiality.

Abbreviations

AD: Alzheimer's disease

ALS: Amyotrophic lateral sclerosis

APC: Antigen presenting cell

ASCL1: Achaete-scute homolog 1

BBB: Blood brain barrier

BCSFB: Blood-cerebrospinal fluid barrier

BDNF: Brain-derived neurotrophic factor

BLMB: Blood-leptomeningeal barrier

BMP: Bone morphogenetic protein

BMSC: Bone marrow-derived stem cell

CCAo: Common carotid artery occlusion

CCL: Chemokine (C-C motif) ligand

CCR: C-C chemokine receptor

CNS: Central nervous system

CNTF: Ciliary neurotrophic factor

CSF: Cerebrospinal fluid

CXCR: C-X-C chemokine receptor

DCs: dendritic cells

DCX: Doublecortin

DG: Dentate gyrus

DGC: Dentate granule cell

Dlx: Distal-less homeobox

d.p.t.: days post transplantation

EAE: Experimental autoimmune encephalomyelitis

EC: Endothelial cell

ES cells: Embryonic stem cells

FACS: Fluorescence-activated cell sorting

FGF: Fibroblast growth factor

http://dx.doi.org/10.5772/55426

GABA: Gamma-aminobutyric acid

GCL: Granule cell layer

GDNF: Glial-derived neurotrophic factor

GFAP: Glial-fibrillary acidic protein

GF-CSF: granulocyte macrophage colony stimulating factor

GMP: Good manufacturing practice

GTP: Good tissue practice

HD: Huntington's disease

hNPC: Human neural stem/precursor cell

Hsp70: Heat shock protein 70

HuCNS-SC: Human CNS stem cell

HVc: hyperstriatum ventrale, pars caudalis

i.a.: Intraartery

i.c.v.: Intracerebroventricular

i.p.c.: Intraparenchyma

i.t.: Intrathecal

i.v.: Intravenous

ICAM: Intercellular adhesion molecule

ICH: Intracerebral hemorrhage

Ig: Immunoglobulin

IGF: Insulin-like growth factor

IL: Interleukin

IML: Inner molecular layer

iN cells: Induced neuronal cells

INF: Interferon

iNOS: inducible nitric oxide synthase

iNSC: Induced neural stem cell

IPC: Intermediate progenitor cell

iPS: Induced pluripotent stem cell

LFA: Leukocyte-function associated antigen

LIF: Leukemia inhibitory factor

LPS: Lipopolysaccharide

MAdCAM: Mucosal addressin cell adhesion molecule

MCAo: Middle cerebral artery occlusion

MCP: Monocyte chemoattractant protein

MHC: Major histocompatibility complex

MMS: Medial migratory stream

MOG: Myelin oligodendrocyte glycoprotein

MPC: Myeloid precursor cell

MRI: Magnetic resonance imaging

MS: Multiple sclerosis

MSC: Mesenchymal stem cell

NCL: Neuronal ceroid lipofuscinose

NeuN: Neuronal nuclei

NF: Neurofilament

NF-kB: Nuclear factor-kB

NGF: Nerve growth factor

NO: Nitric oxide

NPC: Neural stem/precursor cell

NSC: Neural stem cell

OB: Olfactory bulb

OPC: Oligodendrocyte progenitor cells

PD: Parkinson's disease

PGE2: prostaglandine 2

PLP: Proteolipid protein

PSA-NCAM: Polysialylated neural cell adhesion molecule

PSGL: P-selectin glycoprotein ligand

pt: Post-transplantation

RA: Radial astrocyte

RMS: Rostral migratory stream

http://dx.doi.org/10.5772/55426

s.c.: Subcutaneous

SCF: Stem cell factor

SCI: Spinal cord injury

SC: Stem cell

SDF: Stromal cell-derived factor

SE: Status epilepticus

SGZ: Subgranular zone

Shh: Sonic hedgehog

SIDS: Stroke-induced immune depression syndrome

TCR: T cell receptor

TGF: Transforming growth factor

TJ: Tight junction

TLE: Temporal lobe epilepsy

TLR: Toll-like receptor

TNF: Tumor necrosis factor

V-SVZ: Ventricular-subventricular zone

VCAM: Vascular cell adhesion molecule

VEGF: Vascular endothelial growth factor

VLA: Very late antigen

Author details

Matteo Donegà[1,2], Elena Giusto[1,2], Chiara Cossetti[1] and Stefano Pluchino[1*]

*Address all correspondence to: spp24@cam.ac.uk

1 Dept of Clinical Neurosciences, John van Geest Centre for Brain Repair, Wellcome Trust-MRC Stem Cell Institute, University of Cambridge, UK

2 NIHR Biomedical Research Centre, University of Cambridge, UK

References

[1] Altman J, Das GD. Autoradiographic and histological evidence of postnatal hippocampal neurogenesis in rats. The Journal of comparative neurology. 1965 Jun;124(3): 319-35.

[2] Altman J, Das GD. Autoradiographic and histological studies of postnatal neurogenesis. I. A longitudinal investigation of the kinetics, migration and transformation of cells incorporating tritiated thymidine in neonate rats, with special reference to postnatal neurogenesis in some brain regions. The Journal of comparative neurology. 1966 Mar;126(3):337-89.

[3] Goldman SA, Nottebohm F. Neuronal production, migration, and differentiation in a vocal control nucleus of the adult female canary brain. Proceedings of the National Academy of Sciences of the United States of America. 1983 Apr;80(8):2390-4.

[4] Picard-Riera N, Decker L, Delarasse C, Goude K, Nait-Oumesmar B, Liblau R, et al. Experimental autoimmune encephalomyelitis mobilizes neural progenitors from the subventricular zone to undergo oligodendrogenesis in adult mice. Proceedings of the National Academy of Sciences of the United States of America. 2002 Oct 1;99(20): 13211-6.

[5] Pluchino S, Martino G. The therapeutic plasticity of neural stem/precursor cells in multiple sclerosis. Journal of the neurological sciences. 2008 Feb 15;265(1-2):105-10.

[6] Pluchino S, Muzio L, Imitola J, Deleidi M, Alfaro-Cervello C, Salani G, et al. Persistent inflammation alters the function of the endogenous brain stem cell compartment. Brain. 2008 Oct;131(Pt 10):2564-78.

[7] Cossetti C, Alfaro-Cervello C, Donega M, Tyzack G, Pluchino S. New perspectives of tissue remodelling with neural stem and progenitor cell-based therapies. Cell and tissue research. 2012 Jul;349(1):321-9.

[8] Martino G, Pluchino S, Bonfanti L, Schwartz M. Brain regeneration in physiology and pathology: the immune signature driving therapeutic plasticity of neural stem cells. Physiological reviews. 2011 Oct;91(4):1281-304.

[9] Martino G, Franklin RJ, Van Evercooren AB, Kerr DA. Stem cell transplantation in multiple sclerosis: current status and future prospects. Nat Rev Neurol. 2010 May; 6(5):247-55.

[10] Ben-Hur T, Einstein O, Mizrachi-Kol R, Ben-Menachem O, Reinhartz E, Karussis D, et al. Transplanted multipotential neural precursor cells migrate into the inflamed white matter in response to experimental autoimmune encephalomyelitis. Glia. 2003 Jan;41(1):73-80.

[11] Pluchino S, Quattrini A, Brambilla E, Gritti A, Salani G, Dina G, et al. Injection of adult neurospheres induces recovery in a chronic model of multiple sclerosis. Nature. 2003 Apr 17;422(6933):688-94.

[12] Chu K, Kim M, Jeong SW, Kim SU, Yoon BW. Human neural stem cells can migrate, differentiate, and integrate after intravenous transplantation in adult rats with transient forebrain ischemia. Neuroscience letters. 2003 Jun 5;343(2):129-33.

[13] Bottai D, Madaschi L, Di Giulio AM, Gorio A. Viability-dependent promoting action of adult neural precursors in spinal cord injury. Molecular medicine (Cambridge, Mass. 2008 Sep-Oct;14(9-10):634-44.

[14] Lee ST, Chu K, Park JE, Lee K, Kang L, Kim SU, et al. Intravenous administration of human neural stem cells induces functional recovery in Huntington's disease rat model. Neuroscience research. 2005 Jul;52(3):243-9.

[15] Pluchino S, Gritti A, Blezer E, Amadio S, Brambilla E, Borsellino G, et al. Human neural stem cells ameliorate autoimmune encephalomyelitis in non-human primates. Annals of neurology. 2009 Sep;66(3):343-54.

[16] Pluchino S, Zanotti L, Rossi B, Brambilla E, Ottoboni L, Salani G, et al. Neurosphere-derived multipotent precursors promote neuroprotection by an immunomodulatory mechanism. Nature. 2005 Jul 14;436(7048):266-71.

[17] van der Meulen AA, Biber K, Lukovac S, Balasubramaniyan V, den Dunnen WF, Boddeke HW, et al. The role of CXC chemokine ligand (CXCL)12-CXC chemokine receptor (CXCR)4 signalling in the migration of neural stem cells towards a brain tumour. Neuropathology and applied neurobiology. 2009 Dec;35(6):579-91.

[18] Guzman R, De Los Angeles A, Cheshier S, Choi R, Hoang S, Liauw J, et al. Intracarotid injection of fluorescence activated cell-sorted CD49d-positive neural stem cells improves targeted cell delivery and behavior after stroke in a mouse stroke model. Stroke; a journal of cerebral circulation. 2008 Apr;39(4):1300-6.

[19] Martino G, Pluchino S. The therapeutic potential of neural stem cells. Nature reviews. 2006 May;7(5):395-406.

[20] Einstein O, Fainstein N, Vaknin I, Mizrachi-Kol R, Reihartz E, Grigoriadis N, et al. Neural precursors attenuate autoimmune encephalomyelitis by peripheral immunosuppression. Annals of neurology. 2007 Mar;61(3):209-18.

[21] Pluchino S, Zanotti L, Brambilla E, Rovere-Querini P, Capobianco A, Alfaro-Cervello C, et al. Immune regulatory neural stem/precursor cells protect from central nervous system autoimmunity by restraining dendritic cell function. PloS one. 2009;4(6):e5959.

[22] Lee ST, Chu K, Jung KH, Kim SJ, Kim DH, Kang KM, et al. Anti-inflammatory mechanism of intravascular neural stem cell transplantation in haemorrhagic stroke. Brain. 2008 Mar;131(Pt 3):616-29.

[23] Cao W, Yang Y, Wang Z, Liu A, Fang L, Wu F, et al. Leukemia inhibitory factor inhibits T helper 17 cell differentiation and confers treatment effects of neural progenitor cell therapy in autoimmune disease. Immunity. 2011 Aug 26;35(2):273-84.

[24] Takahashi K, Yamanaka S. Induction of pluripotent stem cells from mouse embryonic and adult fibroblast cultures by defined factors. Cell. 2006 Aug 25;126(4):663-76.

[25] Vierbuchen T, Ostermeier A, Pang ZP, Kokubu Y, Sudhof TC, Wernig M. Direct conversion of fibroblasts to functional neurons by defined factors. Nature. 2010 Feb 25;463(7284):1035-41.

[26] Thier M, Worsdorfer P, Lakes YB, Gorris R, Herms S, Opitz T, et al. Direct conversion of fibroblasts into stably expandable neural stem cells. Cell stem cell. 2012 Apr 6;10(4):473-9.

[27] Altman J. Autoradiographic and histological studies of postnatal neurogenesis. 3. Dating the time of production and onset of differentiation of cerebellar microneurons in rats. The Journal of comparative neurology. 1969 Jul;136(3):269-93.

[28] Ming GL, Song H. Adult neurogenesis in the mammalian central nervous system. Annual review of neuroscience. 2005;28:223-50.

[29] Gage FH. Mammalian neural stem cells. Science (New York, NY. 2000 Feb 25;287(5457):1433-8.

[30] Lois C, Alvarez-Buylla A. Proliferating subventricular zone cells in the adult mammalian forebrain can differentiate into neurons and glia. Proceedings of the National Academy of Sciences of the United States of America. 1993 Mar 1;90(5):2074-7.

[31] Kaplan MS, Bell DH. Mitotic neuroblasts in the 9-day-old and 11-month-old rodent hippocampus. J Neurosci. 1984 Jun;4(6):1429-41.

[32] Cameron HA, Woolley CS, McEwen BS, Gould E. Differentiation of newly born neurons and glia in the dentate gyrus of the adult rat. Neuroscience. 1993 Sep;56(2): 337-44.

[33] Goldman SA, Chen Z. Perivascular instruction of cell genesis and fate in the adult brain. Nature neuroscience. 2011 Nov;14(11):1382-9.

[34] Fuchs E, Tumbar T, Guasch G. Socializing with the neighbors: stem cells and their niche. Cell. 2004 Mar 19;116(6):769-78.

[35] Tavazoie M, Van der Veken L, Silva-Vargas V, Louissaint M, Colonna L, Zaidi B, et al. A specialized vascular niche for adult neural stem cells. Cell stem cell. 2008 Sep 11;3(3):279-88.

[36] Shen Q, Wang Y, Kokovay E, Lin G, Chuang SM, Goderie SK, et al. Adult SVZ stem cells lie in a vascular niche: a quantitative analysis of niche cell-cell interactions. Cell stem cell. 2008 Sep 11;3(3):289-300.

[37] Fuentealba LC, Obernier K, Alvarez-Buylla A. Adult neural stem cells bridge their niche. Cell stem cell. 2012 Jun 14;10(6):698-708.

[38] Alvarez-Buylla A, Garcia-Verdugo JM. Neurogenesis in adult subventricular zone. J Neurosci. 2002 Feb 1;22(3):629-34.

[39] Merkle FT, Mirzadeh Z, Alvarez-Buylla A. Mosaic organization of neural stem cells in the adult brain. Science (New York, NY. 2007 Jul 20;317(5836):381-4.

[40] Doetsch F. The glial identity of neural stem cells. Nature neuroscience. 2003 Nov; 6(11):1127-34.

[41] Kriegstein A, Alvarez-Buylla A. The glial nature of embryonic and adult neural stem cells. Annual review of neuroscience. 2009;32:149-84.

[42] Mirzadeh Z, Merkle FT, Soriano-Navarro M, Garcia-Verdugo JM, Alvarez-Buylla A. Neural stem cells confer unique pinwheel architecture to the ventricular surface in neurogenic regions of the adult brain. Cell stem cell. 2008 Sep 11;3(3):265-78.

[43] Sawamoto K, Wichterle H, Gonzalez-Perez O, Cholfin JA, Yamada M, Spassky N, et al. New neurons follow the flow of cerebrospinal fluid in the adult brain. Science (New York, NY. 2006 Feb 3;311(5761):629-32.

[44] Kokovay E, Wang Y, Kusek G, Wurster R, Lederman P, Lowry N, et al. VCAM1 is essential to maintain the structure of the SVZ niche and acts as an environmental sensor to regulate SVZ lineage progression. Cell stem cell. 2012 Aug 3;11(2):220-30.

[45] Lehtinen MK, Zappaterra MW, Chen X, Yang YJ, Hill AD, Lun M, et al. The cerebrospinal fluid provides a proliferative niche for neural progenitor cells. Neuron. 2011 Mar 10;69(5):893-905.

[46] Kopan R, Ilagan MX. The canonical Notch signaling pathway: unfolding the activation mechanism. Cell. 2009 Apr 17;137(2):216-33.

[47] Chapouton P, Jagasia R, Bally-Cuif L. Adult neurogenesis in non-mammalian vertebrates. Bioessays. 2007 Aug;29(8):745-57.

[48] Fernando RN, Eleuteri B, Abdelhady S, Nussenzweig A, Andang M, Ernfors P. Cell cycle restriction by histone H2AX limits proliferation of adult neural stem cells. Proceedings of the National Academy of Sciences of the United States of America. 2011 Apr 5;108(14):5837-42.

[49] Villeda SA, Luo J, Mosher KI, Zou B, Britschgi M, Bieri G, et al. The ageing systemic milieu negatively regulates neurogenesis and cognitive function. Nature. 2011 Sep 1;477(7362):90-4.

[50] Mosher KI, Andres RH, Fukuhara T, Bieri G, Hasegawa-Moriyama M, He Y, et al. Neural progenitor cells regulate microglia functions and activity. Nature neuroscience. 2012 Nov;15(11):1485-7.

[51] Curtis MA, Kam M, Nannmark U, Anderson MF, Axell MZ, Wikkelso C, et al. Human neuroblasts migrate to the olfactory bulb via a lateral ventricular extension. Science (New York, NY. 2007 Mar 2;315(5816):1243-9.

[52] Sanai N, Tramontin AD, Quinones-Hinojosa A, Barbaro NM, Gupta N, Kunwar S, et al. Unique astrocyte ribbon in adult human brain contains neural stem cells but lacks chain migration. Nature. 2004 Feb 19;427(6976):740-4.

[53] Eriksson PS, Perfilieva E, Bjork-Eriksson T, Alborn AM, Nordborg C, Peterson DA, et al. Neurogenesis in the adult human hippocampus. Nature medicine. 1998 Nov; 4(11):1313-7.

[54] Quinones-Hinojosa A, Sanai N, Soriano-Navarro M, Gonzalez-Perez O, Mirzadeh Z, Gil-Perotin S, et al. Cellular composition and cytoarchitecture of the adult human subventricular zone: a niche of neural stem cells. The Journal of comparative neurology. 2006 Jan 20;494(3):415-34.

[55] Ming GL, Song H. Adult neurogenesis in the mammalian brain: significant answers and significant questions. Neuron. 2011 May 26;70(4):687-702.

[56] Yang Z, Ming GL, Song H. Postnatal neurogenesis in the human forebrain: from two migratory streams to dribbles. Cell stem cell. 2011 Nov 4;9(5):385-6.

[57] Guerrero-Cazares H, Gonzalez-Perez O, Soriano-Navarro M, Zamora-Berridi G, Garcia-Verdugo JM, Quinones-Hinojosa A. Cytoarchitecture of the lateral ganglionic eminence and rostral extension of the lateral ventricle in the human fetal brain. The Journal of comparative neurology. 2011 Apr 15;519(6):1165-80.

[58] Sanai N, Nguyen T, Ihrie RA, Mirzadeh Z, Tsai HH, Wong M, et al. Corridors of migrating neurons in the human brain and their decline during infancy. Nature. 2011 Oct 20;478(7369):382-6.

[59] Seri B, Garcia-Verdugo JM, McEwen BS, Alvarez-Buylla A. Astrocytes give rise to new neurons in the adult mammalian hippocampus. J Neurosci. 2001 Sep 15;21(18): 7153-60.

[60] Bonaguidi MA, Song J, Ming GL, Song H. A unifying hypothesis on mammalian neural stem cell properties in the adult hippocampus. Current opinion in neurobiology. 2012 Oct;22(5):754-61.

[61] Seri B, Garcia-Verdugo JM, Collado-Morente L, McEwen BS, Alvarez-Buylla A. Cell types, lineage, and architecture of the germinal zone in the adult dentate gyrus. The Journal of comparative neurology. 2004 Oct 25;478(4):359-78.

[62] Lugert S, Vogt M, Tchorz JS, Muller M, Giachino C, Taylor V. Homeostatic neurogenesis in the adult hippocampus does not involve amplification of Ascl1(high) intermediate progenitors. Nature communications. 2012;3:670.

[63] Ehm O, Goritz C, Covic M, Schaffner I, Schwarz TJ, Karaca E, et al. RBPJkappa-dependent signaling is essential for long-term maintenance of neural stem cells in the adult hippocampus. J Neurosci. 2010 Oct 13;30(41):13794-807.

[64] Lavado A, Lagutin OV, Chow LM, Baker SJ, Oliver G. Prox1 is required for granule cell maturation and intermediate progenitor maintenance during brain neurogenesis. PLoS biology. 2010;8(8).

[65] Song J, Zhong C, Bonaguidi MA, Sun GJ, Hsu D, Gu Y, et al. Neuronal circuitry mechanism regulating adult quiescent neural stem-cell fate decision. Nature. 2012 Sep 6;489(7414):150-4.

[66] Dayer AG, Ford AA, Cleaver KM, Yassaee M, Cameron HA. Short-term and long-term survival of new neurons in the rat dentate gyrus. The Journal of comparative neurology. 2003 Jun 9;460(4):563-72.

[67] Schwartz M, Shechter R. Protective autoimmunity functions by intracranial immunosurveillance to support the mind: The missing link between health and disease. Molecular psychiatry. 2010 Apr;15(4):342-54.

[68] Ziv Y, Ron N, Butovsky O, Landa G, Sudai E, Greenberg N, et al. Immune cells contribute to the maintenance of neurogenesis and spatial learning abilities in adulthood. Nature neuroscience. 2006 Feb;9(2):268-75.

[69] Gould E, Tanapat P. Lesion-induced proliferation of neuronal progenitors in the dentate gyrus of the adult rat. Neuroscience. 1997 Sep;80(2):427-36.

[70] Yang L, Benardo LS, Valsamis H, Ling DS. Acute injury to superficial cortex leads to a decrease in synaptic inhibition and increase in excitation in neocortical layer V pyramidal cells. Journal of neurophysiology. 2007 Jan;97(1):178-87.

[71] Monje ML, Toda H, Palmer TD. Inflammatory blockade restores adult hippocampal neurogenesis. Science (New York, NY. 2003 Dec 5;302(5651):1760-5.

[72] Huehnchen P, Prozorovski T, Klaissle P, Lesemann A, Ingwersen J, Wolf SA, et al. Modulation of adult hippocampal neurogenesis during myelin-directed autoimmune neuroinflammation. Glia. 2011 Jan;59(1):132-42.

[73] Roosendaal SD, Hulst HE, Vrenken H, Feenstra HE, Castelijns JA, Pouwels PJ, et al. Structural and functional hippocampal changes in multiple sclerosis patients with intact memory function. Radiology. 2010 May;255(2):595-604.

[74] Rasmussen S, Imitola J, Ayuso-Sacido A, Wang Y, Starossom SC, Kivisakk P, et al. Reversible neural stem cell niche dysfunction in a model of multiple sclerosis. Annals of neurology. 2011 May;69(5):878-91.

[75] Tepavcevic V, Lazarini F, Alfaro-Cervello C, Kerninon C, Yoshikawa K, Garcia-Verdugo JM, et al. Inflammation-induced subventricular zone dysfunction leads to olfac-

tory deficits in a targeted mouse model of multiple sclerosis. The Journal of clinical investigation. 2011 Dec;121(12):4722-34.

[76] Ransohoff RM. Animal models of multiple sclerosis: the good, the bad and the bottom line. Nature neuroscience. 2012 Aug;15(8):1074-7.

[77] Nait-Oumesmar B, Picard-Riera N, Kerninon C, Decker L, Seilhean D, Hoglinger GU, et al. Activation of the subventricular zone in multiple sclerosis: evidence for early glial progenitors. Proceedings of the National Academy of Sciences of the United States of America. 2007 Mar 13;104(11):4694-9.

[78] Zivadinov R, Zorzon M, Monti Bragadin L, Pagliaro G, Cazzato G. Olfactory loss in multiple sclerosis. Journal of the neurological sciences. 1999 Oct 15;168(2):127-30.

[79] Zhang R, Zhang Z, Zhang C, Zhang L, Robin A, Wang Y, et al. Stroke transiently increases subventricular zone cell division from asymmetric to symmetric and increases neuronal differentiation in the adult rat. J Neurosci. 2004 Jun 23;24(25):5810-5.

[80] Jin K, Wang X, Xie L, Mao XO, Zhu W, Wang Y, et al. Evidence for stroke-induced neurogenesis in the human brain. Proceedings of the National Academy of Sciences of the United States of America. 2006 Aug 29;103(35):13198-202.

[81] Arvidsson A, Collin T, Kirik D, Kokaia Z, Lindvall O. Neuronal replacement from endogenous precursors in the adult brain after stroke. Nature medicine. 2002 Sep; 8(9):963-70.

[82] Yamashita T, Ninomiya M, Hernandez Acosta P, Garcia-Verdugo JM, Sunabori T, Sakaguchi M, et al. Subventricular zone-derived neuroblasts migrate and differentiate into mature neurons in the post-stroke adult striatum. J Neurosci. 2006 Jun 14;26(24): 6627-36.

[83] Thored P, Wood J, Arvidsson A, Cammenga J, Kokaia Z, Lindvall O. Long-term neuroblast migration along blood vessels in an area with transient angiogenesis and increased vascularization after stroke. Stroke; a journal of cerebral circulation. 2007 Nov;38(11):3032-9.

[84] Thored P, Arvidsson A, Cacci E, Ahlenius H, Kallur T, Darsalia V, et al. Persistent production of neurons from adult brain stem cells during recovery after stroke. Stem cells (Dayton, Ohio). 2006 Mar;24(3):739-47.

[85] Nadareishvili Z, Hallenbeck J. Neuronal regeneration after stroke. The New England journal of medicine. 2003 Jun 5;348(23):2355-6.

[86] Kang SS, Keasey MP, Arnold SA, Reid R, Geralds J, Hagg T. Endogenous CNTF mediates stroke-induced adult CNS neurogenesis in mice. Neurobiology of disease. 2012 Aug 31;49C:68-78.

[87] Guerra-Crespo M, Gleason D, Sistos A, Toosky T, Solaroglu I, Zhang JH, et al. Transforming growth factor-alpha induces neurogenesis and behavioral improvement in a chronic stroke model. Neuroscience. 2009 May 5;160(2):470-83.

[88] Sun Y, Jin K, Xie L, Childs J, Mao XO, Logvinova A, et al. VEGF-induced neuroprotection, neurogenesis, and angiogenesis after focal cerebral ischemia. The Journal of clinical investigation. 2003 Jun;111(12):1843-51.

[89] Jin K, Mao XO, Sun Y, Xie L, Greenberg DA. Stem cell factor stimulates neurogenesis in vitro and in vivo. The Journal of clinical investigation. 2002 Aug;110(3):311-9.

[90] Shingo T, Sorokan ST, Shimazaki T, Weiss S. Erythropoietin regulates the in vitro and in vivo production of neuronal progenitors by mammalian forebrain neural stem cells. J Neurosci. 2001 Dec 15;21(24):9733-43.

[91] Weiss S, Dunne C, Hewson J, Wohl C, Wheatley M, Peterson AC, et al. Multipotent CNS stem cells are present in the adult mammalian spinal cord and ventricular neuroaxis. J Neurosci. 1996 Dec 1;16(23):7599-609.

[92] Xu R, Wu C, Tao Y, Yi J, Yang Y, Zhang X, et al. Nestin-positive cells in the spinal cord: a potential source of neural stem cells. Int J Dev Neurosci. 2008 Nov;26(7): 813-20.

[93] Ke Y, Chi L, Xu R, Luo C, Gozal D, Liu R. Early response of endogenous adult neural progenitor cells to acute spinal cord injury in mice. Stem cells (Dayton, Ohio). 2006 Apr;24(4):1011-9.

[94] Vaquero J, Ramiro MJ, Oya S, Cabezudo JM. Ependymal reaction after experimental spinal cord injury. Acta neurochirurgica. 1981;55(3-4):295-302.

[95] Beattie MS, Bresnahan JC, Komon J, Tovar CA, Van Meter M, Anderson DK, et al. Endogenous repair after spinal cord contusion injuries in the rat. Experimental neurology. 1997 Dec;148(2):453-63.

[96] Felix MS, Popa N, Djelloul M, Boucraut J, Gauthier P, Bauer S, et al. Alteration of forebrain neurogenesis after cervical spinal cord injury in the adult rat. Frontiers in neuroscience. 2012;6:45.

[97] Parent JM, Valentin VV, Lowenstein DH. Prolonged seizures increase proliferating neuroblasts in the adult rat subventricular zone-olfactory bulb pathway. J Neurosci. 2002 Apr 15;22(8):3174-88.

[98] Jessberger S, Romer B, Babu H, Kempermann G. Seizures induce proliferation and dispersion of doublecortin-positive hippocampal progenitor cells. Experimental neurology. 2005 Dec;196(2):342-51.

[99] Overstreet-Wadiche LS, Bromberg DA, Bensen AL, Westbrook GL. Seizures accelerate functional integration of adult-generated granule cells. J Neurosci. 2006 Apr 12;26(15):4095-103.

[100] Mathern GW, Leiphart JL, De Vera A, Adelson PD, Seki T, Neder L, et al. Seizures decrease postnatal neurogenesis and granule cell development in the human fascia dentata. Epilepsia. 2002;43 Suppl 5:68-73.

[101] Haughey NJ, Liu D, Nath A, Borchard AC, Mattson MP. Disruption of neurogenesis in the subventricular zone of adult mice, and in human cortical neuronal precursor cells in culture, by amyloid beta-peptide: implications for the pathogenesis of Alzheimer's disease. Neuromolecular medicine. 2002;1(2):125-35.

[102] Yu Y, He J, Zhang Y, Luo H, Zhu S, Yang Y, et al. Increased hippocampal neurogenesis in the progressive stage of Alzheimer's disease phenotype in an APP/PS1 double transgenic mouse model. Hippocampus. 2009 Dec;19(12):1247-53.

[103] Rodriguez JJ, Jones VC, Verkhratsky A. Impaired cell proliferation in the subventricular zone in an Alzheimer's disease model. Neuroreport. 2009 Jul 1;20(10):907-12.

[104] Ermini FV, Grathwohl S, Radde R, Yamaguchi M, Staufenbiel M, Palmer TD, et al. Neurogenesis and alterations of neural stem cells in mouse models of cerebral amyloidosis. The American journal of pathology. 2008 Jun;172(6):1520-8.

[105] Zhang C, McNeil E, Dressler L, Siman R. Long-lasting impairment in hippocampal neurogenesis associated with amyloid deposition in a knock-in mouse model of familial Alzheimer's disease. Experimental neurology. 2007 Mar;204(1):77-87.

[106] Demars M, Hu YS, Gadadhar A, Lazarov O. Impaired neurogenesis is an early event in the etiology of familial Alzheimer's disease in transgenic mice. Journal of neuroscience research. 2010 Aug 1;88(10):2103-17.

[107] Jin K, Peel AL, Mao XO, Xie L, Cottrell BA, Henshall DC, et al. Increased hippocampal neurogenesis in Alzheimer's disease. Proceedings of the National Academy of Sciences of the United States of America. 2004 Jan 6;101(1):343-7.

[108] Parent JM, Vexler ZS, Gong C, Derugin N, Ferriero DM. Rat forebrain neurogenesis and striatal neuron replacement after focal stroke. Annals of neurology. 2002 Dec; 52(6):802-13.

[109] Bonfanti L. From hydra regeneration to human brain structural plasticity: a long trip through narrowing roads. TheScientificWorldJournal. 2011;11:1270-99.

[110] Schwartz M, Shechter R. Systemic inflammatory cells fight off neurodegenerative disease. Nat Rev Neurol. 2010 Jul;6(7):405-10.

[111] Nimmerjahn A, Kirchhoff F, Helmchen F. Resting microglial cells are highly dynamic surveillants of brain parenchyma in vivo. Science (New York, NY. 2005 May 27;308(5726):1314-8.

[112] Hanisch UK, Kettenmann H. Microglia: active sensor and versatile effector cells in the normal and pathologic brain. Nature neuroscience. 2007 Nov;10(11):1387-94.

[113] Lalancette-Hebert M, Gowing G, Simard A, Weng YC, Kriz J. Selective ablation of proliferating microglial cells exacerbates ischemic injury in the brain. J Neurosci. 2007 Mar 7;27(10):2596-605.

[114] Smith JA, Das A, Ray SK, Banik NL. Role of pro-inflammatory cytokines released from microglia in neurodegenerative diseases. Brain research bulletin. 2012 Jan 4;87(1):10-20.

[115] Shechter R, London A, Varol C, Raposo C, Cusimano M, Yovel G, et al. Infiltrating blood-derived macrophages are vital cells playing an anti-inflammatory role in recovery from spinal cord injury in mice. PLoS medicine. 2009 Jul;6(7):e1000113.

[116] Engelhardt B, Ransohoff RM. The ins and outs of T-lymphocyte trafficking to the CNS: anatomical sites and molecular mechanisms. Trends in immunology. 2005 Sep; 26(9):485-95.

[117] Kivisakk P, Imitola J, Rasmussen S, Elyaman W, Zhu B, Ransohoff RM, et al. Localizing central nervous system immune surveillance: meningeal antigen-presenting cells activate T cells during experimental autoimmune encephalomyelitis. Annals of neurology. 2009 Apr;65(4):457-69.

[118] Weller RO, Kida S, Zhang ET. Pathways of fluid drainage from the brain--morphological aspects and immunological significance in rat and man. Brain pathology (Zurich, Switzerland). 1992 Oct;2(4):277-84.

[119] Carare RO, Bernardes-Silva M, Newman TA, Page AM, Nicoll JA, Perry VH, et al. Solutes, but not cells, drain from the brain parenchyma along basement membranes of capillaries and arteries: significance for cerebral amyloid angiopathy and neuroimmunology. Neuropathology and applied neurobiology. 2008 Apr;34(2):131-44.

[120] Moalem G, Leibowitz-Amit R, Yoles E, Mor F, Cohen IR, Schwartz M. Autoimmune T cells protect neurons from secondary degeneration after central nervous system axotomy. Nature medicine. 1999 Jan;5(1):49-55.

[121] Yoles E, Hauben E, Palgi O, Agranov E, Gothilf A, Cohen A, et al. Protective autoimmunity is a physiological response to CNS trauma. J Neurosci. 2001 Jun 1;21(11): 3740-8.

[122] Kipnis J, Mizrahi T, Yoles E, Ben-Nun A, Schwartz M. Myelin specific Th1 cells are necessary for post-traumatic protective autoimmunity. Journal of neuroimmunology. 2002 Sep;130(1-2):78-85.

[123] Schwartz M, Hauben E. T cell-based therapeutic vaccination for spinal cord injury. Progress in brain research. 2002;137:401-6.

[124] Hauben E, Butovsky O, Nevo U, Yoles E, Moalem G, Agranov E, et al. Passive or active immunization with myelin basic protein promotes recovery from spinal cord contusion. J Neurosci. 2000 Sep 1;20(17):6421-30.

[125] Hauben E, Ibarra A, Mizrahi T, Barouch R, Agranov E, Schwartz M. Vaccination with a Nogo-A-derived peptide after incomplete spinal-cord injury promotes recovery via a T-cell-mediated neuroprotective response: comparison with other myelin antigens. Proceedings of the National Academy of Sciences of the United States of America. 2001 Dec 18;98(26):15173-8.

[126] Hohlfeld R, Kerschensteiner M, Stadelmann C, Lassmann H, Wekerle H. The neuroprotective effect of inflammation: implications for the therapy of multiple sclerosis. Journal of neuroimmunology. 2000 Jul 24;107(2):161-6.

[127] Eugster HP, Frei K, Kopf M, Lassmann H, Fontana A. IL-6-deficient mice resist myelin oligodendrocyte glycoprotein-induced autoimmune encephalomyelitis. European journal of immunology. 1998 Jul;28(7):2178-87.

[128] Lacroix S, Chang L, Rose-John S, Tuszynski MH. Delivery of hyper-interleukin-6 to the injured spinal cord increases neutrophil and macrophage infiltration and inhibits axonal growth. The Journal of comparative neurology. 2002 Dec 16;454(3):213-28.

[129] Klusman I, Schwab ME. Effects of pro-inflammatory cytokines in experimental spinal cord injury. Brain Res. 1997 Jul 11;762(1-2):173-84.

[130] Whiteley W, Jackson C, Lewis S, Lowe G, Rumley A, Sandercock P, et al. Inflammatory markers and poor outcome after stroke: a prospective cohort study and systematic review of interleukin-6. PLoS medicine. 2009 Sep;6(9):e1000145.

[131] Okada S, Nakamura M, Mikami Y, Shimazaki T, Mihara M, Ohsugi Y, et al. Blockade of interleukin-6 receptor suppresses reactive astrogliosis and ameliorates functional recovery in experimental spinal cord injury. Journal of neuroscience research. 2004 Apr 15;76(2):265-76.

[132] Serada S, Fujimoto M, Mihara M, Koike N, Ohsugi Y, Nomura S, et al. IL-6 blockade inhibits the induction of myelin antigen-specific Th17 cells and Th1 cells in experimental autoimmune encephalomyelitis. Proceedings of the National Academy of Sciences of the United States of America. 2008 Jul 1;105(26):9041-6.

[133] Gertz K, Kronenberg G, Kalin RE, Baldinger T, Werner C, Balkaya M, et al. Essential role of interleukin-6 in post-stroke angiogenesis. Brain. 2012 Jun;135(Pt 6):1964-80.

[134] Reynolds BA, Weiss S. Generation of neurons and astrocytes from isolated cells of the adult mammalian central nervous system. Science (New York, NY. 1992 Mar 27;255(5052):1707-10.

[135] Aboody KS, Brown A, Rainov NG, Bower KA, Liu S, Yang W, et al. Neural stem cells display extensive tropism for pathology in adult brain: evidence from intracranial gliomas. Proceedings of the National Academy of Sciences of the United States of America. 2000 Nov 7;97(23):12846-51.

[136] Einstein O, Karussis D, Grigoriadis N, Mizrachi-Kol R, Reinhartz E, Abramsky O, et al. Intraventricular transplantation of neural precursor cell spheres attenuates acute

experimental allergic encephalomyelitis. Molecular and cellular neurosciences. 2003 Dec;24(4):1074-82.

[137] Jeong SW, Chu K, Jung KH, Kim SU, Kim M, Roh JK. Human neural stem cell transplantation promotes functional recovery in rats with experimental intracerebral hemorrhage. Stroke; a journal of cerebral circulation. 2003 Sep;34(9):2258-63.

[138] Chu K, Kim M, Jung KH, Jeon D, Lee ST, Kim J, et al. Human neural stem cell transplantation reduces spontaneous recurrent seizures following pilocarpine-induced status epilepticus in adult rats. Brain Res. 2004 Oct 15;1023(2):213-21.

[139] Einstein O, Grigoriadis N, Mizrachi-Kol R, Reinhartz E, Polyzoidou E, Lavon I, et al. Transplanted neural precursor cells reduce brain inflammation to attenuate chronic experimental autoimmune encephalomyelitis. Experimental neurology. 2006 Apr; 198(2):275-84.

[140] Bacigaluppi M, Pluchino S, Peruzzotti-Jametti L, Kilic E, Kilic U, Salani G, et al. Delayed post-ischaemic neuroprotection following systemic neural stem cell transplantation involves multiple mechanisms. Brain. 2009 Aug;132(Pt 8):2239-51.

[141] Sun C, Zhang H, Li J, Huang H, Cheng H, Wang Y, et al. Modulation of the major histocompatibility complex by neural stem cell-derived neurotrophic factors used for regenerative therapy in a rat model of stroke. Journal of translational medicine. 2010;8:77.

[142] Aharonowiz M, Einstein O, Fainstein N, Lassmann H, Reubinoff B, Ben-Hur T. Neuroprotective effect of transplanted human embryonic stem cell-derived neural precursors in an animal model of multiple sclerosis. PloS one. 2008;3(9):e3145.

[143] Zlokovic BV. The blood-brain barrier in health and chronic neurodegenerative disorders. Neuron. 2008 Jan 24;57(2):178-201.

[144] Wolburg H, Lippoldt A. Tight junctions of the blood-brain barrier: development, composition and regulation. Vascular pharmacology. 2002 Jun;38(6):323-37.

[145] Begley DJ. ABC transporters and the blood-brain barrier. Current pharmaceutical design. 2004;10(12):1295-312.

[146] Agrawal S, Anderson P, Durbeej M, van Rooijen N, Ivars F, Opdenakker G, et al. Dystroglycan is selectively cleaved at the parenchymal basement membrane at sites of leukocyte extravasation in experimental autoimmune encephalomyelitis. The Journal of experimental medicine. 2006 Apr 17;203(4):1007-19.

[147] Engelhardt B, Ransohoff RM. Capture, crawl, cross: the T cell code to breach the blood-brain barriers. Trends in immunology. 2012 Aug 24.

[148] Nag S, Venugopalan R, Stewart DJ. Increased caveolin-1 expression precedes decreased expression of occludin and claudin-5 during blood-brain barrier breakdown. Acta neuropathologica. 2007 Nov;114(5):459-69.

[149] Readnower RD, Chavko M, Adeeb S, Conroy MD, Pauly JR, McCarron RM, et al. Increase in blood-brain barrier permeability, oxidative stress, and activated microglia in a rat model of blast-induced traumatic brain injury. Journal of neuroscience research. 2010 Dec;88(16):3530-9.

[150] Jiao H, Wang Z, Liu Y, Wang P, Xue Y. Specific role of tight junction proteins claudin-5, occludin, and ZO-1 of the blood-brain barrier in a focal cerebral ischemic insult. J Mol Neurosci. 2011 Jun;44(2):130-9.

[151] van Assema DM, Lubberink M, Bauer M, van der Flier WM, Schuit RC, Windhorst AD, et al. Blood-brain barrier P-glycoprotein function in Alzheimer's disease. Brain. 2012 Jan;135(Pt 1):181-9.

[152] Errede M, Girolamo F, Ferrara G, Strippoli M, Morando S, Boldrin V, et al. Blood-Brain Barrier Alterations in the Cerebral Cortex in Experimental Autoimmune Encephalomyelitis. Journal of neuropathology and experimental neurology. 2012 Oct; 71(10):840-54.

[153] Hickey WF. Migration of hematogenous cells through the blood-brain barrier and the initiation of CNS inflammation. Brain pathology (Zurich, Switzerland). 1991 Jan;1(2): 97-105.

[154] Engelhardt B, Wolburg-Buchholz K, Wolburg H. Involvement of the choroid plexus in central nervous system inflammation. Microscopy research and technique. 2001 Jan 1;52(1):112-29.

[155] Archambault AS, Sim J, Gimenez MA, Russell JH. Defining antigen-dependent stages of T cell migration from the blood to the central nervous system parenchyma. European journal of immunology. 2005 Apr;35(4):1076-85.

[156] Butcher EC, Picker LJ. Lymphocyte homing and homeostasis. Science (New York, NY. 1996 Apr 5;272(5258):60-6.

[157] Steiner O, Coisne C, Cecchelli R, Boscacci R, Deutsch U, Engelhardt B, et al. Differential roles for endothelial ICAM-1, ICAM-2, and VCAM-1 in shear-resistant T cell arrest, polarization, and directed crawling on blood-brain barrier endothelium. J Immunol. 2010 Oct 15;185(8):4846-55.

[158] Yednock TA, Cannon C, Fritz LC, Sanchez-Madrid F, Steinman L, Karin N. Prevention of experimental autoimmune encephalomyelitis by antibodies against alpha 4 beta 1 integrin. Nature. 1992 Mar 5;356(6364):63-6.

[159] Coisne C, Mao W, Engelhardt B. Cutting edge: Natalizumab blocks adhesion but not initial contact of human T cells to the blood-brain barrier in vivo in an animal model of multiple sclerosis. J Immunol. 2009 May 15;182(10):5909-13.

[160] Man S, Tucky B, Cotleur A, Drazba J, Takeshita Y, Ransohoff RM. CXCL12-induced monocyte-endothelial interactions promote lymphocyte transmigration across an in vitro blood-brain barrier. Science translational medicine. 2012 Feb 1;4(119):119ra14.

[161] Mahad D, Callahan MK, Williams KA, Ubogu EE, Kivisakk P, Tucky B, et al. Modulating CCR2 and CCL2 at the blood-brain barrier: relevance for multiple sclerosis pathogenesis. Brain. 2006 Jan;129(Pt 1):212-23.

[162] Quandt J, Dorovini-Zis K. The beta chemokines CCL4 and CCL5 enhance adhesion of specific CD4+ T cell subsets to human brain endothelial cells. Journal of neuropathology and experimental neurology. 2004 Apr;63(4):350-62.

[163] Alt C, Laschinger M, Engelhardt B. Functional expression of the lymphoid chemokines CCL19 (ELC) and CCL 21 (SLC) at the blood-brain barrier suggests their involvement in G-protein-dependent lymphocyte recruitment into the central nervous system during experimental autoimmune encephalomyelitis. European journal of immunology. 2002 Aug;32(8):2133-44.

[164] Nourshargh S, Hordijk PL, Sixt M. Breaching multiple barriers: leukocyte motility through venular walls and the interstitium. Nat Rev Mol Cell Biol. 2010 May;11(5): 366-78.

[165] Carman CV. Mechanisms for transcellular diapedesis: probing and pathfinding by 'invadosome-like protrusions'. Journal of cell science. 2009 Sep 1;122(Pt 17):3025-35.

[166] Lindvall O, Kokaia Z. Stem cell research in stroke: how far from the clinic? Stroke; a journal of cerebral circulation. 2011 Aug;42(8):2369-75.

[167] Takeuchi H, Natsume A, Wakabayashi T, Aoshima C, Shimato S, Ito M, et al. Intravenously transplanted human neural stem cells migrate to the injured spinal cord in adult mice in an SDF-1- and HGF-dependent manner. Neuroscience letters. 2007 Oct 16;426(2):69-74.

[168] Sandner B, Prang P, Rivera FJ, Aigner L, Blesch A, Weidner N. Neural stem cells for spinal cord repair. Cell and tissue research. 2012 Jul;349(1):349-62.

[169] Shetty AK. Progress in cell grafting therapy for temporal lobe epilepsy. Neurotherapeutics. 2011 Oct;8(4):721-35.

[170] Ahmed AU, Lesniak MS. Glioblastoma multiforme: can neural stem cells deliver the therapeutic payload and fulfill the clinical promise? Expert review of neurotherapeutics. 2011 Jun;11(6):775-7.

[171] Carbajal KS, Schaumburg C, Strieter R, Kane J, Lane TE. Migration of engrafted neural stem cells is mediated by CXCL12 signaling through CXCR4 in a viral model of multiple sclerosis. Proceedings of the National Academy of Sciences of the United States of America. 2010 Jun 15;107(24):11068-73.

[172] Andres RH, Choi R, Pendharkar AV, Gaeta X, Wang N, Nathan JK, et al. The CCR2/CCL2 interaction mediates the transendothelial recruitment of intravascularly delivered neural stem cells to the ischemic brain. Stroke; a journal of cerebral circulation. 2012 Oct;42(10):2923-31.

[173] Peng H, Huang Y, Rose J, Erichsen D, Herek S, Fujii N, et al. Stromal cell-derived factor 1-mediated CXCR4 signaling in rat and human cortical neural progenitor cells. Journal of neuroscience research. 2004 Apr 1;76(1):35-50.

[174] DeGrendele HC, Estess P, Siegelman MH. Requirement for CD44 in activated T cell extravasation into an inflammatory site. Science (New York, NY. 1997 Oct 24;278(5338):672-5.

[175] Rampon C, Weiss N, Deboux C, Chaverot N, Miller F, Buchet D, et al. Molecular mechanism of systemic delivery of neural precursor cells to the brain: assembly of brain endothelial apical cups and control of transmigration by CD44. Stem cells (Dayton, Ohio). 2008 Jul;26(7):1673-82.

[176] Carman CV, Springer TA. A transmigratory cup in leukocyte diapedesis both through individual vascular endothelial cells and between them. The Journal of cell biology. 2004 Oct 25;167(2):377-88.

[177] Mueller A, Mahmoud NG, Strange PG. Diverse signalling by different chemokines through the chemokine receptor CCR5. Biochemical pharmacology. 2006 Sep 14;72(6):739-48.

[178] Weiss N, Deboux C, Chaverot N, Miller F, Baron-Van Evercooren A, Couraud PO, et al. IL8 and CXCL13 are potent chemokines for the recruitment of human neural precursor cells across brain endothelial cells. Journal of neuroimmunology. 2010 Jun; 223(1-2):131-4.

[179] Jaderstad J, Jaderstad LM, Li J, Chintawar S, Salto C, Pandolfo M, et al. Communication via gap junctions underlies early functional and beneficial interactions between grafted neural stem cells and the host. Proceedings of the National Academy of Sciences of the United States of America. Mar 16;107(11):5184-9.

[180] Gnecchi M, Zhang Z, Ni A, Dzau VJ. Paracrine mechanisms in adult stem cell signaling and therapy. Circulation research. 2008 Nov 21;103(11):1204-19.

[181] Gerdes HH, Bukoreshtliev NV, Barroso JF. Tunneling nanotubes: a new route for the exchange of components between animal cells. FEBS letters. 2007 May 22;581(11): 2194-201.

[182] Thery C, Ostrowski M, Segura E. Membrane vesicles as conveyors of immune responses. Nat Rev Immunol. 2009 Aug;9(8):581-93.

[183] Vultur A, Cao J, Arulanandam R, Turkson J, Jove R, Greer P, et al. Cell-to-cell adhesion modulates Stat3 activity in normal and breast carcinoma cells. Oncogene. 2004 Apr 8;23(15):2600-16.

[184] Ben-Hur T, Ben-Menachem O, Furer V, Einstein O, Mizrachi-Kol R, Grigoriadis N. Effects of proinflammatory cytokines on the growth, fate, and motility of multipotential neural precursor cells. Molecular and cellular neurosciences. 2003 Nov;24(3): 623-31.

[185] Pluchino S, Zanotti L, Deleidi M, Martino G. Neural stem cells and their use as therapeutic tool in neurological disorders. Brain research. 2005 Apr;48(2):211-9.

[186] Einstein O, Friedman-Levi Y, Grigoriadis N, Ben-Hur T. Transplanted neural precursors enhance host brain-derived myelin regeneration. J Neurosci. 2009 Dec 16;29(50): 15694-702.

[187] Chu K, Kim M, Park KI, Jeong SW, Park HK, Jung KH, et al. Human neural stem cells improve sensorimotor deficits in the adult rat brain with experimental focal ischemia. Brain Res. 2004 Aug 6;1016(2):145-53.

[188] Martino G, Pluchino S. Neural stem cells: guardians of the brain. Nature cell biology. 2007 Sep;9(9):1031-4.

[189] Capone C, Frigerio S, Fumagalli S, Gelati M, Principato MC, Storini C, et al. Neurosphere-derived cells exert a neuroprotective action by changing the ischemic microenvironment. PloS one. 2007;2(4):e373.

[190] Jiang Q, Zhang ZG, Ding GL, Zhang L, Ewing JR, Wang L, et al. Investigation of neural progenitor cell induced angiogenesis after embolic stroke in rat using MRI. NeuroImage. 2005 Nov 15;28(3):698-707.

[191] Lu P, Jones LL, Snyder EY, Tuszynski MH. Neural stem cells constitutively secrete neurotrophic factors and promote extensive host axonal growth after spinal cord injury. Experimental neurology. 2003 Jun;181(2):115-29.

[192] Lee HJ, Kim KS, Park IH, Kim SU. Human neural stem cells over-expressing VEGF provide neuroprotection, angiogenesis and functional recovery in mouse stroke model. PloS one. 2007;2(1):e156.

[193] Teng YD, Lavik EB, Qu X, Park KI, Ourednik J, Zurakowski D, et al. Functional recovery following traumatic spinal cord injury mediated by a unique polymer scaffold seeded with neural stem cells. Proceedings of the National Academy of Sciences of the United States of America. 2002 Mar 5;99(5):3024-9.

[194] Kim SU, de Vellis J. Stem cell-based cell therapy in neurological diseases: a review. Journal of neuroscience research. 2009 Aug 1;87(10):2183-200.

[195] Lin LF, Doherty DH, Lile JD, Bektesh S, Collins F. GDNF: a glial cell line-derived neurotrophic factor for midbrain dopaminergic neurons. Science (New York, NY. 1993 May 21;260(5111):1130-2.

[196] Zhang WR, Sato K, Iwai M, Nagano I, Manabe Y, Abe K. Therapeutic time window of adenovirus-mediated GDNF gene transfer after transient middle cerebral artery occlusion in rat. Brain Res. 2002 Aug 23;947(1):140-5.

[197] Kobayashi T, Ahlenius H, Thored P, Kobayashi R, Kokaia Z, Lindvall O. Intracerebral infusion of glial cell line-derived neurotrophic factor promotes striatal neurogen-

esis after stroke in adult rats. Stroke; a journal of cerebral circulation. 2006 Sep;37(9): 2361-7.

[198] Chen B, Gao XQ, Yang CX, Tan SK, Sun ZL, Yan NH, et al. Neuroprotective effect of grafting GDNF gene-modified neural stem cells on cerebral ischemia in rats. Brain Res. 2009 Aug 11;1284:1-11.

[199] Nishimura Y, Natsume A, Ito M, Hara M, Motomura K, Fukuyama R, et al. Interferon-beta delivery via human neural stem cell abates glial scar formation in spinal cord injury. Cell transplantation. 2012 Oct 12.

[200] Kim HM, Hwang DH, Lee JE, Kim SU, Kim BG. Ex vivo VEGF delivery by neural stem cells enhances proliferation of glial progenitors, angiogenesis, and tissue sparing after spinal cord injury. PloS one. 2009;4(3):e4987.

[201] Lee HJ, Kim KS, Kim EJ, Choi HB, Lee KH, Park IH, et al. Brain transplantation of immortalized human neural stem cells promotes functional recovery in mouse intracerebral hemorrhage stroke model. Stem cells (Dayton, Ohio). 2007 May;25(5): 1204-12.

[202] Lee HJ, Lim IJ, Lee MC, Kim SU. Human neural stem cells genetically modified to overexpress brain-derived neurotrophic factor promote functional recovery and neuroprotection in a mouse stroke model. Journal of neuroscience research. 2010 Nov 15;88(15):3282-94.

[203] Modo M, Stroemer RP, Tang E, Patel S, Hodges H. Effects of implantation site of stem cell grafts on behavioral recovery from stroke damage. Stroke; a journal of cerebral circulation. 2002 Sep;33(9):2270-8.

[204] Zhang ZG, Jiang Q, Zhang R, Zhang L, Wang L, Zhang L, et al. Magnetic resonance imaging and neurosphere therapy of stroke in rat. Annals of neurology. 2003 Feb; 53(2):259-63.

[205] Kelly S, Bliss TM, Shah AK, Sun GH, Ma M, Foo WC, et al. Transplanted human fetal neural stem cells survive, migrate, and differentiate in ischemic rat cerebral cortex. Proceedings of the National Academy of Sciences of the United States of America. 2004 Aug 10;101(32):11839-44.

[206] Roitberg BZ, Mangubat E, Chen EY, Sugaya K, Thulborn KR, Kordower JH, et al. Survival and early differentiation of human neural stem cells transplanted in a non-human primate model of stroke. Journal of neurosurgery. 2006 Jul;105(1):96-102.

[207] Darsalia V, Kallur T, Kokaia Z. Survival, migration and neuronal differentiation of human fetal striatal and cortical neural stem cells grafted in stroke-damaged rat striatum. The European journal of neuroscience. 2007 Aug;26(3):605-14.

[208] Ben-Hur T. Immunomodulation by neural stem cells. Journal of the neurological sciences. 2008 Feb 15;265(1-2):102-4.

[209] Iadecola C, Anrather J. The immunology of stroke: from mechanisms to translation. Nature medicine. 2011 Jul;17(7):796-808.

[210] Molina-Holgado F, Grencis R, Rothwell NJ. Actions of exogenous and endogenous IL-10 on glial responses to bacterial LPS/cytokines. Glia. 2001 Feb;33(2):97-106.

[211] Croxford JL, Feldmann M, Chernajovsky Y, Baker D. Different therapeutic outcomes in experimental allergic encephalomyelitis dependent upon the mode of delivery of IL-10: a comparison of the effects of protein, adenoviral or retroviral IL-10 delivery into the central nervous system. J Immunol. 2001 Mar 15;166(6):4124-30.

[212] Boyd ZS, Kriatchko A, Yang J, Agarwal N, Wax MB, Patil RV. Interleukin-10 receptor signaling through STAT-3 regulates the apoptosis of retinal ganglion cells in response to stress. Investigative ophthalmology & visual science. 2003 Dec;44(12): 5206-11.

[213] Yang J, Jiang Z, Fitzgerald DC, Ma C, Yu S, Li H, et al. Adult neural stem cells expressing IL-10 confer potent immunomodulation and remyelination in experimental autoimmune encephalitis. The Journal of clinical investigation. 2009 Dec;119(12): 3678-91.

[214] Fainstein N, Vaknin I, Einstein O, Zisman P, Ben Sasson SZ, Baniyash M, et al. Neural precursor cells inhibit multiple inflammatory signals. Molecular and cellular neurosciences. 2008 Nov;39(3):335-41.

[215] Knight JC, Scharf EL, Mao-Draayer Y. Fas activation increases neural progenitor cell survival. Journal of neuroscience research. 2010 Mar;88(4):746-57.

[216] Wang L, Shi J, van Ginkel FW, Lan L, Niemeyer G, Martin DR, et al. Neural stem/ progenitor cells modulate immune responses by suppressing T lymphocytes with nitric oxide and prostaglandin E2. Experimental neurology. 2009 Mar;216(1):177-83.

[217] Ricci-Vitiani L, Casalbore P, Petrucci G, Lauretti L, Montano N, Larocca LM, et al. Influence of local environment on the differentiation of neural stem cells engrafted onto the injured spinal cord. Neurological research. 2006 Jul;28(5):488-92.

[218] Melzi R, Antonioli B, Mercalli A, Battaglia M, Valle A, Pluchino S, et al. Co-graft of allogeneic immune regulatory neural stem cells (NPC) and pancreatic islets mediates tolerance, while inducing NPC-derived tumors in mice. PloS one. 2010;5(4):e10357.

[219] Flugel A, Berkowicz T, Ritter T, Labeur M, Jenne DE, Li Z, et al. Migratory activity and functional changes of green fluorescent effector cells before and during experimental autoimmune encephalomyelitis. Immunity. 2001 May;14(5):547-60.

[220] de Vos AF, van Meurs M, Brok HP, Boven LA, Hintzen RQ, van der Valk P, et al. Transfer of central nervous system autoantigens and presentation in secondary lymphoid organs. J Immunol. 2002 Nov 15;169(10):5415-23.

[221] Mohindru M, Kang B, Kim BS. Functional maturation of proteolipid protein(139-151)-specific Th1 cells in the central nervous system in experimental autoimmune encephalomyelitis. Journal of neuroimmunology. 2004 Oct;155(1-2):127-35.

[222] Guzman R, Choi R, Gera A, De Los Angeles A, Andres RH, Steinberg GK. Intravascular cell replacement therapy for stroke. Neurosurgical focus. 2008;24(3-4):E15.

[223] Takahashi K, Yasuhara T, Shingo T, Muraoka K, Kameda M, Takeuchi A, et al. Embryonic neural stem cells transplanted in middle cerebral artery occlusion model of rats demonstrated potent therapeutic effects, compared to adult neural stem cells. Brain Res. 2008 Oct 9;1234:172-82.

[224] Karimi-Abdolrezaee S, Eftekharpour E, Wang J, Schut D, Fehlings MG. Synergistic effects of transplanted adult neural stem/progenitor cells, chondroitinase, and growth factors promote functional repair and plasticity of the chronically injured spinal cord. J Neurosci. 2010 Feb 3;30(5):1657-76.

[225] Zhang P, Li J, Liu Y, Chen X, Lu H, Kang Q, et al. Human embryonic neural stem cell transplantation increases subventricular zone cell proliferation and promotes peri-infarct angiogenesis after focal cerebral ischemia. Neuropathology. 2011 Aug;31(4): 384-91.

[226] Ziv Y, Avidan H, Pluchino S, Martino G, Schwartz M. Synergy between immune cells and adult neural stem/progenitor cells promotes functional recovery from spinal cord injury. Proceedings of the National Academy of Sciences of the United States of America. 2006 Aug 29;103(35):13174-9.

[227] Wu W, Chen X, Hu C, Li J, Yu Z, Cai W. Transplantation of neural stem cells expressing hypoxia-inducible factor-1alpha (HIF-1alpha) improves behavioral recovery in a rat stroke model. J Clin Neurosci. 2010 Jan;17(1):92-5.

[228] Jin K, Sun Y, Xie L, Mao XO, Childs J, Peel A, et al. Comparison of ischemia-directed migration of neural precursor cells after intrastriatal, intraventricular, or intravenous transplantation in the rat. Neurobiology of disease. 2005 Mar;18(2):366-74.

[229] Lundberg J, Le Blanc K, Soderman M, Andersson T, Holmin S. Endovascular transplantation of stem cells to the injured rat CNS. Neuroradiology. 2009 Oct;51(10): 661-7.

[230] Chu K, Park KI, Lee ST, Jung KH, Ko SY, Kang L, et al. Combined treatment of vascular endothelial growth factor and human neural stem cells in experimental focal cerebral ischemia. Neuroscience research. 2005 Dec;53(4):384-90.

[231] Vendrame M, Cassady J, Newcomb J, Butler T, Pennypacker KR, Zigova T, et al. Infusion of human umbilical cord blood cells in a rat model of stroke dose-dependently rescues behavioral deficits and reduces infarct volume. Stroke; a journal of cerebral circulation. 2004 Oct;35(10):2390-5.

[232] Doeppner TR, Ewert TA, Tonges L, Herz J, Zechariah A, ElAli A, et al. Transduction of neural precursor cells with TAT-heat shock protein 70 chaperone: therapeutic po-

tential against ischemic stroke after intrastriatal and systemic transplantation. Stem cells (Dayton, Ohio). Jun;30(6):1297-310.

[233] Walczak P, Zhang J, Gilad AA, Kedziorek DA, Ruiz-Cabello J, Young RG, et al. Dual-modality monitoring of targeted intraarterial delivery of mesenchymal stem cells after transient ischemia. Stroke; a journal of cerebral circulation. 2008 May;39(5): 1569-74.

[234] Li L, Jiang Q, Ding G, Zhang L, Zhang ZG, Li Q, et al. Effects of administration route on migration and distribution of neural progenitor cells transplanted into rats with focal cerebral ischemia, an MRI study. J Cereb Blood Flow Metab. Mar;30(3):653-62.

[235] Sykova E, Homola A, Mazanec R, Lachmann H, Konradova SL, Kobylka P, et al. Autologous bone marrow transplantation in patients with subacute and chronic spinal cord injury. Cell transplantation. 2006;15(8-9):675-87.

[236] Parr AM, Kulbatski I, Tator CH. Transplantation of adult rat spinal cord stem/progenitor cells for spinal cord injury. Journal of neurotrauma. 2007 May;24(5): 835-45.

[237] Cusimano M, Biziato D, Brambilla E, Donega M, Alfaro-Cervello C, Snider S, et al. Transplanted neural stem/precursor cells instruct phagocytes and reduce secondary tissue damage in the injured spinal cord. Brain. 2012 Feb;135(Pt 2):447-60.

[238] Brundin P, Strecker RE, Londos E, Bjorklund A. Dopamine neurons grafted unilaterally to the nucleus accumbens affect drug-induced circling and locomotion. Experimental brain research Experimentelle Hirnforschung. 1987;69(1):183-94.

[239] Freed CR, Greene PE, Breeze RE, Tsai WY, DuMouchel W, Kao R, et al. Transplantation of embryonic dopamine neurons for severe Parkinson's disease. The New England journal of medicine. 2001 Mar 8;344(10):710-9.

[240] Olanow CW, Goetz CG, Kordower JH, Stoessl AJ, Sossi V, Brin MF, et al. A double-blind controlled trial of bilateral fetal nigral transplantation in Parkinson's disease. Annals of neurology. 2003 Sep;54(3):403-14.

[241] Lindvall O. Why is it taking so long to develop clinically competitive stem cell therapies for CNS disorders? Cell stem cell. 2012 Jun 14;10(6):660-2.

[242] Daley GQ. The promise and perils of stem cell therapeutics. Cell stem cell. 2012 Jun 14;10(6):740-9.

[243] Hyun I, Lindvall O, Ahrlund-Richter L, Cattaneo E, Cavazzana-Calvo M, Cossu G, et al. New ISSCR guidelines underscore major principles for responsible translational stem cell research. Cell stem cell. 2008 Dec 4;3(6):607-9.

[244] Aboody K, Capela A, Niazi N, Stern JH, Temple S. Translating stem cell studies to the clinic for CNS repair: current state of the art and the need for a Rosetta stone. Neuron. 2011 May 26;70(4):597-613.

[245] Amariglio N, Hirshberg A, Scheithauer BW, Cohen Y, Loewenthal R, Trakhtenbrot L, et al. Donor-derived brain tumor following neural stem cell transplantation in an ataxia telangiectasia patient. PLoS medicine. 2009 Feb 17;6(2):e1000029.

[246] Feng Z, Gao F. Stem cell challenges in the treatment of neurodegenerative disease. CNS neuroscience & therapeutics. 2012 Feb;18(2):142-8.

[247] Steiner R, Huhn, S., Koch, T., Al-Uzri, A., Guillaime, D., Sutcliffe, T., Vogel, H., and Selden, N. CNS transplantation of purified human neural stem cells in infantile and late-infantile neuronal ceroid lipofuscinoses: Summary of the Phase I trial. Mol Genet Metab. 2010;99(S35).

[248] Glass JD, Boulis NM, Johe K, Rutkove SB, Federici T, Polak M, et al. Lumbar intraspinal injection of neural stem cells in patients with amyotrophic lateral sclerosis: results of a phase I trial in 12 patients. Stem cells (Dayton, Ohio). 2012 Jun;30(6):1144-51.

[249] Riley J, Federici T, Polak M, Kelly C, Glass J, Raore B, et al. Intraspinal stem cell transplantation in amyotrophic lateral sclerosis: a phase I safety trial, technical note, and lumbar safety outcomes. Neurosurgery. 2012 Aug;71(2):405-16; discussion 16.

[250] Gupta N, Henry RG, Strober J, Kang SM, Lim DA, Bucci M, et al. Neural stem cell engraftment and myelination in the human brain. Science translational medicine. 2012 Oct 10;4(155):155ra37.

[251] Pittenger MF, Mackay AM, Beck SC, Jaiswal RK, Douglas R, Mosca JD, et al. Multilineage potential of adult human mesenchymal stem cells. Science (New York, NY. 1999 Apr 2;284(5411):143-7.

[252] Brazelton TR, Rossi FM, Keshet GI, Blau HM. From marrow to brain: expression of neuronal phenotypes in adult mice. Science (New York, NY. 2000 Dec 1;290(5497): 1775-9.

[253] Woodbury D, Schwarz EJ, Prockop DJ, Black IB. Adult rat and human bone marrow stromal cells differentiate into neurons. Journal of neuroscience research. 2000 Aug 15;61(4):364-70.

[254] Sanchez-Ramos J, Song S, Cardozo-Pelaez F, Hazzi C, Stedeford T, Willing A, et al. Adult bone marrow stromal cells differentiate into neural cells in vitro. Experimental neurology. 2000 Aug;164(2):247-56.

[255] Joyce N, Annett G, Wirthlin L, Olson S, Bauer G, Nolta JA. Mesenchymal stem cells for the treatment of neurodegenerative disease. Regenerative medicine. 2010 Nov; 5(6):933-46.

[256] Gao A, Peng Y, Deng Y, Qing H. Potential therapeutic applications of differentiated induced pluripotent stem cells (iPSCs) in the treatment of neurodegenerative diseases. Neuroscience. 2012 Oct 13.

[257] Yamanaka S. Induced pluripotent stem cells: past, present, and future. Cell stem cell. 2012 Jun 14;10(6):678-84.

[258] Harris VK, Yan QJ, Vyshkina T, Sahabi S, Liu X, Sadiq SA. Clinical and pathological effects of intrathecal injection of mesenchymal stem cell-derived neural progenitors in an experimental model of multiple sclerosis. Journal of the neurological sciences. 2012 Feb 15;313(1-2):167-77.

[259] Yang J, Yan Y, Ma CG, Kang T, Zhang N, Gran B, et al. Accelerated and enhanced effect of CCR5-transduced bone marrow neural stem cells on autoimmune encephalomyelitis. Acta neuropathologica. 2012 Oct;124(4):491-503.

[260] Sher F, Amor S, Gerritsen W, Baker D, Jackson SL, Boddeke E, et al. Intraventricularly injected Olig2-NSCs attenuate established relapsing-remitting EAE in mice. Cell transplantation. 2012 Mar 28.

[261] Jiang Q, Zhang ZG, Ding GL, Silver B, Zhang L, Meng H, et al. MRI detects white matter reorganization after neural progenitor cell treatment of stroke. NeuroImage. 2006 Sep;32(3):1080-9.

Chapter 12

Cell Adhesion Molecules in Neural Stem Cell and Stem Cell-Based Therapy for Neural Disorders

Shan Bian

Additional information is available at the end of the chapter

http://dx.doi.org/10.5772/55136

1. Introduction

Neural stem/progenitor cells (NS/PCs), found in both the developing and the adult mammalian central nervous system (CNS), are a heterogeneous population of multipotent cells with the potential to self-renew by symmetric cell division or to differentiate into neurons, astrocytes and oligodendrocytes through asymmetric cell division (Gage, 2000; Alvarez-Buylla et al., 2001; Temple, 2001; Götz and Huttner, 2005). NS/PCs have been found in almost all regions of the developing mammalian CNS, including the basal forebrain, cerebral cortex, ganglionic eminence, hippocampus, cerebellum, neural crest and spinal cord (Temple, 2001). Throughout development, NS/PCs give rise to neurons and glial cell populations of the CNS. In the adult CNS, NS/PCs are mainly found in the subventricular zone (SVZ) and subgranular layer (SGL) of hippocampal dentate gyrus (DG) (Göritz and Frisén, 2012). The ependymal cells lining the central canal of spinal cord of the adult mouse could be another potential source of adult NS/PCs (Meletis et al., 2008). Because neurogenesis and gliogenesis occur during different stages of mammalian brain development, it was long assumed that neurons and glial cells in the CNS were generated from distinct precursor populations, known as early-embryonic, late-embryonic, and adult NS/PCs. However, abundant evidence has since now demonstrated that embryonic and adult NS/PCs are likely lineage-related. Neuroepithelial cells behaving as NS/PCs during very early developmental stages of the mammalian CNS give rise to radial glial cells around embryonic day 12 (E12). As the progeny of neuroepithelial cells, radial glial cells act as NS/PCs in the fetal and perinatal brain, and develop into astrocyte-like stem cells in the adult brains. Astrocyte-like adult stem cells function as stem cells to generate new nerve cells in the adult mammalian CNS. (Doetsch et al., 1999; Alvarez-Buylla et al., 2001; Merkle et al., 2004; Merkle and Alvarez-Buylla, 2006).

Evidence from recent studies indicates that NS/PC self-renewal and the time course of NSC fate determination are regulated by a combination of nuclear modifications, transcription factors and extrinsic signals from the stem cell microenvironment, also known as stem cell niche (Shi et al., 2008). Cell adhesion molecules (CAMs) located on the cell surface bind to the extracellular matrix (ECM) as well as other cells, thereby connecting cells with their surroundings. As important participants in cell-cell and cell-ECM interactions, CAMs have been shown to play essential roles in NS/PC development, including proliferation, differentiation, and migration, through extrinsic signals from the stem cell niche. Furthermore, many CAMs have been studied for the potential application in repair of CNS and peripheral nervous system (PNS) damage. Based on the role of CAMs in the NS/PC development and on the repair of nervous system, many researchers have investigated the possibility of combining CAMs and stem cell-based transplantation therapy to treat neural disorders and neural injury.

This chapter will discuss the most recent research in CAM functions in NS/PC development, as well as highlighting the applications and future potential of applying CAMs in stem cell-based therapy.

2. Cell adhesion molecules

CAMs are transmembrane proteins located on the cell surface through which cells bind or interact with other cells and the ECM surrounding them. CAMs normally contain three domains: an intracellular domain by which CAMs bind to cytoskeleton proteins, an extracellular domain that interacts with the ECM or other CAMs and a transmembrane domain (Chothia and Jones, 1997). Through these three domains, CAMs can transfer extrinsic signals from microenvironment or other neighboring cells and trigger diverse signaling pathways. Although CAMs can be classified by whether they are calcium-dependent or calcium-independent, most CAMs are categorized into four families: the immunoglobulin superfamily, the cadherins, the integrins, and the selectins (Brackenbury et al., 1981; Shapiro et al., 2007).

The complexity of structure and functions in the nervous system are achieved by the complex interconnections among its component cells. Many types of CAM interactions between various types of neural cells in the nervous system have been reported. Here, the classification and structure-function relationship of CAMs highly expressed in nervous system or play essential roles in nervous system will be introduced and briefly discussed.

2.1. Immunoglobulin superfamily cell adhesion molecules

Immunoglobulin superfamily cell adhesion molecules (IgSF CAMs) are a group of calcium-independent CAMs that share a common structural domain called immunoglobulin (Ig) domain. Members from IgSF CAMs have diverse binding activities and mechanisms of function. Certain IgSF CAMs interact through their extracellular domains with the same type of CAMs, a mechanism referred to as homophilic binding. Conversely, through heterophilic binding, other IgSF CAMs bind to different CAMs or to other cell surface proteins as well as

http://dx.doi.org/10.5772/55136

soluble proteins in the ECM, Due to their structural and binding specificity, different IgSF CAMs have been shown to play distinct roles in the nervous system.

2.1.1. L1CAM subfamily

L1CAM subfamily proteins, which include L1, close homolog of L1 (CHL1), NrCAM, and Neurofascin, are one of the most well known groups of IgSF CAMs and are widely studied in the nervous system. L1CAM proteins can homophilically bind to themselves, and based on their structure, can heterophilically interact with many other IgSF CAMs (e.g. NCAM, TAG-1, and contractin), ECM molecules (e.g. laminian, tenascins, Src and extracellular signal-regulated kinases (Erk)), cytoplasmic proteins, including cytosketeton proteins such as ankyrins, and traffic proteins such as AP-2 (Maness and Schachner, 2007).

The extracellular domain of L1CAM proteins contains six Ig-like domains and four or five fibronectin type III (FNIII) repeats, followed by a transmembrane domain and cytoplasmic domain (Fig. 1A). The Ig1 domain of L1/CHL1 can bind neuropilin-1, an important component of the axonal guidance molecule semaphorin 3A (Sema3A) (Castellani et al., 2002). Similarly, NrCAM has been shown to bind neuropilin-2 to form a complex that mediates Sema3B and Sema3F signaling (Falk et al., 2005). Fluorescent bead aggregation and cell binding assays demonstrate the N-terminal of the L1 Ig1 domain is essential for L1 homophilic binding (Jacob et al., 2002). Through unique motifs in the Ig6 domains, L1 and CHL1 trigger Erk, mitogen-activated protein kinase kinase (MAP2K), phosphatidylinositol-3 kinase (PI3 kinase), and Src signaling pathway by interacting with integrins in some biological processes (Schaefer et al., 1999; Schmid et al., 2000). The intracellular domains of L1CAMs contain a highly conserved motif by which L1CAM proteins bind to ankyrins, a group of adaptor proteins linking integral membrane proteins to the cytoskeleton (Bennett and Baines, 2001). When the cytoplamic domain of Neurofascin is tyrosine phosphorylated, doublecortin (DCX) is recruited and mediates the binding of Neurofascin to microtubules (Kizhatil et al., 2002). An alternative splicing generates a motif YRSL in the cytoplasmic domain of L1 that mediate interactions between L1 and AP-2, a clathrin adaptor that triggers tyrosine-based signals for endocytosis (Kamiguchi et al., 1998). L1 was also shown to directly bind to membrane-cytoskeleton linking ezrin-radixin-moesin (ERM) proteins, through which L1 can bind to the actin cytosketeton and regulate axonal outgrowth and neuronal differentiation (Dickson et al., 2002). These structure-dependent interactions between L1CAMs and other ECM molecules, cytoskeleton proteins and cytoplasmic molecules suggests that are L1CAMs an essential component during the extrinsic signaling transduction regulating neurite outgrowth, axon growth, cell migration and differentiation (Maness and Schachner, 2007). Moreover, the intracellular domain of CHL1 directly interacts with synaptic chaperones Hsc70, CSP and alphaSGT, thereby regulating SNAP25 and VAMP2-induced exocytotic machinery (Andreyeva et al., 2010).

2.1.2. NCAM

Neural cell adhesion molecule (NCAM), also known as CD56, is a glycoprotein expressed on the cell surface of various cell types, including neurons, glial cells, and natural killer cells. NCAM is a unique member of IgSF CAMs because of its 27 alternatively spliced mRNAs and its three major protein isoforms: NCAM-120, NCAM-140, and NCAM-180, so named due to their molecular weights (Reyes et al., 1991).

These three NCAM isoforms share the same extracellular domain, but vary in their transmembrane domains and cytoplasmic region. The extracellular domain of NCAM contains five Ig-like domains and two FNIII repeats, followed by a transmembrane domain in NCAM-140 and NCAM-180 isoforms, but a glycophosphatidyl (GPI) anchor linking to the cell membrane in NCAM-120 (Fig. 1A) (Chothia and Jones, 1997). NCAM-140 has a shorter intracellular domain compared to NCAM-180. The different domains of NCAMs have been reported to play distinct roles in binding activities and biological functions. The Ig-like domains have been shown to be essential for NCAM homophilic binding, and the FNIII repeats are involved in signaling that regulate neurite outgrowth. *Trans*-homophilic binding of NCAMs on different cell surfaces and *cis*-homophilic binding on the same surface have been observed, and both binding models play roles in neurite outgrowth (Walmod et al., 2004). Although the mechanism behind these binding models remains unknown, several studies have suggested that either Ig1 and Ig2 only or all five Ig-like domains are involved in NCAM *trans*-homophilic binding. Both Ig1 and Ig2, and Ig1 and Ig3 have demonstrated important roles in NCAM *cis*-homophilic binding (Frei et al., 1992; Ranheimet al., 1996; Atkins et al., 2004). The FNIII repeats bind fibroblast growth factor receptor (FGFR), which induces NCAM-mediated neurite outgrowth and plasticity. (Kiselyov et al., 2005). A PDZ-like sequence in the cytoplamic domain of NCAM-180 is required for NCAM-180-mediated activation of myosin light chain kinase (MLCK) (Polo-Parada et al., 2005). The intracellular domain of NCAM-180 has also been shown to be essential for the interaction between NCAM-180 and dopamine D2 receptor, by which NCAM regulates dopminergic signaling and behavior (Xiao et al., 2009). Non-receptor tyrosine kinase $p59^{fyn}$ has been found to interact specifically with NCAM-140, which activate $p59^{fyn}$ and focal adhesion kinase (FAK). The subsequent activation of c-Ras1 triggers the Erk signaling (Kolkova et al., 2000). Prion protein (PrP) can recruit and stabilize NCAM-140 on lipid rafts and regulate $p59^{fyn}$ to enhance neurite outgrowth (Santuccione et al., 2005). NCAM-140 also binds to GFRα, the glial-derived neurotrophic factor (GDNF) receptor, which regulates GDNF-mediated Schwann cell migration (Paratcha et al., 2003). Additionaly, the unique carbohydrate motif α-2,8-linked polysialic acid (PSA) on the Ig5 domain, observed only on NCAMs, has been found to negatively influence the homophilic binding of NCAMs as well as their interaction with heparin and heparin sulfate, due to the highly negative charge of PSA (Rutishauser and Landmesser, 1996). Through its inhibition of NCAM homophilic binding and reduction of cell adhesion during cell migration, PSA exhibits functions during many neural processes, including those related to learning and memory, and certain neurological disorders (Becker et al., 1996; Senkov et al., 2006; Stoenica L et al., 2006).

2.1.3. Nectins and Nectin-like molecules

Nectins and Nectin-like molecules (Necls) form another large IgSF CAM family, including four Nectins (Nectin-1, 2, 3, 4) and five Necls (Necl-1, 2, 3, 4, 5), and exhibit cell-cell adhesive functions in a wide range of tissues, including epithelia and neuronal tissue (Takai et al., 2003). All Nectins and Necls share the same structure domains, containing an extracellular

http://dx.doi.org/10.5772/55136

domain with three Ig-like repeats, through which they play their roles in cell-cell adhesion activity, a single transmembrane region, and a cytoplasmic domain (Fig. 1A) (Sakisaka and Takai, 2004). Nectins and Necls can form *cis*-homo-dimers, and subsequently form *trans*-homo-dimers or *trans*-hetero-dimers through their extracellular Ig-like domains. (Sakisaka and Takai, 2004). Nectins can *trans*-interact with c-Src, to activate Rac and Cdc42 signaling (Shimizu and Takai, 2003)., Nectins can bind to F-actin-binding protein afadin through their cytoplasmic domain. Upon binding afadin, Nectins interact with α-catenin and cadherins, playing a role in forming cell-cell adhesion junctions and tight junctions in epithelial cells (Takai and Nakanishi, 2003). Nectins also regulate cell polarization through interacts with Par-3 via their intracellular domain (Takekuni et al., 2003). In addition, Necls interact with many important proteins, such as membrane-associated guanylate kinase Dlg3/MPP3, tumor suppressor DAL-1, CD44, platelet-derived growth factor (PDGF), and Nectins (Yageta et al., 2002; Fukuhara etal., 2003; Takai et al., 2003; Kakunaga S et al., 2004;).

2.1.4. TAG-1

TAG-1, also called TAX-1 (human) or axonin-1, a 135 kDa glycoprotein expressed on the developing axons, belongs to IgSF CAMs superfamily and plays important roles in neurite outgrowth and cell aggregation. TAG-1 has six Ig-like domains followed by four FNIII repeats and is anchored to the cell membrane by a GPI tail (Fig. 1A) (Furley et al., 1990). During development of the central and peripheral nervous system, TAG-1 is transiently expressed both as a soluble form and a GPI-anchored form (Karagogeos et al., 1991). Binding analysis revealed that FNIII repeats but not Ig domains are sufficient for homophilic binding, although both TAG-1 domains types can promote the neurite outgrowth (Tsiotra et al., 1996; Pavlou et al., 2002). The four amino-terminal Ig-like domains have been shown to be important for TAG-1 and neural glial cell adhesion molecule (NgCAM) *cis*-position binding, while the Ig5 and Ig6 domains have a strong inhibitory effect on TAG-1 and NgCAM binding (Buchstaller et al., 1996; Rader et al., 1996). L1 and NrCAM were shown to interact with TAG-1 through its Ig-like domains in the *trans*-position (Pavlou et al., 2002).

2.1.5. MAG

Myelin-associated glycoprotein (MAG) is expressed on the surface of oligodendrocytes in the CNS and Schwann cells in the PNS, and has been implicated in neuron-oligodendrocyte and oligodendrocyte-oligodendrocyte interactions in the CNS and glia-glia interaction in the PNS (Sternberger et al., 1979; Quarles, 1984; Martini and Schachner, 1986; Poltorak et al., 1987). Two MAG isoforms have been identified as small MAG (S-MAG) and large MAG (L-MAG). Similar to other IgSF CAMs, MAGs contain an identical extracellular domains composed of five Ig-like domains, a transmembrane segment, and two distinct cytoplasmic domains, which distinguish the S-MAG and L-MAG (Fig. 1A) (Salzer et al., 1987). MAG binds gangliosides, the most abundant sialylated glycoconjugates in the nervous system, which are important for neuron-oligodendrocyte interaction (Schnaar et al., 1998). MAG was shown to interact with the leucine-rich repeat (LRR)-containing GPI-linked Nogo-66 receptor (NgR), inducing the inhibition of neurite outgrowth (Fournier et al., 2001; Domeniconi et al., 2002).

There are many other IgSF CAMs, not to be introduced in this chapter, found in other mammalian body systems, such as intercellular cell adhesion molecule (ICAM-1, expressed on endothelial cells and cells of the immune system), vascular cell adhesion molecule (VCAM-1, expressed on blood vessels), platelet-endothelial cell adhesion molecule (PECAM-1, expressed on the surface of platelets, monocytes, neutrophils, and some types of T-cells).

2.2. Cadherins

Cadherins, a family of calcium-dependent CAMs, are the major architectural molecules mediating cell-cell adhesion of extracellular domains at intercellular junctions. The cadherin superfamily consists of four subgroups, including classic cadherins, protocadherins, desmosomal and unconventional cadherins, which share a similar structure with extracellular Ca^{2+}-binding domains, known as cadherin repeats (Angst et al., 2001). Although the homophilic binding of specific cadherin subtypes has been described frequently, several cadherin subtypes were also found to interact heterophilically (Ahrens et al., 2002).

2.2.1. Classic cadherins

In vertebrates, classic cadherins, including epithelial cadherin, neural cadherin, and placental cadherin and so on, have five extracellular cadherin repeats, a transmembrane domain, and a highly conserved intracellular domain (Fig. 1B) (Tepass et al., 2000). By binding to Ca^{2+} on the boundary between cadherin repeats, the extracellular domain is stabilized and undergoes homophilic interactions, which is essential for the adhesion function of cadherins (Nagar et al., 1996). Both *cis*- and *trans*-homophilic interactions were observed in classic cadherin-mediated cell adhesion (Zhang et al., 2009). Classic cadherins bind to two cytoplasmic proteins, p120-catenin and β-catenin, through the conserved intracellular motif (Tepass et al., 2000). The cadherin-β-catenin complex interacts with α-catenin and subsequently associates with actin, forming an adherin junction and a cell-cell signaling centre together with cytoskeletal and signaling molecules (Wheelock and Johnson, 2003; Drees et al., 2005). In addition to catenin proteins, classic cadherins have been observed to interact with many other molecules from various signaling pathways (Erez et al, 2005). For example, cadherin-mediated cell adhesion is involved in Wnt, Hedgehog, Ras, and RhoGTPase signaling (Stepniak et al., 2009; Watanabe et al., 2009; Heuberger and Birchmeier, 2010). Cadherin-catenin complex also interacts with receptor-type tyrosine kinases, such as FGFRs and epithelial growth factor receptors (EGFRs) (Mason, 1994; Perrais et al., 2007). Such crosstalk between cadherin system and other signaling pathways allow cadherins to play critical roles in diverse cell biological behaviors. In humans, there are 18 cadherins in the classic cadherin superfamily, of which epithelial cadherin (E-cadherin, or cadherin-1), is the most well studied member. As its name indicates, E-cadherin is expressed in developmental and adult epithelial tissues. The cytoplasmic tail of E-cadherin, containing a highly phosphorylated region, is essential for the function of E-cadherin in the formation cell-cell junction (Tepass et al., 2000). Neural cadherin (N-cadherin or cadherin-2) is broadly expressed in neuroepithelial cells during early embryonic and neonatal development, particularly in the nuclei and laminae, and neuroanatomical connections during late embryonic stages and early postnatal development (Redies and Takeichi, 1993). N-cadherin has been reported to play critical roles in the establishment of left-right asymmetry and synaptogenesis as well as catenin-mediated processes related to learning and memory (Benson and Tanaka, 1998; García-Castro et al., 2000). Placental cadherin (P-cadherin or Cadherin-3) is highly expressed in undifferentiated cells in epithelial tissues and human embryonic stem cells, and is a putative stem and precursor cell maker (Raymond et al., 2009).

2.2.2. Protocadherins

With more than 100 having been identified in mammals, protocadherins are the largest subfamily of cadherins (Hirano and Takeichi, 2012). Protocadherins can be further sorted into two groups, based on their genomic distribution: clustered protocadherins, whose coding genes are located on human chromosome 5 in tandem order, and consist of three gene subclusters Pcdhα, Pcdhβ and Pcdhγ, and non-clustered protocadherins, whose genes are distributed among different chromosomes and are divided into several subgroups, incuding δ-protocadherins (Vanhalst et al., 2005). The extracellular domains of protocadherins contain more than five cadherin motifs, which differ from the characteristic features of classical cadherins (Fig. 1B) (Sano et al., 1993). Unlike the highly conserved cytoplasmic tails of classic cadherins, the intracellular domains of protocadherins are variable, suggesting their diverse functions. Although protocadherins appear to display weaker cell-cell adhesion than classic cadherins, protocadherins show diverse biological functions in the CNS, including roles in neuronal differentiation and synaptogenesis (Hirano and Takeichi, 2012).

2.3. Integrins

Integrins are a group of cell surface receptors that form a large CAM subfamily. Integrins are heterodimeric glycoproteins consisting of two distinct subunits: an α-subunit and β-subunit that each penetrates the cell membrane once and has a short intracellular tail which is typically 40 to 70 amino acids long. Thus far, 18 α-subunits and 8 β-subunits have been identified and 24 αβ combinations have been observed (Hynes, 2002). The extracellular domain of α-subunits, which generally are larger than 900 amino acids, is divided into four subdomains containing a ligand-binding N-terminal region including a sevenfold repeat, among which repeats 5, 6, and 7 contain a Ca^{2+}- binding structure. Half of the α-subunits contain an extra I-domain, which contributes a divalent cation-binding site and facilitates an interaction with IgSF CAMs (Fig. 1C) (Landis, et al., 1994). The ectodomains of integrin β-subunits are typically larger than 600 amino acids and contain 8 subdomains, including an N-terminal signal region as well as a metal-binding site that is directly involved in ligand-integrin interactions (Fig. 1C). The extracellular N-terminal of α-chain and β-chain form a ligand-binding αβ headpiece (Fig. 1C). Through interactions between their extracellular domain and ECM ligands and other CAM family members, integrins can perform outside-in signaling to mediate cell response to its surrounding environment. Moreover, through their intracellular tails, integrins can also perform inside-out signaling, thereby relaying the intercellular cell state to the extracellular environment. Using both outside-in and inside-out signal transduction models, integrins not only aid in facilitating cell-ECM interactions, but also play roles in many other biological activities, including cell cycling as well as cell growth, survival, and differentiation.

2.4. Selectins

The selectin family of CAMs is a group of calcium-dependent transmembrane glycoproteins. Three selectins have been identified, including E-selectin (found on endothelial cells), L-selectin (found on lymphocytes), and P-selectin (found on platelets and endothelial cells). All identified selectins share a similar extracellular structure, composed of an N-terminal sequence, a calcium-binding lectin domain, an EGF-like domain, in addition to a non-conserved transmembrane domain, a non-conserved cytoplasmic tail and a differing number of consensus repeats (Fig. 1D). Selectins have been reported to play important roles in the immune system associated with inflammation and cancer progression (Barthel et al., 2007).

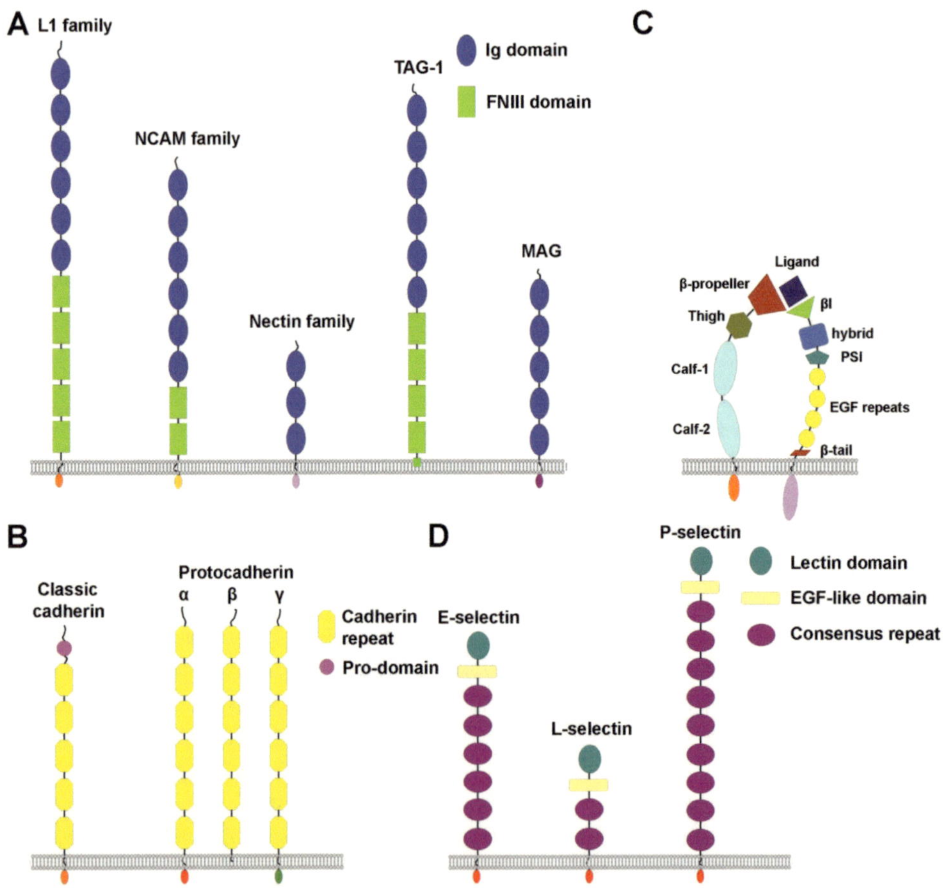

Figure 1. Schematic diagram of diverse domain structure of different CAM superfamilies. A. IgSF CAM proteins. B. Cadherins. C. Integrins. D. Selectins

http://dx.doi.org/10.5772/55136

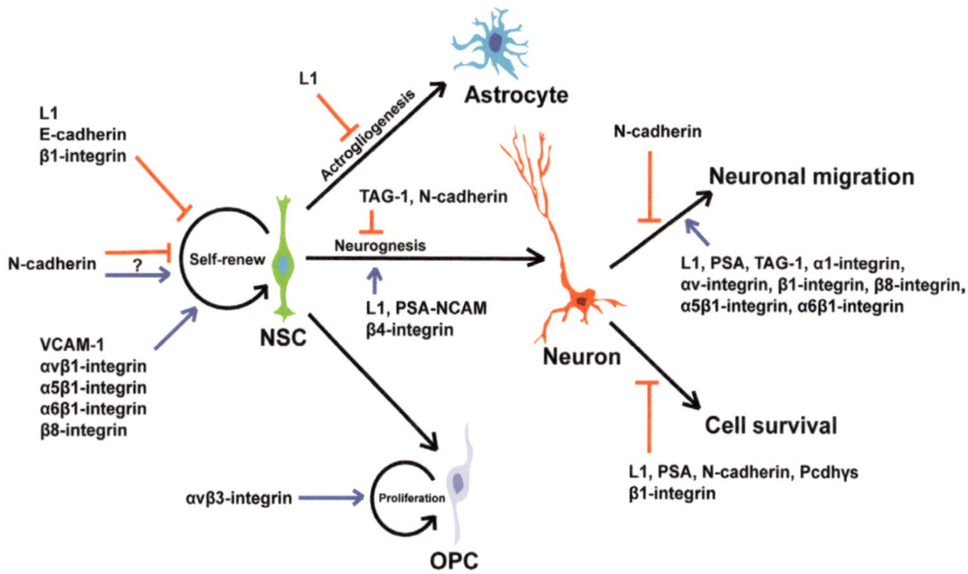

Figure 2. Roles of cell adhesion molecules in neural stem/precursor cell self-renewal, differentiation, cell survival and migration.

3. Cell adhesion molecules and neural stem/progenitor cell development

In the mammalian embryonic CNS, NS/PCs are found in almost all regions, and give rise to every type of neural cells, including various subtypes of neurons, astrocytes, and oligoden drocytes, forming the most complex and functional organ of the body (Temple, 2001). In the adult CNS, astrocyte-like type-B stem cells reside only in certain regions, namely as the subventricular zone (SVZ), from which stem cells migrate into olfactory bulb (OB) and differentiate into mature neurons, and the SGL of hippocampal DG, from which stem cells migrate and extend processes into the molecular layer of the hippocampus (Gage, 2000). Ependymal cells lining the central canal of adult mouse spinal cord can exhibit dramatic proliferation after spinal cord injury and give rise to newborn glial cells in the injured spinal cord (Göritz1 and Frisén, 2012). Many CAMs from different subfamilies are expressed in these neurogenic regions, although their functions have not been fully understood. Increasingly evidences suggest that CAMs are not only important for maintenance of the architecture and shape of stem cell niche, but also play essential roles in signaling transduction from the stem cell niche to regulate cell survival, proliferation, differentiation, and migration. Recent progress in NS/PC cell fate research indicates that stem cell maintenance and cell fate determination are regulated by a combination of epigenetic modulation, intrinsic transcription factors and extrinsic signals from the NS/PC niche (Shi et al., 2008). Transcriptional factors play essential roles in NS/PC fate determination by regulating multiple downstream gene expression in NS/PCs. However, the functions of these intrinsic factors are regulated by extrinsic cues through various signaling pathways mediated by cell surface receptors and down-stream cellular signaling elements. Modulation of cell fate by such extrinsic cues provides an excellent mechanism for neural progenitor cells to adapt to the environment. As bridging molecules between extrinsic signals from neighbor cells or the ECM and intracellular transcriptional regulation, CAMs play critical roles during these processes.

3.1. Cell adhesion molecules regulate neural stem/progenitor cell self-renewal and proliferation

Self-renewal is essential for NS/PCs to perpetuate and maintain their population in an undifferentiated state, which is critical for the CNS development, learning and memory-related plasticity in adult animals as well as replacement for dead cell following injury. Recent studies showed that the NS/PC niche, comprised of ECM molecules, soluble factors, and cell surface molecules, is essential for NS/PC identification, maintenance of stem cell population, and preventing contact with differentiation stimuli and cell apoptotic signals (Doetsch, 2003; Miller and Gauthier-Fisher, 2009; Moore and Lemischka, 2006). Among the NS/PC niche molecules, CAMs play a critical role as transducers that mediate signaling between stem cells and their niche.

IgSF CAMs play diverse roles in the regulation of NS/PC self-renewal and proliferation through specific signaling pathways. L1 is highly expressed in the developing neurons of the cerebral cortex, hippocampus, and corpus callosum, with expression declining to low levels with maturation (Demyanenko et al., 1999). Reduced hippocampal neurons were observed in L1-deficient mice, suggesting that L1 may have a potential role in neurogenesis (Demyanenko et al., 1999). Although undifferentiated NS/PCs do not express L1, substrate-coated L1 inhibits the proliferation of cultured NS/PCs *in vitro* (Dihné et al., 2003). Further investigation using L1-deficient precursors suggests the mechanism by which L1 affects NS/PC proliferation is through heterophilic binding with unknown receptors (Dihné et al., 2003). Ectopic expression of L1 using human GFAP promoter in NS/PCs, or overexpression of L1 on the ESC-derived neural precursors resulted in an impaired proliferation *in vitro* (Bernreuther et al., 2006; Cui et al., 2010; Cui et al., 2011; Xu et al., 2011). Interestingly, L1 isoform 2 is expressed on the surface of human ESCs, in which it positively affects proliferation. Deletion of stable L1 decreased hESC proliferation, while L1 overexpression enhanced proliferation, suggesting the diverse roles of L1 on different cell types by different mechanisms (Son et al., 2011). In the adult CNS, VCAM-1 is expressed on the apical processes of type B astrocyte-like adult stem cells in the adult mouse SVZ, and has been reported to be essential for SVZ cytoarchitecture, maintenance of adult NS/PCs, and neuroblast chain migration (Kokovay et al., 2012). Blocking VCAM1 expression with VCAM1 antibodies disrupts position and cytoarchetecture of ependymal cells and type B cells in at ventricular surface. Further investigation demonstrated that VCAM1 maintains adult NS/PCs through NOX2 production of reactive oxygen species (Kokovay et al., 2012).

Neuroepithelial attachments at adherens junctions are critical for NS/PC maintenance and self-renewal. As essential molecules during the nervous system development, cadherins have been reported to play roles in neural tube, neuroepithelial layer and boundary formation, synaptogenesis, axon guidance and regulation of NS/PC behaviors. E-cadherin is expressed in the

http://dx.doi.org/10.5772/55136

embryonic and adult ventricles, where NS/PCs reside, and in clonal stem cell colonies *in vitro*. Deletion of E-cadherin results in a reduction of NSC self-renewal, but an increase of neural precursor proliferation in vivo (Karpowicz et al., 2009). *In vitro* culture experiments provide further evidence that loss of E-cadherin reduces NSC self-renewal, while overexpressing E-cadherin results in an increased NSC number, due to the adhesion function of E-cadherin (Karpowicz et al., 2009). N-cadherin, which is highly expressed in the nervous system, appears to be important for maintenance of NS/PCs and cell proliferation. During neural tube development in zebrafish, N-cadherin was shown to restrict proliferation of neural precursors in the dorsal neural tube by regulating the cell cycle length (Chalasani and Brewster, 2011). Increased cell division and shortened cell cycle length, mediated by ligand-independent activation of Hedgehog signaling, were observed in the N-cadherin mutant, (Chalasani and Brewster, 2011). However, other studies have found opposing effects of N-cadherin. Knock-down of N-cadherin expression by *in utero* electroporation of shRNA targeting N-cadherin resulted in reduced β-catenin signaling, leading to increased cell cycle exit and enhanced premature neuronal differentiation. These effects could be rescued by introducing a stabilized active form of β-catenin (Zhang et al., 2010). Another *in vitro* study using fusion protein N-cadhsion-Fc (including extracellular domain of N-cadherin and Fc domain of IgG)-coated culture conditions demonstrated that N-cadherin-Fc promotes maintenance of the neural precursor cell lines P19 and MEB5, and promotes proliferation (Yue et al., 2010). In a more recent study in spinal cord, forkhead transcription factors Foxp2 and Foxp4 suppressed N-cadherin to promote the detachment of differentiating neurons from the neuroepithelium. Blocking Foxp2 and Foxp4 function in both mouse and chick enhances progenitor maintenance, implying that N-cadherin plays diverse in different systems and signaling pathways (Rousso et al., 2012).

Many integrins are expressed in neurogenic regions of the developing and adult mammalian CNS. Cultured neurospheres from newborn mouse forebrain express five major integrins, including $\alpha5\beta1$-, $\alpha6A\beta1$-, $\alpha v\beta1$-, $\alpha v\beta5$-, $\alpha v\beta8$-, and low levels of $\alpha6B\beta1$-integrin. Antibody inhibition of $\alpha5\beta1$- and $\alpha v\beta1$-integrins reduces NS/PC proliferation (Jacques et al., 1998). With the use of fluorescence-activated cell sorting, $\alpha5\beta1$-integrin was shown to be highly expressed in multipotent NS/PCs and downregulated during neuronal differentiation, suggesting a role in maintenance of characteristic NS/PC features (Yoshida et al., 2003). $\alpha6\beta1$-integrin, which is highly expressed in human NS/PCs, is used as a neural precursor cell marker, and was found to form a functional complex with laminin $\gamma1$, and netrin-4 on the NS/PC surface, that promotes cell proliferation (Hall et al., 2006; Staquicini et al., 2009). Antibody inhibition of β1-integrin signaling results in an increased population of abventricular dividing precursors, a phenomenon also observed in the neocortex of mice deficient in Laminin $\alpha2$, a ligand of β1-integrin (Loulier et al., 2009). Although diminished proliferation was observed in cultured β1-integrin-deficient neural precursors, a small percentage of β1-integrin-negative cells survived and formed neurospheres normally, suggesting that β1-integrin is not required for maintenance of all NS/PCs (Leone et al., 2005). Further investigation has revealed that β1-integrin contributes to the maintenance of NS/PCs by activating MAPK signaling (Campos et al., 2004; Wang et al., 2011). β1-integrin interacts with Notch and EGFR signalling pathways, suggesting that ECM molecule or growth factor involvement may be important in β1-integrin-mediated regulation of NS/PCs (Campos et al., 2006). β1-integrin is also involved in basic FGF (bFGF) and EGF-mediated proliferation of neuroepithelial cells (Suzuki et al., 2010). Moreover, the carbohydrate-binding protein Galectin-1 regulates proliferation of adult NS/PC through interactions with β1-integrin (Sakaguchi et al., 2010). Additionally, the FNIII domain 6-8 of ECM molecule tenascin-R inhibits NS/PC proliferation through interactions with β1-integrin, an effect that is eliminated with antibody inhibition of β1-integrin (Liao et al., 2008). PDGF stimulates proliferation of oligodendrocyte precursor cells (OPCs) through activation of αvβ3-integrin, which in turn activates PI3 kinase-dependent signaling pathway (Baron et al., 2002). Neurovascularization, regulated by αvβ8-integrin, plays essential roles during the CNS development (Proctor et al., 2005). Analysis of adult β8-integrin-deficient mouse brains shows abnormalities in the SVZ and RMS, with a smaller OB. Reduced proliferation of cells in the SVZ and RMS in β8-integrin-deficient brains and smaller neurospheres formed by β8-integrin-deficient NS/PCs reveal the important role of β8-integrin on NS/PC maintenance, potentially through transforming growth factor β (TGFβ) signaling (Mobley et al., 2009).

3.2. Cell adhesion molecules regulate neural stem/progenitor cell differentiation

In the mammalian CNS, different neural cell types arise and migrate in a precise temporospatial manner. During mouse brain development, neurons first arise around embryonic day 12 (E12), with neurogenesis peaking at E14 and ceaseing around E18. Astrocytes appear at approximately E18, with their numbers peaking in the neonatal period, and oligodendrocytes are generated after birth when the neurogenesis has largely subsided (Bayer and Altman, 1991).

In vitro cultured neural precursors showed enhanced neuronal differentiation, associated with a reduced astrogliogenesis and unchanged oligodendrocyte genesis whether cultured on substrate-coated L1, or co-cultured with L1-overexpressing fibroblasts (Dihné et al., 2003). Further analysis of different neuronal subtypes that differentiated from NS/PCs shows that L1 inhibits cholinergic neuron differentiation, but promotes differentiation of GABAergic neurons (Dihné et al., 2003). Under an *in vitro* culture system, ectopic expression of L1 in NS/PCs, or overexpression of L1 on the ESC-derived neural precursors promotes neuronal differentiation, but inhibits astrocytic differentiation (Bernreuther et al., 2006; Cui et al., 2010; Cui et al., 2011; Xu et al., 2011). L1 overexpression also resulted in an increased yield of GABAergic neurons after transplant into lesioned striatum (Bernreuther et al., 2006). Another important IgSF CAM PSA-NCAM, identified as a neural precursor marker, is expressed by NS/PCs in neurogenic regions in the developing and certain regions of the adult CNS, such as SVZ, hippocampus, and rostral migratory stream (RMS) (Seki and Arai, 1993; Bonfanti and Theodosis, 1994; Doetsch and Alvarez-Buylla, 1996; Alonso, 1999). Polysialic acid (PSA)-directed migration and differentiation of neural precursors are essential for mouse brain development. SVZ precursors migrate along the RMS towards the OB, where they differentiate into interneurons that do not express PSA. Interestingly, specific deletion of PSA from SVZ precursors using endoneuraminidase-N (Endo N) results in failure of migration and premature differentiation. Nestin-positive cells fail to migrate and develop neuronal cell features, such as long neurites. Furthermore, these cells are positive tyrosine hydroxylase (TH) (Petridis

http://dx.doi.org/10.5772/55136

et al., 2004). Further investigation revealed that PSA-dependent differentiation requires cell-cell contact, facilitated by NCAM triggering p59fyn and MAPK p44/42 phosphorylation (Beggs et al., 1997; Petridis et al., 2004). Through a double-knockout of two PSA synthases ST8SiaII and ST8SiaIV, PSA was deleted, resulting in an increased number of glial fibrillary acidic protein (GFAP)-positive astrocytes both *in vitro* and *in vivo* (Angata et al., 2007). TAG-1 plays a role in inhibiting neurogenesis by interacting with amyloid β precursor protein (APP). TAG-1 and APP are co-localized in the neurogenic niche and neural precursors, and can bind to each other to release APP intracellular domain (AICD), triggering the APP signaling pathway to negatively regulate neurogenesis (Ma et al., 2008). Fe65 is also expressed in the neurogenic VZ and has been identified as a downstream element of TAG-1-APP signaling during regulation of neurogenesis (Ma et al., 2008).

The roles of cadherins on neural differentiation are still not well understood. Although neural differentiation appears to occur normally in N-cadherin-deficient mice (Kadowaki et al., 2007), a recent study knocking down N-cadherin expression by *in utero* eletroporation resulted in an enhanced neuronal differentiation, likely through β-catenin signaling as it can be rescued by introducing a stabilized active form of β-catenin (Zhang et al., 2010). Another study showed that bone morphogenetic protein 4 (BMP4) induces astrocytic differentiation of NS/PCs through PI3 kinase-mediated upregulation of N-cadherin (Kim et al., 2010).

β4-integrin plays an essential role in NS/PC differentiation. Under *in vitro* culture system, knockdown of endogenous β4-integrin inhibits cell differentiation and reduces of FGFR2 expression, while overexpression of β4-integrin in NS/PCs promotes differentiation (Su et al., 2009). The FNIII domain 6-8 of ECM molecule tenascin-R, through interactions with β1-integrin, promotes astrocytic differentiation, while preventing differentiation into neurons and oligodendrocytes; Conversely, epidermal growth factor-like (EGFL) domain of tenascin-R promotes neuronal differentiation, while reducing differentiation into astrocytes and oligodendrocytes, through interactions with β1-integrin that can be inhibited with application of β1-integrin antibody (Liao et al., 2008).

3.3. Cell adhesion molecules regulate neuronal migration

When NS/PCs give birth to neurons or glial cells, the differentiating or differentiated cells will migrate away from the stem cell niche to their appropriate location. In the developing CNS, specific neurons migrate in a specific pathway to reach their final destination in the brain. For example, in the developing cerebral cortex, the newborn neurons generated by radial glial cells in the embryonic VZ undergo radial migration along the long processes of radial glial cells, forming the different layers of cortical plate. Newborn interneurons generated from the ganglionic eminence also migrate in tangential path into the cortex without interacting with radial glial cells (Ghashghaei et al., 2007). In the adult CNS, NS/PCs from SVZ migrate along the rostral migratory stream towards the OB and from the SGL of hippocampus migrate a short distance into molecular layer (Ghashghaei et al., 2007). As cell surface molecules interacting with the surrounding environment, CAMs play essential roles in cell migration..

Overexpression of L1 using human GFAP promoter in NS/PCs promotes neuronal migration in *in vitro* neurosphere migration assays, and also enhances migration of transplanted cells in the injured spinal cord (Xu et al., 2011). L1-overexpression in the neural precursors derived from ESCs also promotes migration into the lesioned CNS tissues (Bernreuther et al., 2006; Cui et al., 2011; Xu et al., 2011). The NS/PCs born in the adult SVZ express high levels of PSA-NCAM, and migrate along the RMS towards the OB and differentiate into interneurons. Increasingly evidence has demonstrated the importance of PSA in controlling NS/PC migration pattern, although PSA is not required for radial migration of interneurons within OB (Hu et al., 1996). SVZ precursors are unable to migrate when PSA expression is abolished, resulting in a smaller OB (Cremer et al., 1994; Hu et al., 1996; Angata et al., 2007). Expressed by hippocampal precursors in SGL, PSA is also required for the migration of newborn granule cells (Burgess et al., 2008). Although the mechanism by which PSA guides the NS/PCs is still not clear, it has been hypothesized that PSA regulates directed migration towards guidance cues. A recent study observing the effect of PSA on migration of OPCs showed that PSA-positive OPCs polarize and directly migrate towards concentration gradients of PDGF, while loss of the PSA tail of NCAM causes an altered migration pattern in response to PDGF gradients, indicating the PSA is involved in a regulatory network requiring environmental cues (Zhang et al., 2004). TAG-1 is expressed by cortical GABAergic interneurons and mediates their migration from the ganglionic eminence towards the developing cortex. Inhibition of TAG-1 function results in a remarkable reduction of GABAergic interneurons in the cortex (Denaxa et al., 2001).

Recently it was demonstrated that N-cadherin is involved in radial neuronal migration during cortical development. In N-cadherin conditional knockout mice, neuroepithelial and radial glial cells can not expand their cell bodies and processes to span the distance from the ventricular surface to the pial surface, which is essential for neuronal migration and cortical lamination, resulting in disorganization of the entire cortex (Kadowaki et al., 2007). Knock-down of N-cadherin expression by *in utero* electroporation reduced β-catenin signaling, causing enhanced premature neuronal migration, which can be rescued by introducing a stabilized active form of β-catenin (Zhang et al., 2010). A recent study showed that Rab GTPases-dependent endocytic pathways are critical for radial migration during the cortical development, through N-cadherin trafficking (Kawauchi et al., 2010). Deficiency in N-cadherin also causes mispositioning of neurons in the zebrafish neural tube (Lele et al., 2002). N-cadherin was also shown to regulate the directional chain migration of cerebellar granule neurons in zebrafish by continuously coordinating cell-cell contacts and cell polarity through the remodeling of adherens junctions (Rieger et al., 2009). Classic cadherins were also shown to regulate tangential migration of precerebellar neurons in the caudal hindbrain. N-cadherin, Cadherin-6, Cadherin-8, and Cadherin-11 are expressed in the migratory stream of lateral reticular nucleus (LRN) and neurons of external cuneate nucleus (ECN). Overexpression of two dominant negative constructs, a membrane-bound form and a cytoplasmic form, but not full length N-cadherin and Cadherin-11, inhibits LRN/ECN neuron migration, suggesting classic cadherins regulate contact-dependent tangential migration probably through their adhesive functions (Taniguchi et al., 2006). Additionaly, cadherins are also required for neural crest cells migration (Monier-Gavelle and Duband, 1997; Nakagawa and Takeichi, 1998; Borchers et al., 2001; Shoval et al., 2007).

http://dx.doi.org/10.5772/55136

Integrins were also shown to control migration of neural precursors. During migration along the RMS, antibodies for specific integrins, such as α1-, β1-, and αv-integrins, inhibit neuronal migration during the stages at which they are expressed (Murase and Horwitz, 2002). β1-integrins-dificient progenitor cells exhibit impaired migration in different ECM substrates (Leone et al., 2005). Antibody inhibition of neurosphere-expressed $\alpha6\beta1$-integrins results in inhibition of tangential chain migration of NS/PC *in vitro* (Jacques et al., 1998). The $\alpha6\beta1$-integrin, laminin $\gamma1$, and netrin-4 complex was shown to be a promoter of NS/PC migration (Staquicini et al., 2009). Another study analyzing adult neural precursors from the striatum reveals that $\alpha6\beta1$- and $\alpha5\beta1$-integrins promote cell migration (Tate et al., 2004). β8-integrin has also been shown to be important for neural precursor migration. Selective deletion of β8-integrin expression in neuroblasts causes abnormal chain migration and a reduction in OB size (Mobley et al., 2009; Mobley and McCarty, 2011).

3.4. Cell adhesion molecules regulate cell survival

During embryonic and early postnatal development, approximately 50% of newborn neurons die due to apoptosis in almost every CNS regions. In adult the CNS, cell death mostly occurs in the regions of neurogenesis. For example, in the adult hippocampus, about 70% of newborn neurons die within three weeks after failing to form the functional connections with the existing neural circuits. CAMs have been shown to play essential roles during apoptosis signaling.

In the *in vitro* culture system, mouse or human L1Fc fusion proteins can protect cerebellar granule neurons from cell death induced by growth factor deprivation, staurosporine treatment, and oxidative stress (Loers et al., 2005). With the use of specific inhibitors of signal transduction molecules, PI3-kinase activity was shown to be important for the neuroprotective effect of L1. Moreover, protein kinase PDK1 and Akt were found to be potential downstream targets (Loers et al., 2005). In other *in vitro* assays, coated L1 had no effect on the extent of precursor cell death, measured by TUNEL staining (Dihné et al., 2003). Ectopic expression of L1 also showed no effect on cell death of ESC-derived neural precursors, suggesting diverse roles of L1 on different cell types (Bernreuther et al., 2006). PSA-NCAM is also involved in neuronal survival. After removal or inactivation of PSA using Endo N, impaired survival was observed in cultured cortical neurons. This effect of PSA has been shown to be involved the BDNF signaling pathway (Vutskits et al., 2001). In NCAM-deficient mice, increased cell apoptosis was observed in the SVZ and the RMS. Interestingly, increased cell death occurred in the PSA$^+$NCAM$^+$ neuroblasts, but not in PSA$^-$NCAM$^+$ astrocyte, suggesting that PSA-NCAM but not NCAM is important for cell survival (Gascon et al., 2007).

Cadherins, such N-cadherin and Pcdhγs, have also been found to be involved in neural cell survival. N-cadherin can enhance cell survival of both mouse spinal cord neurons and rat hippocampal neurons *in vitro* through pro-apoptotic protein Bim-related signaling pathway (Lelièvre et al., 2012). A cyclic peptide including a functional binding motif HAVDI in the extracellular domain 1 of N-cadherin promote *in vitro* survival of different population of CNS neurons, by binding to and clustering N-cadherin in neurons and thereby activating N-cadherin/FGFR signaling cascade (Skaper et al., 2004). Pcdhγs is proven to be involved in neural cell survival. Abolishing Pcdhγs gene cluster results in the apoptosis of spinal interneurons, and retinal interneurons and ganglion cells, but the mechanism of Pcdhγs in cell survival is still unclear (Wang et al., 2002; Lefebvre et al., 2008).

Integrins also play a role regulation of cell survival. Cultured β1-integrin-deficient neural progenitors display high levels of cell death (Leone et al., 2005). Inhibition of phosphatidu-choline-specific phospholipase C in cultured neural cells results in a reduction of cell survival, associated with upregulation of β4-integrin and Rb protein (Lv et al., 2006).

4. Applications of cell adhesion molecules in stem cell-based regenerative therapy

Many CAMs have been investigated for pre-clinical studies in the treatment of neural injury and neurodegenerative disorders as their roles in CNS development and regeneration after CNS injury have been revealed. L1 has been reported to play important roles in neuronal survival and migration as well as neurite outgrowth and extension, axonal guidance and synaptic plasticity *in vitro* and *in vivo* (Hortsch, 1996). Thus, it has been studied as a target for the treatment of neurodegeneration and neural injuries. Application of soluble L1Fc to injured rat spinal cords significantly improved locomotor recovery compared to both the PBS-treatment group and the IgM antibody control group (Roonprapunt et al., 2003). Further analysis using biotinylated dextran amine for corticospinal axon labeling revealed that L1Fc promoted axonal growth (Roonprapunt et al., 2003). Another study using adeno-associated virus (AAV)-mediated L1 expression also promoted functional recovery including stepping ability and muscle coordination after mouse spinal cord injury (Chen et al., 2007). Increased regeneration of corticospinal tract axons and levels of 5-hydroxytryptamine (5-HT), an important neurotransmitter, were observed in the AAV-L1-treated spinal cord, along with a reduction of astrocytes and reduced expression of the neurite outgrowth-inhibitor and chondroitin sulfate proteoglycan NG2. Increased expression level of cyclic AMP, phosphorylated CREB, Rac1 and MAPK and a reduction of GTP-RhoA expression were observed in AAV-L1-treated spinal cord, revealing that exogenous L1 promotes functional recovery by triggering multiple signaling pathways (Chen et al., 2007). The reparative effects of L1 on spinal cord injury were achieved after acute application of L1. A separate study applying AAV-L1 alone, Chondroitinase ABC (ChaseABC) alone, which degrades chondroitin sulfates that inhibit neurite outgrowth, or a combination of AAV-L1 and ChaseABC to treat the sub-chronic spinal cord injury (Lee et al., 2012). Results revealed that when compared to the AAV-green fluorescent protein (GFP) injection control group, AAV-L1 treatment improves voluntary movements, while ChaseABC application enhances body weight support. Injection of the combination of AAV-L1 and ChaseABC results in improvement in both parameters, with increased densities of cholinergic and GABAergic terminals in motor neuron and enhanced synaptic rearrangements (Lee et al., 2012).

Polysialic acid (PSA) is a long linear homopolymer glycan carried by NCAMs that plays essential roles in PSA-NCAM-mediated activities, including cell-cell interaction and cell

http://dx.doi.org/10.5772/55136

migration (Hu et al., 1996; Johnson et al., 2005). Numerous studies have used PSA to promote adult CNS or PNS repair. Expression of PSA at the lesion site can loosen scar tissue and reduce inhibitory interactions with growth cones. Thus, engineered overexpression of PSA on the astrocyte scar enhanced Purkinje cell axonal regeneration in the lesioned cerebellum of growth related genes L1/GAP-43 double transgenic mice (Zhang et al., 2007). Similarly, induced expression of PSA in the glial scar of injured spinal cords promoted regeneration of sensory axons (Zhang et al., 2007). PSA glycomimetics has also been reported to promote plasticity and functional recovery after spinal cord injury in mice (Mehanna, 2010). PSA was also applied in the study of regeneration following peripheral nerve injury. PSA glycomimetic promotes myelination and functional recovery after peripheral nerve injury (Mehanna et al., 2009). PSA-mimetic enhances Schwann cell proliferation and process elongation in vitro, which may be mediated by interaction with Schwann cell-expressed NCAM and FGFR (Mehanna et al., 2009).

Human natural killer cell glycan (HNK-1) is found on glycolipids and glycoproteins, including many CAMs, such as L1, NCAM, MAG, TAG-1 in the nervous system, and has been shown to play roles in cell recognition and adhesion (Morita et al., 2008). Application of HNK-1 mimic peptide in injured peripheral nerves resulted in larger motor neuron somata and enhanced axonal remyelination resulting in better functional recovery compared to the mice treated with a scrambled peptide (Simova et al., 2006). *In vitro* assays demonstrated that HNK-1 mimic peptide enhances neurite outgrowth and survival of motor neurons (Simova et al., 2006).

Due to the role of CAMs in the NS/PC development and repair of nervous system injury, many researchers have investigated the possibility of combining CAMs and stem cell-based transplantation therapy to treat neural disorder and neural injury models. As the most widely studied CAM associated with neural regeneration, L1 has been investigated as a potential candidate for stem cell-based therapeutic strategy for treatment of neurodegenerative diseases or regeneration after neural injury. Overexpresion of L1 in NS/PCs results in a reduced proliferation, enhanced neuronal migration and differentiation, as well as decreased astrogenesis using an *in vitro* culture system and enhanced migration and survival of grafted cell after transplantation into the injured spinal cord (Xu et al., 2011). Transplanted L1-overexpressing NS/PCs increased soma size and enhanced synaptic input to host motor neuron in the lesion site. Moreover, an increase in axons expressing tyrosine hydroxylase—a enzyme vital for the synthesis of catecholaminergic neurotransmitters—distal to the lesion site was observed in mice recipients of L1-expressing NS/PCs transplantation (Xu et al., 2011). Overall a significantly improved locomotor functional recovery was observed in mice transplanted with L1-expressing precursors when compared to mice transplanted with wild-type NS/PCs after spinal cord injury (Xu et al., 2011). The L1-overexpressing neural precursor derived from ESCs exhibits decreased cell proliferation *in vitro*, enhanced neuronal differentiation both *in vitro* and *in vivo* resulting in a diminished astrocytic differentiation *in vivo* without affecting cell death compared to wild-type cells (Bernreuther et al., 2006; Cui et al., 2010; Cui et al., 2011). Transplantation of L1-overexpressing ESC-derived neural precursors at the lesion site after spinal cord injury reduces glial scar volume, enhances graft size, promotes neuronal migration, and decreases the microglial/macrophage reaction in the lesion site, and thereby improves locomotor functional recovery (Cui et al., 2011). In addition to application in treatment for spinal cord injury, L1-expressing NS/PCs have also been studied as potential candidates in neurodegenerative disease models, such as Huntington's and Parkinson's disease. In a quinolinic acid-induced Huntington's disease mouse model, L1-overexpressing neural precursors derived from ESCs generated more graft-derived GABAergic neurons in the lesioned striatum and improved locomotor functional recovery in rotation behavior test compared to the control mice transplanted wild-type cells (Bernreuther et al., 2006). In the MPTP-induced mouse model of Parkinson's disease, transplantation of L1-overexpressing ESC-derived neural precursors led to a better functional recovery in apomorphine-induced rotation test when compared to the mice treated with wild-type cells and vehicle-injected mice (Cui et al., 2010). Further morphological analysis revealed an increased number of dopaminergic neurons, leading to increased dopamine level in the striatum ipsilateral to the transplantation region of L1-expressing neural precursors (Cui et al., 2010). Due to the L1-mediated promotion of neurite outgrowth and neuronal survival by homophilic binding, NS/PCs overexpressing trimerized L1 extracellular domain and full length L1 were transplanted in to the injured spinal cord. The injured mice transplanted with NS/PCs overexpressing trimerized L1 extracellular domain and full length L1 exhibit improved functional recovery in locomotor behavior after spinal cord injury when compared to the group transplanted with only L1-overexpressing NS/PCs (He et al., 2012). The trimer-L1-expressing stem cells displayed enhanced reduction in glial scar volume in the lesion site and expression of chondroitin sulfates, preventing degeneration of corticospinal axons, promoting remyelination and enhancing regrowth of serotonergic axons (He et al., 2012). These results further demonstrated the potential application of L1-expressing NS/PC transplantation and could be of great therapeutic value. In cell replacement therapy, the transplanted cells need not only to be differentiated into proper cell types, but also to migrate into the injured region. In light of the effects of PSA on cell migration, researchers were motivated to attempt to overexpress PSA in transplanted stem cells. Ectopic expression of PSA on the ESC-derived neural precursors results in enhanced migratory ability after transplantation in the rodent striatum (Glaser et al., 2007).

Stem cell-based therapeutic applications have attracted a great deal of attention as multiple potential stem cell types were developed, particularly with the establishment of induced pluripotent stem cells (iPSCs). These stems cells seemed to provide near limitless potential in treating many human diseases. However, stem cell-based therapeutic applications for neurological disorders have faced many obstacles and setbacks, such as immunorejection, tumor formation, and low efficiency resulting from low cell survival, and failure to migrate and form functional neural connections with existing neural circuitry. Moreover, the molecular mechanisms that underlie NS/PC proliferation and differentiation into distinct cell types remain unclear. Nonetheless, advances in stem cell research and advantages of the combination of CAMs and stem cells in pre-clinical research, like enhanced cell survival, promoted cell migration, and increased neuronal differentiation, are encouraging. All these advantages suggest that CAMs have tremendous potential for application in stem cell-based cell replacement therapy for neurodegenerative diseases and spinal injuries.

http://dx.doi.org/10.5772/55136

Acknowledgements

We thank Dr. Tao Sun and Miss Aisha Abdullah for providing thoughtful comments. Owing to space limitations, I apologize for being unable to cite many excellent papers in this field. This work was supported by the Ellison Medical Foundation (T. S.), an award from the Hirschl/ Weill-Caulier Trust (T. S.) and an R01-MH083680 grant from the NIH/NIMH (T. S.).

Author details

Shan Bian

Department of Cell and Developmental Biology, Cornell University Weill Medical College, USA

References

[1] Ahrens, T, Pertz, O, Haussinger, D, Fauser, C, Schulthess, T, & Engel, J. (2002). Analysis of heterophilic and homophilic interactions of cadherins using the c-Jun/c-Fos dimerization domains. *J Biol Chem*, , 277, 19455-60.

[2] Alonso, G. (1999). Neuronal progenitor-like cells expressing polysialylated neural cell adhesion molecule are present on the ventricular surface of the adult rat brain and spinal cord. *J Comp Neurol*, , 414, 149-66.

[3] Alvarez-buylla, A, García-verdugo, J. M, & Tramontin, A. D. (2001). A unified hypothesis on the lineage of neural stem cells. *Nat Rev Neurosci*, , 2, 287-93.

[4] Andreyeva, A. Leshchyns'ka I, Knepper M, Betzel C, Redecke L, Sytnyk V, Schachner M. ((2010). CHL1 is a selective organizer of the presynaptic machinery chaperoning the SNARE complex. *PLoS One*, 5: e12018.

[5] Angata, K, Huckaby, V, Ranscht, B, Terskikh, A, Marth, J. D, & Fukuda, M. (2007). Polysialic acid-directed migration and differentiation of neural precursors are essential for mouse brain development. *Mol Cell Biol*, , 27, 6659-68.

[6] Angst, B. D, Marcozzi, C, & Magee, A. I. (2001). The cadherin superfamily: diversity in form and function. *J Cell Sci*, , 114, 629-41.

[7] Atkins, A. R, Gallin, W. J, Owens, G. C, Edelman, G. M, & Cunningham, B. A. (2004). Neural cell adhesion molecule (N-CAM) homophilic binding mediated by the two N-terminal Ig domains is influenced by intramolecular domain-domain interactions. *J Biol Chem*, , 279, 49633-43.

[8] Barthel, S. R, Gavino, J. D, Descheny, L, & Dimitroff, C. J. (2007). Targeting selectins and selectin ligands in inflammation and cancer. *Expert Opin Ther Targets*, , 11, 1473-91.

[9] Bayer, S. A, & Altman, J. (1991). Neocortical Development. *New York: Raven Press.*

[10] Becker, C. G, Artola, A, Gerardy-schahn, R, Becker, T, Welzl, H, & Schachner, M. (1996). The polysialic acid modification of the neural cell adhesion molecule is involved in spatial learning and hippocampal long-term potentiation. *J Neurosci Res*, , 45, 143-52.

[11] Beggs, H. E, Baragona, S. C, Hemperly, J. J, & Maness, P. F. (1997). NCAM140 interacts with the focal adhesion kinase fak) and the SRC-related tyrosine kinase p59(fyn). *J Biol Chem*, 272: 8310-9., 125.

[12] Bennett, V, & Baines, A. J. (2001). Spectrin and ankyrin-based pathways: metazoan inventions for integrating cells into tissues. *Physiol Rev*, , 81, 1353-92.

[13] Benson, D. L, & Tanaka, H. (1998). N-cadherin redistribution during synaptogenesis in hippocampal neurons. *J Neurosci*, , 18, 6892-904.

[14] Bernreuther, C, Dihné, M, Johann, V, Schiefer, J, Cui, Y, Hargus, G, Schmid, J. S, Xu, J, Kosinski, C. M, & Schachner, M. (2006). Neural cell adhesion molecule L1-transfected embryonic stem cells promote functional recovery after excitotoxic lesion of the mouse striatum. *J Neurosci*, , 26, 11532-9.

[15] Bonfanti, L, & Theodosis, D. T. (1994). Expression of polysialylated neural cell adhesion molecule by proliferating cells in the subependymal layer of the adult rat, in its rostral extension and in the olfactory bulb. *Neuroscience*, , 62, 291-305.

[16] Borchers, A, David, R, & Wedlich, D. (2001). Xenopus cadherin-11 restrains cranial neural crest migration and influences neural crest specification. *Development*, , 128, 3049-60.

[17] Brackenbury, R, Rutishauser, U, & Edelman, G. M. (1981). Distinct calcium-independent and calcium-dependent adhesion systems of chicken embryo cells. *Proc Natl Acad Sci U S A*, , 78, 387-91.

[18] Buchstaller, A, Kunz, S, Berger, P, Kunz, B, Ziegler, U, Rader, C, & Sonderegger, P. (1996). Cell adhesion molecules NgCAM and axonin-1 form heterodimers in the neuronal membrane and cooperate in neurite outgrowth promotion. *J Cell Biol*, , 135, 1593-607.

[19] Burgess, A, Wainwright, S. R, Shihabuddin, L. S, Rutishauser, U, Seki, T, & Aubert, I. (2008). Polysialic acid regulates the clustering, migration, and neuronal differentiation of progenitor cells in the adult hippocampus. *Dev Neurobiol*, , 68, 1580-90.

[20] Campos, L. S, Decker, L, Taylor, V, & Skarnes, W. (2006). Notch, epidermal growth factor receptor, and beta1-integrin pathways are coordinated in neural stem cells. *J Biol Chem*, , 281, 5300-9.

[21] Campos, L. S, Leone, D. P, Relvas, J. B, Brakebusch, C, Fässler, R, Suter, U, & Ffrench-constant, C. (2004). Beta1 integrins activate a MAPK signalling pathway in neural stem cells that contributes to their maintenance. *Development*, , 131, 3433-44.

[22] Castellani, V, De Angelis, E, Kenwrick, S, & Rougon, G. (2002). Cis and trans interactions of L1 with neuropilin-1 control axonal responses to semaphorin 3A. *EMBO J*, , 21, 6348-57.

[23] Chalasani, K, & Brewster, R. M. (2011). N-cadherin-mediated cell adhesion restricts cell proliferation in the dorsal neural tube. *Mol Biol Cell*, , 22, 1505-15.

[24] Chen, J, Wu, J, Apostolova, I, Skup, M, Irintchev, A, Kügler, S, & Schachner, M. (2007). Adeno-associated virus-mediated L1 expression promotes functional recovery after spinal cord injury. *Brain*, , 130, 954-69.

[25] Chothia, C, & Jones, E. Y. (1997). The molecular structure of cell adhesion molecules. *Annu Rev Biochem*, , 66, 823-62.

[26] Cui, Y. F, Hargus, G, Xu, J. C, Schmid, J. S, Shen, Y. Q, Glatzel, M, Schachner, M, & Bernreuther, C. (2010). Embryonic stem cell-derived L1 overexpressing neural aggregates enhance recovery in Parkinsonian mice. *Brain*, , 133, 189-204.

[27] Cui, Y. F, Xu, J. C, Hargus, G, Jakovcevski, I, Schachner, M, & Bernreuther, C. (2011). Embryonic stem cell-derived L1 overexpressing neural aggregates enhance recovery after spinal cord injury in mice. *PLoS One*, 6: e17126.

[28] Dahlin-huppe, K, Berglund, E. O, Ranscht, B, & Stallcup, W. B. (1997). Mutational analysis of the L1 neuronal cell adhesion molecule identifies membrane-proximal amino acids of the cytoplasmic domain that are required for cytoskeletal anchorage. *Mol Cell Neurosci*, , 9, 144-56.

[29] Demyanenko, G. P, Tsai, A. Y, & Maness, P. F. (1999). Abnormalities in neuronal process extension, hippocampal development, and the ventricular system of L1 knockout mice. *J Neurosci*, , 19, 4907-20.

[30] Denaxa, M, Chan, C. H, Schachner, M, Parnavelas, J. G, & Karagogeos, D. (2001). The adhesion molecule TAG-1 mediates the migration of cortical interneurons from the ganglionic eminence along the corticofugal fiber system. *Development*, , 128, 4635-44.

[31] Dickson, T. C, Mintz, C. D, Benson, D. L, & Salton, S. R. (2002). Functional binding interaction identified between the axonal CAM L1 and members of the ERM family. *J Cell Biol*, , 157, 1105-12.

[32] Dihné, M, Bernreuther, C, Sibbe, M, Paulus, W, & Schachner, M. (2003). A new role for the cell adhesion molecule L1 in neural precursor cell proliferation, differentiation, and transmitter-specific subtype generation. *J Neurosci*, , 23, 6638-50.

[33] Doetsch, F. (2003). A niche for adult neural stem cells. *Curr Opin Genet Dev*, , 13, 543-50.

[34] Doetsch, F, & Alvarez-buylla, A. (1996). Network of tangential pathways for neuronal migration in adult mammalian brain. *Proc Natl Acad Sci U S A*, , 93, 14895-900.

[35] Doetsch, F, Caille, I, Lim, D. A, Garcia-verdugo, J. M, & Alvarez-buylla, A. (1999). Subventricular zone astrocytes are neural stem cells in the adult mammalian brain. *Cell*,*97*, 703-16.

[36] Domeniconi, M, Cao, Z, Spencer, T, Sivasankaran, R, Wang, K, Nikulina, E, Kimura, N, Cai, H, Deng, K, Gao, Y, He, Z, & Filbin, M. (2002). Myelin-associated glycoprotein interacts with the Nogo66 receptor to inhibit neurite outgrowth. *Neuron*, , 35, 283-90.

[37] Drees, F, Pokutta, S, Yamada, S, Nelson, W. J, & Weis, W. I. (2005). Alpha-catenin is a molecular switch that binds E-cadherin-beta-catenin and regulates actin-filament assembly. *Cell*, , 123, 903-15.

[38] Erez, N, Bershadsky, A, & Geiger, B. (2005). Signaling from adherens-type junctions. *Eur J Cell Biol*, , 84, 235-44.

[39] Falk, J, Bechara, A, Fiore, R, Nawabi, H, Zhou, H, Hoyo-becerra, C, Bozon, M, Rougon, G, Grumet, M, Püschel, A. W, Sanes, J. R, & Castellani, V. (2005). Dual functional activity of semaphorin 3B is required for positioning the anterior commissure. *Neuron*, , 48, 63-75.

[40] Fournier, A. E. GrandPre T, Strittmatter SM. ((2001). Identification of a receptor mediating Nogo-66 inhibition of axonal regeneration. *Nature*, , 409, 341-6.

[41] Frei, T. von Bohlen und Halbach F, Wille W, Schachner M. ((1992). Different extracellular domains of the neural cell adhesion molecule (N-CAM) are involved in different functions. *J Cell Biol*, , 118, 177-94.

[42] Fukuhara, H, Masuda, M, Yageta, M, Fukami, T, Kuramochi, M, Maruyama, T, Kitamura, T, & Murakami, Y. (2003). Association of a lung tumor suppressor TSLC1 with MPP3, a human homologue of Drosophila tumor suppressor Dlg. *Oncogene*, , 22, 6160-5.

[43] Furley, A. J, Morton, S. B, Manalo, D, Karagogeos, D, Dodd, J, & Jessell, T. M. (1990). The axonal glycoprotein TAG-1 is an immunoglobulin superfamily member with neurite outgrowth-promoting activity. *Cell*, , 61, 157-70.

[44] Gage, F. H. (2000). Mammalian neural stem cells. *Science*, , 287, 1433-38.

[45] García-castro, M. I, Vielmetter, E, & Bronner-fraser, M. (2000). N-Cadherin, a cell adhesion molecule involved in establishment of embryonic left-right asymmetry. *Science*, , 288, 1047-51.

[46] Gascon, E, Vutskits, L, Jenny, B, Durbec, P, & Kiss, J. Z. (2007). PSA-NCAM in postnatally generated immature neurons of the olfactory bulb: a crucial role in regulating expression and cell survival. *Development*, 134: 1181-90., 75.

[47] Ghashghaei, H. T, Lai, C, & Anton, E. S. (2007). Neuronal migration in the adult brain: are we there yet? *Nat Rev Neurosci*, , 8, 141-51.

[48] Glaser, T, Brose, C, Franceschini, I, Hamann, K, Smorodchenko, A, Zipp, F, Dubois-dalcq, M, & Brüstle, O. (2007). Neural cell adhesion molecule polysialylation enhances the sensitivity of embryonic stem cell-derived neural precursors to migration guidance cues. *Stem Cells*, , 25, 3016-25.

[49] Göritz, C, & Frisén, J. (2012). Neural stem cells and neurogenesis in the adult. *Cell Stem Cell,10*, 657-9.

[50] Götz, M, & Huttner, W. B. (2005). The cell biology of neurogenesis. *Nat Rev Mol Cell Biol,6*, 777-88.

[51] Hall, P. E, Lathia, J. D, Miller, N. G, Caldwell, M. A, & Ffrench-constant, C. (2006). Integrins are markers of human neural stem cells. *Stem Cells*, , 24, 2078-84.

[52] He, X, Knepper, M, Ding, C, Li, J, Castro, S, Siddiqui, M, & Schachner, M. (2012). Promotion of spinal cord regeneration by neural stem cell-secreted trimerized cell adhesion molecule l1. *PLoS One*, 7: e46223.

[53] Heuberger, J, & Birchmeier, W. (2010). Interplay of cadherin-mediated cell adhesion and canonical Wnt signaling. *Cold Spring Harb Perspect Biol*, 2: a002915.

[54] Hirano, S, & Takeichi, M. (2012). Cadherins in brain morphogenesis and wiring. *Physiol Rev*, , 92, 597-634.

[55] Hortsch, M. family of neural cell adhesion molecules: old proteins performing new tricks. *Neuron*, , 17, 587-93.

[56] Hu, H, Tomasiewicz, H, Magnuson, T, & Rutishauser, U. (1996). The role of polysialic acid in migration of olfactory bulb interneuron precursors in the subventricular zone. *Neuron*, , 16, 735-43.

[57] Hynes, R. O. (2002). Integrins: bidirectional, allosteric signaling machines. *Cell*, , 110, 673-87.

[58] Jacob, J, Haspel, J, Kane-goldsmith, N, & Grumet, M. (2002). L1 mediated homophilic binding and neurite outgrowth are modulated by alternative splicing of exon 2. *J Neurobiol*, , 51, 177-89.

[59] Jacques, T. S, Relvas, J. B, Nishimura, S, Pytela, R, Edwards, G. M, Streuli, C. H, & Ffrench-constant, C. (1998). Neural precursor cell chain migration and division are regulated through different beta1 integrins. *Development*, , 125, 3167-77.

[60] Johnson, C. P, Fujimoto, I, Rutishauser, U, & Leckband, D. E. (2005). Direct evidence that neural cell adhesion molecule (NCAM) polysialylation increases intermembrane repulsion and abrogates adhesion. *J Biol Chem*, , 280, 137-45.

[61] Kadowaki, M, Nakamura, S, Machon, O, Krauss, S, Radice, G. L, & Takeichi, M. (2007). N-cadherin mediates cortical organization in the mouse brain. *Dev Biol*, , 304, 22-33.

[62] Kamiguchi, H, Long, K. E, Pendergast, M, Schaefer, A. W, Rapoport, I, Kirchhausen, T, & Lemmon, V. (1998). The neural cell adhesion molecule L1 interacts with the AP-2 adaptor and is endocytosed via the clathrin-mediated pathway. *J Neurosci*, , 18, 5311-21.

[63] Kakunaga, S, Ikeda, W, Shingai, T, Fujito, T, Yamada, A, Minami, Y, Imai, T, & Takai, Y. (2004). Enhancement of serum- and platelet-derived growth factor-induced cell proliferation by Necl-5/Tage4/poliovirus receptor/CD155 through the Ras-Raf-MEK-ERK signaling. *J Biol Chem*, , 279, 36419-25.

[64] Karagogeos, D, Morton, S. B, Casano, F, Dodd, J, & Jessell, T. M. (1991). Developmental expression of the axonal glycoprotein TAG-1: differential regulation by central and peripheral neurons in vitro. *Development*, , 112, 51-67.

[65] Karpowicz, P, Willaime-morawek, S, Balenci, L, Deveale, B, Inoue, T, & Van Der Kooy, D. (2009). E-Cadherin regulates neural stem cell self-renewal. *J Neurosci*, , 29, 3885-96.

[66] Kawauchi, T, Sekine, K, Shikanai, M, Chihama, K, Tomita, K, Kubo, K, Nakajima, K, Nabeshima, Y, & Hoshino, M. (2010). Rab GTPases-dependent endocytic pathways regulate neuronal migration and maturation through N-cadherin trafficking. *Neuron*, , 67, 588-602.

[67] Kim, M. Y, Kaduwal, S, Yang, D. H, & Choi, K. Y. (2010). Bone morphogenetic protein 4 stimulates attachment of neurospheres and astrogenesis of neural stem cells in neurospheres via phosphatidylinositol 3 kinase-mediated upregulation of N-cadherin. *Neuroscience*, , 170, 8-15.

[68] Kiselyov, V. V, Soroka, V, Berezin, V, & Bock, E. (2005). Structural biology of NCAM homophilic binding and activation of FGFR. *J Neurochem*, , 94, 1169-79.

[69] Kizhatil, K, Wu, Y. X, Sen, A, & Bennett, V. (2002). A new activity of doublecortin in recognition of the phospho-FIGQY tyrosine in the cytoplasmic domain of neurofascin. *J Neurosci*, , 22, 7948-58.

[70] Kolkova, K, Novitskaya, V, Pedersen, N, Berezin, V, & Bock, E. (2000). Neural cell adhesion molecule-stimulated neurite outgrowth depends on activation of protein kinase C and the Ras-mitogen-activated protein kinase pathway. *J Neurosci*, , 20, 2238-46.

[71] Kokovay, E, Wang, Y, Kusek, G, Wurster, R, Lederman, P, Lowry, N, Shen, Q, & Temple, S. (2012). VCAM1 is essential to maintain the structure of the SVZ niche and acts as an environmental sensor to regulate SVZ lineage progression. *Cell Stem Cell*, , 11, 220-30.

[72] Landis, R. C, Mcdowall, A, Holness, C. L, Littler, A. J, Simmons, D. L, & Hogg, N. (1994). Involvement of the "I" domain of LFA-1 in selective binding to ligands ICAM-1 and ICAM-3. *J Cell Biol*, , 126, 529-37.

[73] Lee, H. J, Bian, S, Jakovcevski, I, Wu, B, Irintchev, A, & Schachner, M. (2012). Delayed applications of L1 and chondroitinase ABC promote recovery after spinal cord injury. *J Neurotrauma*, 2012, 29: 1850-63.

[74] Lefebvre, J. L, Zhang, Y, Meister, M, Wang, X, & Sanes, J. R. (2008). gamma-Protocadherins regulate neuronal survival but are dispensable for circuit formation in retina. *Development*, , 135, 4141-51.

[75] Lele, Z, Folchert, A, Concha, M, Rauch, G. J, Geisler, R, Rosa, F, Wilson, S. W, Hammerschmidt, M, & Bally-cuif, L. (2002). parachute/n-cadherin is required for morphogenesis and maintained integrity of the zebrafish neural tube. *Development*, , 129, 3281-94.

[76] Lelièvre, E. C, Plestant, C, Boscher, C, Wolff, E, Mège, R. M, & Birbes, H. (2012). N-cadherin mediates neuronal cell survival through Bim down-regulation. *PLoS One*, 7: e33206.

[77] Leone, D. P, Relvas, J. B, Campos, L. S, Hemmi, S, Brakebusch, C, Fässler, R, Ffrench-constant, C, & Suter, U. (2005). Regulation of neural progenitor proliferation and survival by beta1 integrins. *J Cell Sci*, , 118, 2589-99.

[78] Liao, H, Huang, W, Schachner, M, Guan, Y, Guo, J, Yan, J, Qin, J, Bai, X, & Zhang, L. (2008). Beta 1 integrin-mediated effects of tenascin-R domains EGFL and FN6-8 on neural stem/progenitor cell proliferation and differentiation in vitro. *J Biol Chem*, , 283, 27927-36.

[79] Loers, G, Chen, S, Grumet, M, & Schachner, M. (2005). Signal transduction pathways implicated in neural recognition molecule L1 triggered neuroprotection and neuritogenesis. *J Neurochem*, , 92, 1463-76.

[80] Loulier, K, Lathia, J. D, Marthiens, V, Relucio, J, Mughal, M. R, Tang, S. C, Coksaygan, T, Hall, P. E, Chigurupati, S, Patton, B, Colognato, H, Rao, M. S, Mattson, M. P, Haydar, T. F, & Ffrench-constant, C. (2009). beta1 integrin maintains integrity of the embryonic neocortical stem cell niche. *PLoS Biol*, 7: e1000176.

[81] Lv, X, Wang, N, Su, L, Zhang, S, & Miao, J. (2006). Inhibition of PC-PLC blocked the survival of mouse neural cells by up-regulating the expression of integrin beta4 and Rb. *Dev Neurosci*, , 28, 499-504.

[82] Ma, Q. H, Futagawa, T, Yang, W. L, Jiang, X. D, Zeng, L, Takeda, Y, Xu, R. X, Bagnard, D, Schachner, M, Furley, A. J, Karagogeos, D, Watanabe, K, Dawe, G. S, & Xiao, Z. C. APP signalling pathway through Fe65 negatively modulates neurogenesis. *Nat Cell Biol*, , 10, 283-94.

[83] Maness, P. F, & Schachner, M. (2007). Neural recognition molecules of the immunoglobulin superfamily: signaling transducers of axon guidance and neuronal migration. *Nat Neurosci*, , 10, 19-26.

[84] Martini, R, & Schachner, M. (1986). Immunoelectron microscopic localization of neural cell adhesion molecules (L1, N-CAM, and MAG) and their shared carbohydrate epitope and myelin basic protein in developing sciatic nerve. *J Cell Biol*, , 103, 2439-48.

[85] Mason, I. (1994). Cell signalling. Do adhesion molecules signal via FGF receptors? *Curr Biol*, , 4, 1158-61.

[86] Mehanna, A, Jakovcevski, I, Acar, A, Xiao, M, Loers, G, Rougon, G, Irintchev, A, & Schachner, M. (2010). Polysialic acid glycomimetic promotes functional recovery and plasticity after spinal cord injury in mice. *Mol Ther*, , 18, 34-43.

[87] Mehanna, A, Mishra, B, Kurschat, N, Schulze, C, Bian, S, Loers, G, Irintchev, A, & Schachner, M. (2009). Polysialic acid glycomimetics promote myelination and functional recovery after peripheral nerve injury in mice. *Brain*, , 132, 1449-62.

[88] Merkle, F. T, & Alvarez-buylla, A. (2006). Neural stem cells in mammalian development. *Curr Opin Cell Biol*, , 18, 704-9.

[89] Merkle, F. T, Tramontin, A. D, Garcia-verdugo, J. M, & Alvarez-buylla, A. (2004). Radial glia give rise to adult neural stem cells in the subventricular zone. *Proc Natl Acad Sci U S A,101*, 17528-32.

[90] Miller, F. D, & Gauthier-fisher, A. Home at last: neural stem cell niches defined. *Cell Stem Cell*, , 4, 507-10.

[91] Mobley, A. K, & Mccarty, J. H. (2011). integrin is essential for neuroblast migration in the rostral migratory stream. *Glia*, , 59, 1579-87.

[92] Mobley, A. K, Tchaicha, J. H, Shin, J, Hossain, M. G, & Mccarty, J. H. (2009). Beta8 integrin regulates neurogenesis and neurovascular homeostasis in the adult brain. *J Cell Sci*, , 122, 1842-51.

[93] Monier-gavelle, F, & Duband, J. L. (1997). Cross talk between adhesion molecules: control of N-cadherin activity by intracellular signals elicited by beta1 and beta3 integrins in migrating neural crest cells. *J Cell Biol*, , 137, 1663-81.

[94] Moore, K. A, & Lemischka, I. R. (2006). Stem cells and their niches. *Science*, , 311, 1880-5.

[95] Morita, I, Kizuka, Y, Kakuda, S, & Oka, S. (2008). Expression and function of the HNK-1 carbohydrate. *J Biochem*, , 143, 719-24.

[96] Murase, S, & Horwitz, A. F. (2002). Deleted in colorectal carcinoma and differentially expressed integrins mediate the directional migration of neural precursors in the rostral migratory stream. *J Neurosci*, , 22, 3568-79.

[97] Nagar, B, Overduin, M, Ikura, M, & Rini, J. M. (1996). Structural basis of calcium-induced E-cadherin rigidification and dimerization. *Nature*, , 380, 360-4.

[98] Nakagawa, S, & Takeichi, M. (1998). Neural crest emigration from the neural tube depends on regulated cadherin expression. *Development*, , 125, 2963-71.

[99] Paratcha, G, Ledda, F, & Ibáñez, C. F. (2003). The neural cell adhesion molecule NCAM is an alternative signaling receptor for GDNF family ligands. *Cell*, , 113, 867-79.

[100] Pavlou, O, Theodorakis, K, Falk, J, Kutsche, M, Schachner, M, Faivre-sarrailh, C, & Karagogeos, D. (2002). Analysis of interactions of the adhesion molecule TAG-1 and its domains with other immunoglobulin superfamily members. *Mol Cell Neurosci*, , 20, 367-81.

[101] Perrais, M, Chen, X, Perez-moreno, M, & Gumbiner, B. M. (2007). E-cadherin homophilic ligation inhibits cell growth and epidermal growth factor receptor signaling independently of other cell interactions. *Mol Biol Cell*, , 18, 2013-25.

[102] Petridis, A. K, Maarouf, A, & Rutishauser, U. (2004). Polysialic acid regulates cell contact-dependent neuronal differentiation of progenitor cells from the subventricular zone. *Dev Dyn*, , 230, 675-84.

[103] Polo-parada, L, Plattner, F, Bose, C, & Landmesser, L. T. (2005). NCAM 180 acting via a conserved C-terminal domain and MLCK is essential for effective transmission with repetitive stimulation. *Neuron*, , 46, 917-31.

[104] Poltorak, M, Sadoul, R, Keilhauer, G, Landa, C, Fahrig, T, & Schachner, M. (1987). Myelin-associated glycoprotein, a member of the L2/HNK-1 family of neural cell adhesion molecules, is involved in neuron-oligodendrocyte and oligodendrocyte-oligodendrocyte interaction. *J Cell Biol*, , 105, 1893-9.

[105] Proctor, J. M, Zang, K, Wang, D, Wang, R, & Reichardt, L. F. (2005). Vascular development of the brain requires beta8 integrin expression in the neuroepithelium. *J Neurosci*, , 25, 9940-8.

[106] Quarles, R. H. (1984). Myelin-associated glycoprotein in development and disease. *Dev Neurosci*, , 6, 285-303.

[107] Rader, C, Kunz, B, Lierheimer, R, Giger, R. J, Berger, P, Tittmann, P, Gross, H, & Sonderegger, P. (1996). Implications for the domain arrangement of axonin-1 derived from the mapping of its NgCAM binding site. *EMBO J*, , 15, 2056-68.

[108] Ranheim, T. S, Edelman, G. M, & Cunningham, B. A. (1996). Homophilic adhesion mediated by the neural cell adhesion molecule involves multiple immunoglobulin domains. *Proc Natl Acad Sci U S A*, , 93, 4071-5.

[109] Raymond, K, Deugnier, M. A, Faraldo, M. M, & Glukhova, M. A. (2009). Adhesion within the stem cell niches. *Curr Opin Cell Biol*, , 21, 623-9.

[110] Redies, C, & Takeichi, M. (1993). Expression of N-cadherin mRNA during development of the mouse brain. *Dev Dyn*, , 197, 26-39.

[111] Rieger, S, Senghaas, N, Walch, A, & Köster, R. W. (2009). Cadherin-2 controls directional chain migration of cerebellar granule neurons. *PLoS Biol*, 7: e1000240.

[112] Reyes, A. A, Small, S. J, & Akeson, R. (1991). At least 27 alternatively spliced forms of the neural cell adhesion molecule mRNA are expressed during rat heart development. *Mol Cell Biol*, , 11, 1654-61.

[113] Roonprapunt, C, Huang, W, Grill, R, Friedlander, D, Grumet, M, Chen, S, Schachner, M, & Young, W. (2003). Soluble cell adhesion molecule L1-Fc promotes locomotor recovery in rats after spinal cord injury. *J Neurotrauma*, , 20, 871-82.

[114] Rousso, D. L, Pearson, C. A, Gaber, Z. B, Miquelajauregui, A, Li, S, Portera-cailliau, C, Morrisey, E. E, & Novitch, B. G. (2012). Foxp-mediated suppression of N-cadherin regulates neuroepithelial character and progenitor maintenance in the CNS. *Neuron*, , 74, 314-30.

[115] Rutishauser, U, & Landmesser, L. (1996). Polysialic acid in the vertebrate nervous system: a promoter of plasticity in cell-cell interactions. *Trends Neurosci*, , 19, 422-7.

[116] Sakaguchi, M, Imaizumi, Y, Shingo, T, Tada, H, Hayama, K, Yamada, O, Morishita, T, Kadoya, T, Uchiyama, N, Shimazaki, T, Kuno, A, Poirier, F, Hirabayashi, J, Sawamoto, K, & Okano, H. (2010). Regulation of adult neural progenitor cells by Galectin-1/beta1 Integrin interaction. *J Neurochem*, , 113, 1516-24.

[117] Sakisaka, T, & Takai, Y. (2004). Biology and pathology of nectins and nectin-like molecules. *Curr Opin Cell Biol*, , 16, 513-21.

[118] Salzer, J. L, Holmes, W. P, & Colman, D. R. (1987). The amino acid sequences of the myelin-associated glycoproteins: homology to the immunoglobulin gene superfamily. *J Cell Biol*, , 104, 957-65.

[119] Sano, K, Tanihara, H, Heimark, R. L, Obata, S, & Davidson, M. St John T, Taketani S, Suzuki S. ((1993). Protocadherins: a large family of cadherin-related molecules in central nervous system. *EMBO J*, , 12, 2249-56.

[120] Santuccione, A, & Sytnyk, V. Leshchyns'ka I, Schachner M. ((2005). Prion protein recruits its neuronal receptor NCAM to lipid rafts to activate and to enhance neurite outgrowth. *J Cell Biol*, 169: 341-54., 59fyn.

[121] Schaefer, A. W, Kamiguchi, H, Wong, E. V, Beach, C. M, Landreth, G, & Lemmon, V. (1999). Activation of the MAPK signal cascade by the neural cell adhesion molecule L1 requires L1 internalization. *J Biol Chem*, , 274, 37965-73.

[122] Schmid, R. S, Pruitt, W. M, & Maness, P. F. (2000). A MAP kinase-signaling pathway mediates neurite outgrowth on L1 and requires Src-dependent endocytosis. *J Neurosci*, , 20, 4177-88.

[123] Schnaar, R. L, Collins, B. E, Wright, L. P, Kiso, M, Tropak, M. B, Roder, J. C, & Crocker, P. R. (1998). Myelin-associated glycoprotein binding to gangliosides. Structural specificity and functional implications. *Ann N Y Acad Sci*, , 845, 92-105.

[124] Seki, T, & Arai, Y. (1993). Distribution and possible roles of the highly polysialylated neural cell adhesion molecule (NCAM-H) in the developing and adult central nervous system. *Neurosci Res*, , 17, 265-90.

[125] Senkov, O, Sun, M, Weinhold, B, Gerardy-schahn, R, Schachner, M, & Dityatev, A. (2006). Polysialylated neural cell adhesion molecule is involved in induction of long-term potentiation and memory acquisition and consolidation in a fear-conditioning paradigm. *J Neurosci*, , 26, 10888-98.

[126] Shapiro, L, Love, J, & Colman, D. R. (2007). Adhesion molecules in the nervous system: structural insights into function and diversity. *Annu Rev Neurosci.*, 30, 451-74.

[127] Shi, Y, Sun, G, & Stewart, R. (2008). Neural stem cell self-renewal. *Crit Rev Oncol Hematol,65*, 43-53.

[128] Shimizu, K, & Takai, Y. (2003). Roles of the intercellular adhesion molecule nectin in intracellular signaling. *J Biochem*, , 134, 631-6.

[129] Shoval, I, Ludwig, A, & Kalcheim, C. (2007). Antagonistic roles of full-length N-cadherin and its soluble BMP cleavage product in neural crest delamination. *Development*, , 134, 491-501.

[130] Simova, O, Irintchev, A, Mehanna, A, Liu, J, Dihné, M, Bächle, D, Sewald, N, Loers, G, & Schachner, M. (2006). Carbohydrate mimics promote functional recovery after peripheral nerve repair. *Ann Neurol*, , 60, 430-37.

[131] Skaper, S. D, Facci, L, Williams, G, Williams, E. J, Walsh, F. S, & Doherty, P. (2004). A dimeric version of the short N-cadherin binding motif HAVDI promotes neuronal cell survival by activating an N-cadherin/fibroblast growth factor receptor signalling cascade. *Mol Cell Neurosci*, , 26, 17-23.

[132] Son, Y. S, Seong, R. H, Ryu, C. J, Cho, Y. S, Bae, K. H, Chung, S. J, Lee, B, Min, J. K, & Hong, H. J. (2011). Brief report: L1 cell adhesion molecule, a novel surface molecule

of human embryonic stem cells, is essential for self-renewal and pluripotency. *Stem Cells*, , 29, 2094-9.

[133] Staquicini, F. I, Dias-neto, E, Li, J, Snyder, E. Y, Sidman, R. L, Pasqualini, R, & Arap, W. (2009). Discovery of a functional protein complex of netrin-4, laminin gamma1 chain, and integrin alpha6beta1 in mouse neural stem cells. *Proc Natl Acad Sci U S A*, , 106, 2903-8.

[134] Stepniak, E, Radice, G. L, & Vasioukhin, V. (2009). Adhesive and signaling functions of cadherins and catenins in vertebrate development. *Cold Spring Harb Perspect Biol*, 1: a002949.

[135] Sternberger, N. H, Quarles, R. H, Itoyama, Y, & Webster, H. D. (1979). Myelin-associated glycoprotein demonstrated immunocytochemically in myelin and myelin-forming cells of developing rat. *Proc Natl Acad Sci U S A*, , 76, 1510-4.

[136] Stoenica, L, Senkov, O, Gerardy-schahn, R, Weinhold, B, Schachner, M, & Dityatev, A. (2006). In vivo synaptic plasticity in the dentate gyrus of mice deficient in the neural cell adhesion molecule NCAM or its polysialic acid. *Eur J Neurosci*, , 23, 2255-64.

[137] Su, L, Lv, X, Xu, J, Yin, D, Zhang, H, Li, Y, Zhao, J, Zhang, S, & Miao, J. (2009). Neural stem cell differentiation is mediated by integrin beta4 in vitro. *Int J Biochem Cell Biol*, , 41, 916-24.

[138] Suzuki, Y, Yanagisawa, M, Yagi, H, Nakatani, Y, & Yu, R. K. (2010). Involvement of beta1-integrin up-regulation in basic fibroblast growth factor- and epidermal growth factor-induced proliferation of mouse neuroepithelial cells. *J Biol Chem*, , 285, 18443-51.

[139] Takai, Y, Irie, K, Shimizu, K, Sakisaka, T, & Ikeda, W. (2003). Nectins and nectin-like molecules: roles in cell adhesion, migration, and polarization. *Cancer Sci*, , 94, 655-67.

[140] Takai, Y, & Nakanishi, H. (2003). Nectin and afadin: novel organizers of intercellular junctions. *J Cell Sci*, , 116, 17-27.

[141] Takekuni, K, Ikeda, W, Fujito, T, Morimoto, K, Takeuchi, M, Monden, M, & Takai, Y. (2003). Direct binding of cell polarity protein PAR-3 to cell-cell adhesion molecule nectin at neuroepithelial cells of developing mouse. *J Biol Chem*, , 278, 5497-500.

[142] Taniguchi, H, Kawauchi, D, Nishida, K, & Murakami, F. (2006). Classic cadherins regulate tangential migration of precerebellar neurons in the caudal hindbrain. *Development*, , 133, 1923-31.

[143] Tate, M. C, García, A. J, Keselowsky, B. G, Schumm, M. A, & Archer, D. R. LaPlaca MC. ((2004). Specific beta1 integrins mediate adhesion, migration, and differentiation of neural progenitors derived from the embryonic striatum. *Mol Cell Neurosci*, , 27, 22-31.

[144] Temple, S. (2001). The development of neural stem cells. *Nature,414*, 112-17.

[145] Tepass, U, Truong, K, Godt, D, Ikura, M, & Peifer, M. (2000). Cadherins in embryonic and neural morphogenesis. *Nat Rev Mol Cell Biol*, , 1, 91-100.

[146] Tsiotra, P. C, Theodorakis, K, Papamatheakis, J, & Karagogeos, D. (1996). The fibronectin domains of the neural adhesion molecule TAX-1 are necessary and sufficient for homophilic binding. *J Biol Chem*, , 271, 29216-22.

[147] Vanhalst, K, Kools, P, Staes, K, Van Roy, F, & Redies, C. (2005). delta-Protocadherins: a gene family expressed differentially in the mouse brain. *Cell Mol Life Sci*, , 62, 1247-59.

[148] Vutskits, L, Djebbara-hannas, Z, Zhang, H, Paccaud, J. P, Durbec, P, Rougon, G, Muller, D, & Kiss, J. Z. (2001). PSA-NCAM modulates BDNF-dependent survival and differentiation of cortical neurons. *Eur J Neurosci*, , 13, 1391-402.

[149] Walmod, P. S, Kolkova, K, Berezin, V, & Bock, E. (2004). Zippers make signals: NCAM-mediated molecular interactions and signal transduction. *Neurochem Res*, , 29, 2015-35.

[150] Wang, Y, Yao, M, Zhou, J, Zheng, W, Zhou, C, Dong, D, Liu, Y, Teng, Z, Jiang, Y, Wei, G, & Cui, X. (2011). The promotion of neural progenitor cells proliferation by aligned and randomly oriented collagen nanofibers through β1 integrin/MAPK signaling pathway. *Biomaterials*, , 32, 6737-44.

[151] Wang, X, Weiner, J. A, Levi, S, Craig, A. M, Bradley, A, & Sanes, J. R. (2002). Gamma protocadherins are required for survival of spinal interneurons. *Neuron*, , 36, 843-54.

[152] Watanabe, T, Sato, K, & Kaibuchi, K. (2009). Cadherin-mediated intercellular adhesion and signaling cascades involving small GTPases. *Cold Spring Harb Perspect Biol*, 1: a003020.

[153] Wheelock, M. J, & Johnson, K. R. (2003). Cadherin-mediated cellular signaling. *Curr Opin Cell Biol*, , 15, 509-14.

[154] Yageta, M, Kuramochi, M, Masuda, M, Fukami, T, Fukuhara, H, Maruyama, T, Shibuya, M, & Murakami, Y. (2002). Direct association of TSLC1 and DAL-1, two distinct tumor suppressor proteins in lung cancer. *Cancer Res*, , 62, 5129-33.

[155] Yoshida, N, Hishiyama, S, Yamaguchi, M, Hashiguchi, M, Miyamoto, Y, Kaminogawa, S, & Hisatsune, T. (2003). Decrease in expression of alpha 5 beta 1 integrin during neuronal differentiation of cortical progenitor cells. *Exp Cell Res*, , 287, 262-71.

[156] Yue, X. S, Murakami, Y, Tamai, T, Nagaoka, M, Cho, C. S, Ito, Y, & Akaike, T. (2010). A fusion protein N-cadherin-Fc as an artificial extracellular matrix surface for maintenance of stem cell features. *Biomaterials*, , 31, 5287-96.

[157] Xiao, M. F, Xu, J. C, Tereshchenko, Y, Novak, D, Schachner, M, & Kleene, R. (2009). Neural cell adhesion molecule modulates dopaminergic signaling and behavior by regulating dopamine D2 receptor internalization. *J Neurosci*, , 29, 14752-63.

[158] Xu, J. C, Bernreuther, C, Cui, Y. F, Jakovcevski, I, Hargus, G, Xiao, M. F, & Schachner, M. expressing radial glia and astrocytes enhance recovery after spinal cord injury. *J Neurotrauma*, , 28, 1921-37.

[159] Zhang, H, Vutskits, L, Calaora, V, Durbec, P, & Kiss, J. Z. (2004). A role for the polysialic acid-neural cell adhesion molecule in PDGF-induced chemotaxis of oligodendrocyte precursor cells. *J Cell Sci*, , 117, 93-103.

[160] Zhang, J, Woodhead, G. J, Swaminathan, S. K, Noles, S. R, Mcquinn, E. R, Pisarek, A. J, Stocker, A. M, Mutch, C. A, Funatsu, N, & Chenn, A. (2010). Cortical neural precursors inhibit their own differentiation via N-cadherin maintenance of beta-catenin signaling. *Dev Cell*, , 18, 472-9.

[161] Zhang, Y, Ghadiri-sani, M, Zhang, X, Richardson, P. M, Yeh, J, & Bo, X. (2007). Induced expression of polysialic acid in the spinal cord promotes regeneration of sensory axons. *Mol Cell Neurosci*, , 35, 109-19.

[162] Zhang, Y, Sivasankar, S, Nelson, W. J, & Chu, S. (2009). Resolving cadherin interactions and binding cooperativity at the single-molecule level. *Proc Natl Acad Sci U S A*, , 106, 109-14.

[163] Zhang, Y, Zhang, X, Yeh, J, Richardson, P, & Bo, X. (2007). Engineered expression of polysialic acid enhances Purkinje cell axonal regeneration in L1/GAP-43 double transgenic mice. *Eur J Neurosci*, Figure 1. Schematic diagram of diverse domain structure of different CAM superfamilies. A. IgSF CAM proteins. B. Cadherins. C. Integrins. D. Selectins, 25, 351-61.

Chapter 13

Neuroinflammation on the Epigenetics of Neural Stem Cells

Nando Dulal Das and Young Gyu Chai

Additional information is available at the end of the chapter

http://dx.doi.org/10.5772/54981

1. Introduction

Under physiological conditions, neuronal stem cells (NSCs) can undergo both self-renewal and differentiation stages. The formation of new neurons, neurogenesis, is a vital process by which the brain maintains its lifelong plasticity in response to extrinsic and intrinsic changes. However, the exact mechanisms that regulate NSC self-renewal and differentiation are largely unknown. NSCs become stimulated after neuronal injury and can migrate at pathological sites (Nakatomi et al., 2002; Russo et al., 2011) that dictate the potential of NSCs therapeutic use in pathological conditions of the central nervous system. In this chapter, we describe the effect of neuroinflammation in NSCs and discuss whether the inflammatory mediators can epigenetically affect the capacity of NSCs and alter their proliferation and differentiation ability. The mechanism by which the inflammatory environment influences the NSC niche and thus, alters the self-renewal, survival, migration, and differentiation of the NSCs is currently unknown (Martino and Pluchino, 2006). Several studies have focused the effects of inflammation on the regenerative capacity of NSCs subjected to microglial activation after an acute injury or after LPS treatment. Overall, the connection between brain inflammation and NSC neurogenesis and the role of the niche in the modulation of neuronal differentiation under alternative conditions are under intense investigation.

To gain further insight into these phenomena, we describe epigenetic mechanisms, including DNA methylation and histone modification in NSCs inflammation. DNA methylation and histone modification are known to play significant roles in the modulation of stem cell proliferation and differentiation (Li and Zhao, 2008). Regarding DNA methylation, methylated CpG-binding protein (MBD) deficiency results the suppression of NSCs differentiation. Therefore, to identify the downstream target genes of MBDs has potential in NSCs differentiation study. Histone modifications are another important epigenetic mark. There are many types of post-translational modifications of the residues at histone tails, including methylation of lysines and arginines, acetylation, phosphorylation, ubiquitination, SUMOylation, and ADP-ribosylation. Among the histone modifications, histone H3 lysine (K) methylation is a central epigenetic modification with both activating and repressive roles in eukaryotic chromatin (Reinberg et al., 2004).

Next, we will focus on epigenetic involvement in neurodegenerative diseases and NSCs. Actually, inflammatory stimuli induce beneficial effects (e.g., phagocytosis of debris and apoptotic cells), and inflammation is linked to tissue repair processes, uncontrolled inflammation may result in production of neurotoxic factors that amplify underlying disease states and pathogenesis of Alzheimer's disease (AD), Parkinson's disease (PD), amyotrophic lateral sclerosis (ALS), multiple sclerosis (MS), and a growing number of other nervous system pathologies (Glass et al., 2010). Here, endogenous NSCs cannot fully compensate the neuronal loss in such neurodegenerative diseases. The possible reasons include the lack of trophic support and inhibitory signals within the brain microenvironment (Croft and Przyborski, 2009), indicative of oxidative stress (Kelly et al., 2011) and age-related neuroinflammation. In summary, the recent development of stem cells technology open new areas of research aimed at stimulating neuronal regeneration in the brain during aging, neuroinflammation and neurodegenerative diseases (Russo et al., 2011). Here, we overview the therapeutic approach of NSC and how these stem cells are responsible for brain homeostasis, induction of neurogenesis in several diseased states. Finally, this chapter indicates the possibility of combination therapy of epigenetic drug with NSC transplantation in these neurodegenerative diseases.

2. Epigenetics and neuroinflammation

Alterations in cell signaling by environmental changes can remodel epigenetic marks (Borrelli et al., 2008; Weaver et al., 2007). Epigenetics thus presents potential explanations for sustained changes in transcriptional activity associated with cell differentiation, learning and memory, age-related neurodegeneration and effects of early experience, repeated drug exposure, chronic stress, and environmental toxins. The implicit hypothesis is that environmental signals alter chromatin modifications, which then serve as the mechanism for the transcriptional 'plasticity' that mediates sustained variation in neural function (Meaney and Ferguson-Smith, 2010). Although most extensively studied in embryonic stem cells, such 'bivalent' domains, which are also found in the adult brain (Sanz et al., 2008), suggest a developmentally 'poised' state awaiting environmental direction. Indeed, such states may mark the potential for plasticity. The same epigenetic mark can recruit effectors that activate as well as others that repress transcription. We describe the detail epigenetic changes in NSC later in this chapter. We summarize the recent evidence that physiological and environmental signals influence adaptive transcriptional responses in neurons through the epigenetic modification of chromatin. We highlight to the regulation of histone modifications and DNA methylation in response to neuroinflammation and related signaling. In addition, mechanisms that induce chromatin modifications in association with multiprotein complexes on neuronal gene promoters are mentioned.

http://dx.doi.org/10.5772/54981

3. NSC and inflammation

To maintain brain homeostasis, NSCs are highly controlled under physiological conditions in which the stem cell niche is vital for the NSC self-renewal, proliferation, differentiation, and migration. NSCs become activated after neuronal injury and migrate to the site of injury, indicating that some regulators at the injury site can guide the migration of precursor cells. Damaged neurons can be repaired by the activation of endogenous neuronal stem cells, which migrate to regions of the brain injury, differentiate into neuronal cells, and integrate into neuronal circuits (Belmadani et al., 2006; Russo et al., 2011) The mechanism by which the inflammatory environment influences the NSC niche and thus, alters the self-renewal, survival, migration, and differentiation of the NSCs is currently unknown (Martino and Pluchino, 2006). Alterations of NSC functions either pro-neurogenic or anti-neurogenic in inflammation may depend on the NSC niche and activation of brain microglial cells. It is reported that activated microglia in inflammatory conditions can inhibit neurogenesis (Butovsky et al., 2006). On the contrary, activated microglia also showed helpful for neurogenesis (Hanisch and Kettenmann, 2007). Inflammatory cytokines and nitric oxide (NO) released by microglial cells can inhibit the adult neurogenesis. Activation of microglia with LPS results the production of inflammatory mediators in vitro, including TNF-α and IL-6, that inhibit the generation of neurons from NSCs (Monje et al., 2003). However, modification of microglial status by other cytokines, such as IL-4 or low dose interferon-γ (IFN-γ) changes their phenotype to strongly promote neurogenesis (Butovsky et al., 2006). However, the positive effects are at least partly dependent on microglia production of insulin-like growth factor-1 (IGF-1), a potent proneurogenic growth factor. Though controversial, this raises the possibility that some types of controlled inflammation may be exploited in CNS regeneration or in combating neurological diseases that have pronounced chronic proinflammatory components (Rolls et al., 2009)

Several studies have focused the effects of inflammation on the regenerative capacity of NSCs subjected to microglial activation after an acute injury or after LPS treatment. It is reported that TLR4 is expressed by NSCs, and LPS suppresses the proliferation of NSCs under culture conditions via an NF-κB-dependent mechanism (Monje et al., 2003; Rolls et al., 2007). In addition, TLR4 can directly modulate the self-renewal and cell-fate decision of neuronal progenitor cells (Rolls et al., 2007). Overall, the effects of proinflammatory signaling on NSCs go beyond simple changes in the abundance of new neurons (Carpentier and Palmer, 2009). It is shown that the neurons generated during the period of inflammation are morphologically normal, with normal cell body location, polarity, and branching, yet they display an accentuated inhibitory or excitatory responses in immature versus mature neurons, respectively (Jakubs et al., 2008). So, the functions of new neurons are severely affected by immune signaling. Moreover, the connection between brain inflammation and NSC neurogenesis and the role of the niche in the modulation of neuronal differentiation under alternative conditions are under intense investigation.

4. Epigenetic significance in NSC-inflammation

Epigenetic refers to any heritable influence (in the progeny of cells or individuals) on chromosome or gene function that is not accompanied by a change in DNA sequence (Yoder et al., 1997). It includes processes such as DNA methylation, histone modification and noncoding RNA expression. Appropriate gene function either activation or repression at inflammatory stages of NSC progression could be achieved by such epigenetic regulation. Here, we cover recent reports involving the role of epigenetic mechanisms in NSC-inflammation and its fate on NSC mechanisms. One of the important epigenetic mechanisms, DNA methylation in the genome is established by a family of DNA methyltransferases (DNMTs). Maintenance of methylation patterns is achieved by a function of DNMT1 during DNA replication, while de novo methylation is primarily catalyzed by DNMT3a and DNMT3b. DNA methylation is responsible for the regulation of gene expression, where two mechanisms are involved. First, methylation of CpG dinucleotides affects DNA structure and can directly interfere with the binding of TFs to their target sequences(Takizawa et al., 2001); second, a more pervasive effect, methyl-CpG-binding domain (MBD)-containing protein family members can bind to genes with methylated CpG dinucleotides, thereby suppressing the genes' expression (Nan et al., 1997) Though, DNA methylation is actively involved in the acquisition of multipotentiality in NSC from early-, mid- to late-gestation. Here, we mainly focus on two well-studied pathways that act synergistically to promote astrocytic differentiation of NSC are those activated by the interleukin-6 (IL-6) family of cytokines such as leukemia inhibitory factor (LIF), ciliary neurotrophic factor (CNTF) and cardiotrophin-1 (CT-1) and bone morphogenetic protein (BMP) signaling (Juliandi et al., 2010). In early- and mid-gestational NSCs, astrocytic gene promoters such as glial fibrillary acidic protein (GFAP) are hypermethylated, a status that impedes binding of the STAT3-p300/CBP-SMADs complex to its target sequence and thus prevents these NSCs from differentiating into astrocytes even when the cells are stimulated by astrocyte-inducing cytokines (Takizawa et al., 2001). These IL-6 family cytokines have been shown to be expressed in NSCs and neurons in the fetal mouse brain (Barnabe-Heider et al., 2005); but how DNA methylation of IL-6 does affect of NSC differentiation in inflammation is not disclosed. On the contrary, the STAT3 binding site-containing GFAP promoter in NSCs at late gestation is barely methylated, so that upon LIF stimulation these NSCs can differentiate into astrocytes (Takizawa et al., 2001). Overall, the genome-wide DNA methylation status of NSCs in well-defined inflammatory conditions may give us possible clue of gene specific methylation status of NSC whether DNA methylation can play an important role in defining the NSC fate from neurogenesis to astrocytogenesis in inflammatory conditions. Notch signaling is a conserved pathway from insects to mammals, which contributes to cell-to-cell communication (Louvi and Artavanis-Tsakonas, 2006) and controls cell fate determination in the CNS (Lundkvist and Lendahl, 2001). Upon Notch activation by its ligand, the Notch intracellular domain (NICD) is released from the plasma membrane and is translocated into the nucleus, where it converts a particular repressor complex into an activator complex (Nakayama et al., 2008). It is confirmed that Notch ligands are indeed expressed in neuronally committed NPCs and young neurons, and that these ligands activate Notch signaling in the residual NSCs. Further, forced expression of NICD in midgestational NSCs induced the upregulation of nuclear factor 1A (NF1A), which in turn accelerated demethylation of astrocytic gene promoters by preventing DNMT1 from binding to them and thus allowed precocious astrocytic differentiation in response to LIF stimulation (Namihira et al., 2009).

It has shown that methyl binding domain (MBD) proteins expressed predominantly in neurons, and not in astrocytes or oligodendrocytes, in the CNS (Kishi and Macklis, 2004); may regulate in NSC differentiation. It was found that exon1 of GFAP are hypermethylated in all neural cell types and that only in neurons, methyl-CpG-binding protein 2 (MeCP2), a member of the MBD family, is highly expressed and binds to this methylated exon1 region (Setoguchi et al., 2006) that is linked to block the astrocyte differentiation. Indeed, ectopic expression of MeCP2 directs NSCs to become neurons and inhibits astrocytic differentiation, even in the presence of astrocyte-inducing cytokines such as LIF and BMP2 (Tsujimura et al., 2009). MBD1-deficient NSCs generate fewer neurons than do wild type NSCs, suggesting an important role for MBD1 in neuronal fate specification (Zhao et al., 2003).

Histone proteins within the chromosome play a significant role in chromatin structure, gene transcription and epigenetic information. Multiple modifications decorate each histone tail within the nucleosome, and some amino acids on the histone tail can be modified in several different ways. Covalent modifications of histone taile include methylation, acetylation, phosphorylation, ubiquitylation, sumoylation, glycosylation, biotinylation, carbonylation and adenosine diphosphate (ADP)-ribosylation (Strahl and Allis, 2000). Among these, modifications by histone acetylation and methylation are the most common. Acetylation and deacetylation of lysine residue in histone tails is mediated by histone acetyl transferases (HATs) and histone deacetylases (HDACs), respectively (Hsieh and Gage, 2005). Histone acetylation by HATs is responsible for open chromatin (euchromatin) formation that leads to transcriptional activation. Conversely, HDACs result decrease of histone acetylation and formation of condensed chromatin (heterochromatin) that causes transcriptional silencing. Adult hippocampal-derived NSCs differentiate predominantly into neurons, at the expense of astrocytes and oligodendrocytes, when treated by the antiepileptic and HDAC inhibitor valproic acid (VPA) in vitro, even in conditions that favor glia-specific differentiation (Hsieh et al., 2004). VPA-mediated HDAC inhibition upregulates the neuron-specific gene NeuroD, a neurogenic basic helix-loop-helix transcription factor (TF), is resulting in the induction and suppression, respectively, of neuronal and glial differentiation. In the developing rat brain and in cultured E14 NSCs, VPA treatment has also been shown to promote neurogenesis by activating the Ras-ERK pathway (Jung et al., 2008).

Histone methylation is involved in the regulation of a variety of nuclear processes dedicated to the maintenance of active and silent states of gene expression, which is essential for cellular regulation, homeostasis and fate determination (Cloos et al., 2008). There are five lysine residues in the histone N termini that are prominently methylated. H3K4 and H3K36 methylation primarily transduce activating functions, whereas H3K9, H3K27, and H4K20 methylation is mainly associated with repressed chromatin. Histone lysine methylationcan result in mono-, di-, or trimethyl states and each distinct methylation state confers different biological read-outs. Histone lysine trimethyl states, particularly those with repressive functions, appear relatively robust because they are stably propagated during several cell divisions (Lachner et al., 2004). Among the histone modifications, histone H3 lysine (K) methylation is a central epigenetic modification with both activating and repressive roles in eukaryotic chromatin (Reinberg et al., 2004). JmjC domain proteins demethylate histone lysine and arginine residues in an oxidative reaction that requires Fe (II) and α-ketoglutarate as cofactors. Depending on their target specificity, JmjC domain proteins promote transcriptional repression or activation, thereby impacting important processes such as hormone response, stem cell renewal, germ cell development, and cellular proliferation and differentiation (Beyer et al., 2008). Interestingly, a range of JmjC proteins is induced in different cancers and has been linked to cell proliferation (Cloos et al., 2006) and the suppression of senescence (Pfau et al., 2008). Members of the JMJD2 family that target H3K9me3/me2 and H3K36me3/me2 are highly expressed in prostate cancer (Wissmann et al., 2007).

Recently, we reported the effect of lipopolysaccharides (LPS) on NSCs epigenetics, where we used an immortalized neuroectodermal stem cell line, NE-4C. The NE-4C cell line was cloned from the anterior brain vesicles of E9 mouse embryos lacking functional p53 (Livingstone et al., 1992). Non-induced NE-4C cells grow as homogeneous, epithelial-like populations, and in response to all-trans retinoic acid (RA) treatment, they differentiate into neurons on a highly reproducible schedule (Jelitai et al., 2004). We found that histone demethylase, Jmjd2b is functional in long-term LPS treatment and regulates the histone demethylation of the promoters of its target genes that may be crucial in multiple signaling pathways and biological processes in murine NSCs (NE-4C cells). MetaCore pathway analysis revealed the gene networks and canonical pathways affected in Jmjd2b-attenuated NE-4C cells that involved neurophysiological processes (receptor-mediated axon growth repulsion, GABA-A receptor life cycle), the Notch1-mediated pathway for NF-κB activity modulation, and TGF-β-dependent induction. Several extrinsic factors affect the histone methylation status of NSCs. In the postnatal mouse brain, MLL1 is required for neurogenesis and its deficiency in NSCs in the subventricular zone (SVZ) leads to a glial lineage preference. One of the key downstream regulators of SVZ neurogenesis, Dlx2, is not expressed in MLL1-deficient NSCs. This is due to a change in histone methylation of Dlx2, from a single high level of H3K4 trimethylation (H3K4me3) to a bivalent poised state marked by both activating H3K4me3 and repressive H3K27me3 (Lim et al., 2009).

We found that Jmjd2b is functional in long-term LPS treatment and regulates the histone demethylation of the promoters of its target genes that may be crucial in multiple signaling pathways and biological processes in NE-4C cells. Jmjd2b is a newly identified member of the histone demethylase Jmjd2 family that is characterized by the catalytic Jumonji C (JmjC) domain. Jmjd2b specifically targets the trimethylated lysine 9 of histone H3 (H3K9) for demethylation at pericentric heterochromatin and euchromatin (Fodor et al., 2006). It is reported that JMJD2B is critical to breast cancer cell survival under conditions of normoxia and hypoxia, which occurs partially via the regulation of cell cycle progression, is highly expressed in ERα-positive primary breast cancers, and is an adverse prognostic factor in hypoxic breast cancers (Yang et al., 2010). In this study, MetaCore pathway analysis was used to reveal the gene networks and canonical pathways affected in Jmjd2b-kd cells. Among the network, generation of neurons, neurogenesis, cell differentiation, and cellular developmental processes were most significantly affected in Jmjd2b-attenuated NE-4C cells. The significantly downregulated genes were clustered in different networks and canonical pathways. We found that Jmjd2b-kd NE-4C cells downregulated various key genes involved in neurophysiological processes (receptor-mediated axon growth repulsion, GABA-A receptor life cycle), the Notch1-mediated pathway for NF-κB activity modulation, and TGF-β-dependent induction. Jmjd2b encodes a histone demethylase that has been recently shown to be a HIF-1α target gene (Yang et al., 2009). Jmjd2b attenuation significantly inhibited p65, iNOS, Bcl2 and TGF-β expression in JMJD3-kd NE-4C cells. A GeneGo analysis of Jmjd2b-kd NE-4C cells revealed that Jmjd2b attenuation affected the generation of neurons, neurogenesis, system development, cell differentiation and cellular development processes. Several genes involved in the receptor-mediated axon growth repulsion (semaphorin 3a, pleiotrpphin-OSF1, ephrin A receptor 2), the GABA-A receptor life cycle (GABA-A receptor beta 2), the NOTCH1-mediated pathway for NF-κB activity modulation (c-Rel, Jagged 1, p65/p52) and the TGF-β-dependent induction (TGF-β2, Jagged1, N-cadherin, Lef1) were directly or indirectly affected by Jmjd2b attenuation. We predict that Jmjd2b recruitment may be necessary for the expression of regulated genes from several pathways that are crucial for various neurological functions. These results suggest that LPS has an inflammatory effect on NE-4C cells via epigenetic modulation.

It has also been reported that the mRNA expression of NeuroD, a neural progenitor cell marker, was significantly decreased in the hippocampus of aged mice compared with that in young mice. In light of previous results, we examined the presence of H3K9me3 at the NeuroD promoter but did not observe a reduction of the H3K9me3 level in Jmjd2b-kd NE-4C cells. We predicted that other histone modifications might be involved at the promoter site of NeuroD for its expression. However, the functions of most histone demethylases, including Jmjd2b, are not clear under inflammatory conditions, and the mechanism by which Jmjd2b epigenetically regulates gene expression in NSC inflammation has not been well shown. Therefore, the clarification of the function of Jmjd2b may help to identify novel therapeutic targets for brain inflammation.

5. Epigenetic regulations of proinflammatory cytokines in NSC

Cytokines are the secreted molecules that mediate communication between immune cells and between immune system and host. Cytokines encompass a broad class of signaling molecules that have the potential to influence an immense variety of signals that regulate NSC function, including growth factor production, electrical activity, synaptic function, and axonal path finding (Carpentier and Palmer, 2009). We will focus our discussion on the epigenetic regulations of inflammatory cytokines in NSC. Though, several recent reports shown that important cytokines include TNF-α, IL-6, and IL-1β have prominent inhibitory effect on adult neurogenesis in vivo. TNF-α can induce apoptosis in NSCs or newborn neurons via TNFR1. TNFR1 signaling, but not that of TNFR2, has been demonstrated to inhibit neurogenesis in the normal hippocampus (Iosif et al., 2006). In addition, neurogenesis is severely affected by another strong inflammatory mediator, NO. It has been reported that the SVZ cell proliferation rate is significantly increased after the inhibition of neuronal NOS activity (Sun et al., 2005). Notably, the pathological concentration of NO has a skewing effect on NSC differentiation when the pro-astroglial fate is very dominant (Covacu et al., 2006). At present, not many studies have been reported regarding the epigenetic involvement for cytokine regulations in NSC; recently, we reported LPS could affect NSC in vitro via epigenetic regulation (Das et al., 2012). The in vitro treatment of NE-4C cells with LPS (1 μg/ml for 96 h) significantly increased Jmjd2b expression and decreased the levels of H3K9me3. It has been reported that IL-1β suppresses the proliferation of hippocampal progenitor cells (Koo and Duman, 2008).The decreased proliferation of neural stem cells is responsible for decreased neural differentiation, and increased proliferation could correspond to the promotion of neurogenesis. We predicted that H3K9me3 is involved in Jmjd2b-attenuated NE-4C cells. A ChIP analysis showed that Jmjd2b-attenuated samples experienced an increase in the H3K9me3 on inflammatory signaling-mediated genes. An induced presence of H3K9me3 has been observed at the promoters of the Notch1, IL-1β, and IL-2 genes in Jmjd2b-kd NE-4C cells, suggesting that Jmjd2b can fine-tune the local chromatin state to enhance the transcription of these genes (Das et al., 2012).

6. Potential of NSC in neurodegenerative diseases

Neurogenesis by endogenous NSC cannot fully overcome the neuronal loss observed in neurodegenerative diseases. One reason for this limited response is the lack of trophic support and inhibitory signals within the brain microenvironment (Croft and Przyborski, 2009), indicative of oxidative stress and age-related neuroinflammation. These observations stimulated a search for agents that could increase neurogenesis and enhance neuroprotection (Russo et al., 2011). Now we will discuss the various neurodegenerative diseases including Alzheimer'sdisease (AD), Parkinson's disease (PD), amyotrophic lateral sclerosis (ALS), multiple sclerosis (MS), epilepsy, and stroke in inflammatory contexts. Each of the neurodegenerative diseases considered here is distinguished by a disease-specific mechanism for induction of inflammatory responses. The distinct pathways for production of inducers of inflammation—such as Ab, a-synuclein, mutant SOD1, and myelin peptide mimetic—and the specific anatomical locations at which these processes occur are likely determinants of the specific pathologicial features of each disease. In particular, TLRs and other pattern recognition receptors expressed on microglia and astrocytes are likely to play significant roles in initiating inflammatory responses. Later the downstream signal transduction pathways like NF-κB and AP-1 appear to play general roles in mediating the production of amplifiers and effector molecules, such as cytokines (e.g., TNF-α, IL-1β, and IL-6), ROS, and NO which involving in neurotoxicity for all of the neurodegenerative diseases (Glass et al., 2010).

Now we will focus on epigenetic involvement for such neurodegenerative diseases. Genes that are epigenetically regulated in Alzheimer's disease are *S100A2* (a member of the S100 family of calcium-binding proteins) and *SORBS3* (a sorbin and SH3 domain containing the cell-adhesionProtein) that display significant different level of DNA methylation (Siegmund et al., 2007). S100A2 has been previously identified as a metastatic inductor in non-small-cell lung cancer (Bulk et al., 2009), but its role in Alzheimer's disease pathogenesis remains unknown.

Most importantly, S100B, another member of the S100 family, which acts as a neurotrophic and pro-survival neuronal factor, might have a role in Alzheimer's disease pathogenesis and how does the exogenous and endogenous NSC express and epigenetically regulate such neurotrophic factors is still unknown. Among other epigenetic regulations in the pathogenesis of PD; DNA hypomethylation of TNF-α can directly lead to specific vulnerability of the substantianigra could be the direct consequence of PD (Pieper et al., 2008) and that may potentiate why the cytokine mediate inflammation is one of the major causes for PD. It is also reported that TNF-α overexpression induces apoptosis in neuronal cells and TNF-α levels are high in the CSF of patients with Parkinson's disease (Mogi et al., 1996). Multiple sclerosis is an inflammatory chronic disease characterized by a demyelinating process, which is followed by neurodegeneration. Although little is known about the epigenetics of this disease, some evidence suggests hypomethylation was proven at the promoter region of *PADl2* (peptidyl arginine deiminase, type II), also found to be overexpressed in multiple sclerosis. PADl2 catalyzes the citrullination of myelin basic protein that can change the properties of myelin (Mastronardi et al., 2007; Urdinguio et al., 2009). Epilepsy is described as a common chronic neurological disorder characterized by recurrent spontaneous seizures. Sporadic epilepsy can arise as a result of traumatic brain injury, stroke, abnormalities in brain wiring, toxic-metabolic etiologies, inflammation, autoimmunity, or an imbalance in the ratio of inhibitory to excitatory synaptic transmission (Hwang et al, 2012; Berg et al, 2010). Spontaneous seizures activate REST and promote deacetylation of core histone protein H4 (a mark of gene repression) at the RE1 site of the gria2 promoter (gene encoding the AMPAR subunit GluA2) recruits mSin3A and CoREST, HDACs-1/2, G9a and MeCP2, while promoting an increase in acetylation of H4 (a mark of open chromatin) at the promoter of brain-derived neurotrophic factor BDNF (Tsankova et al, 2004). Although, GluA2 expression was decreased, leading to an increase in GluA2-lacking, Ca2+-permeable AMPARs at CA3 synapses and neuronal death in CA3. Alterations of these proteins contribute to the pathophysiology of recurrent seizures. In epileptic adult rats transplanted fetal NSC (E14 rat) cells differentiated into neurons (13%, mostly GABAergic) and astrocytes (57%) and showed a reduction of motor seizure by 43% and severe convulsive seizure by 90% (Waldau et al., 2010). But how does the epigenetic regulators in exogenous NSC play crucial role in epilepsy yet to disclose. Ischemic insults also trigger activation of REST in mature hippocampal neurons destined to die and that the increase in REST correlates with a decrease in histone acetylation and gene silencing of GluA2. This is significant in that the GluA2 subunit prevents Ca2+ influx via AMPA receptors (AMPARs), is essential to synapatogenesis, long lasting forms of synaptic plasticity and neuronal death (Hwang et al., 2012; Liu and Zukin, 2007). Since, REST is a master transcriptional regulator of neuronal genes in pluripotent stem cells and neural progenitors and that loss of REST during the late stages of neural differentiation by ubiquitin based proteosomal degradation is required for acquisition of the neural phenotype (Hwang et al., 2012). The more study needs to answer how does REST perform at transplanted NSC in stroke model and whether other synaptic proteins correlate with REST for neuronal death in a clinically relevant ischemic stroke model.

Finally, more research will be required to understand the epigenetic mechanisms that underlie the neuroprotective roles of NSC in neurodegenerative diseases.

7. Conclusion

Due to self-renewal ability and differentiation to various neural cell types, NSC has great potential for clinical treatment of neurological diseases and dysfunctions. This regenerative capacity of NSC hold a great promise to open new areas of research aimed at stimulating neuronal regeneration in the brain during aging, neuroinflammation and neurodegenerative diseases. Epigenetic regulation along with other mechanisms can control these properties of NSCs. However, our knowledge about the precise mechanisms that control NSC function in neuroinflammation is still in its infancy and many avenues remain to be explored. The acute innate proinflammatorysignaling cascade strongly suppresses the production and retentionof new neurons in the adult brain. Here, the related immune signaling and epigenetic role might involve that must be addressed. We are just at the beginning of understanding the field. There are not reproducible comprehensive profiles of the DNA methylomes and histone modifications of NSCs in proper inflammatory stages that could generate some biomarkers to test in disease-associated conditions. Although more neurodevelopmental diseases caused by mutations in epigenetic genes are being identified, we still do not understand how the disturbance of DNA methylation and histone modification would directly affect NSC fate except the regulation of some neurotrophic factors. Moreover, a clear epigenetic interpretation that control the stimulation of neurogenesis during neuroinflammation, and the integration of NSC in diseased brain could assist to develop novel therapeutic approaches with a potential application in neuroinflammatory diseases.

Acknowledgements

This work was supported by the National Research Foundation of Korea Grant funded by the Korean Government (No.2012-0009212).

Author details

Nando Dulal Das* and Young Gyu Chai

*Address all correspondence to: nando.hu@gmail.com; ygchai@hanyang.ac.kr

Division of Molecular & Life Science, Hanyang University, Gyeonggi-do, Ansan, Korea

The authors declare that they have no conflict of interest.

References

[1] Barnabe-Heider, F., Wasylnka, J.A., Fernandes, K.J., Porsche, C., Sendtner, M., Kaplan, D.R., and Miller, F.D. (2005). Evidence that embryonic neurons regulate the onset of cortical gliogenesis via cardiotrophin-1. Neuron *48*, 253-265.

[2] Belmadani, A., Tran, P.B., Ren, D., and Miller, R.J. (2006). Chemokines regulate the migration of neural progenitors to sites of neuroinflammation. The Journal of neuroscience : the official journal of the Society for Neuroscience *26*, 3182-3191.

[3] Berg, A.T, Berkovic, S.F., Brodie, M.J., Buchhalter, J., Cross, J.H., Van Emde, B.W. *et al.* (2010). Revised terminology and concepts for organization of seizures and epilepsies: report of the ILAE commission on classification and terminology, 2005–2009. Epilepsia *51*, 676–685.

[4] Beyer, S., Kristensen, M.M., Jensen, K.S., Johansen, J.V., and Staller, P. (2008). The histone demethylases JMJD1A and JMJD2B are transcriptional targets of hypoxia-inducible factor HIF. The Journal of biological chemistry *283*, 36542-36552.

[5] Borrelli, E., Nestler, E.J., Allis, C.D., and Sassone-Corsi, P. (2008). Decoding the epigenetic language of neuronal plasticity. Neuron *60*, 961-974.

[6] Bulk, E., Sargin, B., Krug, U., Hascher, A., Jun, Y., Knop, M., Kerkhoff, C., Gerke, V., Liersch, R., Mesters, R.M., *et al.* (2009). S100A2 induces metastasis in non-small cell lung cancer. Clinical cancer research : an official journal of the American Association for Cancer Research *15*, 22-29.

[7] Butovsky, O., Ziv, Y., Schwartz, A., Landa, G., Talpalar, A.E., Pluchino, S., Martino, G., and Schwartz, M. (2006). Microglia activated by IL-4 or IFN-gamma differentially induce neurogenesis and oligodendrogenesis from adult stem/progenitor cells. Molecular and cellular neurosciences *31*, 149-160.

[8] Carpentier, P.A., and Palmer, T.D. (2009). Immune influence on adult neural stem cell regulation and function. Neuron *64*, 79-92.

[9] Cloos, P.A., Christensen, J., Agger, K., and Helin, K. (2008). Erasing the methyl mark: histone demethylases at the center of cellular differentiation and disease. Genes & development 22, 1115-1140.

[10] Cloos, P.A., Christensen, J., Agger, K., Maiolica, A., Rappsilber, J., Antal, T., Hansen, K.H., and Helin, K. (2006). The putative oncogene GASC1 demethylates tri- and dimethylated lysine 9 on histone H3. Nature *442*, 307-311.

[11] Covacu, R., Danilov, A.I., Rasmussen, B.S., Hallen, K., Moe, M.C., Lobell, A., Johansson, C.B., Svensson, M.A., Olsson, T., and Brundin, L. (2006). Nitric oxide exposure diverts neural stem cell fate from neurogenesis towards astrogliogenesis. Stem cells *24*, 2792-2800.

[12] Croft, A.P., and Przyborski, S.A. (2009). Mesenchymal stem cells expressing neural antigens instruct a neurogenic cell fate on neural stem cells. Experimental neurology *216*, 329-341.

[13] Das, N.D., Choi, M.R., Jung, K.H., Park, J.H., Lee, H.T., Das, A., Kim, S.H., and Chai, Y.G. (2012). Functional Analysis of Histone Demethylase Jmjd2b on Lipopolysaccharide-Treated Murine Neural Stem Cells (NSCs). Neurotoxicity research.

[14] Glass, C.K., Saijo, K., Winner, B., Marchetto, M.C., and Gage, F.H. (2010). Mechanisms underlying inflammation in neurodegeneration. Cell *140*, 918-934.

[15] Hanisch, U.K., and Kettenmann, H. (2007). Microglia: active sensor and versatile effector cells in the normal and pathologic brain. Nature neuroscience *10*, 1387-1394.

[16] Hsieh, J., and Gage, F.H. (2005). Chromatin remodeling in neural development and plasticity. Current opinion in cell biology *17*, 664-671.

[17] Hsieh, J., Nakashima, K., Kuwabara, T., Mejia, E., and Gage, F.H. (2004). Histone deacetylase inhibition-mediated neuronal differentiation of multipotent adult neural progenitor cells. Proceedings of the National Academy of Sciences of the United States of America *101*, 16659-16664.

[18] Hwang, J-Y., Aromolaran, K.A., Zukin, R.S., (2012). Epigenetic Mechanisms in Stroke and Epilepsy Neuropsychopharmacology Reviews, 1–16.

[19] Iosif, R.E., Ekdahl, C.T., Ahlenius, H., Pronk, C.J., Bonde, S., Kokaia, Z., Jacobsen, S.E., and Lindvall, O. (2006). Tumor necrosis factor receptor 1 is a negative regulator of progenitor proliferation in adult hippocampal neurogenesis. The Journal of neuroscience : the official journal of the Society for Neuroscience *26*, 9703-9712.

[20] Jakubs, K., Bonde, S., Iosif, R.E., Ekdahl, C.T., Kokaia, Z., Kokaia, M., and Lindvall, O. (2008). Inflammation regulates functional integration of neurons born in adult brain. The Journal of neuroscience : the official journal of the Society for Neuroscience *28*, 12477-12488.

[21] Jelitai, M., Anderova, M., Marko, K., Kekesi, K., Koncz, P., Sykova, E., and Madarasz, E. (2004). Role of gamma-aminobutyric acid in early neuronal development: studies with an embryonic neuroectodermal stem cell clone. Journal of neuroscience research *76*, 801-811.

[22] Juliandi, B., Abematsu, M., and Nakashima, K. (2010). Epigenetic regulation in neural stem cell differentiation. Development, growth & differentiation *52*, 493-504.

[23] Jung, G.A., Yoon, J.Y., Moon, B.S., Yang, D.H., Kim, H.Y., Lee, S.H., Bryja, V., Arenas, E., and Choi, K.Y. (2008). Valproic acid induces differentiation and inhibition of proliferation in neural progenitor cells via the beta-catenin-Ras-ERK-p21Cip/WAF1 pathway. BMC cell biology *9*, 66.

[24] Kelly, L., Grehan, B., Chiesa, A.D., O'Mara, S.M., Downer, E., Sahyoun, G., Massey, K.A., Nicolaou, A., and Lynch, M.A. (2011). The polyunsaturated fatty acids, EPA

http://dx.doi.org/10.5772/54981

and DPA exert a protective effect in the hippocampus of the aged rat. Neurobiology of aging *32*, 2318 e2311-2315.

[25] Kishi, N., and Macklis, J.D. (2004). MECP2 is progressively expressed in post-migratory neurons and is involved in neuronal maturation rather than cell fate decisions. Molecular and cellular neurosciences *27*, 306-321.

[26] Koo, J.W., and Duman, R.S. (2008). IL-1beta is an essential mediator of the antineurogenic and anhedonic effects of stress. Proceedings of the National Academy of Sciences of the United States of America *105*, 751-756.

[27] Lachner, M., Sengupta, R., Schotta, G., and Jenuwein, T. (2004). Trilogies of histone lysine methylation as epigenetic landmarks of the eukaryotic genome. Cold Spring Harbor symposia on quantitative biology *69*, 209-218.

[28] Lim, D.A., Huang, Y.C., Swigut, T., Mirick, A.L., Garcia-Verdugo, J.M., Wysocka, J., Ernst, P., and Alvarez-Buylla, A. (2009). Chromatin remodelling factor Mll1 is essential for neurogenesis from postnatal neural stem cells. Nature *458*, 529-533.

[29] Liu, S.J., Zukin, R.S. (2007). Ca2+-permeable AMPA receptors in synaptic plasticity and neuronal death. Trends in Neuroscience *30*, 126–134.

[30] Livingstone, L.R., White, A., Sprouse, J., Livanos, E., Jacks, T., and Tlsty, T.D. (1992). Altered cell cycle arrest and gene amplification potential accompany loss of wild-type p53. Cell *70*, 923-935.

[31] Louvi, A., and Artavanis-Tsakonas, S. (2006). Notch signalling in vertebrate neural development. Nature reviews Neuroscience *7*, 93-102.

[32] Lundkvist, J., and Lendahl, U. (2001). Notch and the birth of glial cells. Trends in neurosciences *24*, 492-494.

[33] Martino, G., and Pluchino, S. (2006). The therapeutic potential of neural stem cells. Nature reviews Neuroscience *7*, 395-406.

[34] Mastronardi, F.G., Noor, A., Wood, D.D., Paton, T., and Moscarello, M.A. (2007). Peptidyl argininedeiminase 2 CpG island in multiple sclerosis white matter is hypomethylated. Journal of neuroscience research *85*, 2006-2016.

[35] Meaney, M.J., and Ferguson-Smith, A.C. (2010). Epigenetic regulation of the neural transcriptome: the meaning of the marks. Nature neuroscience *13*, 1313-1318.

[36] Mogi, M., Harada, M., Narabayashi, H., Inagaki, H., Minami, M., and Nagatsu, T. (1996). Interleukin (IL)-1 beta, IL-2, IL-4, IL-6 and transforming growth factor-alpha levels are elevated in ventricular cerebrospinal fluid in juvenile parkinsonism and Parkinson's disease. Neuroscience letters *211*, 13-16.

[37] Monje, M.L., Toda, H., and Palmer, T.D. (2003). Inflammatory blockade restores adult hippocampal neurogenesis. Science *302*, 1760-1765.

[38] Nakatomi, H., Kuriu, T., Okabe, S., Yamamoto, S., Hatano, O., Kawahara, N., Tamura, A., Kirino, T., and Nakafuku, M. (2002). Regeneration of hippocampal pyramidal neurons after ischemic brain injury by recruitment of endogenous neural progenitors. Cell *110*, 429-441.

[39] Nakayama, K., Nagase, H., Hiratochi, M., Koh, C.S., and Ohkawara, T. (2008). Similar mechanisms regulated by gamma-secretase are involved in both directions of the bidirectional Notch-Delta signaling pathway as well as play a potential role in signaling events involving type 1 transmembrane proteins. Current stem cell research & therapy *3*, 288-302.

[40] Namihira, M., Kohyama, J., Semi, K., Sanosaka, T., Deneen, B., Taga, T., and Nakashima, K. (2009). Committed neuronal precursors confer astrocytic potential on residual neural precursor cells. Developmental cell *16*, 245-255.

[41] Nan, X., Campoy, F.J., and Bird, A. (1997). MeCP2 is a transcriptional repressor with abundant binding sites in genomic chromatin. Cell *88*, 471-481.

[42] Pfau, R., Tzatsos, A., Kampranis, S.C., Serebrennikova, O.B., Bear, S.E., and Tsichlis, P.N. (2008). Members of a family of JmjC domain-containing oncoproteins immortalize embryonic fibroblasts via a JmjC domain-dependent process. Proceedings of the National Academy of Sciences of the United States of America *105*, 1907-1912.

[43] Pieper, H.C., Evert, B.O., Kaut, O., Riederer, P.F., Waha, A., and Wullner, U. (2008). Different methylation of the TNF-alpha promoter in cortex and substantia nigra: Implications for selective neuronal vulnerability. Neurobiology of disease *32*, 521-527.

[44] Reinberg, D., Chuikov, S., Farnham, P., Karachentsev, D., Kirmizis, A., Kuzmichev, A., Margueron, R., Nishioka, K., Preissner, T.S., Sarma, K., *et al.* (2004). Steps toward understanding the inheritance of repressive methyl-lysine marks in histones. Cold Spring Harbor symposia on quantitative biology *69*, 171-182.

[45] Rolls, A., Shechter, R., London, A., Ziv, Y., Ronen, A., Levy, R., and Schwartz, M. (2007). Toll-like receptors modulate adult hippocampal neurogenesis. Nature cell biology *9*, 1081-1088.

[46] Rolls, A., Shechter, R., and Schwartz, M. (2009). The bright side of the glial scar in CNS repair. Nature reviews Neuroscience *10*, 235-241.

[47] Russo, I., Barlati, S., and Bosetti, F. Effects of neuroinflammation on the regenerative capacity of brain stem cells. J Neurochem *116*, 947-956.

[48] Russo, I., Barlati, S., and Bosetti, F. (2011). Effects of neuroinflammation on the regenerative capacity of brain stem cells. Journal of neurochemistry *116*, 947-956.

[49] Sanz, L.A., Chamberlain, S., Sabourin, J.C., Henckel, A., Magnuson, T., Hugnot, J.P., Feil, R., and Arnaud, P. (2008). A mono-allelic bivalent chromatin domain controls tissue-specific imprinting at Grb10. The EMBO journal *27*, 2523-2532.

[50] Setoguchi, H., Namihira, M., Kohyama, J., Asano, H., Sanosaka, T., and Nakashima, K. (2006). Methyl-CpG binding proteins are involved in restricting differentiation plasticity in neurons. Journal of neuroscience research *84*, 969-979.

[51] Siegmund, K.D., Connor, C.M., Campan, M., Long, T.I., Weisenberger, D.J., Biniszkiewicz, D., Jaenisch, R., Laird, P.W., and Akbarian, S. (2007). DNA methylation in the human cerebral cortex is dynamically regulated throughout the life span and involves differentiated neurons. PloS one *2*, e895.

[52] Strahl, B.D., and Allis, C.D. (2000). The language of covalent histone modifications. Nature *403*, 41-45.

[53] Sun, Y., Jin, K., Childs, J.T., Xie, L., Mao, X.O., and Greenberg, D.A. (2005). Neuronal nitric oxide synthase and ischemia-induced neurogenesis. Journal of cerebral blood flow and metabolism : official journal of the International Society of Cerebral Blood Flow and Metabolism *25*, 485-492.

[54] Takizawa, T., Nakashima, K., Namihira, M., Ochiai, W., Uemura, A., Yanagisawa, M., Fujita, N., Nakao, M., and Taga, T. (2001). DNA methylation is a critical cell-intrinsic determinant of astrocyte differentiation in the fetal brain. Developmental cell *1*, 749-758.

[55] Tsankova, N.M., Kumar, A., Nestler, E.J. (2004). Histone modifications at gene promoter regions in rat hippocampus after acute and chronic electroconvulsive seizures. Journal of Neuroscience *24*, 5603–5610.

[56] Tsujimura, K., Abematsu, M., Kohyama, J., Namihira, M., and Nakashima, K. (2009). Neuronal differentiation of neural precursor cells is promoted by the methyl-CpG-binding protein MeCP2. Experimental neurology *219*, 104-111.

[57] Urdinguio, R.G., Sanchez-Mut, J.V., and Esteller, M. (2009). Epigenetic mechanisms in neurological diseases: genes, syndromes, and therapies. Lancet neurology *8*, 1056-1072.

[58] Waldau, B., Hattiangady, B., Kuruba, R., Shetty, A.K. (2010). Medial ganglionic eminencederived neural stem cell grafts ease spontaneous seizures and restore GDNF expression in a rat model of chronic temporal lobe epilepsy. Stem Cells *28*, 1153–1164.

[59] Weaver, I.C., D'Alessio, A.C., Brown, S.E., Hellstrom, I.C., Dymov, S., Sharma, S., Szyf, M., and Meaney, M.J. (2007). The transcription factor nerve growth factor-inducible protein a mediates epigenetic programming: altering epigenetic marks by immediate-early genes. The Journal of neuroscience : the official journal of the Society for Neuroscience *27*, 1756-1768.

[60] Wissmann, M., Yin, N., Muller, J.M., Greschik, H., Fodor, B.D., Jenuwein, T., Vogler, C., Schneider, R., Gunther, T., Buettner, R., *et al.* (2007). Cooperative demethylation

by JMJD2C and LSD1 promotes androgen receptor-dependent gene expression. Nature cell biology *9*, 347-353.

[61] Yang, J., Jubb, A.M., Pike, L., Buffa, F.M., Turley, H., Baban, D., Leek, R., Gatter, K.C., Ragoussis, J., and Harris, A.L. (2010). The histone demethylase JMJD2B is regulated by estrogen receptor alpha and hypoxia, and is a key mediator of estrogen induced growth. Cancer research *70*, 6456-6466.

[62] Yang, J., Ledaki, I., Turley, H., Gatter, K.C., Montero, J.C., Li, J.L., and Harris, A.L. (2009). Role of hypoxia-inducible factors in epigenetic regulation via histone demethylases. Annals of the New York Academy of Sciences *1177*, 185-197.

[63] Yoder, J.A., Walsh, C.P., and Bestor, T.H. (1997). Cytosine methylation and the ecology of intragenomic parasites. Trends in genetics : TIG *13*, 335-340.

[64] Zhao, X., Ueba, T., Christie, B.R., Barkho, B., McConnell, M.J., Nakashima, K., Lein, E.S., Eadie, B.D., Willhoite, A.R., Muotri, A.R., *et al.* (2003). Mice lacking methyl-CpG binding protein 1 have deficits in adult neurogenesis and hippocampal function. Proceedings of the National Academy of Sciences of the United States of America *100*, 6777-6782.

Chapter 14

Primary Neural Stem Cell Cultures from Adult Pig Brain and Their Nerve-Regenerating Properties: Novel Strategies for Cell Therapy

Olivier Liard and Emmanuel Moyse

Additional information is available at the end of the chapter

http://dx.doi.org/10.5772/55726

1. Introduction

Neurons are one among the very few differentiated cell types which are strictly post-mitotic and therefore non-renewable by sister cell divisions. The production of new neurons, or neurogenesis, does though occur in limited areas of adult mammalian nervous system, due to the existence of intrinsic stem cells which has been formally demonstrated by a primary culture approach on adult brain tissues from mouse [1], man [2,3] and other species [4]. These "neural stem cells" satisfy the general stemness criteria, i.e.: illimited proliferative activity, self-renewal capacity, multipotency (i.e. capacity to generate all the differentiated cell types of their host tissue, which for neural stem cells encompasses neurons, astrocytes, oligodendrocytes), proliferative reactivity to tissue lesions and physiological signals [5]. This discovery has raised great promise for clinical repair in neurology and led to coin a novel principle in cell therapy, which consists in replacing lost neurons by grafting *in vitro* purified neural stem cells [6,7,8]. This theoretical principle has been challenged by the putative tumour-promoting potential of grafted proliferative cells. Such risk can reasonably be ruled out by the recent demonstration that adult brain-derived neural stem cells are highly resilient to tumorigenesis and senescence, much more than embryonic neuroblasts, embryonic stem cells, induced pluripotent stem cells or any other stem cell type from adult mammalian tissues [9]. However, the use of homotypic neural stem cell transplantation in human clinics seems ethically difficult, since it requires initial sampling of alive brain tissue. In order to overcome this limitation, we have developed primary neural stem cell culture from adult porcine brain. The pig is indeed the non-primate species that is immunologically closest to human [10], which already allowed successful protocols of xenotransplantation in human clinics using *in vitro*-expanded primary cell preparations from several adult pig tissues – but not from the nervous system.

2. General properties of neural stem cells in adult mammals

Neural stem cells are identified and isolated in primary cultures of adult nervous tissues in the presence of epithelial growth factor (EGF) and basic fibroblast growth factor (bFGF), by the growth of cellular masses called "neurospheres" which include the three neural lineages: neurons, astrocytes, oligodendrocytes [1]. This primary culture is termed "neurosphere assay" (NSA) and allows to generate and expand great numbers of proliferative neurogenic cells through successive passages [11]. Recent methodological refinements allowed to isolate and distinguish neural stem cells *per se* from downstream progenitors, i.e. proliferative cells being committed to a single lineage although yet undifferentiated [12].

Beside the two initially recognized and most active neurogenic structures of adult mammalian brain (the subventricular zone of telencephalon and the subgranular layer of hippocampal dentate gyrus), several topographically restricted regions of the central nervous system were shown to yield neurospheres in EGF-bFGF-treated primary culture and therefore to contain neural stem cells [13,14]. Spatial frequency among tissue and *in vivo* yield of neuronal versus glial progenies are highly variable, ranging among: 95% in hippocampus and hypothalamus, 60-70% in SVZ-olfactory bulb system, 40% in dorsal vagal complex, 0% in spinal cord despite as high a proliferation rate as in SVZ [15]. However, *in vitro*-expanded neurospheres display strikingly reproducible properties, regarding both growth dynamics (10-12 days for the primary generation, 6-7 days for the subsequent ones) and lineage proportions after differentiation commitment by growth factor removal: 25% neurons, 70% astrocytes, 5% oligodendrocytes [11].

Neural stem cells generate differentiated progenies, like other types of adult stem cells, through several intermediate stages which can now be distinguished by their respective marker combinations – or "molecular profiles" or "transcriptomic signatures" (Figure 1). The biological properties of neural stem cells from adult nervous tissues are reminiscent of those characterizing neural progenitors from embryos, especially *in vitro*. However, accumulated data in the past decade have revealed important differences between these two categories of neural precursor cells, which led to consider them as distinct cell types. The differences include indeed intrinsic molecular mechanisms of proliferation and differentiation, as well as receptivity to extracellular signals [16,17,18]. Most interestingly, neural stem cells from adults are much less teratogenic than embryonic stem cells and foetal neural progenitors [9,19, 20,21,22]. Primary cultures of neural stem cells from the archetypical subventricular zone of rodent, which is the most highly proliferative of endogenous neurogenic niches in mammalian brain, could indeed be maintained over more than 70 successive *in vitro* generations (or "passages" in the neurosphere assay) without any alteration of growth kinetics, self-renewal capacity and lineage potential [9]. *In vivo* ectopic grafts of *in vitro*-expanded neurospheres never triggered tumors in experimental rodents [9], by contrast to primary cultures of other stem cell categories from adult mammals [19,20,21]. In the recent years, some typical stem cells have been formally characterized in a variety of mammalian tumors, including neural ones; however, these "cancer stem cells" are distinct lineages from intrinsic stem cells of healthy tissues and seem to arise *de novo* as a consequence of DNA damage independently of tissue stem cells [23]. In the case of neural tissues, it must be recalled that beyond early post-natal ages (4-years in man) brain cancers are exclusively gliomas [24,25]. This fact renders most unlikely that adult neural cancers might derive from neural stem cells. Indeed, in a recent extensive survey of above 2000 brain metastases from 500 patients, neural stem cells were never found affected [22]. Regarding neurological repair aims for human clinics, the specially high resilience of adult neural stem cells against tumorigenesis is a determinant advantage in view of cell therapy applications.

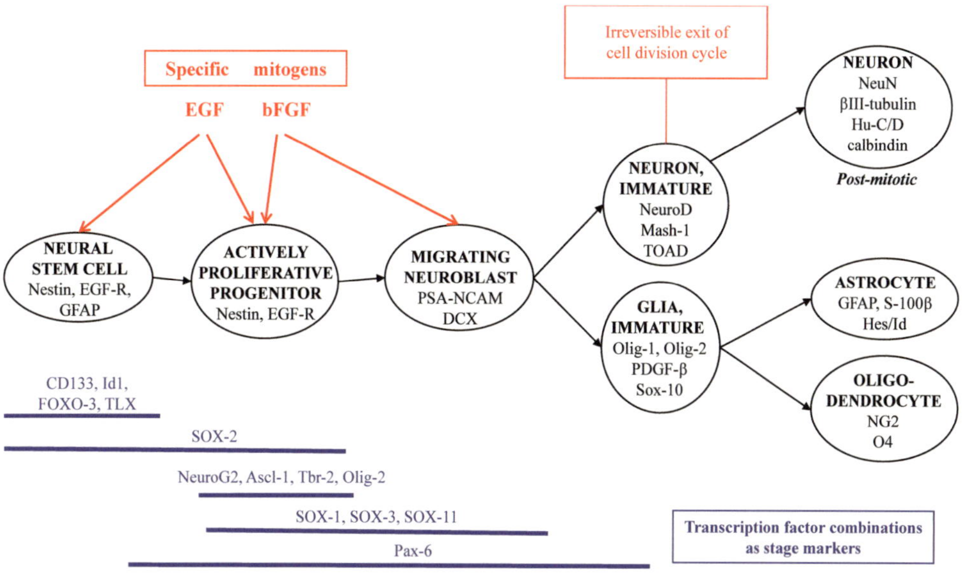

Figure 1. The "neuropoietic system" of adult neurogenesis. Modelized from references 15,38.

3. Neurosphere isolation from adult pig brain subventricular zone

Neural progenitors have been isolated from foetal porcine brain by several laboratories in the last years, by using the neurosphere assay. We decided to develop isolation of adult brain-derived neural stem cells, because of their higher safety than other available proliferative neurogenic cell types in terms of oncogenic potential (cf above). Neural stem cells are found only in rare and discrete structures, the best characterized and most productive of which is the subventricular zone (SVZ) of the forebrain in rodents. We have isolated and characterized for the first time the neural stem cells of adult pig SVZ, by transposing the "neurosphere assay" from routine protocols of rodent studies [26].

3.1. The key to success for the "neurosphere assay": Adequate tissue sampling

This particular primary culture is known to be hampered by two pitfalls: i) delay between *in vivo* blood supply arrest (upon either sacrifice of experimental animals or tissue explantation) and tissue immersion into appropriate fresh survival medium like the Hank's one or artificial cerebrospinal fluid (see below); ii) excess of non-competent tissue around the NSC-containing area within the microdissected sample.

In order to minimize the duration of brain tissue sampling, which is lengthened in adult pig by the requirement of dorsal skull sawing, we performed the surgical approach of telencephalic subventricular region in deeply anesthetized pig instead of immediately after sacrifice. A rough 10mm-thick, transverse section of midbrain was explanted and immediately transferred into a large Petri dish filled with fresh Hank's medium, and the subventricular zone was then microdissected out of this thick slice on the basis of macroscopical neuroanatomical landmarks by reference to a detailed pig brain atlas [27]. Despite this precaution, our first neurosphere assay attempts on adult pig SVZ failed, which was subsequently attributed to improper tissue sampling. Indeed, the telencephalic subventricular zone is proportionally much larger in adult pig than in adult rodents, and proliferative activity is already known as highly heterogeneous among the whole anatomical extent of SVZ in rodent brain [28]. We performed therefore the subregional mapping of proliferative activity within adult pig SVZ, by Ki-67 immunohistochemistry using a routine protocol [29,30]. The Ki-67 protein is indeed a phylogenetically conserved marker which is expressed exclusively during the cell division cycle of eukaryotic cells, but throughout all its successive phases (G1-S-G2-M); it is therefore widely used in clinical anatomo-pathology and oncology. Proliferative cells, appearing *in situ* as Ki-67 immunoreactive nuclei, were almost exclusively grouped in a narrow ventral subfield of the Nissl-stained subventricular zone, lining the ventral border of lateral ventricle of adult pig brain (Figure 2). The major lateral territory of adult pig SVZ was almost devoid of Ki-67 labelling (Figure 2). We have confirmed the anatomical restriction of the "niche" to such a small subfield of neuroanatomically defined SVZ, by specific labelling of the radial glia which has been extensively characterized in rodent brain SVZ, using double vimentin-glial fibrillary acidic protein (GFAP) immunohistochemistry [26]. *In vitro* culture of neural stem cells from adult pig brain became successful when the SVZ tissue explants were microdissected according to the anatomical lineaments of specific Ki-67 immunoreactive labelling.

3.2. Primary neurospheres from adult pig brain as a vehicle of neural stem cells

For primary cell culture ("neurosphere assay"), a piece of SVZ tissue was microdissected from the pig forebrain slice in low-calcium artificial cerebrospinal fluid (aCSF: 124 mM NaCl, 5 mM KCl, 3.2 mM $MgCl_2$, 0.1 mM $CaCl_2$, 26 mM $NaHCO_3$, 100 mM glucose, pH 7.38, according to Weiss et al., 1996). As previously described [30], the tissue sample was rinsed twice with aCSF, and digested first in 40 U cystein-EDTA-βmercaptoethanol-preactivated papain (Sigma, L'Isle D'Abeau, France) for 10 min at 37°C, then in 250 μL undiluted TrypLE™ Express solution (heat-resistant, microbially-produced, purified trypsin-like enzyme, Gibco cat 12604-013, Invitrogen, Cergy-Pontoise, France) for 10 min at 37°C. After the addition of 750 μL of fresh aCSF and centrifugation for 8 min at 400g at room temperature, the cell pellet was resuspended in 1mL culture medium (Dulbecco's modified Eagle medium [DMEM, Sigma], B27 supplement [Gibco Invitrogen], 200 U/mL penicillin and 200 μg/mL streptomycin [Gibco Invitrogen]) containing 20ng/mL Epidermal Growth Factor (EGF, Gibco Invitrogen) and 20 ng/mL basic

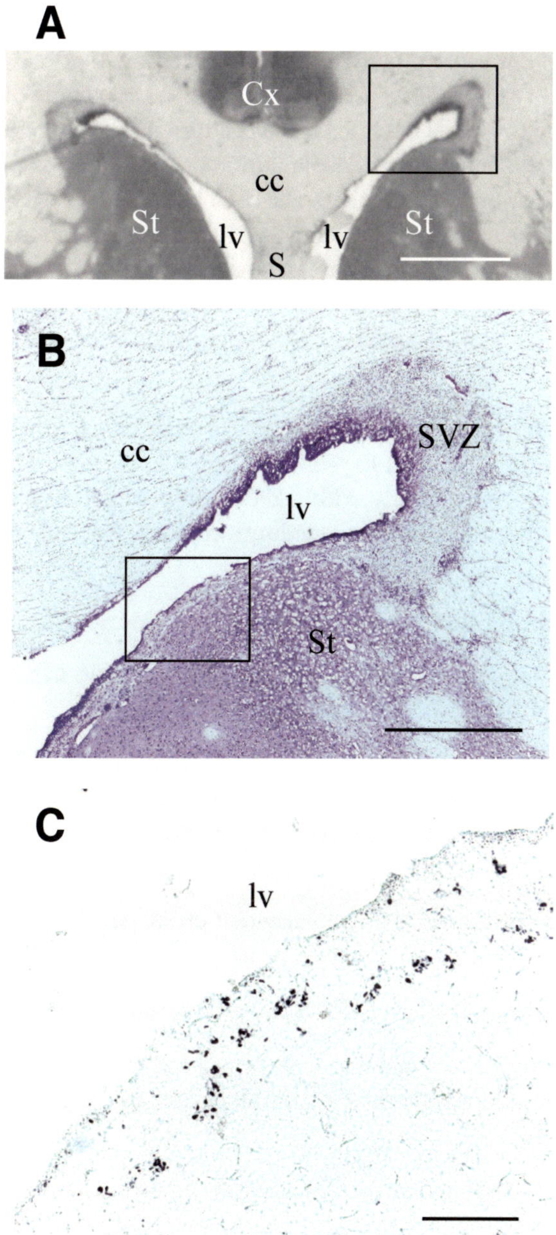

Figure 2. *In situ* localization of the proliferative cells of the adult pig SVZ (modified from original data of reference 26). (A) Macroscopic aspect of Nissl-stained coronal section of adult pig forebrain. (B) Light microscopic aspect of the SVZ region in a Nissl-stained pig brain section. (C) Ki-67 immunohistochemistry on adjacent section of the previous one, with high magnification on the boxed area of B. Note the high density of Ki-67-positive nuclei in this area, as reavealed by brown precipitate, which contrasts with the lateral field of SVZ devoid of Ki-67 immunoreactivity. Legend: *cc: corpus callosum, Cx: cerebral cortex, lv: lateral telencephalic ventricle, S: septum pellucidum, St: striatum, SVZ: subventricular zone.*

Fibroblast Growth Factor (bFGF, Gibco Invitrogen). The cells were dissociated gently with a 26G steel needle mounted on a disposable 1mL syringe, counted on a Malassez slide, and seeded at various densities according to purported utilization. Optimal results were obtained with either Corning 24-well plates or with Falcon 6-well ones. Seedings for basic cell biology kinetics were performed at 3,000 cells per mL of culture medium (subclonal concentration). Maximal neurosphere expansion for transplantation or molecular assays was obtained with seeding at 30,000 cells per mL. Cultures must be monitored daily to follow the morphological growth of the neurospheres; when their diameter reaches 150μm indeed, growth slows down while sphere surface apparently thickens and hardens, and culture proves unable to be expanded any more. Passage (i.e. sphere dissociation, cell resuspension and re-seeding as for primary culture) was performed therefore when the majority of spheres was 100-120μm in diameter.

For passage, the primary spheres were collected in sterile tubes, incubated for 45-60 min at 37°C in 250 μL undiluted TrypLE™ Express solution (Gibco Invitrogen) per 6 mL culture-derived pellet, and dissociated gently with a 26G steel syringe; dispersed cells were centrifuged, resuspended in fresh proliferative medium, counted ant their dilution adjusted to be seeded as above.

Both primary and passaged spheres were prone to two kinds of utilization:

- for neural stem cell expansion, subsequent passages were run in proliferative conditions, i.e. in the presence of 20ng/mL EGF and 20ng/mL bFGF;
- for cell differentiation, spheres were picked with a Pasteur pipette, seeded in wells containing poly-D-lysine (Sigma) precoated sterile coverslips and cultured in the above culture medium (DMEM, B27, penicillin-streptomycin) in the absence of EGF and bFGF for 7-10 days.

All culture media were renewed by the replacement of 500 μL of medium per well every 2-3 days.

Stemness criteria fulfilment by adult pig SVZ-derived spheres in the presence of EGF and bFGF.

We have observed in our primary neurospheres from adult pig SVZ the properties which have been established as the seminal criteria of neural stem cells since their pioneering discovery in adult murine brain [1,11].

The existence of EGF/bFGF-dependent proliferative cells in the ventro-medial SVZ from adult pig brain was demonstrated by the growth of spherical masses from dissociated pieces of this tissue [26] which displayed similar morphology and size distribution to neurospheres obtained from the rat SVZ [30]. Each pair of SVZ samples from 1 pig brain yielded routinely 1 million dispersed cells for seeding, i.e. the total number of cells collected per pig SVZ was about 10 times that of mouse SVZ [11,31]. The pig SVZ spheres appeared though after a longer delay in culture (9-12 days in vitro [DIV]) than primary neurospheres from adult rat SVZ (3-4 DIV: 30). Each pair of SVZ from one pig generated 3000 primary spheres after 2 weeks in the present culture conditions and at initial seeding densities of 30,000 cells per mL. These primary cells could be expanded linearly on this base through several passages (up to 6 with no change of growth slope) [26].

The self-renewal capacity of primary sphere-generating cells from adult pig SVZ was established by the obtention of novel spheres from dissociation of mature primary Spheres. Re-seeding of dispersed neurosphere cells in the same proliferative medium after 12-14 DIV led to novel "secondary" spheres that were morphologically similar to the primary spheres and grew slightly faster than primary spheres. A second passage at 8-10 DIV led to tertiary spheres that displayed similar morphology and growth rate to secondary spheres. In two independent cultures from 2 pigs each, 5-6 subsequent passages could be performed with similar aspects and cell number amplifications of respective generations. The second passage regularly triggered a 3-time-magnification of total cell number, while other passage increments (I, III and next ones) averaged 20-30% [26].

The multipotentiality and neural character of EGF/bFGF-dependent proliferative cells of adult pig SVZ were demonstrated on mature spheres which were deprived of the 2 mitogens after having reached proliferative maturity, by immunocytochemical detection of neuronal and neuroglial markers in three distinct subpopulations of each sphere – according to the classical lineage test of neurosphere assay [1,11,30]. Quantitative analysis with confocal microscopy indicated that all neurospheres from adult pig SVZ yielded similar proportions of the three neural lineages: 25% neurons, 70% astrocytes, 5% oligodendrocytes. A discrete distribution of the three cell lineages was also reproducibly observed inside the neurospheres, with oligodendrocytes being buried in the core of the spheres while both astrocytes and neurons were concentrated at the periphery [26].

3.3. Transcriptomic signature of neural stem cells in adult pig SVZ neurospheres

We have assayed the expression of phenotypic markers by RT-PCR on total RNA from proliferative, undifferentiated SVZ neurospheres of adult pig SVZ, which were collected at the second or third generation as described above. Pig SVZ neurospheres displayed positive expression of RNAs encoding for:

- nestin, an intermediate filament which is systematically expressed by neural stem cells and embryonic neural progenitors;
- the proliferation marker cyclin D1;
- the neural stem cell proteins Sox-2, Hes-1, $p75^{NTR}$ (the low-affinity receptor of neurotrophins), Lhx2;
- the neuronal proteins βIII-tubulin, doublecortin (DCX), neuron-specific enolase (NSE), Internexin α, Pbx1, Mash1, NeuroD1;
- the oligodendrocyte markers CNPase, Olig-2, PDGF-Rα (Figure 3).

The astrocyte marker glial fibrillary acidic protein (GFAP) has been detected on differentiated neurospheres from adult pig SVZ by immunocytochemistry using the antibody against cow

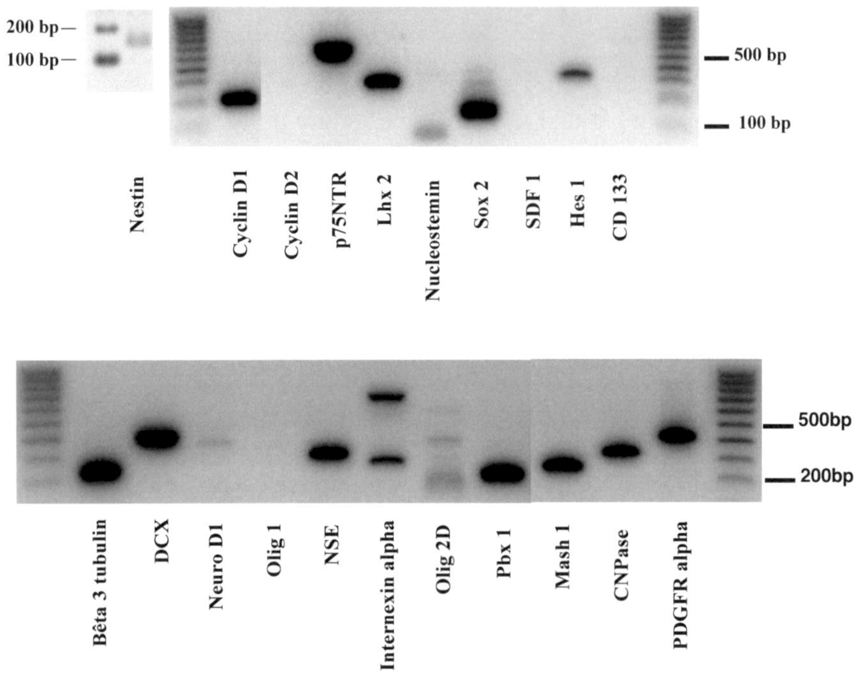

Figure 3. Transcriptomic profile of proliferative neurospheres from adult pig SVZ (modified from original data of reference 26). Agarose gel electrophoretic analysis of amplicons from RT-PCR on proliferative neurosphere RNA with primers designed on the basis of human and/or pig genomic sequences. Stem cell markers: Nestin, Hes-1, Lhx-2, p75NTR, Sox-2, Cyclin D1, Cyclin D2. Committed progenitor markers: Pbx1, βIII-tubulin, internexin-α, NeuroD, NSE, DCX, Mash-1, CNPase, PDGFR-α. PCR products are shown beside co-electrophoresed commercial DNA ladders.

GFAP (cf above) but its transcript could not be detected in the proliferative neurosphere RNAs (probably due to unknown specificity of porcine gene sequence and primers therefore).

The lack of Olig-1 transcripts is expected at this pro-differentiating stage since this gene is expressed in the late phase of oligodendroglial lineage maturation [32,33]; it is also in keeping with Hes-1 detection in the present proliferative spheres, since Hes-1 overexpression was shown to correlate with Olig-1 down-regulation during differentiation [34,35].

In the persistent absence of any specific marker of either neural stem cell or neural committed progenitors, this series of transcripts fits best with the molecular profile of neural stem/precursor cells which has been characterized in rodents and human [36,37,38]. On these bases, we may assume that the major cell type in our adult pig SVZ-derived neurospheres correspond to neural multipotent progenitors, i.e. a stage of the neural lineage that is slightly more mature than the primitive neural stem cells but equally multipotent [26]. This interpretation is supported by additional RT-PCR data in our neurospheres: the absence of the stemness gene transcripts CD133 and SDF-1 [39], the weakness of nucleostemin expression [40], and the lack of detectable cyclin D2, which was demonstrated to specifically drive neural stem cells into division onset by assaying *in vivo* adult neurogenesis in knockout mice for each of the three cyclins D [41]. Cyclin D1, which was readily expressed in the present material, conversely operates in actively dividing neural progenitors like in most of proliferating cell types in adult organisms.

The present data can be compared with published molecular profiling of neural progenitor cultures from the pig foetal brain [42,43] or adult retina [44,45]. Among the 13 markers that have been addressed by RT-PCR on mRNA extracts of proliferating neurospheres, nine yielded similar results between foetal and adult SVZ, i.e. either negativity (CD133, SDF-1) or positivity (nestin, Sox-2, Hes-1, Pbx-1, β-III-tubulin, doublecortin, Nogo-A). Two differences were encountered between expanded neural cells from adult and foetal brains. The first of these was the detectable expression of cyclin D2 in foetal, but not adult, pig neurospheres. This indicates that neural stem cell expansion is higher in foetal than in adult neurosphere cultures [42]. Furthermore, the lack of Ki-67 detection in pre-differentiation neurospheres from foetal [42], but not adult (present data) supports the same interpretation, since primitive neural stem cells are much less actively dividing than neural precursors at the immediately subsequent maturation stage [46]. Interestingly, the published molecular profile of *in vitro* expanded human embryonic neural stem cells (hNSC) is very close, but not identical, to those of pig neural precursor cells. The major difference was the expression of the stemness gene CD133 in hNSCs that otherwise expressed both Ki-67 and GFAP like adult and foetal pig cells, respectively [42].

In the future, production of neural stem/progenitor cells from the adult pig by the present procedure can promote cell therapy improvements, first in animal models, but also in clinical attempts to repair neural lesions in the use of adult human. By analogy with other tissues [47, 48], adult pig neural precursor cells should be feasible for transplantation into diseased humans without immune rejection [49,50]. The present study thus provides the substratum for an alternative source to human allografts for transplantation in neurodegenerative disorders or injuries, which is now open to experimentation.

3.4. *In vitro* lineage labelling

In order to track *in vivo* the fate of transplanted neural stem cells and their progenies, *in vitro*-expanded neurospheres must be labeled prior to transplantation with a tag which will be retained permanently and transmitted through division without affecting survival and differentiation of labeled cells.

The easiest and long-established tagging method for living cell tracking after transplantation, is incubation of dividing cells with a labeled analog of DNA biosynthesis precursors: nucleotides. Two tools are available: tritiated thymidine (^{3}H-Thy) and bromo-deoxy-uridine (BrdU), and can be detected *in situ* by autoradiography and immunohistochemistry respectively. These tags are easy to apply and detect, but they have the major inconvenient to fade across subsequent rows of cell division because of the semi-replicative mechanism of DNA replication. Practically, this nuclear labeling becomes undetectable after 2 successive divisions of initially labeled cells.

Another, more recent procedure allows true lineage-labeling without fading across successive rows of division: it consists in applying a viral vector of the green fluorescent protein (GFP) gene, either by *in vivo* stereotaxic surgery into the stem cell-containing niche [51] or *in vitro* on expanded primary neurospheres [52]. In collaboration with a retrovirology laboratory (U421 INSERM, Ecole Normale Supérieure, Lyon, France), we have chosen the latter approach [53] and used a lentiviral vector of green fluorescent protein gene (LV-GFP) which had already been validated for human dendritic cells [54]. LV-GFP was freshly synthetized, titrated and tested as previously described [55] stored at -80°C and unfrozen just before use. Adult pig SVZ-derived neurospheres were initially expanded through at least 2 passages. Then, 4-5 days after the last passage, the growing neurospheres were preincubated 30 minutes in the presence of 1mM polybrene (Sigma) and void lentiviral particles (VLP for optimal integration yield; 55) at 0.1 x MOI (multiplicity of infection), and further incubated 2h with fresh culture medium containing 1mM polybrene and LV-GFP at 0.3x MOI; incubation medium was then replaced by fresh standard culture medium. LV-GFP-infected neurospheres were allowed 3 days culture in standard conditions for optimal lineage labeling (Figure 4).

4. Improvement of post-lesional neural outcome with primary pig neurosphere transplantation inside a venous autograft

Cell therapy for the nervous system has encountered a major empirical obstacle: in adult mammalian brain, the neurogenic potential of neural stem cells is inhibited by intrinsic tissue microenvironment except for the very few naturally neurogenic areas, which has been formally demonstrated by heterotopic transplantation studies [56]. To overcome this problem, we developed a novel paradigm in order to improve post-lesion regeneration in peripheral nerves: transplantation of *in vitro*-expanded, proliferating neurospheres from adult pig subventricular zone inside a homotypic venous graft which is sutured at both ends onto the lesioned nerve ends.

Bridging a long nerve gap with a homotypic venous graft is already used in human neurology in order to ameliorate post-lesional recovery of peripheral nerves [57,58,59,60], but it was barely explored for neuronal cell therapy applications. The disabilities resulting from peripheral nerve lesions in adults typically display no neuron loss but pathway disruption, upon disappearance of distal axonal segments, which was extensively documented to lead to restricted or aberrant synaptic reconnexion of surviving neurons [61]. We postulated that neural stem cell grafting on the lesion site might overcome this limitation, by generating chains of interconnected neurons that would functionally replace the lost nerve substance more efficiently than existing severed neuronal axons.

To test this hypothesis, we attempted transplantation of the proliferating neurospheres from adult pig subventricular zone (SVZ) which we had characterized above, inside an autologous venous graft, following surgical transsection of nervis cruralis with 30mm-long gap in adult pig [53]. The following section summarizes the hallmarks of this study, methods and results [53].

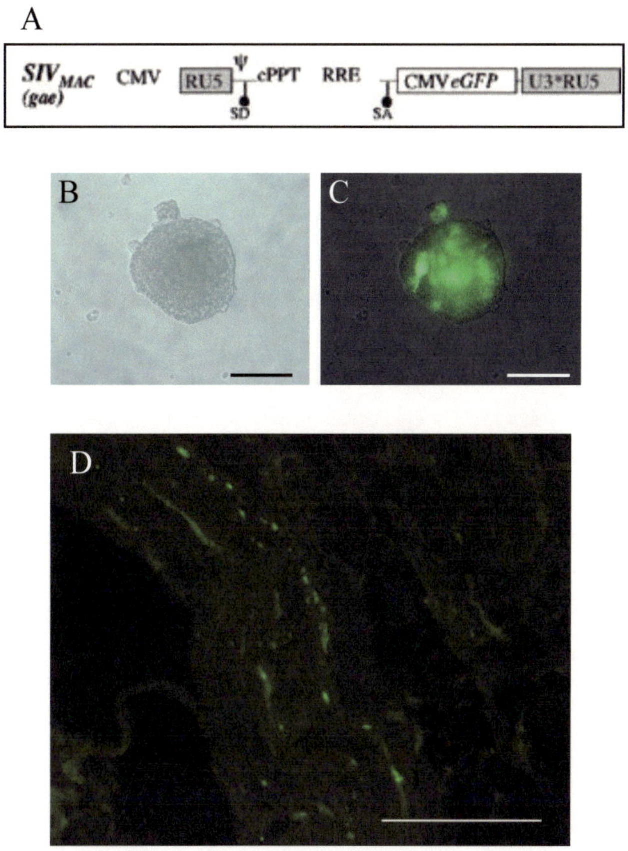

Figure 4. Lineage labelling of neurospheres from adult pig SVZ by *in vitro* viral transfer of GFP gene (modelized from references 54, 55]. The functional structure of the lentiviral GFP vector is schematized in (A) (see 54 for more details). (B,C) Labelled neurospheres 3 days after *in vitro* infection with the lentiviral vector of GFP, as observed under a photonic microscope with natural light (B) or green fluorescence (C). (D) *In situ* detection of neurosphere-derived GFP in a transplanted venous bridge at 8 months after lesion, by fluorescent light microscopy. Legend: *CMV: cytomegalovirus; gag: infection competency coding sequence; GFP: green fluorescent protein; PPT: purine-rich element (to increase exon recognition); RRE: RNA export responsive element; SIV: Simian immunodeficiency virus.* Scale bars: 50 µm (B,C), 0.5mm (D).

4.1. The experimental cell therapy paradigm and its functional outcome

Our lesion model was unilateral transsection of nervis cruralis with 30 mm-long gap in adult pig, under anesthesia. This lesion induces a major motor defect, consisting in the loss of the right leg extension over the thigh, which can be quantified by electromyography of the crural-innervated muscle quadriceps. Our experimental cell therapy paradigm consisted in bridging the nerve gap immediately after damage with an autologous venous segment (sampled from vena mammalian externalis) which was sutured at the proximal end over the perinevre of the severed nerve, filled through the opposite end of the venous graft with freshly prepared suspension of *in vitro*-expanded neurospheres from adult pig SVZ (300 µL, 1000 neurospheres) which we had previously characterized (see section II) and sutured the venous neuroguide over the distal end of the severed nerve (Figure 5). Control animals did not receive neurospheres in the venous shaft before saturation. The transplanted neurospheres had been expanded *in vitro* from adult pig SVZ through 2-3 passages and have been labeled *in vitro* with either BrdU (for short-term post-lesional survival times) or lentivirally-transferred fluorescent protein gene (see above, section II.4) prior to transplantation (Liard et al 2012).

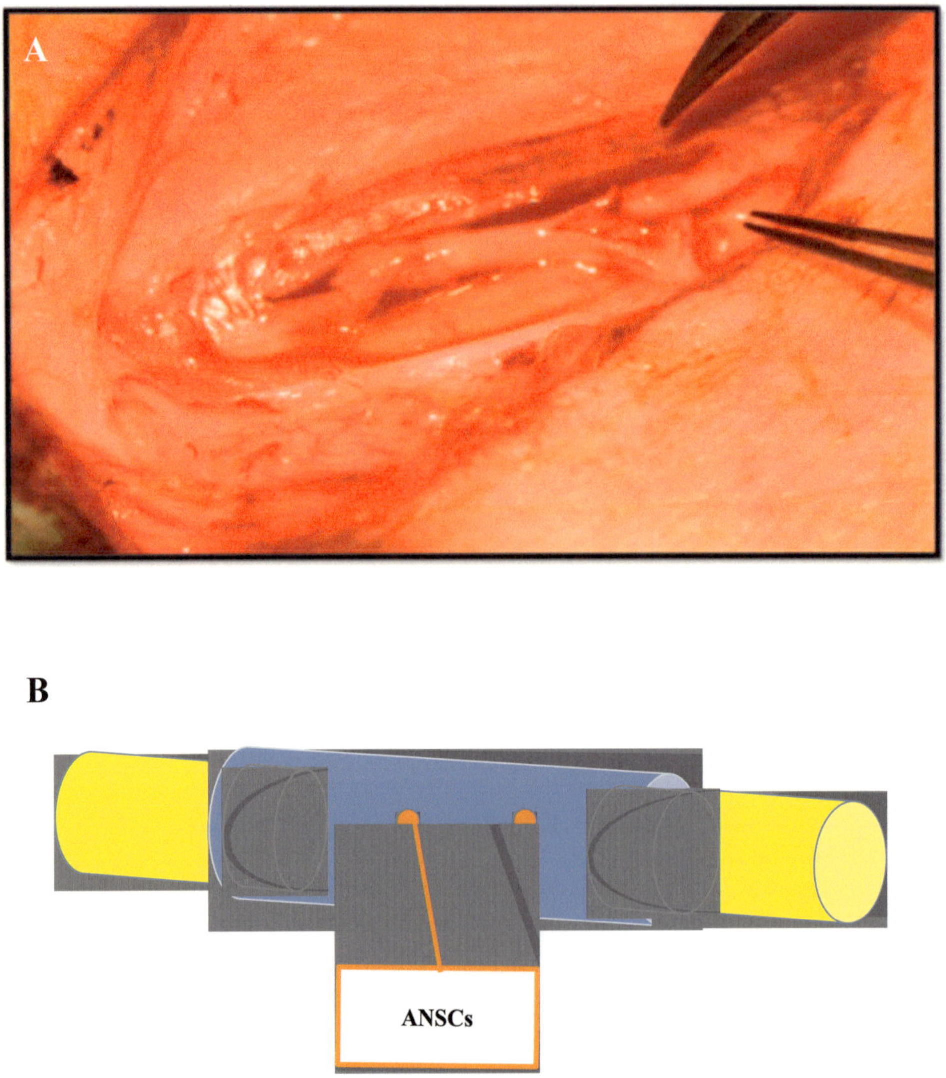

Figure 5. Surgical paradigm for adult pig neural stem cell (ANSCs) transplantation inside a venous graft (modelized from reference 53]. (A) Surgical bridging of lesioned nerve by a venous graft (below the forceps tips; the arrow points to the distal, unsutured end of the graft). (B) Procedure schematization.

http://dx.doi.org/10.5772/55726

Lesion-induced loss of leg extension on the thigh became definitive in controls (up to 8 months after lesion) but was reversed by 90-180 days after neurosphere-filled vein grafting. Prior to surgery, electromyograms of muscle quadriceps vast internal displayed amplitudes of 4.9 to 6.1 mV and post-stimulus latencies of 0.89 to 3.7 ms. Immediately after surgical realization of nervis cruralis substance loss, electromyograms of muscle quadriceps were negative for all experimental animals. Electromyography showed stimulo-detection recovery in neurosphere-transplanted pigs which was partial at 180 days after lesion and almost complete by 240 days, while electromyograms of controls were still negative (Liard et al 2012). Interestingly, in neurosphere-transplanted pigs, post-stimulus latencies returned earlier to normal (i.e., at 180 days post-lesion) than electromyogram amplitude (which was fairly detectable although much lower than on intact leg).

Therefore, transplanting *in vitro* purified-expanded adult neural stem cells inside a venous neuroguide on the site of a peripheral nerve lesion promoted efficient functional recovery in adult pig [26]. We further addressed the underlying mechanisms by post-mortem immunohistochemical analysis of the bridged nerve (see below).

4.2. Exclusively neuronal fate of grafted neurosphere cells

In order to assess the fate of transplanted neurosphere cells [53], we euthanized the experimental animals at various post-lesion intervals, sampled the experimental vein-bridged nerve segment and processed it for immunohistochemistry (fixation by 24-hour-immersion at 4°C into a buffered 4% paraformaldehyde solution, cryoprotection in 30% sucrose-containing phosphate buffer). After snap-freezing in isopentane at -45°C, serial sagittal sections were collected from each frozen vein-bridged nerve segment in a cryostat and processed for phenotypic marker immunohistofluorescence. Fluorescent dyes were chosen so as to be compared with either intrinsic green fluorescent protein (GFP) or with BrdU immunohistofluorescence, depending on the method for neurosphere cell labeling prior to transplantation [53].

At 8 days after lesion-transplantation, BrdU-positive cells were localized inside the venous tube as flattened spherical groups and were all immunoreactive for the specific marker of immature migration-prone neurons: doublecortin (DCX) [53]. At longer survival delays assessed (45 to 240 days) all neurosphere-derived cells, whether labeled with BrdU or virally-transferred GFP, expressed the specific marker of mature neurons: neuronal nuclear antigen (NeuN) or neurofilament protein NF-68 (Figure 4; modified from ref 53). By contrast, none of labeled neurosphere-derived cells coexpressed any of glial markers assessed: CNPase (for myelinating glia, i.e. oligodendrocytes or Schwann cells), S-100β and GFAP (for astrocytes), at either of post-lesional delays. Therefore *adult neural stem cell-derived progenies of grafted neurospheres inside the venous bridge survived and differentiated into neurons exclusively* [53]. Moreover, newly-formed neurons distributed inside the venous graft along the longitudinal axis of the severed nerve, which suggests that new neurons would have created gap-filling interconnected chains in-between the proximal and distal ends of the severed nerve. However, such issue remains to be investigated by using a more resolutive approach (electron microscopy).

Our finding contrasts with numerous previous attempts to improve post-lesional repair of peripheral with putatively regenerating cell transplantation. The latter strategy, or cell therapy, has been attempted with primary cultures of Schwann cells [62,63], olfactory bulb enshealting cells [64], or various types of stem cells [65]. In all these studies, progenies of grafted stem cells mostly proved to be glial cells, which indirectly enhance axonal regeneration [65,66,67]. Since none of these paradigms used vascular bridge for stem cell transplantation, our results suggest that the venous wall *per se* stimulates neuronal differentiation of stem cell progenies. It is in keeping with the recent demonstrations that vascular walls favor neurogenesis from adult neural stem cells [68,69,70], which is mediated in rodent models by the vascular endothelium-derived growth factor (VEGF) [71,72,73,74]. Our results suggest that choosing a venous trunk versus an artificial neuroguide to bridge a nerve gap is an interesting solution because of intrinsic property. Furthermore, all surgeons are able to take a vein trunk on superficial vein network and this is much less expensive than neurotubes in emergency conditions.

4.3. Graft-induced activation of intrinsic Schwann cells in the lesioned nerve

At 180 and 240 days after nerve lesion and GFP-expressing neurosphere transplantation, CNPase immunohistofluorescent labeling was much higher than in vein-bridged lesioned controls which had not received neurosphere transplantation [53]. Both the number of CNPase-immunoreactive Schwann cell processes and their labeling intensities, as quantified by computerized image analysis, were significantly higher in neurosphere-grafted nerves than in controls [26]. Our result is in keeping with a previous report showing that a venous graft favored Schwann cell proliferation after nerve lesion [75]. In our paradigm though, since controls have received a neurosphere-devoid venous graft, our results indicated that neurosphere cells emit diffusible signals that stimulate Schwann cells. Therefore, *adult neural stem cell-derived progenies of grafted neurospheres inside the venous bridge promoted activation of intrinsic myelinating Schwann cells*. This result is in keeping with accumulating reports demonstrating that primary neurospheres from adult brain tissues do secrete *in vitro* some neurotrophic factors [76,77].

Altogether, our novel cell therapy paradigm in adult pig promotes efficient functional recovery of a peripheral nerve after lesion with long substance loss, correlatively with genesis of new neurons in-between the lesional gap and activation of intrinsic myelinating cells.

Similar results have been obtained in adult rat by using a similar strategy (Xu et al 2012) or mesenchymal stem cells combined with acellular conduits (Zhao et al 2012, Jia et al 2012). From both our and others' studies though (as reviewed in ref 66), a pending issue remains unaddressed: which is the long-term fate of ectopic graft-derived new neurons? The putative formation of ectopic multisynaptic neuronal nets bridging the gap between the two sides of the lesional gap, as suggested by the topographical distribution of new neurons (see above), is made plausible by previous experimental demonstrations of synaptic connectivity between ectopic stem cell-derived new neurons and host neural net, both in neurogenic [56] and non-neurogenic (Lu et al 2012) adult structures of mammalian central nervous system. However, this issue requires to be directly addressed by means of electron microscopy.

Another question arising from our results, concerns the potential long term negative effects of grafted stem cell-derived ectopic neurons. Such transplantation might indeed result into neuromas: either neoplastic neuromas deriving from the grafted proliferative stem cells

(Johnson et al 2012), or non-neoplastic neuromas consisting in aberrant axonal swellings (Rajput et al 2012).

5. Future applications of these new paradigms

The results above (sections II & III) bring about a proof-of-concept for future improvements of neurological clinics. However, before it can be tested in human patients, our paradigm deserves some improvements.

5.1. Alternative source of adult neural stem cells

The direct transposition of the present cell therapy strategy to human patients would raise ethical problems, regarding either autologous sampling of patient's subventricular zone as a source of neurogenic neurospheres, or xenotransplantation of neurospheres from adult pig SVZ due to currently discussed virological risks of porcine tissues [10]. A convenient alternative can be provided by the recently discovered neurogenic stem cells of biopsied human olfactory mucosa [84]. This tissue has been demonstrated indeed to harbor a population of mesenchymal stem cells which displays the same neurogenic properties as original neural stem cells after isolation and expansion in primary culture. To this aim, olfactory mucosa biopsies are enzymatically and mechanically separated from the olfactory neuroepithelium, and the cell suspension from the resulting tissue (olfactory chorion) is seeded in a Dulbecco's medium (DMEM) supplemented with F12 supplement and 10% fetal bovine serum. This primary culture yields adherent proliferating cells that are self-renewable under passage into the same medium. Upon seeding into serum-free DMEM supplemented with insulin, transferrin, selenium, EGF and bFGF, olfactory chorion-derived cells grow into typical neurospheres that exhibit the same morphology, growth kinetics and multipotentiality as neurospheres from SVZ [3,85,86,87]. These *in vitro*-expanded neurogenic neurospheres were then demonstrated to restore function and plasticity after brain lesions in adult rodents [88]. This cell preparation would therefore represent a convenient alternative for clinical application of our cell therapy strategy for post-lesional repair of peripheral nerves. The long-term safety of this novel cell type should however be checked precisely in future assays, since cell therapies with mesenchymal stem cells were recently documented to elicit neuromas in human patients [89].

5.2. Alternative neuroguide

Another issue in our cell therapy strategy concerns the use of autologous venous shaft as a neuroguide. Although largely used in human clinics [57, 58], such procedure has indeed been reported to generate inflammatory or necrotic outcomes. This risk could be avoided by using a synthetic neuroguide which has already been shown to favor neurogenesis from transplanted neural stem cells [59,60,90,91]. The basic neuroguide structure can be manufactured by diverse methods: spinning mandrel technology, sheet rolling, injection-molding, freeze-drying, and electro-spinning.

It will be interesting to study this cell enhancement with FDA- and CE-approved neuroguides [60, 90]:

- Nerve tubes from biodegradable and biological materials:
- type 1 collagen =Integra NeuraGen ® ; Neuro-matrix and Neurolac [91,92,93] with retrospective studies [94];
- Nerve tubes from biodegradable and synthetic materials :
- polyglycolic acid PGA = **NEUROTUBE®** [95,96] ;
- poly-lactic-glycolic acid (PLGA) ;
- poly-L-lactide-caprolactone (PLCL) **Neurolac®** [97];
- polyvinyl alcohol hydrogel SaluBridge.
- Nerve guide with internal multitubular architecture, which is required for bridging long nerve gaps. If we take this point of view, acellular nerve graft Axogen'Avance ® is promissing and obtained by cryocongelation and chemical treatment from human nerves [93,98]. The graft can be handled like an autograft, held by epineurium and bridge gaps up to 50 mm.

With synthetic neurotube, three-dimensional scaffold can be obtained by electrospinning and progress with nanotechnologies. This approach could be optimized with extracellular matrix (ECM) proteins or peptides (laminin-1) and neurotrophic growth factors like basic fibroblast growth factor (bFGF), brain derived neurotrophic factor (BDNF) or nerve growth factor (NGF) [99].

Above all maybe, the use of acellular or totally synthetic conduits for nerve gap bridging is a *priori* likely to reduce the risk of post-operative neuromas. More systematic and extensive studies are required to evaluate properly the risk over benefit probability ratios of all available combinations. For instance, a novel cell therapy strategy which differs from ours merely by using an acellular vein graft, was recently reported to produce only 35-75% of neural cells, including glial and proliferative cells along with neurons [100]. The highly positive pro-repair impact of autologous vein wall, which we have empirically demonstrated, may prove difficult to mimic with man-controlled substitutes.

6. Conclusion

Thus, transplantation of neural stem cells from adult mammalian brain inside an autologous venous graft provides an efficient repair strategy, in the pig model. Our study provides the proof-of-concept for further study in human clinics. It also provides progress in basic cell biology, since the grafted exogenous neurospheres were shown to promote nerve regeneration through 2 distinct original mechanisms: i) indirect activation of intrinsic myelinating glial cells (as recently reviewed in 66), and ii) genesis of new neurons aligned in-between the two ends of the severed nerve. The latter one is totally novel in the field of post-lesional plasticity and repair of peripheral nerves. It deserves though further analysis, by deciphering the neurotransmitter phenotype of neurosphere-derived new neurons and their ultrastructural pattern of connectivity.

http://dx.doi.org/10.5772/55726

Author details

Olivier Liard[1] and Emmanuel Moyse[2*]

*Address all correspondence to: emmanuel.moyse@univ-tours.fr

1 Maxillo-facial Surgery, Clinique Toulouse-Lautrec, 2 rue Jacques Monod, Albi, France

2 Centre INRA of Tours, Unit PRC (Physiology of Reproduction and Behaviour, UMR-7247 INRA-University of Tours), Nouzilly, France

References

[1] Reynolds BA, Weiss S. Generation of neurons and astrocytes from isolated cells of the adult mammalian central nervous system. Science 1992; 255:1707-1710.

[2] Eriksson PS, Perfilieva E, Björk-Erikkson T, Alborn AM, Nordborg C, Peterson DA, Gage FH. Neurogenesis in the adult human hippocampus. Nat. Med. 1998; 4(11): 1313-1317.

[3] Murrell W, Féron F, Wetzig A, Cameron N, Splatt K, Bellette B, Bianco J, Perry C, Lee G, Mackay-Sim A. Multipotent stem cells from adult olfactory mucosa. Dev. Dyn. 2005; 233(2):496-515.

[4] Bonfanti L. From Hydra regeneration to human brain structural plasticity: a long trip through narrowing processes. Sci. World J. 2011; 11:1270-1299.

[5] McKay R. Stem cells in the central nervous system. Science 1997; 276:66-71.

[6] Lindvall O, Kokaia Z. Stem cells for the treatment of neurological disorders. Nature 2006; 441:1094-1096.

[7] Féron F. Current cell therapy strategies for repairing the central nervous system. Rev. Neurol. (Paris) 2007; 163(Spec.N.1): 3S23-30.

[8] Gincberg G, Arien-Zakay H, Lazarovici P, Lelkes PI. Neural stem cells: therapeutic potential for neurodegenerative diseases. Br. Med. Bull. 2012; 1-13.

[9] Foroni C, Galli R, Cipeletti B, Caumo A, Alberti S, Fiocco R, Vescovi A. Resilience to transformation and inherent genetic and functional stability of adult neural stem cells ex vivo. Cancer Res. 2007; 67(8):3725-3733.

[10] Ekser B, Ezzelarab M, Hara H, van der Windt DJ, Wijkstrom M, Bottino R, Trucco M, Cooper DK. Clinical xenotransplantation: the next medical revolution? Lancet 2012; 379(9816):672-683.

[11] Siebzehnrubl FA, Vedam-Mai V, Azari H, Reynolds BA, Deleyrolle LP. Isolation and characterization of adult neural stem cells. Methods Mol. Biol. 2011; 750:61-77.

[12] Azari H, Sharififar S, Fortin JM, Reynolds BA. The neuroblast assay: an assay for the generation and enrichment of neuronal progenitor cells from differentiating neural stem cell progeny using flow cytometry. J. Vis. Exp. 2012; doi: 10.3791/3712.

[13] Weiss S, Dunne C, Hewson J, Wohl C, Wheatley M, Peterson AC, Reynolds BA. Multipotent CNS stem cells are present in the adult mammalian spinal cord and ventricular neuroaxis. J. Neurosci. 1996; 16:7599-7609.

[14] Migaud M, Batailler M, Segura S, Duittoz A, Franceschini I, Pillon D. Emerging new sites for adult neurogenesis in the mammalian brain: a comparative study between the hypothalamus and the classical neurogenic zones. Eur. J. Neurosci. 2010; 32(12): 2042-52.

[15] Moyse E, Segura S, Liard O, Mahaut S, Mechawar N. Control of proliferation and lineage commitment in adult neural stem cells by tissular microenvironment in healthy and post-lesional conditions. Current Stem Cell Research and Therapy 2008; 3(3): 163-184.

[16] Aiba K, Sharov AA, Carter MG, Foroni C, Vescovi AL, Ko MSH. Defining a developmental path to neural fate by global expression profiling of mouse embryonic stem cells and adult neural stem/progenitor cells. Stem Cells 2006; 24:889-895.

[17] Marei HES. Transcription factors expressed in embryonic and adult olfactory bulb neural stem cells reveal distinct proliferation, differentiation and epigenetic control. Genomics 2012; doi: dx.doi.org/10.1016/j.ygeno.2012.09.006.

[18] Yang J, Gu P, Menges S, Klassen H. Quantitative changes in gene transcription during induction of differentiation in porcine neural progenitor cells. Mol. Vis. 2012; 18:1484-1504.

[19] Fong CY, Gauthaman Km Bongso A. Teratomas from pluripotent stem cells: a clinical hurdle. J. Cell Biochem. 2010; 111:769-781.

[20] Shih CC, Forman SJ, Chu P, Slovak M. Human embryonic stem cells are prone to generate primitive, undifferentiated tumors in engrafted human fetal tissues in severe combined immunodeficient mice. Stem Cells Dev. 2007; 16(6):893-902.

[21] Miura K, Okada Y, Aoi T, Okada A, Takahashi K, Okita K, Nakagawa M, Koyanagi M, Tanabe K, Ohnuki M, Ogawa D, Ikeda E, Okano H, Yamanaka S. Variation in the safety of induced pluripotent stem cell lines. Nat. Biotechnol. 2009; 27:743-745.

[22] Wan JF, Zhang SJ, Wang L, Zhao KL. Implications for preserving neural stem cells in whole brain radiotherapy and prophylactic cranial irradiation: a review of 2270 metastases in 488 patients. J. Radiat. Res. 2012; doi: 10.109/jrr/rrs085.

[23] Valent P, Bonnet D, DeMaria R, Lapidot T, Copland M, Melo JV, Chomienne C, Ishikawa F, Schuringa JJ, Stassi G, Huntly B, Herrmann H, Soulier J, Roesch A, Schuur-

huis GJ, Wöhrer S, Arock M, Zuber J, Cerny-Reiterer S, Johnsen HE, Andreeff M, Eaves C. Cancer stem cells definitions and terminology: the devil is in the details. Nat. Rev. Cancer 2012; doi: 10.1038/nrc3368.

[24] Vescovi AL, Galli R, Reynolds BA. Brain tumour stem cells. Nat. Rev. Cancer 6:425-436.

[25] Deleyrolle LP, Harding A, Cato K, Siebzehnrubl FA, Rahman M, Azari H, Olson S, Gabrielli B, Osborne G, Vescovi A, Reynolds BA. Evidence for label-retaining tumour-initiating cells in human glioblastoma. Brain 2011; 134:1331-1334.

[26] Liard O*, Segura S*, Pascual A, Gaudreau P, Fusai T, Moyse E. *In vitro* isolation of neural precursor cells from the adult pig subventricular zone. J. Neurosci. Methods 2009; 182(2):172-179.

[27] Felix B, Léger MA, Albe-Fessard D, Marcilloux JC, Rampin O, Laplace JP. Stereotaxic atlas of the pig brain. Brain Res Bull 1999; 49(1-2):1-137.

[28] Doetsch F, Garcia-Verdugo JM, Alvarez-Buylla A. Cellular composition and three dimensional organization of the subventricular germinal zone in the adult mammalian brain. J. Neurosci. 1997; 17: 5046-5061.

[29] Scholzen T, Gerdes J. The Ki-67 protein: from the known and the unknown. J. Cell Physiol. 2000; 182:311-322.

[30] Charrier C, Coronas V, Fombonne J, Roger M, Jean A, Krantic S, Moyse E. Characterization of neural stem cells in the dorsal vagal complex of adult rat by *in vivo* proliferation labeling and *in vitro* neurosphere assay. Neuroscience 2006; 138: 5-16.

[31] Reynolds BA, Rietze RL. Neural stem cells and neurospheres – reevaluating the relationship. Nat. Methods 2005; 2:333-6.

[32] Bouvier C, Bartoli C, Aguirre-Cruz L, Virard I, Colin C, Fernandez C, Gouvernet J, Figarella-Branger D. Shared oligodendrocyte lineage gene expression in gliomas and oligodendrocyte progenitor cells. J. Neurosurg. 2003; 99:344-50.

[33] Bradl M, Lassmann H. Oligodendrocytes: biology and pathology. Acta Neuropathol. 2010; 119:37-53.

[34] Wu Y, Lin Y, Levine EM, Rao MS. Hes1 but not Hes5 regulates an astrocyte versus oligodendrocyte fate choice in glial restricted precursors. Dev. Dyn. 2003; 226(4): 675-89.

[35] Liu J, Casaccia P. Epigenetic regulation of oligodendrocyte identity. Trends Neurosci. 2010; 33(4):193-201.

[36] Kemperman G. Adult neurogenesis. Oxford University Press 2006.

[37] Lowell S, Benchoua A, Heavey B, Smith AG. Notch promotes neural lineage entry by pluripotent embryonic stem cells. PLoS Biology 2006; 4(5):e121.

[38] Hsieh J. Orchestrating transcriptional control of adult neurogenesis. Genes Dev. 2012; 26(10):1010-1021.

[39] Tiveron MC, Rossel M, Moepps B. Tiveron MC, Rossel M, Moepps B, Zhang YL, Seidenfaden R, Favor J, König N, Cremer H. Molecular interaction between projection neuron precursors and invading interneurons via stromal-derived factor 1 (CXCL12)/CXCR4 signaling in the cortical subventricular zone/intermediate zone. J. Neurosci. 2006; 26(51):13273-13278.

[40] Kafienah W, Mistry S, William Ch, Hollander AP. Nucleostemin is a marker of proliferating stromal stem cells in adult human bone marrow. Stem Cells 2006; 24(4): 1113-1120.

[41] Kowalczyk A, Filipowski RK, Rylski M, Wilczynski GM, Konopacki FA, Jaworski J, Ciemeryck MA, Sicinski P, Kaczmarek L. The critical role of cyclin D2 in adult neurogenesis. J. Cell Biol. 2004; 167(2):209-13.

[42] Schwartz PH, Nethercott H, Kirov H, Ziacian B, Young MJ, Klassen H. Expression of neurodevelopmental markers by cultured porcine neural precursor cells. Stem Cells 2005; 23:1286-1294.

[43] Harrower TP, Tyers P, Hooks Y, Barker RA. Long-term survival and integration of porcine expanded neural precursor cell grafts in a rat model of Parkinson's disease. Exp. Neurol. 2006; 197:56-69.

[44] Klassen H, Kiilgaard JF, Zahir T, Ziaeian B, Kirov II, Schefig E, Warfvinge K, Young MJ. Progenitor cells from the porcine neural retina express photoreceptor markers after transplantation to the subretinal space of allorecipients. Stem Cells 2007; 25(5): 1222-1230.

[45] Gu P, Harwood LJ, Zhang X, Wylie M, Curry WJ, Cogliati T. Isolation of retinal progenitor and stem cells from the porcine eye. Mol. Vis. 2007; 13:1045-57.

[46] Von Bohlen und Halbach O. Immunohistochemical markers for staging neurogenesis in hippocampus. Cell Tissue Res. 2007; 329:409-20.

[47] Bucher P, Morel P, Buhler LH. Xenotransplantation: an update on recent progress and future perspectives. Transpl. Int. 2005; 18:894-901.

[48] Vodicka P, Smetana K Jr, Dvorankova B, Emerick T, Xu YZ, Ourednik J, Ourednik V, Motlik J. The miniature pig as an animal model in biomedical research. Ann N Y Acad Sci. 2005; 1049:161-171.

[49] Armstrong RJ, Harrower TP, Hurelbrink CB, Mc Laughin M, Ratcliffe EL, Tyers P, Richards A, Dunnett SB, Rosser AE, Barker RA. Porcine neural xenografts in the immunocompetent rat: immune response following grafting of expanded neural precursor cells. Neuroscience 2001; 106: 201-206.

[50] Yan J, Xu L, Welsh A.M, Chen D, Hazel T. Johe K, Koliatsos VE. Combinated immunosuppressive agents or CD4 antibodies prolong survival of human neural stem cell

grafts and improve disease outcomes in ALS transgenic mice. Stem Cells 2006; 24:1976-85.

[51] Suhonen J, Ray J, Blömer U, Gage FH, Kaspar B. Ex vivo and in vivo gene delivery to the brain. Curr. Protoc. Hum. Genet. Chapter 13: Unit 13.3.

[52] Franceschini IA, Feigenbaum-Lacombe V, Casanova P, Lopez-Lastra M, Darlix JL, Dalcq MD. Efficient gene transfer in mouse neural precursors with a bicistronic retroviral vector. J. Neurosci. Res. 2001; 65(3):208-219.

[53] Liard O, Segura S, Sagui E, Nau A, Pascual A, Cambon M, Darlix JL, Fusai T, Moyse E. Adult brain-derived neural stem cells grafting into a vein bridge increases post-lesional recovery and regeneration in a peripheral nerve of adult pig. Stem Cell International 2012; 2012:128732.

[54] Mangeot PE, Duperrier K, Nègre D, Boson B, Rigal D, Cosset FL, Darlix JL. High levels of transduction of human dendritic cells with optimised SIV vectors. Mol. Ther. 2002; 5(3): 283-90.

[55] Goujon C, Jarrosson-Wuillème L, Bernaud J, Rigal D, Darlix JL, Cimarelli A. With a little help from a friend: increasing HIV transduction of monocyte-derived dendritic cells with virion-like particles of SIV(MAC). Gene Ther. 2006; 13(12): 991-4.29)

[56] Shihabuddin LS, Horner PJ, Ray J, Gage FH. Adult spinal cord stem cells generate neurons after transplantation in the adult dentate gyrus. J. Neurosci. 2000; 20(23): 8727-8735.

[57] Chiu DTW, Strauch B. A prospective clinical evaluation of autologous vein grafts used as a conduit for distal sensory nerve defects of 3cm or less. Plastic and Reconstructive Surgery 1990; 86(5):928-934.

[58] Cooley BC. History of vein grafting. Microsurgery 1998; 18(4):234-236.

[59] Deal DN, Griffin JW, Hogan MV. Nerve conduits for nerve repair or reconstruction. J. Am. Acad. Orthop. Surg. 2012; 20(2):63-68.

[60] Kehoe S, Zhang XF, Boyd D. FDA approved guidance conduits and wraps for peripheral nerve injury: a review of materials and efficacy. Injury 2012; 43(5):553-572.

[61] Kouyoumdjian JA. Peripheral nerve injuries: a retrospective survey of 456 cases. Muscle Nerve 2006; 34(6):785-788.

[62] Timmer M, Robben S, Muller-Ostermeyer F. Axonal regeneration across long gaps in silicone chambers filled with Schwann cells overexpressing high molecular weight FGF-2. Biomed. Mater. Eng. Cell Transplant. 2003; 12 (3): 265-77.

[63] Hood B, Levene HB, Levi AD. Transplantation of autologous Schwann cells for the repair of segmental peripheral nerve defects. Neurosurg. Focus 2009; 26(2):E4.

[64] Verdu E, Narvarro X, Gudino-Cabrera G, Rodriguez F.J, Ceballos D, Valero A, Nieto-Sampedro M. Olfactory bulb ensheathing cells enhance peripheral nerve regeneration, Neuroreport 1999; 10:1097–1101.

[65] Walsh S, Midha R. Use of stem cells to augment nerve injury repair. Neurosurgery 2009; 65(4 Suppl): A80-86.

[66] Ren Z, Wang Y, Peng J, Zhao Q, Lu S. Role of stem cells in the regeneration and repair of peripheral nerves. Rev. Neurosci. 2012; 23(2):135-143.

[67] Walsh SK, Kumar R, Grochmal JK, Kemp SW, Forden J, Midha R. Fate of stem cell transplants in peripheral nerves. Stem Cell Res. 2012; 8(2):226-238.

[68] Tavazoie M, Van der Vaken L, Silva-Vargas V, Louissaint M, Colonna L, Zaidi B, Garcia-Verdugo JM, Doetsch F. A specialized vascular niche for adult neural stem cells. Cell Stem Cell 2008; 3(3):279-288.

[69] Bennett L, Yang M, Enikolopov G, Iacovitti L. Circumventricular organs: a novel site of neural stem cells in the adult brain. Mol. Cell Neurosci. 2009; 41(3):337-47.

[70] Goldberg JS, Hirschi KH. Diverse roles of the vasculature within the neural stem cell niche. Regen. Med. 2009; 4(6):879-97.

[71] Ferrara N, Gerber HP, LeCouter J. The biology of VEGF and its receptors. Nature Med 2003; 9: 669–676.

[72] Schänzer A, Wachs FP, Wilhelm D, Acker T, Cooper-Kuhn C, Beck H, Winkler J, Aigner L, Plate KH, Kuhn HG. Direct stimulation of adult neural stem cells in vitro and neurogenesis in vivo by vascular endothelial growth factor. Brain Pathol. 2004;14(3):137-148.

[73] Wada T, Haigh JJ, Ema M, Hitoshi S, Chaddah R, Rossant J, Nagi A, van der Kooy D. Vascular endothelial factor directly inhibits primitive neural stem cell survival but promotes definitive neural stem cell survival. J. Neurosci. 2006; 26:6803-6812.

[74] Calvo CF, Fontaine RH, Soueid J, Tammela T, Makinen T, Alfaro-Cervello C, Bonnaud F, Miguez A, Benhaim L, Xu Y, Barallobre MJ, Moutkine I, Lyytikkä J, Tatlisumak T, Pytowski B, Zalc B, Richardson W, Kessaris N, Garcia-Verdugo JM, Alitalo K, Eichmann A, Thomas JL. Vascular endothelial growth factor receptor 3 directly regulates murine neurogenesis. Genes Dev. 2011; 25:831-844.

[75] Tseng CY, Hu G, Ambron RT, Chiu DT. Histologic analysis of Schwann cell migration and peripheral nerve regeneration in the autogenous venous nerve conduit (AVNC). J Reconstr Microsurg. 2003; 19(5): 331-40.

[76] Xiao M, Klueber KM, Lu C, Guo Z, Marshall CT, Wang H, Roisen FJ. Human adult olfactory neural progenitors rescue axotomized rodent rubrosphinal neurons and promote functional recovery. Exp. Neurol. 2005; 194(1):12-30.

[77] Chirasani SR, Sternjak A, Wend P, Momma S, Campos B, Hermann IM, Graf D, Mitsiadis T, Herold-Mende C, Besser D, Synowitz M, Kettenmann H, Glass R. Bone mor-

phogenetic protein-7 release from endogenous neural precursor cells suppresses the tumourigenicity of stem-like glioblastoma cells. Brain 2010; 133(Pt 7): 1961-72.

[78] Xu L, Zhou S, Feng GY, Zhang LP, Zhao DM, Sun Y, Liu Q, Huang F. Neural stem cells enhance nerve regeneration after sciatic nerve injury in rats. Mol. Neurobiol. 2012; 46(2):265-274.

[79] Zhao Z, Wang Y, Peng J, Ren Z, Zhang L, Guo Q, Xu WS, Lu S. Improvement in nerve regeneration through a decellularized nerve graft by supplementation with bone marrow stromal cells-in-fibrin. Cell transplant 2012; in press.

[80] Jia H, Wang Y, Tong XJ, Liu GB, Li Q, Zhang LX, Sun XH. Sciatic nerve repair by acellular nerve xenografts implanted with BMSCs in rats. Synapse 2012; 66(3): 256-269.

[81] Lu P, Wang Y, Graham L, McHale K, Gao M, Wu D, Brock J, Blesch A, Rosenzweig ES, Havton LA, Zheng B, Conner JM, Marsala M, Tuszynski MH. Long-distance growth and connectivity of neural stem cells after severe spinal cord injury. Cell 2012; 150(6):1264-1273.

[82] Johnson TS, O'Neill AC, Motarjem PM, Nazzal J, Randolph M, Winograd JM. Tumor formation following murine neural precursor cell transplantation in a rat peripheral nerve injury model. Journal of Reconstructive Microsurgery 2008; 24(8):545-550.

[83] Rajput K, Reddy S, Shankar H. Painful neuromas. Clin. J. Pain 2012; 28(7):639-645.

[84] Girard SD, Devéze A, Nivet E, Gepner B, Roman FS, Féron F. Isolating nasal olfactory stem cells from rodents or humans. J. Vis. Exp. 2011; 54: doi:pii. 10.3791/2762.

[85] Murrell W, Wetzig A, Donnellan M, Féron F, Burne T, Meedeniya A, Kesby J, Bianco J, Perry C, Silburn P, Mackay-Sim A. Olfactory mucosa is a potential source for autologous stem cell therapy for Parkinson's disease. Stem Cells 2008; 26(8):2183-2192.

[86] Delorme B, Nivet E, Gaillard J, Häupl T, Ringe J, Devèze A, Magnan J, Sohier J, Khrestchatisky M, Roman FS, Charbord P, Sensebé L, Layrolle P, Féron F. The human nose harbors a niche of olfactory ectomesenchymal stem cells displaying neurogenic and osteogenic properties. Stem Cells Dev. 2010; 19(6):853-866.

[87] Boone N, Loriod B, Bergon A, Sbai O, Formisano-Tréziny C, Gabert J, Khrestchatisky M, Nguyen C, Féron F, Axelrod FB, Ibrahim EC. Olfactory stem cells, a new cellular model for studying molecular mechanisms underlying familial dysautonomia. PLoS one 2010; 5(12):e15590.

[88] Nivet E, Vignes M, Girard SD, Pierrisnard C, Devèze A, Magnan J, Lanté F, Khrestchatisky M, Féron F, Roman FS. Engraftment of human nasal olfactory stem cells restores neuroplasticity in mice with hippocampal lesions. J.Clin. Invest. 2011; 121(7): 2808-2820.

[89] Kraus PD, Wolff D, Grauer O, Angstwurm, Jarius S, Wandiger KP, Holler E, Schulte-Mattler W, Kleiter I. Muscle cramps and neuropathies in patients with allogeneic

hematopoietic stem cell transplantation and graft-versus-host disease. PLoS one 2012; 7(9):e44922.

[90] 84 Meek MF, Coert JH. US Food and Drug Administration / Conformit Europe-approved absorbable nerve conduits for clinical repair of peripheral and cranial nerves. Ann. Plast. Surg. 2008; 60(4):466-472.

[91] Lin YC, Marra KG. Injectable systems and implantable conduits for peripheral nerve repair. Biomed. Mater. 2012; 7(2):024102.

[92] Farole A, Jamal BT. A bioabsorbable collagen nerve cuff (NeuraGen) for repair of lingual ans inferior alveolar nerve injuries: a case series. J. Oral Maxillofac. Surg. 2008; 66(10):2058-2062.

[93] Whitlock EL, Tuffaha SH, Luciano JP, Yan Y, Hunter DA, Magill CK, Moore AM, Tong AY, Mackinnon SE, Borschel GH. Processed allografts and type I collagen conduits for repair of peripheral nerve gaps. Muscle Nerve 2009; 39(6):787-799.

[94] Wangensteen KJ, Kalliainen LK. Collagen tube conduits in peripheral nerve repair: a retrospective analysis. Hand (NY) 2009; 5(3):273-277.

[95] Mackinnon SE, Dellon AL. Clinical nerve reconstruction with a bioabsorbable polyglycolic acid tube. Plast. Reconstr. Surg. 1990; 85(3):419-424.

[96] Weber RA, Breidenbrach WC, Brown RE, Jabaley ME, Mass DP. A randomized prospective study of polyglycolic acid conduits for digital nerve reconstruction in humans. Plast. Reconstr. Surg. 2000; 106(5):1036-1045.

[97] Bertleff MJ, Meek MF, Nicolai JP. A prospective clinical evaluation of biodegradable neurolac nerve guides for sensory nerve repair in the hand. J. Hand Surg. Am. 2005; 30(3):513-518.

[98] Johnson PJ, Newton P, Hunter DA, Mackinnon SE. Nerve endoneurial microstructure facilitates uniform distribution of regenerative fibers: a post hoc comparison of midgraft nerve fiber densities. J. Reconstr. Microsurg. 2011; 27(2):83-90.

[99] Madduri S, Gander B (2012) Growth factor delivery systems and repair strategies for damaged peripheral nerves. J Control Release 2012; 161(2): 274-282.

[100] Hassan NH, Sulong AF, Ng MH, Htwe O, Idrus RB, Roohi S, Naicker AS, Abdullah S. Neural-differentiated mesenchymal stem cells incorporated into muscle stuffed vein scaffold forms a stable living nerve conduit. J. Orthop. Res. 2012; 30(10): 1674-1681.